AI-Enhanced Plant Omics

CABI BIOTECHNOLOGY SERIES

Biotechnology, in particular the use of transgenic organisms, has a wide range of applications including agriculture, forestry, food and health. There is evidence that it could make a major impact in producing plants and animals that are able to resist stresses and diseases, thereby increasing food security. There is also potential to produce pharmaceuticals in plants through biotechnology, and provide foods that are nutritionally enhanced. Genetically modified organisms can also be used in cleaning up pollution and contamination. However, the application of biotechnology has raised concerns about biosafety, and it is vital to ensure that genetically modified organisms do not pose new risks to the environment or health. To understand the full potential of biotechnology and the issues that relate to it, scientists need access to information that not only provides an overview of and background to the field, but also keeps them up to date with the latest research findings.

This series, which extends the scope of CABI's successful "Biotechnology in Agriculture" series, addresses all topics relating to biotechnology including transgenic organisms, molecular analysis techniques, molecular pharming, in vitro culture, public opinion, economics, development and biosafety. Aimed at researchers, upper-level students and policy makers, titles in the series provide international coverage of topics related to biotechnology, including both a synthesis of facts and discussions of future research perspectives and possible solutions.

Titles Available

1. Animal Nutrition with Transgenic Plants
   Edited by G. Flachowsky
2. Plant-derived Pharmaceuticals: Principles and Applications for Developing Countries
   Edited by K.L. Hefferon
3. Transgenic Insects: Techniques and Applications
   Edited by M.Q. Benedict
4. Bt Resistance: Characterization and Strategies for GM Crops Producing Bacillus thuringiensis Toxins
   Edited by Mario Soberón, Yulin Gao and Alejandra Bravo
5. Plant Gene Silencing: Mechanisms and Applications
   Edited by Tamas Dalmay
6. Ethical Tensions from New Technology: The Case of Agricultural Biotechnology
   Edited by Harvey James
7. GM Food Systems and their Economic Impact
   Tatjana Brankov and Koviljko Lovre
8. Endophyte Biotechnology: Potential for Agriculture and Pharmacology
   Edited by Alexander Schouten
9. Forest Genomics and Biotechnology
   Richard Meilan and Matias Kirst
10. Transgenic Insects: Techniques and Applications, 2nd Edition
    Edited by M.Q. Benedict and Maxwell J. Scott
11. Plant Omics: Advances in Big Data Biology
    Edited by Hajime Ohyanagi, Eiji Yamamoto, Ai Kitazumi and Kentaro Yano
12. Next-generation Sequencing and Agriculture
    Edited by Phillip Bayer and Dave Edwards

13. Aquaculture and Fisheries Biotechnology: Genetic Approaches
Rex A Dunham

14. Algal Biotechnology
Edited by Qiang Wang

15. Genetic Engineering of Vegetable Crops
Edited by Pritam Kalia

16. Plant Breeding Technology: Future Trends and Challenges
Edited by Zeba Khan

17. Stress-Resilient Crops: Coordinated Omics-CRISPR-Nanotechnology Strategies
Edited by Jen-Tsung Chen

18. High Throughput Plant Metabolomics
Edited by Jen-Tsung Chen

19. Plant Biotechnology and Sustainable Agriculture: Bridging the Gap for Global Food Security
Edited by Tariq Aftab

20. Biotechnology of Orchids and Other Ornamentals
Edited by Pablo Bolaños-Villegas, Jian-Zhi Huang

21. Deep Learning in Plant Science and Agricultural Biotechnology
Edited by Jen-Tsung Chen

22. AI-Enhanced Plant-omics
Edited by Jen-Tsung Chen

AI-Enhanced Plant Omics

Edited by

Jen-Tsung Chen

CABI

CABI is a trading name of CAB International

CABI
Nosworthy Way
Wallingford
Oxfordshire OX10 8DE
UK

Tel: +44 (0)1491 832111
E-mail: info@cabi.org
Website: www.cabi.org

CABI
200 Portland Street
Boston
MA 02114
USA

T: +1 (617)682-9015
E-mail: cabi-nao@cabi.org

A catalogue record for this book is available from the British Library, London, UK.

ISBN-13: 9781800628625 (hardback)
9781800628632 (ePDF)
9781800628649 (ePub)

DOI: 10.1079/9781800628649.0000

Commissioning Editor: Ward Cooper
Editorial Assistant: Emma McCann
Production Editor: James Bishop

Typeset by Exeter Premedia Services Pvt Ltd, Chennai, India
Printed and bound in the UK by CPI Group (UK) Ltd, Croydon, CR0 4YY

Contents

Contributors

Maryam Afzaal, Department of Biochemistry Government College University Faisalabad -38000, Punjab, Pakistan; Email: maryamchattha88@gmail.com

Yogesh K. Ahlawat, Centre of Research Impact and Outcome, Chitkara University Institute of Engineering and Technology, Rajpura, 140401, Punjab, India; Email: ykahlawa@mtu.edu

Ajaz Ahmed, Directoriate of Non timber Forest products Khyber Pakhtunkhwa Forest Department Chitral-17200, Pakistan; Email: ajaz.ntfp@hotmailcom

Mohammad Ali, Department of Computer Science, Aligarh Muslim University, Aligarh, 202002, India; Email: mohdali9065@gmail.com

Abid Mehmood Alvi, Department of Plant Protection, Ghazi University, Dera Ghazi Khan-32200, Punjab, Pakistan; Email: aalvi@gudgk.edu.pk

Jameel M. Al-Khayri, Department of Agricultural Biotechnology, College of Agriculture and Food Sciences, King Faisal University, Al-Ahsa 31982, Saudi Arabia; Email: jkhayri@kfu.edu.sa

Shumaila Anam, Department of Botany, University of Agriculture Faisalabad-38000, Punjab, Pakistan; Email: shumailaanam44@gmail.com

Arumugam Vijaya Anand, Department of Human Genetics and Molecular Biology, Bharathiar University, Coimbatore 641 046, Tamil Nadu, India; Email: avahgmb@buc.edu.in

Mohd Anas, Department of Botany, Faculty of Life Sciences, Aligarh Muslim University, Aligarh, 202002, India; Email: mohdanas147177@gmail.com

Esakkimuthu Balaji, Prashant Fertility and Research Centre, Chennai 600031, Tamil Nadu, India; Email: balaji9994651754@gmail.com

Loulaq Banday, Department of Biotechnology, University of Kashmir, Srinagar, UT J&K, India; Email: loulaqanwar@gmail.com

Sana Basri, Proteomics and Plant Biotechnology Lab, Department of Botany, Faculty of Life Sciences, Aligarh Muslim University, Aligarh, 202002, India; Email: sana.basri8270@gmail.com

Kuppusamy Bhagavathi, Department of Human Genetics and Molecular Biology, Bharathiar University, Coimbatore 641 046, Tamil Nadu, India; Email: bhagavathikuppusamy@gmail.com

Maske Neha Chandrakant, Department of Agricultural Biotechnology, Assam Agricultural University, Jorhat-785013, Assam, India; Email: maskeneha1250@gmail.com

Neha Chaurasiya, Proteomics and Plant Biotechnology Lab, Department of Botany, Faculty of Life Sciences, Aligarh Muslim University, Aligarh, 202002, India; Email: chaurasiyaneha143@gmail.com

Jen-Tsung Chen, Department of Life Sciences, National University of Kaohsiung, Nanzih District, Kaohsiung 811, Taiwan; Email: jentsung@nuk.edu.tw

Siddhartha Das, Department of Plant Pathology, M.S Swaminathan School of Agriculture, Centurion University of Technology and Management, Paralakhemundi, Odisha, India; Email: siddhartha.das10@gmail.com

Tabasum Mohi Ud Din, Department of Botany, Amar Singh College, Srinagar, UT J&K, India; Email: tabsumji@gmail.com

Monisha Domesa, Department of Biochemistry, Bharathiar University, Coimbatore 641 046, Tamil Nadu, India; Email: monishadomesa172002@gmail.com

Pradip Kumar Averi Dubey, Department of Human Genetics and Molecular Biology, Bharathiar University, Coimbatore 641 046, Tamil Nadu, India; Email: averidubey.12@gmail.com

Suman Dutta, School of Agriculture and Rural Development, Ramakrishna Mission Vivekananda Educational and Research Institute, Belur Math, Howrah, West Bengal; Email: duttasumansiliguri@gmail.com

Jemaa Essemine, Jiangsu Key Laboratory of Crop Genetics and Physiology, Agricultural College of Yangzhou University, Yangzhou, 225009, China; Bioinformatics Section, Department of Informatics, University of Quebec at Montreal, Pavillon President-Kennedy, Montreal, QC H2X 3Y7, Canada; Email: jemaa.essemine@gmail.com

Sahul Ameed Nihaal Fathima, Department of Human Genetics and Molecular Biology, Bharathiar University, Coimbatore 641 046, Tamil Nadu, India; Email: nihaal2822@gmail.com

Xianzhong Feng, Key Laboratory of Soybean Molecular Design Breeding, Northeast Institute of Geography and Agroecology, Chinese Academy of Sciences, Changchun, 130102, PR China; Email: fengxianzhong@iga.ac.cn

Shanmugam Gavaskar, Department of Computer Application, Bharathiar University, Coimbatore 641 046, Tamil Nadu, India; Email: gavaskar@buc.edu.in

Subham Ghosh, Discipline of Bioinformatics, ICAR-Indian Agricultural Statistics Research Institute, Pusa, New Delhi-110012

Agnishwar Girigoswami, Faculty of Allied Health Sciences, Chettinad Hospital and Research Institute, Chettinad Academy of Research and Education, Kelambakkam, Chennai 602 103, Tamil Nadu, India; Email: dragnishwar@care.edu.in

Koyeli Girigoswami, Chettinad Academy of Research and Education, Kelambakkam, Chennai 602 103, Tamil Nadu, India; Email: koyelig@care.edu.in

Pemula Gowtham, Chettinad Academy of Research and Education, Kelambakkam, Chennai 602 103, Tamil Nadu, India; Email: g.pemula96@gmail.com

Mokhtar Guerfel, Department of Biological Sciences, College of Education Afif, Shaqra University, Ar-Riyad, Saudi Arabia; Email: mokhtar_guerfel@su.edu.sa

Ishfaq Majid Hurrah, Plant Biotechnology Division, CSIR-Indian Institute of Integrative Medicine, Sanat Nagar Srinagar, UT J&K, India; Email: eshfaqmajeed@gmail.com

Sasikumar Janani, Department of Human Genetics and Molecular Biology, Bharathiar University, Coimbatore 641 046, Tamil Nadu, India; Email: jananisasi45@gmail.com

Esha Kainat, Department of Botany, University of Gujrat-50700, Punjab, Pakistan; Email: eshameerkainat@gmail.com

Muhammad Kamran, Department of Plant Protection, Ghazi University, Dera Ghazi Khan-32200, Punjab, Pakistan; Email: muhkamran@gudgk.edu.pk

Kamal Kashyap, Department of Zoology, Rabindranath Tagore University, Hojai-782435, Assam, India; Email: kamalkashyap753@gmail.com

Mehrab Khalil, Zoology, Wildlife and Fisheries, University of Agriculture, Faisalabad-38000, Punjab, Pakistan; Email: mehrabkhalil8@gmail.com

Priti Kumari, Department of Human Genetics and Molecular Biology, Bharathiar University, Coimbatore 641 046, Tamil Nadu, India; Email: pt211000@gmail.com

Muhammad Majeed, Department of Botany, University of Gujrat-50700, Punjab, Pakistan; Email: m.majeed@uog.edu.pk

Marimuthu Malarvizhi, PG and Research Department of Zoology, Government Arts College, Coimbatore 641 018, Tamil Nadu, India; Email: pavimalar333@gmail.com

Sayanti Mandal, Department of Chemistry and Biochemistry, Sharda School of Engineering & Sciences, Sharda University, Greater Noida, Uttar Pradesh, India; Email: mandalsayanti@gmail.com

Mani Manoj, Department of Human Genetics and Molecular Biology, Bharathiar University, Coimbatore 641 046, Tamil Nadu, India; Email: manojinthravathi@gmail.com

Ganesan Megha, Department of Biochemistry, Bharathiar University, Coimbatore 641 046, Tamil Nadu, India; Email: meghaganesan123@gmail.com

Manoj Kumar Mishra, Department of Biotechnology, SR Institute of Management & Technology, Lucknow, Uttar Pradesh, India; Email: manoj30biotech@gmail.com

Isack Ibrahim Mrisho, College of Chinese Medicinal Material, Jilin Agricultural University, Changchun, 130118, PR China; Email: issackmson98@gmail.com

Elshan Musazade, Key Laboratory of Soybean Molecular Design Breeding, Northeast Institute of Geography and Agroecology, Chinese Academy of Sciences, Changchun, 130102, PR China; Email: elshan.musazade1@gmail.com

Pragya Pallavi, Center for Global Health Research, Saveetha Institute of Medical and Technical Sciences (SIMATS), Chennai 602 105, Tamil Nadu, India; Email: pragyapallavi56@gmail.com

Priyanka Paul, Department of Agronomy and Agroforestry, M.S Swaminathan School of Agriculture, Centurion University of Technology and Management, Paralakhemundi, Odisha, India; Email: priyankapaul85090@gmail.com

Sivaji S Pillai, Department of Human Genetics and Molecular Biology, Bharathiar University, Coimbatore 641 046, Tamil Nadu, India; Email: sivajispillai@gmail.com

Jeyabal Philomenathan Antony Prabhu, Department of Human Genetics and Molecular Biology, Bharathiar University, Coimbatore 641 046, Tamil Nadu, India; Email: jantony.prabhu@gmail.com

Mingnan Qu, Jiangsu Key Laboratory of Crop Genetics and Physiology, Agricultural College of Yangzhou University, Yangzhou, 225009, China; Email: qmn@yzu.edu.cn

Manoharan Rajesh, Department of Biochemistry, Bharathiar University, Coimbatore 641 046, Tamil Nadu, India; Email: rajeshmanoh@gmail.com

Iram Rasool, Plant Biotechnology Division, CSIR-Indian Institute of Integrative Medicine, Sanat Nagar Srinagar, UT J&K, India; Email: khaniram523@gmail.com

Asirvatham Alwin Robert, Department of Endocrinology and Diabetes, Prince Sultan Military Medical City, Ministry of Defense, Riyadh 11159, Kingdom of Saudi Arabia; Email: aalwinrobert@gmail.com

Biswajit Roy, Information Technology Group CSIR-Centre for Cellular & Molecular Biology, Habsiguda, Uppal Road, Hyderabad - 500 007, Telangana, India; Email: broy@ccmb.res.in

Bhaswati Sarma, Department of Botany, ADP College, Haiborgaon, Nagaon-782002, Assam, India; Email: bswati12@gmail.com

Sangeeta Sarma, Department of Botany, Jagiroad College, Jagiroad, Morigaon-782410, Assam, India; Email: sangeeta.sarma8@gmail.com

Abhik Sarkar, Discipline of Bioinformatics, ICAR-Indian Agricultural Statistics Research Institute, Pusa, New Delhi-110012

Varun Kumar Sharma, Department of Biotechnology & Microbiology, School of Sciences, Noida International University, Gautam Budh Nagar-201308, Uttar Pradesh, India; Email: varungenetics@gmail.com

Madhavan Shenbagam, Department of Biochemistry and Biotechnology, Annamalai University, Chidambaram 608 002, India Email: shenbu15@gmail.com

Salvinder Singh, Department of Agricultural Biotechnology, Assam Agricultural University, Jorhat-785013, Assam, India; Email: salvinder.singh@aau.ac.in

Dhanawantari L. Singha, Department of Botany, Rabindranath Tagore University, Hojai-782435, Assam, India; dhanawantarisingha@gmail.com

Dipro Sinha, Discipline of Bioinformatics, ICAR-Indian Agricultural Statistics Research Institute, Pusa, New Delhi-110012

Shilpy Singh, Department of Biotechnology & Microbiology, School of Sciences, Noida International University, Gautam Budh Nagar-201308, Uttar Pradesh; Email: shilpy.singh24@gmail.com

Mahankali Sravanalakshmi, Department of Human Genetics and Molecular Biology, Bharathiar University, Coimbatore 641 046, Tamil Nadu, India; Email: sravani99991@gmail.com

Macharla Sreenidhi, Department of Human Genetics and Molecular Biology, Bharathiar University, Coimbatore 641 046, Tamil Nadu, India; Email: macharlasreenidhi@gmail.com

Hannah Elizabeth Stephen Babu, Department of Human Genetics and Molecular Biology, Bharathiar University, Coimbatore 641 046, Tamil Nadu, India; Email: hannahelizahanu07@gmail.com

Alagarsamy Sumithra, Department of Computer Science and Engineering, SNS College of Technology, Coimbatore 641 035, Tamil Nadu, India; Email: sumithra.a.cse@snsct.org

Kempanna Sushmitha, Department of Biochemistry, Bharathiar University, Coimbatore 641 046, Tamil Nadu, India; Email: sushmithak099@gmail.com

Paneerselvam Swarnamughi, Department of Genetics, University of Madras, Chennai- 600113, Tamil Nadu, India; Email: swarnamughi313@gmail.com

Arpita Talukdar, Department of Botany, Rabindranath Tagore University, Hojai-782435, Assam, India; Email: arpitatalukdar1234@gmail.com

Aftab Umar, Department of Botany, University of Agriculture Faisalabad-38000, Punjab, Pakistan; Email: aftabumar538@gmail.com

Shanmugam Velayuthaprabhu, Department of Biotechnology, Bharathiar University, Bharathiar University, Coimbatore 641 046, Tamil Nadu, India Email: velayuthaprabhu@buc.edu.in

Kannan Vijayarani, Department of Education, Bharathiar University, Bharathiar University, Coimbatore 641046, Tamil Nadu, India; Email: kvijayarani@yahoo.com

Zhan Xu, Yazhouwan National Laboratory, No. 8 Huanjin Road, Yazhou District, Sanya City, Hainan Province 572024, China; Email: ricexuzhan@hotmail.com

Anurag Yadav, Department of Microbiology, C.P. College of Agriculture, Sardarkrushinagar Dantiwada Agricultural University, Sardarkrushinagar, Dist. Banaskantha, Gujarat-385506, India; Email: anuragyadav123@gmail.com

Kusum Yadav, Department of Biochemistry, University of Lucknow, Lucknow, Uttar Pradesh-226007, India; Email: anukusum@gmail.com

Preface

In the era of multiple omics (multi-omics), the ways to deal with the resulting big data, i.e. bioinformatics, became one of the leading technologies for uncovering complicated cellular processes and biological networks to construct spatiotemporal and high-dimensional biology. Among bioinformatics approaches, in recent years, artificial intelligence (AI) algorithms such as machine learning, deep learning, and most recently generative AI, have been applied to studying and integrating with plant-omics, including genomics, transcriptomics, proteomics, metabolomics, epigenomics, and so on, which are expected to solve biological problems by modeling, simulation, prediction, detection, identification, etc. Without a doubt, nowadays, AI technology has become a game changer in all fields, including plant biological research.

Plant biologists explore multi-omics to enable the release of the potential from genomes to develop crops with better quality, higher yield, and more tolerance to stressors. In the scenario of global climate change and increasing human population, omics-assisted crop breeding is promising to support the sustainable development goal (SDG) of "zero hunger" globally by achieving sustainable agriculture. AI technology is indispensable throughout the entire strategy, particularly for its contribution to efficiency and saving labor. For instance, AI technology can advance agriculture to become smarter by integrating with sensors and robotics to construct dynamic three-dimensional plant phenotyping and further digital phenomics, not only to be applied to agricultural production but also for assisting crop breeding, plant disease identification, detection, and protection.

This book summarizes current achievements in AI tools such as machine learning, deep learning, and generative AI, and their applications in plant-omics approaches, such as functional genomics, transcriptomics, proteomics, metabolomics, and epigenomics, highlighting the knowledge concerning the interactions between significant crops and environmental stressors, pests, and pathogens. This book includes AI-based strategies for constructing digital plant-omics with single-cell and spatial resolution and three-dimensional plant biology, which are valuable for future precision crop breeding and climate-smart agriculture.

The Editor appreciates all contributors for their innovative chapters and the publisher for the instruction and assistance.

Jen-Tsung Chen
National University of Kaohsiung, Taiwan

1 Artificial Intelligence: An Introduction to Technical Advancements and Current Biological Applications in Plants

Mani Manoj[1], Kuppusamy Bhagavathi[1], Sahul Ameed Nihaal Fathima[1], Sivaji S. Pillai[1], Marimuthu Malarvizhi[2], Madhavan Shenbagam[3], Shanmugam Velayuthaprabhu[4] and Arumugam Vijaya Anand*

[1]Department of Human Genetics and Molecular Biology, Bharathiar University, Coimbatore, Tamil Nadu, India; [2]PG and Research Department of Zoology, Government Arts College, Coimbatore, Tamil Nadu, India; [3]Department of Biochemistry and Biotechnology, Annamalai University, Chidambaram, India; [4]Department of Biotechnology, Bharathiar University, Coimbatore, Tamil Nadu, India

Abstract

Artificial intelligence (AI) is rapidly transforming the landscape of plant sciences by offering innovative tools and techniques that enhance our understanding of complex biological systems. This chapter presents a comprehensive introduction to the role of AI in plant research, covering its integration with plant-omics technologies and its diverse applications in genomics, transcriptomics, proteomics, metabolomics, phenomics, and imaging. It explores the use of machine learning and deep learning models for trait analysis, disease prediction, and gene function annotation. Furthermore, the chapter delves into the transformative role of AI in crop improvement, precision breeding, and sustainable agriculture. By automating phenotyping, optimizing resource use, and facilitating smart pest management, AI is paving the way for climate-resilient and high-yield agricultural practices. Despite its immense potential, the deployment of AI in plant sciences faces challenges such as data quality, model interpretability, and ethical concerns. This chapter concludes by reflecting on current limitations and highlighting future directions for AI-driven innovations that promise to revolutionize plant biology and agriculture.

Keywords: Artificial intelligence, plant-omics, machine learning, deep learning, crop improvement

1.1 Introduction to Artificial Intelligence in Plant Sciences

Artificial intelligence (AI) is the design and development of computer systems that can execute tasks that normally require human intelligence. These tasks are visual perception, speech recognition, decision making, and language translation (Russell and Norvig, 2016). AI in the present times is a revolutionary factor in most industries, and its use in plant sciences is on the cusp of revolutionizing our understanding, cultivation, and sustainable management of plant life (Singh *et al.*, 2021). AI has transcended its traditional uses, affected a wide range of industries, and left its mark on

*Corresponding author: avahgmb@buc.edu.in

 AI-Enhanced Plant Omics (ed. Jen-Tsung Chen)
DOI: 10.1079/9781800628649.0001

industries globally (Williamson *et al.*, 2021). This includes the domain of plant science, which is a very fertile ground for the use of AI, revolutionizing revered practices. The use of AI in plant science brings a wave of change, with new solutions to old challenges facing agriculture, ecology, and environmental sustainability (Bhoi and Mahali, 2024). As AI technologies develop, the capacity to interpret complex biological systems, improve agricultural productivity, and improve sustainable practices is enhanced. From precision agriculture through machine learning (ML) for soil health prediction to intelligent sensors that allow real-time monitoring of plant growth, the synergy of AI and plant species seeks not only to improve efficiency but also to solve some of humanity's most significant issues, such as food shortages and climate change (Van Dijk *et al.*, 2021). Plant science encompasses disciplines such as molecular biology, genetics, ecology, and agronomy, involving the intricate interactions between plants and their environment (Bhoi and Mahali, 2024). Understanding these dynamics is important to improve crop productivity, food security, and climate change mitigation (Harfouche *et al.*, 2019). AI integration enhances sustainable agriculture by using practices like agroforestry, precision agriculture, and organic farming (Patil and Kumar, 2022). Promoting AI-based irrigation systems, drought-tolerant crops, and soil conservation can improve water efficiency and retain soil health. AI technologies can also contribute to the development of high-yielding, disease-resistant crops that are responsive to adapting to changing climatic conditions (Rani *et al.*, 2023). Agricultural robots or agrobots are transforming agriculture by sowing seeds and applying fertilizer automatically. This stimulates plant development and minimizes the use of fertilizer, with beneficial implications for the environment. Sensor-equipped robots collect vital information on soil, humidity, and nutrients, enabling farmers to make informed decisions and extract maximum yield (Rani *et al.*, 2023). Crop and soil monitoring is critical in contemporary agriculture, enabling farmers to make informed decisions, maximize resource utilization, and elevate yields sustainably. Advanced sensors like satellites, drones, and Internet of Things (IoT) devices offer vital information on crop health, soil moisture, and nutrient levels, enabling the application of targeted fertilization and irrigation. These technologies, collectively, improve data analysis and monitoring efficiency (Wang *et al.*, 2015).

Intelligent spraying or variable-rate spraying is a new-generation agricultural practice that utilizes advanced technology to maximize pesticide application. This practice seeks to lower chemical consumption while maximizing efficacy against pests and diseases. Sensor- and GPS-equipped intelligent sprayers identify variations in the field, creating detailed maps to mark treatment zones (Kumar *et al.*, 2023). AI has a far-reaching impact on crop production at the farm gate through a range of mechanisms. This includes optimization of seed, fertilizer, and chemical application; automation of irrigation scheduling; monitoring of soil, crop, and animal health; monitoring of disease and pests; and decision support for the placement of agricultural machinery in fields, illustrated in Fig. 1.1 (Sampath *et al.*, 2023). Human activities are interfering with ecological balance and compromising biodiversity, making it essential to preserve plant species, which act as habitats for many organisms. Plants supply crucial resources for humankind, which offer a wide range of things including food, medicine, and shelter. Trees and forests are crucial for the production of oxygen and the reduction of toxic greenhouse gas emissions. By applying AI, scientists can improve the efforts undertaken to save endangered plant species and ensure a healthy ecosystem (Amara *et al.*, 2017). AI enables rapid assessment of plant traits and real-time monitoring of the environment using robotics. The use of plant phenotyping enhances our understanding of plant physiology, resulting in the production of better seed varieties and reducing the use of pesticides and fertilizers, thereby encouraging a more sustainable agriculture model (Van Dijk *et al.*, 2021). Exotic diseases and pests have the capability to cause extensive damage to plant growth and productivity. So early detection is important to enable appropriate interventions. Recent developments in computer vision have increasingly enabled effective detection of these pests (Renuka, 2022).

ML is a crucial component in the identification of genes responsible for stress resistance in crops. Significant research includes the use of support vector machines (SVMs) in the identification of drought-related genes. In order

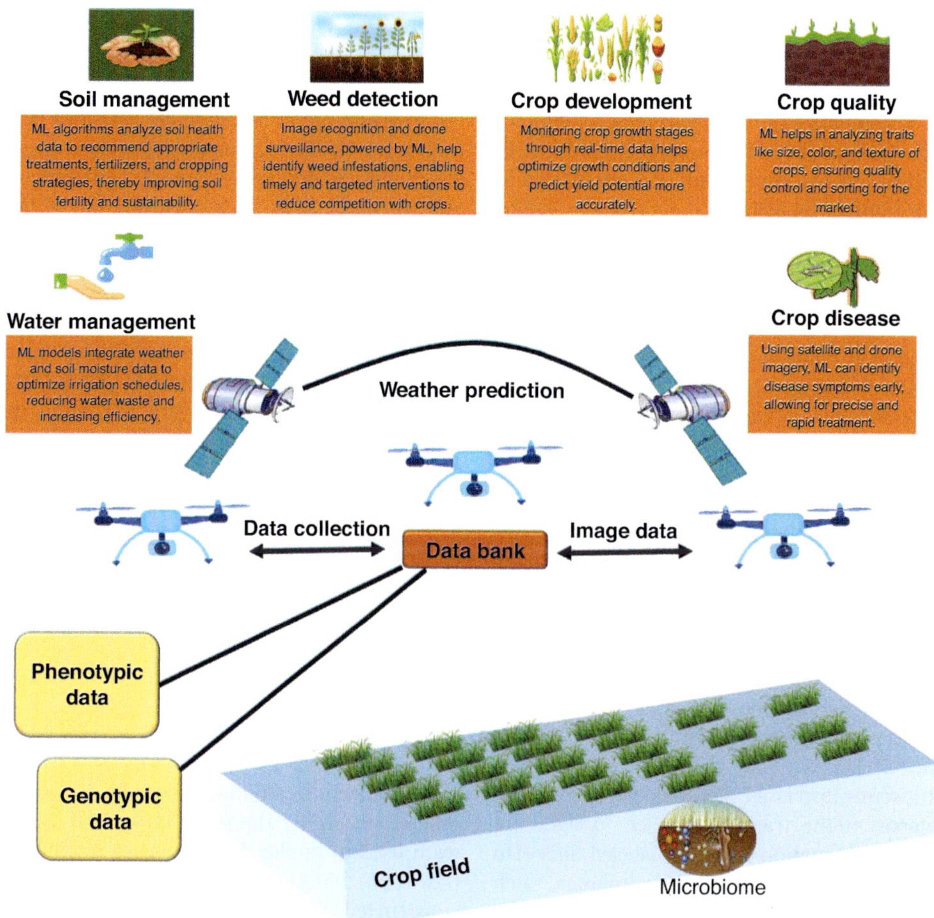

Fig. 1.1. Artificial intelligence in plant biology. (Author's own image.)

to assess the salt tolerance of maize, Liang *et al.* (2011) and Shikha *et al.* (2017) employed neural networks and SVM to identify the single nucleotide polymorphisms (SNPs) that contribute to the salt tolerance (Liang *et al.*, 2011; Ravari *et al.*, 2015; Shikha *et al.*, 2017). Additionally, *Arabidopsis thaliana* was shown to have significant microRNAs, with an R^2 value of 0.96 when using SVM techniques to generate predictions for abiotic stress (Asefpour Vakilian, 2020). ML methods indicated common responses to stress in rice (Shaik and Ramakrishna, 2014). Using a clustering algorithm, 216 plants were classified systematically as per their incomplete metabolite profiles and both nutritional and taxonomic correlations were found (Liu *et al.*, 2017). ML methods were also employed in ascertaining the impact of abiotic stress on *Arabidopsis*: 23 proteins were found as markers of responses to stress, and it was inferred that biochemical regulation plays an important role in varying environmental conditions (Fürtauer *et al.*, 2018).

1.2 Machine Learning and Deep Learning in Plant-Omics

Big data analysis using AI in the form of ML has four main components: system, user, domain, and data. These components interact bidirectionally; for instance, data are fed to ML, and outputs generate new data. Users provide feedback and domain information for enhancing ML predictions, while the domain supplies contextual information for model deployment. The system holds the hardware and software to execute ML algorithms (Zhou *et al.*, 2017). ML has three classes: supervised, unsupervised, and reinforcement learning (Greener *et al.*, 2022). Supervised learning employs datasets with features alongside labels, where the ML model can learn patterns to predict labels and hence deploy them to new datasets (Jiang *et al.*, 2020). Moreover, outputs of unsupervised methods, i.e. clusters of hierarchical clustering and principal components of principal component analysis (PCA), can be provided as inputs of supervised methods to decrease feature space (Zhou *et al.*, 2017). Semi-supervised methods are handy when handling datasets with few labeled data, as is the most frequent situation in plant biology research, where few genes of interest may have functional annotations. Such methods attempt to improve the performance of models by discarding the unlabeled data and switching to supervised learning or discounting labels to carry out unsupervised learning (Daniels *et al.*, 2021). With the assistance of unlabeled data, semi-supervised methods can learn from patterns in the training dataset, in addition, to infer information from the labeled data (Hu *et al.*, 2022). Unsupervised techniques, such as hierarchical clustering and PCA, are also used on unlabeled datasets to identify patterns in the features to determine the number of classes inherent in the dataset. The process can be highly robust and effective, especially where sample labels may be inaccurate or non-existent, as it is focused on clustering features (Scheurer and Slager, 2020). For example, a latent feature like protein structure may drive patterns in sequence data; even without labels for the protein structure, an unsupervised model can be capable of identifying patterns in the sequence data that are related to their structural properties (Rives *et al.*, 2021). The reinforcement learning algorithms, just like unsupervised learning, do not depend on input–output pairings. Instead, reinforcement learning is based on experience feedback, acting on rewards or penalties for responses rather than on explicit errors for unwanted outcomes (Averbeck and O'Doherty, 2022).

In scenarios like crop yield prediction from multi-omics data, quality in feature extraction can significantly influence outcomes, and both supervised and unsupervised techniques can fail to give a direct mapping between raw data and crop yield. Reinforcement learning is a good alternative under such scenarios (Elavarasan and Vincent, 2020). Deep learning (DL) consists of neural networks with several hidden layers, which can be trained using both supervised and unsupervised techniques. Each of the hidden layers learns sequentially to recognize data patterns based on its preceding layers (Kamilaris and Prenafeta-Boldú, 2018). The structure of the hidden layers, including the number and the size, can be modified in accordance with the specific objectives of the research (Dong *et al.*, 2022). For instance, feed-forward neural networks with "thin" hidden layers are generally used to process sequential data like DNA sequences, whereas convolutional neural networks (CNNs) are optimally used in datasets with two or more dimensions, i.e. multi-omics data (Camacho *et al.*, 2018). Nevertheless, one of the chief shortcomings of DL methods is that they require large datasets, which are usually restricted in the majority of areas of plant molecular studies, especially in mono-omics research. High-accuracy DL models were used to predict blood glucose trajectories (Fox *et al.*, 2018). An iterative DL network was constructed for protein secondary structure prediction (Novakovsky *et al.*, 2023). Moreover, a DL model was used for tumor suppressor gene and oncogene prediction (Tavanaei *et al.*, 2017). DL models have also been found to be able to predict DNA methylation states at the single-cell level (Angermueller *et al.*, 2017). DL models were also used to predict the viability of drug-treated cancer cell lines (Menden *et al.*, 2013). In a recent study, DL models with a densely connected network architecture were compared on nine large-scale genomic datasets, including environment–genomic interactions, and the results were contrasted with the genomic best linear unbiased prediction (GBLUP) model (Montesinos-López *et al.*, 2018). The results showed that the DL models were good, outperforming GBLUP in six of the nine datasets, especially in cases where the genotype considerations and environmental interaction terms were not included. Some of the most commonly used architectures in neural networks include densely connected networks (multilayer perceptrons), convolutional networks, and recurrent networks (Gulli and Pal, 2017).

In the current work, a dense connectivity network was utilized, for example, an input layer, a few hidden layers, and an output layer, with the connections between the neurons being given weights as a representation of gained experience. Each layer constitutes a data filter that progressively refines the information pertaining to the intended task. Training the network is done by exposing the dataset to the network for a specified number of epochs, which are then broken down into smaller batches to improve efficiency. Model fitting is influenced by the number of epochs utilized, and the application of regularization methods, for example, dropout regularization, reduces the possibility of overfitting by temporarily disabling a portion of neurons during training. Optimal hyperparameter selection for the model is a challenging but essential exercise (Montesinos-López *et al.*, 2018). AI-based models have demonstrated tremendous potential in disease resistance prediction in different crop species. Through the application of DL and ML methods, the models improve the accuracy and efficiency in the prediction of diseases, a critical component of the enhancement of crop yields and ensuring sustainability. The combination of genomic data and image-based evaluations has particularly been useful in disease-resistant trait detection and plant disease diagnosis.

1.3 AI Applications in Genomics and Transcriptomics

Genomics is going towards an era of data-intensive science. AI, in the form of DL algorithms, is at the forefront of knowledge and pattern extraction from this abundance of genomic data. A deep convolutional neuronal network called DeepVariant could detect gene variation within a contemporaneous next generation sequencing (NGS) read (Poplin *et al.*, 2018). This was done by learning statistical correlations of images of read pileups that surround putative variants and genuine genotype calls from the model. One of the greatest challenges faced by genomics is to call nucleotide variation in an individual's genome as compared to a reference sequence, a task named 'variant calling'. Efficient yet accurate variant calling is important in order to discern genomic variation for phenotypic variation and disease. Clairvoyante overcomes this challenge by predicting variant type, zygosity, alternative alleles, and indel length. Clairvoyante ignores sample information and is able to identify variants in under 2 hour on a normal server. Clairvoyante is optimized for leveraging long-read sequencing data from platforms such as Pacific Biosciences (PacBio) and Oxford Nanopore Technology (ONT) but is flexible enough to be widely applied to short-read datasets too (Luo *et al.*, 2019; Chen *et al.*, 2023; Ahsan *et al.*, 2024). DeepPVP (PhenomeNet Variant Predictor) excels in identifying variants from whole-exome and whole-genome sequence data (Trieu *et al.*, 2020), whereas ExPecto excels at predicting the tissue-specific transcriptional consequences of mutation and functional SNPs accurately (Zhou *et al.*, 2018). AI-driven whole-genome sequencing can transform the field of genomics and further patient outcomes with integration into the world of multi-omics data.

Gene annotation (GA) is the process of identifying and interpreting the functional elements encoded in a genome. GA relies heavily on automated methods, which include ML (Eraslan *et al.*, 2019; Zou *et al.*, 2019; Mahood *et al.*, 2020) as well as other computational methods, such as rule-based and heuristic methods (Altschul *et al.*, 1990; Stanke *et al.*, 2006). A range of methodologies has been developed for the categorization of nonsynonymous variants (Tang and Thomas, 2016). The combined annotation-dependent depletion approach (CADD) combines many predictive features in a machine-learning framework to estimate the deleteriousness of genetic variants (Kircher *et al.*, 2014). PrimateAI, which uses CNNs trained on variants with known pathogenicity and incorporates data augmentation through cross-species information, has shown superiority to previous methodologies when trained directly on sequence alignments (Sundaram *et al.*, 2018). This network learns important protein domains, conserved amino acid positions, and sequence dependencies directly from a training dataset of around 120,000 human samples.

PrimateAI is a major improvement in variant classification tools that can potentially decrease the reliance of clinical reporting on previously known knowledge. SpliceAI, a 32-layer deep neural network, can predict both canonical and non-canonical splicing directly from exon–intron junction sequence data (Jaganathan *et al.*, 2019). DL approaches have also greatly increased our ability to find regulatory elements and predict the effect of genetic variation on these elements (Alipanahi *et al.*, 2015; Kelley *et al.*, 2018). With the creation of increasingly vast amounts of data at low expense, and improvement in complementary hardware, new types of AI enable a range of prediction tasks, including annotating the regulatory genome and classifying single-cell data (Angermueller *et al.*, 2017; Ching *et al.*, 2018; Eraslan *et al.*, 2019; Zou *et al.*, 2019). DL models can identify complex patterns. For example, in predicting the DNA sequences in the human genome that regulate gene transcription in a given cell type, these models can learn features such as sequence composition (e.g. GC content), presence or absence of specific motif patterns like transcription factors (TFs) binding sites, local chromatin accessibility, biophysical properties of DNA (e.g. bendability), and positional variations among sequence features. Deep neural network (DNN) models became the most common predictive modeling paradigm in the regulatory genomics field (Toneyan *et al.*, 2022). These models take in a hypothesized regulatory DNA sequence (typically between 100 and 10,000 base pairs) and attempt to predict some dynamic features (specifically, cell or context specificity) of the activity of that sequence. For example, a model may predict whether a specific TF binds to that sequence in a specific cell type based on a chromatin immunoprecipitation followed by a sequencing (ChIP-seq) experiment (Zhou and Troyanskaya, 2015; Kelley *et al.*, 2016; Kim *et al.*, 2021; Avsec *et al.*, 2021). Additionally, ML algorithms have been utilized to detect distinct features, including protein-coding genes, microRNAs (miRNAs), long non-coding RNAs (Sun *et al.*, 2015), polyadenylation sites (Gao *et al.*, 2018), DNAse I hypersensitive sites (Lyu *et al.*, 2018), cis-regulatory elements (CREs), and chromatin states (Ernst and Kellis, 2017). The workflow for predicting gene function using AI models is illustrated in Fig. 1.2.

The detection of these genetic features can utilize ML in a classification model: genetic features can be detected as "protein-coding genes" or not, "miRNA" or not, or "CRE" or not. Thus, methods like SVMs, random forests (RFs), and CNNs can be used for genomic feature prediction. SVMs have been shown to be effective for both binary classification problems (e.g. presence or absence of a gene) and multi-class problems (e.g. gene, miRNA, pseudogene, transposon) (Mathur and Foody, 2008); however, compared to RFs, they need tuning of multiple parameters. Indeed, RFs can provide better performance on most tasks (Caruana and Niculescu-Mizil, 2006). However, despite the rapid proliferation of various algorithms since the 1990s, hidden Markov models (HMMs) remain the most effective ML approach for a wide range of structural feature prediction tasks in plant biology. HMM algorithms can be trained iteratively in order to predict the functional category (e.g. splice site, exon, enhancer) of single nucleotides from the categories of neighboring nucleotides. The study of gene expression is essential to the development of both biological and medical research fields. The identification of patterns in gene expression profiles gives valuable molecular signatures that allow the prediction, diagnosis, and forecasting the development of various diseases. Gene expression analysis is significantly enhanced by data science methods, particularly through the use of ML algorithms. Nevertheless, with the increasing complexity and size of genomic data, there is a need for analytical methods that scale well (Moreno *et al.*, 2022). The use of transcriptome-based toxicity prediction is increasingly being viewed as a valuable supplement to conventional chemical structure-based methods. Similar to quantitative structure–activity relationship (QSAR) models, which investigate the connection between molecular structure and biological activity (Bradbury, 1994; Cronin and Dearden, 1995), transcriptomic models examine gene expression profiles to deduce toxicological effects. Although early ML techniques, such as k-nearest neighbors (k-NNs), were employed to classify toxicity, they were confronted with the issue of handling asymmetric datasets—a problem also prevalent in transcriptomics due to the imbalance of toxic and non-toxic expression profiles (Dunn III, 1988). Contemporary

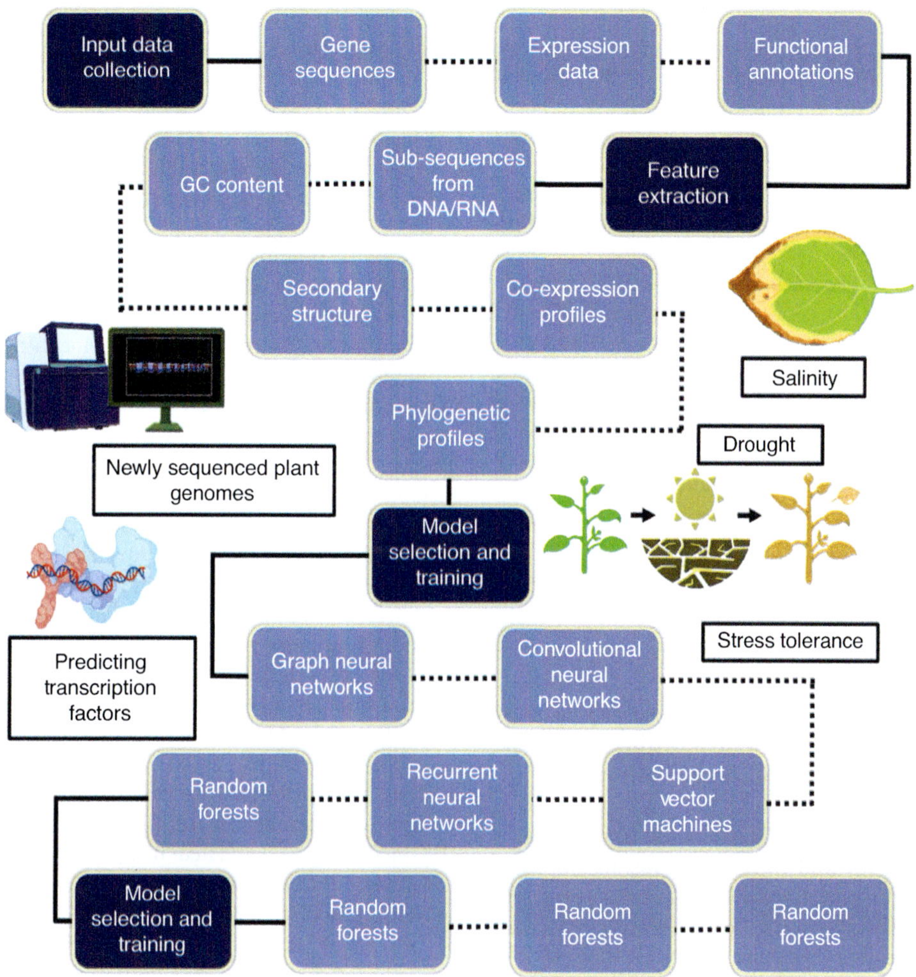

Fig. 1.2. Predicting gene function using AI models. (Author's own image.)

algorithms, such as artificial neural networks (ANNs) and decision tree-based models such as RFs, provide enhanced generalization performance, particularly in the case of imbalanced transcriptomic datasets (Kumar and Anitha, 2014). DNNs have demonstrated superior performance when optimized with advanced loss functions, even under conditions of extreme data imbalance (Myint *et al.*, 2012). When integrated with feature representations derived from transcriptomic fingerprints or pathway-level descriptors, these models exhibit robust predictive capabilities. Their performance is often comparable to that of models using structural fingerprints, such as ECFP6 and MACCS, in quantitative structure–activity relationship (QSAR) studies (Myint and Xie, 2014).

In addition, enhanced descriptors, such as those obtained from element-specific persistent homology (ESPH) and other biological markers, which were originally applied in structural models (Wu and Wei, 2018), can also be applied in transcriptomics by employing topological and interaction-level gene data to improve the predictive capability of toxicity. The incorporation of transcriptomic data into toxicity prediction models has significantly enhanced the accuracy of potential toxicant identification. Compared to traditional approaches that rely on chemical structural descriptors alone, gene expression profile analysis provides a dynamic picture of cellular response to chemical exposures. For instance, Schwartz *et al.* (2015) treated 3D-cultured human pluripotent stem cell-derived neural cells with a series of compounds and employed RNA-seq to capture whole-genome expression profiles. They subsequently applied SVM to predict the chemicals based on their toxicity, with a remarkable average AUC value of 0.91 (Murray *et al.*, 2018). Yamane *et al.* (2016) conducted chemical treatments on human embryonic stem cells and analyzed the resulting transcriptomes to construct gene interaction networks. By classifying chemicals into functional categories such as neurotoxins and carcinogens, they enhanced the accuracy of toxicity prediction models, including the detection of delayed or late-onset toxic effects (Yamane *et al.*, 2016). These reports underscore the potential of integrating transcriptomic data with ML algorithms to advance the science of toxicology.

1.4 Proteomics and Metabolomics: AI-Based Approaches

Prediction of three-dimensional (3D) structures of proteins and their functional mechanisms offers revolutionary potential in plant proteomics, especially in the improvement of essential crop-related agricultural traits (Ceasar and Ebeed, 2024). One significant barrier lies in predicting the structures of proteins without known homologs. Although template-free protein structure prediction (PSP), spurred by ML and optimization methods based on search strategies have shown potential, exhibiting high precision in prediction for complex proteins, especially those relevant to drug discovery, remains challenging (Mufassirin *et al.*, 2023). Basic protein modeling methods, including homology modeling, ab initio modeling, and threading, have been the cornerstone of the research. This research area has witnessed a rapid transformation from the discovery of DL-based approaches to the development of highly accurate AI models like AlphaFold (including AlphaFold, AlphaFold2, and AlphaFold3), RosettaFold, and ProteinBERT described in Table 1.1 (Chen *et al.*, 2024). The integration of AI with large amounts of biological data enables genetic dissection, identification of functional genes, regulatory element prediction, and phenotypic prediction. One revolutionary breakthrough in this field is intelligent precision design breeding (IPDB), which uses AI for the systematic design of crops (Zhu *et al.*, 2025).

Notwithstanding such progress, use of AI and optimization algorithms (OA) in plant biology is constrained by convoluted computational definitions and algorithmic demands. Thus, a review to dispel the mystery surrounding modeling strategies and optimization approaches is needed by plant biologists (Hesami and Jones, 2020). In genomics, although DNA assembly and sequencing are experimentally dominated, gene annotation is computationally driven, involving predictive models combining statistical evidence and biological knowledge (Gremme *et al.*, 2005). AI and ML are also revolutionizing prediction of plant secondary metabolites production. Khanna *et al.* (2024) showed how deep neural networks, ensemble learning, and k-nearest neighbor algorithms could infer intricate relationships between macronutrient availability and secondary metabolite production. The models were validated using computer-simulated environments to determine robustness (Khanna *et al.*, 2024).

Fathi *et al.* (2025) utilized artificial neural networks, radial basis function (RBF) and multilayer perceptron (MLP), to predict concentrations of key secondary metabolites, e.g. carotenoids, proline, and stress-induced soluble sugars, hence offering an alternative to traditional chemical analyses. Likewise, AI-assisted metabolomics analysis offers informative information on the mechanisms of disease, the process of drug discovery, and potential metabolic interventions (Fathi *et al.*, 2025; Hasan *et al.*, 2025). In pathway research, AI is utilized across the discovery, design, and optimization phases. Techniques such as ML, DL, natural language processing, network analysis, and data mining are utilized to integrate omics data, enabling more informative biology (Liao *et al.*, 2025). Murmu *et al.* (2024) identified the use of AI in omics-based plant defense research, despite existing limitations in existing approaches.

Table 1.1. Deep learning approaches for plant protein prediction. (Author's own table.)

AI model	Description	Plant protein applications	References
AlphaFold	AlphaFold, created by DeepMind, accurately predicts 3D protein structures from sequences of amino acids.	Used to forecast structures of many plant proteins, such as those responsible for photosynthesis, stress, and development.	Baek *et al*., 2021; Elnaggar *et al*., 2021; Jumper *et al*., 2021
AlphaFold 2	A more powerful version of AlphaFold with increased accuracy and performance, employing cutting-edge neural network architecture.	Utilized for precise structural predictions of intricate plant proteins to help understand their function and interactions.	Baek *et al*., 2021; Elnaggar *et al*., 2021; Jumper *et al*., 2021
AlphaFold 3	The newest version, which is also able to predict protein structures and interactions with other molecules.	Enables thorough modeling of plant protein complexes, such as those in metabolic pathways and signaling.	Baek *et al*., 2021; Elnaggar *et al*., 2021; Jumper *et al*., 2021
RoseTTAFold	This model, developed by the Baker Lab, combines deep learning and Rosetta energy functions for structure prediction.	Used for modeling plant proteins where structures are not available experimentally, assisting with functional annotation and engineering.	Baek *et al*., 2021; Elnaggar *et al*., 2021; Jumper *et al*., 2021
ProtTrans	Uses transformer-based language models trained on large protein sequence datasets for diverse predictive tasks.	Aids in the prediction of structural features and functions of plant proteins, particularly those with limited homology to known structures.	Baek *et al*., 2021; Elnaggar *et al*., 2021; Jumper *et al*., 2021

1.5 AI in Plant Phenomics and Imaging Technologies

The union of AI and high-throughput phenotyping (HTP) technologies has revolutionized crop improvement through the rapid and accurate evaluation of crop traits at a large scale. AI algorithms are responsible for processing large volumes of data, detecting meaningful associations between phenotypic traits and genetic factors, thereby facilitating the selection of desirable traits for breeding programs (Sheikh *et al.*, 2024). DL has immensely enhanced image analysis software, such that it is now more accessible for HTP use in plant biology. Thus, researchers are now able to answer basic questions in reproductive biology through the interrogation of complex image data (Warman and Fowler, 2021). Computer vision techniques, facilitated by ML, have been extensively used in high-throughput plant phenotyping, delivering non-destructive ways of visualizing plant growth and physiological reactions that elucidate gene functions regarding trait development (Mochida *et al.*, 2019). Vision-based phenotyping approaches allow for the non-destructive measurement of plant traits, ranging from single-trait measurements to overall measurements of crop canopies using different imaging modalities (Kolhar and Jagtap, 2023). On top of this, automated machine learning (AutoML) frameworks are being implemented in image-based phenotyping for the development of predictive models of various plant traits, automating data interpretation and analysis (Koh *et al.*, 2021).

High-resolution, 3D phenotyping of plants is made possible by sophisticated field robots equipped with laser and camera sensors, transforming physical plants into digital twins that can give estimates of plant height and leaf area (Esser *et al.*, 2023). Similarly, unmanned aerial vehicles (UAVs) with onboard image sensors combined with ML approaches have made crop breeding and precision agriculture activities more streamlined (Gano *et al.*, 2024). High-throughput image analysis makes monitoring plant growth automatic, significantly reducing labor and enhancing productivity (Jiang and Li, 2020). Using ML algorithms, such as RF, enhances the precision of trait extraction from images, enabling accurate, non-destructive monitoring of plant growth (Lee *et al.*, 2018; Das Choudhury *et al.*, 2019). Intermittent image capture makes analysis of dynamic growth patterns over time possible, giving insights into developmental stages and environment responses (Du *et al.*, 2020). Multimodal imaging modalities like RGB, thermal, and hyperspectral imaging further enrich phenotyping datasets, enabling more detailed analysis of plant health and development (Koirala *et al.*, 2019).

AI, in the form of DL methods, facilitates early detection of plant stress through the identification of minute changes in images that cannot be detected by the human eye. Early detection is critical for intervention and prevention of adverse effects. DL architectures, such as CNNs and long short-term memory (LSTM) networks, have been highly accurate in detecting and classifying different plant stressors, such as drought, nutrient deficiency, and disease (Azimi *et al.*, 2021). Multispectral and hyperspectral imaging further improves this capacity by detecting physiological and biochemical changes associated with plant stress, which, with AI, allows for more accurate detection (Mahlein, 2016; Mahlein *et al.*, 2024). The application of interpretability techniques, such as saliency maps and layer-wise relevance propagation, ensures that AI systems focus on biologically relevant features, thus improving understanding of their decision-making process (Nagasubramanian *et al.*, 2020). Additionally, time-series analysis of plant images using AI models allows for monitoring of stress development, gaining insight into the dynamics of stress and allowing for the development of timely management strategies (Azimi *et al.*, 2021). Despite these improvements, the performance of AI models, especially those based on DL, is highly dependent on the availability of large, annotated datasets, which are generally challenging to obtain for a range of plant species and stress conditions. Transfer learning addresses this by allowing pre-trained models (previously trained on a larger dataset) to be transferred to new, smaller datasets, thus improving performance and reducing training time (Nagasubramanian *et al.*, 2019).

1.6 Role of AI in Crop Improvement and Breeding

The use of AI in crop improvement and breeding is increasingly common, offering new avenues for enhancing the efficiency and sustainability of agriculture. The ability of AI to examine and process extensive data is of paramount importance in addressing the complex issues of plant breeding, particularly in determining the relationship between genotype and phenotype, optimizing breeding approaches, and enhancing the stress tolerance of plants. In addition, the integration of AI with multi-omics technology, such as genomics and phenomics, is revolutionizing the field by enabling precise and rapid gene identification, thus accelerating crop improvement schemes. One of the major contribution areas of AI is high-throughput phenotyping and genotyping. AI enhances the speed and accuracy of processing complex datasets, thus enhancing the knowledge on the genetic foundation of key agronomic traits (Khan *et al.*, 2022).

ML algorithms, including DL models, are common in genomic selection and feature selection, thus making accurate predictions of crop performance under different environmental conditions (Negus *et al.*, 2024). Such technologies have become essential to modern breeders seeking to accelerate selection cycles and increase predictability of traits. AI is also leading the charge in enhancing stress tolerance and precision breeding. ML algorithms are used to enhance plant resistance to abiotic and biotic stresses, thus contributing to improved productivity and adaptability. This attribute becomes more essential because of the effects of climate change, where crops have to adapt to irregular and extreme environmental conditions (Sharma *et al.*, 2025). The precision breeding facilitated by AI also supports the optimization of gene-editing systems and prediction of the effect of gene variants, which are essential in breeding varieties that are adapted to future environmental conditions (Farooq *et al.*, 2024). When it comes to genomic predictions and crop modeling, AI is once again crucial. Advanced models like SoyDNGP are utilized for genomic predictions, which facilitate intelligent parent selection and precise trait forecasting crucial ingredients for successful breeding programs (Feng *et al.*, 2024). Moreover, AI-facilitated crop growth modeling and computer vision methods are being utilized to enhance phenomics, thus facilitating enhanced monitoring and management of plant growth and development throughout the entire life of the crop (Negus *et al.*, 2024). In spite of the colossal potential of AI in enhancing crop improvement, some challenges exist. The construction of explainable AI models is still a core priority in bringing transparency and interpretability to breeding choices. Also, the integration of environmental variation into genomic predictions is a challenging but essential task in the construction of resilient and dynamic breeding programs. Addressing these challenges will be essential to achieving the transformative promise of AI in the domain of sustainable agriculture (Feng *et al.*, 2024).

AI-assisted marker selection in plant breeding is revolutionizing the way breeders identify and select plants with favorable traits. The process employs molecular markers and AI together to significantly enhance the efficiency and accuracy of breeding programs. With the use of ML, AI technologies are combined with marker-assisted selection (MAS) to bypass the shortcomings of conventional breeding methods, hence enabling more efficient selection procedures, particularly for complex traits like drought tolerance in crops like winter wheat. One of the innovations in this area is HTP, which employs remote sensing technologies, e.g. UAVs with the use of ML algorithms, to enable efficient and non-invasive solutions to conventional phenotyping techniques. The innovation is central to MAS because it accelerates the detection of plants that express favorable traits through the rapid and accurate collection of data (Chang-Brahim *et al.*, 2024). Molecular markers such as simple sequence repeats (SSR), restriction fragment length polymorphism (RFLP), and SNPs form the backbone of MAS, enabling identification and selection of economically significant traits with increased speed and accuracy (Shrestha *et al.*, 2020). Combining MAS with genomic selection (GS) introduces an all-encompassing approach by employing genome-wide marker data to predict the genetic potential of breeding lines. The combination enhances the improvement of functional traits to enable breeders to make data-driven decisions at earlier stages of the breeding cycle (Verma *et al.*, 2024). AI-assisted MAS also finds application in other areas such as germplasm characterization, genetic diversity analysis, and trait stacking, all of which contribute to accelerating crop improvement activities (Kumawat *et al.*, 2020). However, despite all its advantages, MAS is martyred by numerous challenges. One of the principal problems is the need for stable and precise relations between markers and traits, particularly for polygenic traits controlled by intricate genetic architectures. AI can overcome these problems by providing interpretable models that can handle large, high-dimensional data sets and identify subtle genetic patterns (Rani *et al.*, 2014; Chang-Brahim *et al.*, 2024).

Although AI-aided marker selection promises revolutionary potential for contemporary plant breeding, its limitations must be recognized. Its effectiveness relies on the quality and quantity of data, the interpretability of AI predictions, and the ease of incorporating AI tools into conventional breeding protocols. The translation of scientific breakthroughs into useful technology is also a critical bottleneck. Breeding programs need to invest in capacity development and digital infrastructure to maximize the advantages of AI-aided MAS and achieve sustainable gains in crop productivity and resilience.

1.6.1 AI-based prediction of crop yield and performance

The use of AI in plant performance evaluation and crop yield forecasting is revolutionizing modern agriculture. This new technique utilizes the power of ML and DL in processing big and heterogeneous datasets, such as climatic patterns, soil health parameters, past yield data, and satellite images, to provide highly accurate yield forecasting estimates. These predictions drastically surpass the efficiency of traditional manual methods, which are time-consuming, labor-intensive, and focused. AI has several advantages, such as automation, real-time processing, and the ability to scan large agricultural fields. Most important is the synergy in using AI with remote sensing technologies, such as high-resolution satellite images from missions such as Sentinel-2. Such data sets include critical information about the health of plants, growth stages, and stress responses, enabling a deeper understanding of crop behavior.

A wide variety of AI techniques has been applied in this domain. ML models, such as decision trees, SVMs, and ANNs, are widely used to represent complex, non-linear relationships in agricultural data (Goel and Pandey, 2024). In addition, hybrid architectures, such as fuzzy-based AI systems incorporating fuzzy logic and ML methods, have been reported to be effective in dealing with uncertainty and variability in agricultural systems (Kumar *et al.*, 2024). These techniques allow for more precise and robust predictions of yields even under uncertain environmental conditions. Remote sensing data, especially vegetation indices and spectral reflectance measurements derived from Sentinel-2 data, are widely utilized for crop health monitoring. Models such as RF and CNNs are effective in processing such data, providing high spatial and temporal resolution estimates of crop yield potential (Aslan *et al.*, 2024). However, several challenges remain. One of the key limitations is the uneven quality and availability of agricultural data across regions. Variability in data acquisition methods and the lack of standardized datasets can reduce the generality of AI models (Goel and Pandey, 2024). Further, the development of models with the ability to adapt to different crop types, varying climatic zones, and a variety of agricultural practices is a serious challenge. AI tools have a tendency to require localized training and tuning to be effective across different regions.

However, great strides are being made. For instance, the Extra Trees regression model was better at predicting yield than conventional models such as RF and conventional ANNs (Sharma *et al.*, 2024). Additionally, uses of AI are extending to early disease detection in crops, crop suitability assessment based on site-specific data, and real-time plant stress monitoring. DL algorithms, especially CNNs, have shown high accuracy in these areas, allowing for improved agronomic decision making and optimization of yield potential (Sadashiv *et al.*, 2021). To realize the full potential of AI in agriculture, data standardization, model interoperability, and explainability issues need to be addressed. Future studies should focus on creating transparent and transferable AI models, investigate hybrid and ensemble methods, and increase cross-regional collaborations for open data sharing. Such innovations will allow for the development of more resilient, productive, and sustainable farming systems globally.

1.6.2 Genomic selection and precision breeding using AI

The intersection of AI with genomic selection and precision breeding approaches is revolutionizing the landscape of modern plant breeding. By facilitating rapid, data-driven decision-making processes, AI bridges the genotype–phenotype gap, thereby enabling the creation of crops characterized by high yield, stress tolerance, and climate change tolerance. Such developments are acutely necessary in light of the urgent need to guarantee global food security in the face of growing populations and environmental pressures. In genomic selection, AI and ML algorithms are employed to analyze large genomic datasets, identifying patterns and correlations between genetic markers and desirable phenotypic traits. Such developments enable earlier and more precise predictions of trait expression, thereby shortening the breeding cycle (Sharma *et al.*, 2025). Relative to conventional MAS, AI-based models are capable of capturing the complex interactions between multiple genes and environmental factors, thereby maximizing accuracy and predictive capability. AI also increases the precision and effectiveness of gene-editing tools, including CRISPR-Cas9. Through the prediction of off-target effects and simulation of the phenotypic impact of particular genetic changes, AI enables breeders to optimize crop traits with greater confidence (Patil, 2023; Farooq *et al.*, 2024). Such developments enable the creation of superior cultivars that have superior yield potential, enhanced nutritional quality, and improved resistance to both biotic and abiotic stresses.

Precision breeding is also being further augmented through AI-based simulations forecasting plant performance under future environmental conditions. These models enable the creation of cultivars tailored to anticipated climate conditions, thus ensuring long-term agricultural resilience (Farooq *et al.*, 2024). Methods like fluorescence-based hybridization and high-resolution imaging, when combined with AI, provide deep phenotyping insights into cellular and developmental processes, enabling breeders to select genotypes with improved adaptability (Fang, 2024). Besides trait selection, AI is also critical in ensuring agricultural sustainability by optimizing breeding methods, cropping systems, and resource allocation. For instance, AI can identify optimal intercropping configurations or suggest effective crop rotations for improving soil health and overall productivity. Additionally, AI helps to extract useful insights from large genetic repositories stored within global gene banks, unveiling hidden traits and enabling biodiversity conservation (Farooq *et al.*, 2024). A notable project in this area is CropGPT, a global project committed to leveraging AI and biological big data for the improvement of precision breeding. CropGPT is a good example of an international collaborative effort in integrating computational and agricultural sciences to grow crops that are ready for future challenges (Zhu *et al.*, 2024). Nevertheless, the application of AI in plant breeding raises important ethical and regulatory challenges. Ensuring equitable access to AI technologies, protecting indigenous crop varieties, and maintaining genetic diversity are critical concerns. Additionally, there is a need for open data governance and inclusive policy mechanisms to support the responsible use of AI in agriculture (Patil, 2023).

1.7 AI in Plant Disease Detection and Pest Management

ML has been a valuable method for early detection of plant diseases, with vast potential to improve agricultural yield and global food security. Different ML methods, such as RF, k-Nearest Neighbors, Linear Regression, Naive Bayes, Neural Networks, and SVM, have been used to develop predictive models that enable early disease detection. Predictive models are generally developed from high-resolution optical images taken through drones, which are processed using image processing algorithms to extract disease detection-related features (Ahmed and Yadav, 2023). Among the different methods, DL through CNNs has been highly effective in separating plant diseases in large datasets, allowing real-time evaluation and management of plant health problems. In addition, ensemble methods through different ML algorithms have been found to have enhanced predictive potential. For instance, a 97.81% accuracy rate has been demonstrated for a RF Classifier for leaf disease detection (Nayak *et al.*, 2024).

Despite this, there are still a number of challenges where agricultural environments are constrained by limited resources; infrastructural limitations prevent the effective use of such advanced technology. However, cost-effective and flexible solutions, including the use of open-access data and effective feature extraction methods such as the histogram of oriented gradient (HOG), provide promising alternatives to effective disease surveillance (Nagila and Mishra, 2024). To date, research has been concentrated on disease detection; however, there is increasing justification that research on AI use in plant pathology should also be directed toward its prevention, management, and recovery of infection (Omaye *et al.*, 2024). AI has also revolutionized pathogen detection in agriculture from being extremely time-consuming and labor-intensive to using automated, accurate, and scalable systems. DL models in conjunction with the IoT and computer vision enable real-time disease detection using images taken by smartphones and drones (Lee *et al.*, 2024). Such models, particularly CNNs, provide scalability and flexibility in various agricultural environments (Anand, 2024).

One of the most important abilities of AI models is their capacity to identify infectious pathogens, i.e. viruses, bacteria, and fungi, which continue to pose gigantic threats to global agricultural production (Garg *et al.*, 2024). However, improving generalizability is a high priority, particularly in the context of emerging diseases, challenges in domain adaptation due to differences in data distributions, and large-scale spatial modeling. Future research has to address these gaps to create more efficient disease-management models

(Dwivedi *et al.*, 2024). Intelligent pest management is another revolutionary application of AI. AI-based tools, e.g. machine vision, enable precise identification of pests, monitoring, and forecasting, which are essential to reducing crop loss ranging from 20% to 40% globally. Applications like Plantix and Bioleaf demonstrate the practical use of AI in pest management, enabling farmers to obtain timely diagnoses and treatment advice (Rahman and Ravi, 2022). Advanced pest management systems merge sensor data, satellite imagery, and meteorological forecasts to optimize pesticide application and minimize ecotoxic risks (Mittu *et al.*, 2025). Economically, these AI-based systems reduce reliance on chemical inputs, optimize resource allocation, and provide a high return on investment, making enabling policies a prerequisite to ensure widespread adoption (Rathnayake *et al.*, 2024). Advanced solutions, e.g. edge AI-based pest monitoring systems, are particularly beneficial in resource-constrained settings. For example, one such system in orchards demonstrated an accuracy level of more than 70% in the detection of brown marmorated stink bug infestations using CNN-based image analysis, indicating the potential role of AI in integrated pest management (Kargar *et al.*, 2022).

1.8 Artificial Intelligence in Sustainable Agriculture and Smart Farming

Sustainable agriculture focuses on farming methods that enhance long-term ecological resilience and productivity, thus enabling future generations to satisfy their dietary needs without depleting natural resources (Alzubi and Galyna, 2023). As the world continues to grow in population and food security is disrupted by climate change, resource utilization, and environmental degradation, the need for sustainable practices is on the rise. Intelligent farming has become part of sustainable agriculture, and AI is a major factor in this shift. Through the utilization of AI technologies like computer vision and ML, new systems can identify pest infestations, disease, and nutrient deficiencies through image inspection of crops. These AI-based tests enable early intervention, reduce crop loss, and optimize input use efficiency.

AI is actually a branch of precision agriculture, wherein data gathered using sensors, drones, and satellites are processed to evaluate soil health, weather conditions, and crop growth. Based on this information, farmers can wisely utilize water, fertilizers, and pesticides, thus reducing wastage and optimizing crop yield. Furthermore, AI has immense potential to solve the urgent issues of traditional farming, especially in resource-poor rural areas. By improving decision-making support, reducing risks, and providing labor-efficient options, AI fosters the resilience of agricultural practices and facilitates equitable access to technological innovations. Long-term agricultural sustainability is increasingly dependent on the application of data-driven technologies. AI facilitates the creation of climate-smart agricultural practices, optimizing the efficiency of resource utilization while reducing the negative impacts of climate change on food production. This revolutionary potential makes AI an essential element of innovative and sustainable agricultural practices (Gryshova *et al.*, 2024).

1.9 Challenges and Future Perspectives of AI in Plant-Omics

Integrating plant-omics data analysis with AI is difficult in many aspects, mainly because of the complexity and volume of the data to be handled. Omics tools such as genomics, transcriptomics, proteomics, and metabolomics produce large, heterogeneous data that need to be analyzed using advanced AI algorithms. One of the greatest challenges is harmonizing data collection protocols and ensuring platform compatibility. Then, choosing the right AI models and imaging sensors is essential but challenging due to the heterogeneity of biological data (Murmu *et al.*, 2024). Interpreting AI models is another huge challenge in the integration. DL and other advanced algorithms are "black boxes," and it becomes challenging to know how conclusions are drawn. Overfitting, where models are good at training data but poor at new, unseen data, is another technical challenge. Moreover, AI-based predictions need to be experimentally validated, which is time- and resource-consuming. All of these obstacles prevent AI technology from being widely used in plant-omics (Ali and Mohammed, 2024). To counter these challenges, researchers are suggesting a number of strategic solutions. The creation of easy-to-use bioinformatics platforms can enable plant biologists to make the best use of AI tools even without computational skills. The use of common data standards and open-source repositories will further make data accessible and interoperable. Interdisciplinary collaboration is the key to unifying plant science, computer science, and statistics researchers and making the successful adoption of AI in complex biological systems a reality. DL techniques, when implemented with thoughtful design and validation, can also enhance the interpretation of stress symptoms and predictive modeling accuracy.

In the years to come, the future of AI in plant-omics is bright. DL and ML already augment our knowledge of plant stress responses and defense strategies. With multi-omics data integration, AI allows researchers to deconstruct molecular pathways and create effective crop protective strategies against biotic and abiotic stresses. Integrating these, particularly in horticulture, is revolutionary as AI-supported phenotyping platforms provide high-throughput, non-invasive methods rivaling conventional observation methods (Zhang *et al.*, 2024). In addition, AI is opening new avenues in bioinformatics, facilitating ease in the identification of stress-responsive genes and the mapping of regulatory networks vital for crop resilience. All these are fundamental to precision breeding, environmentally friendly crop management, and climate-resilient agriculture. Although issues like data complexity and model interpretability persist, the integration of AI in plant-omics is set to transform agricultural science, playing a significant role in global food security and environmental sustainability (Nabwire *et al.*, 2021).

1.10 Conclusion

ML and DL, among AI technologies, are driving seismic breakthroughs in genomic annotation, phenotype prediction, disease diagnosis, and precision agriculture. These breakthroughs not only simplify research but also aid world food security by improving crop resilience, enhancing breeding techniques, and promoting sustainable agriculture. However, this revolution is not without challenges; challenges like data heterogeneity, ethics, and computational limitations continue to constrain the overall potential of AI in plant-omics. Overcoming these through collaborative research, robust data infrastructures, and ethics guidelines will be the order of the day. As AI technologies continue to develop, their judicious integration into plant sciences promises to usher in a revolutionary era of innovation, where data-intensive approaches can tackle agriculture's burning needs in the face of climatic variability. Moreover, the interdisciplinary nature of AI in plant biology promotes collaboration among computational scientists, biotechnologists, geneticists, and agronomists, thus ensuring holistic approaches to solving practical agricultural problems. As AI systems become increasingly interpretable and transparent, their integration into decision making will provide actionable insights for researchers and farmers. Continued investment in AI training, developing infrastructure, and ethical use of AI will be crucial in realizing its impact. Lastly, the judicious and creative use of AI in plant sciences not only enhances crop productivity and resilience but also ensures global efforts towards food sustainability, environmental integrity, and climate change adaptation.

References

Ahmed, I. and Yadav, P.K. (2023) Plant disease detection using machine learning approaches. *Expert Systems* 40(5), e13136. DOI: 10.1111/exsy.13136.

Ahsan, M.U., Gouru, A., Chan, J., Zhou, W. and Wang, K. (2024) A signal processing and deep learning framework for methylation detection using Oxford Nanopore sequencing. *Nature Communications* 15(1), 1448. DOI: 10.1038/s41467-024-45778-y.

Ali, A.M. and Mohammed, M.A. (2024) A comprehensive review of artificial intelligence approaches in omics data processing: Evaluating progress and challenges. *International Journal of Mathematics, Statistics, and Computer Science* 2, 114–167. DOI: 10.59543/ijmscs.v2i.8703.

Alipanahi, B., Delong, A., Weirauch, M.T. and Frey, B.J. (2015) Predicting the sequence specificities of DNA- and RNA-binding proteins by deep learning. *Nature Biotechnology* 33(8), 831–838. DOI: 10.1038/nbt.3300.

Altschul, S.F., Gish, W., Miller, W., Myers, E.W. and Lipman, D.J. (1990) Basic local alignment search tool. *Journal of Molecular Biology* 215(3), 403–410. DOI: 10.1016/S0022-2836(05)80360-2.

AlZubi, A.A. and Galyna, K. (2023) Artificial intelligence and internet of things for sustainable farming and smart agriculture. *IEEE Access* 11, 78686–78692. DOI: 10.1109/ACCESS.2023.3298215.

Amara, J., Bouaziz, B. and Algergawy, A. (2017) A deep learning-based approach for banana leaf diseases classification. In: *Proceedings of the Datenbanksysteme für Business, Technologie und Web (BTW '17) - Workshopband.*

Anand, A. (2024) Leaf disease detection system through advanced AI and machine learning integration. *SSRN Electronic Journal.* DOI: 10.2139/ssrn.5018301.

Angermueller, C., Lee, H.J., Reik, W. and Stegle, O. (2017) DeepCpG: Accurate prediction of single-cell DNA methylation states using deep learning. *Genome Biology* 18(1), 67. DOI: 10.1186/s13059-017-1189-z.

Asefpour Vakilian, K. (2020) Machine learning improves our knowledge about miRNA functions towards plant abiotic stresses. *Scientific Reports* 10(1), 3041. DOI: 10.1038/s41598-020-59981-6.

Aslan, M.F., Sabanci, K. and Aslan, B. (2024) Artificial intelligence techniques in crop yield estimation based on sentinel-2 data: A comprehensive survey. *Sustainability* 16(18), 8277. DOI: 10.3390/su16188277.

Averbeck, B. and O'Doherty, J.P. (2022) Reinforcement-learning in fronto-striatal circuits. *Neuropsychopharmacology* 47(1), 147–162. DOI: 10.1038/s41386-021-01108-0.

Avsec, Ž., Weilert, M., Shrikumar, A., Krueger, S., Alexandari, A. *et al.* (2021) Base-resolution models of transcription-factor binding reveal soft motif syntax. *Nature Genetics* 53(3), 354–366. DOI: 10.1038/s41588-021-00782-6.

Azimi, S., Wadhawan, R. and Gandhi, T.K. (2021) Intelligent monitoring of stress induced by water deficiency in plants using deep learning. *IEEE Transactions on Instrumentation and Measurement* 70, 1–13. DOI: 10.1109/TIM.2021.3111994.

Baek, M., DiMaio, F., Anishchenko, I., Dauparas, J., Ovchinnikov, S. *et al.* (2021) Accurate prediction of protein structures and interactions using a three-track neural network. *Science* 373(6557), 871–876. DOI: 10.1126/science.abj8754.

Bhoi, C. and Mahali, S.K. (2024) The role of artificial intelligence used in plant science: A review. *Asian Journal of Research in Botany* 7(2), 263–272. DOI: 10.47814/ijssrr.v7i3.1925.

Bradbury, S.P. (1994) Predicting modes of toxic action from chemical structure: An overview. *SAR and QSAR in Environmental Research* 2(1–2), 89–104. DOI: 10.1080/10629369408028842.

Camacho, D.M., Collins, K.M., Powers, R.K., Costello, J.C. and Collins, J.J. (2018) Next-generation machine learning for biological networks. *Cell* 173(7), 1581–1592. DOI: 10.1016/j.cell.2018.05.015.

Caruana, R. and Niculescu-Mizil, A. (2006) An empirical comparison of supervised learning algorithms. In: *ICML '06: Proceedings of the 23rd international conference on Machine learning*, Pittsburgh, Pennsylvania, pp. 161–168. DOI: 10.1145/1143844.1143865.

Ceasar, S.A. and Ebeed, H.T. (2024) *The Present State and Impact of AI-Driven Computational Tools for Predicting Plant Protein Structures*. Protein & Peptide Letters.

Chang-Brahim, I., Koppensteiner, L.J., Beltrame, L., Bodner, G., Saranti, A. *et al.* (2024) Reviewing the essential roles of remote phenotyping, GWAS and explainable AI in practical marker-assisted selection for drought-tolerant winter wheat breeding. *Frontiers in Plant Science* 15, 1319938. DOI: 10.3389/fpls.2024.1319938.

Chen, L., Li, Q., Nasif, K.F.A., Xie, Y., Deng, B. *et al.* (2024) AI-driven deep learning techniques in protein structure prediction. *International Journal of Molecular Sciences* 25(15), 8426. DOI: 10.3390/ijms25158426.

Chen, N.-C., Kolesnikov, A., Goel, S., Yun, T., Chang, P.-C. *et al.* (2023) Improving variant calling using population data and deep learning. *BMC Bioinformatics* 24(1), 197. DOI: 10.1186/s12859-023-05294-0.

Ching, T., Himmelstein, D.S., Beaulieu-Jones, B.K., Kalinin, A.A., Do, B.T. *et al.* (2018) Opportunities and obstacles for deep learning in biology and medicine. *Journal of the Royal Society Interface* 15(141), 20170387. DOI: 10.1098/rsif.2017.0387.

Cronin, M.T.D. and Dearden, J.C. (1995) QSAR in toxicology. 1. Prediction of aquatic toxicity. *Quantitative Structure-Activity Relationships* 14(1), 1–7. DOI: 10.1002/qsar.19950140102.

Daniels, M.W., Dvorkin, D., Powers, R.K. and Kechris, K. (2021) Semi-supervised learning using hierarchical mixture models: Gene essentiality case study. *Mathematical and Computational Applications* 26(2), 40. DOI: 10.3390/mca26020040.

Das Choudhury, S., Samal, A. and Awada, T. (2019) Leveraging image analysis for high-throughput plant phenotyping. *Frontiers in Plant Science* 10, 508. DOI: 10.3389/fpls.2019.00508.

Dong, J., Zhao, M., Liu, Y., Su, Y. and Zeng, X. (2022) Deep learning in retrosynthesis planning: Datasets, models and tools. *Briefings in Bioinformatics* 23(1), bbab391. DOI: 10.1093/bib/bbab391.

Du, J., Lu, X., Fan, J., Qin, Y., Yang, X. *et al.* (2020) Image-based high-throughput detection and phenotype evaluation method for multiple lettuce varieties. *Frontiers in Plant Science* 11, 563386. DOI: 10.3389/fpls.2020.563386.

Dunn, W.J., III (1988) QSAR approaches to predicting toxicity. *Toxicology Letters* 43(1–3), 277–283. DOI: 10.1016/0378-4274(88)90033-1.

Dwivedi, M., Pandey, B. and Saxena, V. (2024) Analyzing the frontier of AI-based plant disease detection: Insights and perspectives. In: *Microbial Data Intelligence and Computational Techniques for Sustainable Computing*. Springer Nature Singapore, Singapore, pp. 195–204. DOI: 10.1007/978-981-99-9621-6_13.

Elavarasan, D. and Vincent, P.M.D. (2020) Crop yield prediction using deep reinforcement learning model for sustainable Agrarian applications. *IEEE Access* 8, 86886–86901. DOI: 10.1109/ACCESS.2020.2992480.

Elnaggar, A., Heinzinger, M., Dallago, C., Rehawi, G., Wang, Y. *et al.* (2021) ProtTrans: Toward understanding the language of life through self-supervised learning. *IEEE Transactions on Pattern Analysis and Machine Intelligence* 44(10), 7112–7127. DOI: 10.1109/TPAMI.2021.3095381.

Eraslan, G., Avsec, Ž., Gagneur, J. and Theis, F.J. (2019) Deep learning: New computational modelling techniques for genomics. *Nature Reviews Genetics* 20(7), 389–403. DOI: 10.1038/s41576-019-0122-6.

Ernst, J. and Kellis, M. (2017) Chromatin-state discovery and genome annotation with chromHMM. *Nature Protocols* 12(12), 2478–2492. DOI: 10.1038/nprot.2017.124.

Esser, F., Rosu, R.A., Cornelißen, A., Klingbeil, L., Kuhlmann, H. *et al.* (2023) Field robot for high-throughput and high-resolution 3D plant phenotyping: Towards efficient and sustainable crop production. *IEEE Robotics & Automation Magazine* 30(4), 20–29. DOI: 10.1109/MRA.2023.3321402.

Fang, J. (2024) Breeding 5.0: AI-driven revolution in designed plant breeding. *Molecular Plant Breeding* 15(1), 27–33. DOI: 10.5376/mpb.2024.15.0004.

Farooq, M.A., Gao, S., Hassan, M.A., Huang, Z., Rasheed, A. *et al.* (2024) Artificial intelligence in plant breeding. *Trends in Genetics* 40(10), 891–908. DOI: 10.1016/j.tig.2024.07.001.

Fathi, S., Movlodzadeh, R., Heidari, S. and Najafian, S. (2025) Application of machine learning models to predict secondary metabolites for the first time in the valuable medicinal plant (*Dracocephalum moldavica* L.). *In Review*. DOI: 10.21203/rs.3.rs-6105806/v1.

Feng, W., Gao, P. and Wang, X. (2024) AI breeder: Genomic predictions for crop breeding. *New Crops* 1, 100010. DOI: 10.1016/j.ncrops.2023.12.005.

Fox, I., Lynn, A., Mamta, J., Rodica, P.-B. and Jenna, W. (2018) Deep multi-output forecasting: Learning to accurately predict blood glucose trajectories. In: *Proceedings of the 24th ACM SIGKDD International Conference on Knowledge Discovery & Data Mining (KDD '18)*, ACM, August 19–23, 2018. DOI: 10.1145/3219819.3220102.

Fürtauer, L., Pschenitschnigg, A., Scharkosi, H., Weckwerth, W. and Nägele, T. (2018) Combined multivariate analysis and machine learning reveals a predictive module of metabolic stress response in *Arabidopsis thaliana*. *Molecular Omics* 14(6), 437–449. DOI: 10.1039/c8mo00095f.

Gano, B., Bhadra, S., Vilbig, J.M., Ahmed, N., Sagan, V. *et al.* (2024) Drone-based imaging sensors, techniques, and applications in plant phenotyping for crop breeding: A comprehensive review. *The Plant Phenome Journal* 7(1), e20100. DOI: 10.1002/ppj2.20100.

Gao, X., Zhang, J., Wei, Z. and Hakonarson, H. (2018) DeepPolyA: A convolutional neural network approach for polyadenylation site prediction. *IEEE Access* 6, 24340–24349. DOI: 10.1109/ACCESS.2018.2825996.

Garg, T., Dwivedi, P., Mishra, M.K., Joshi, N.C., Shrivastava, N. *et al.* (2024) Artificial intelligence in plant disease identification: Empowering agriculture. In: *Methods in Microbiology*, Vol. 55. Academic Press, pp. 179–193. DOI: 10.1016/bs.mim.2024.05.007.

Goel, M. and Pandey, M. (2024) Crop yield prediction using AI: A review. In: *2024 2nd International Conference on Disruptive Technologies (ICDT)*, IEEE, Greater Noida, India, pp. 1547–1553. DOI: 10.1109/ICDT61202.2024.10489432.

Greener, J.G., Kandathil, S.M., Moffat, L. and Jones, D.T. (2022) A guide to machine learning for biologists. *Nature Reviews Molecular Cell Biology* 23(1), 40–55. DOI: 10.1038/s41580-021-00407-0.

Gremme, G., Brendel, V., Sparks, M.E. and Kurtz, S. (2005) Engineering a software tool for gene structure prediction in higher organisms. *Information and Software Technology* 47(15), 965–978. DOI: 10.1016/j.infsof.2005.09.005.

Gryshova, I., Balian, A., Antonik, I., Miniailo, V., Nehodenko, V. *et al.* (2024) Artificial intelligence in climate smart in agricultural: Toward a sustainable farming future. *Access Journal - Access to Science, Business, Innovation in the Digital Economy* 5(1), 125–140. DOI: 10.46656/access.2024.5.1(8).

Gulli, A. and Pal, S. (2017) *Deep learning with Keras: Implementing Deep Learning Methods and Neural Networks with the Power of Python*. Packt Publishing Ltd.

Harfouche, A.L., Jacobson, D.A., Kainer, D., Romero, J.C., Harfouche, A.H. *et al.* (2019) Accelerating climate resilient plant breeding by applying next-generation artificial intelligence. *Trends in Biotechnology* 37(11), 1217–1235. DOI: 10.1016/j.tibtech.2019.05.007.

Hasan, M.E., Osman, M.A., Hemeida, A.A. and ElHefnawi, M. (2025) Bioinformatics and artificial intelligence approaches in metabolic pathways. In: *In Metabolic Dynamics in Host-Microbe Interaction*. Springer Nature Singapore, Singapore, pp. 17–32.

Hesami, M. and Jones, A.M.P. (2020) Application of artificial intelligence models and optimization algorithms in plant cell and tissue culture. *Applied Microbiology and Biotechnology* 104(22), 9449–9485. DOI: 10.1007/s00253-020-10888-2.

Hu, L., Li, J., Peng, X., Xiao, J., Zhan, B. *et al.* (2022) Semi-supervised NPC segmentation with uncertainty and attention guided consistency. *Knowledge-Based Systems* 239, 108021. DOI: 10.1016/j.knosys.2021.108021.

Jaganathan, K., Kyriazopoulou Panagiotopoulou, S., McRae, J.F., Darbandi, S.F., Knowles, D. *et al.* (2019) Predicting splicing from primary sequence with deep learning. *Cell* 176(3), 535–548. DOI: 10.1016/j.cell.2018.12.015.

Jiang, Y. and Li, C. (2020) Convolutional neural networks for image-based high-throughput plant phenotyping: A review. *Plant Phenomics* 2020, 4152816. DOI: 10.34133/2020/4152816.

Jiang, T., Gradus, J.L. and Rosellini, A.J. (2020) Supervised machine learning: A brief primer. *Behavior Therapy* 51(5), 675–687. DOI: 10.1016/j.beth.2020.05.002.

Jumper, J., Evans, R., Pritzel, A., Green, T., Figurnov, M. *et al.* (2021) Highly accurate protein structure prediction with alphafold. *Nature* 596(7873), 583–589. DOI: 10.1038/s41586-021-03819-2.

Kamilaris, A. and Prenafeta-Boldú, F.X. (2018) Deep learning in agriculture: A survey. *Computers and Electronics in Agriculture* 147, 70–90. DOI: 10.1016/j.compag.2018.02.016.

Kargar, A., Wilk, M.P., Zorbas, D., Gaffney, M.T. and Q'Flynn, B. (2022) A novel resource-constrained insect monitoring system based on machine vision with edge AI. In: *2022 IEEE 5th International Conference on Image Processing Applications and Systems (IPAS)*, IEEE, Genova, Italy, pp. 1–6. DOI: 10.1109/IPAS55744.2022.10052895.

Kelley, D.R., Snoek, J. and Rinn, J.L. (2016) Basset: Learning the regulatory code of the accessible genome with deep convolutional neural networks. *Genome Research* 26(7), 990–999. DOI: 10.1101/gr.200535.115.

Kelley, D.R., Reshef, Y.A., Bileschi, M., Belanger, D., McLean, C.Y. *et al.* (2018) Sequential regulatory activity prediction across chromosomes with convolutional neural networks. *Genome Research* 28(5), 739–750. DOI: 10.1101/gr.227819.117.

Khan, M.H.U., Wang, S., Wang, J., Ahmar, S., Saeed, S. *et al.* (2022) Applications of artificial intelligence in climate-resilient smart-crop breeding. *International Journal of Molecular Sciences* 23(19), 11156. DOI: 10.3390/ijms231911156.

Khanna, A., Jain, S., Choudhury, T., Choudhury, T. and Kotecha, K. (2024) Predicting secondary metabolites from diverse macronutrients using machine learning and AI. In: *2024 OPJU International Technology Conference (OTCON) on Smart Computing for Innovation and Advancement in Industry 4.0*, IEEE, Raigarh, India, pp. 1–6. DOI: 10.1109/OTCON60325.2024.10688121.

Kim, D.S., Risca, V.I., Reynolds, D.L., Chappell, J., Rubin, A.J. *et al.* (2021) The dynamic, combinatorial cis-regulatory lexicon of epidermal differentiation. *Nature Genetics* 53(11), 1564–1576. DOI: 10.1038/s41588-021-00947-3.

Kircher, M., Witten, D.M., Jain, P., O'Roak, B.J., Cooper, G.M. *et al.* (2014) A general framework for estimating the relative pathogenicity of human genetic variants. *Nature Genetics* 46(3), 310–315. DOI: 10.1038/ng.2892.

Koh, J.C.O., Spangenberg, G. and Kant, S. (2021) Automated machine learning for high-throughput image-based plant phenotyping. *Remote Sensing* 13(5), 858. DOI: 10.3390/rs13050858.

Koirala, A., Walsh, K.B., Wang, Z. and McCarthy, C. (2019) Deep learning – method overview and review of use for fruit detection and yield estimation. *Computers and Electronics in Agriculture* 162, 219–234. DOI: 10.1016/j.compag.2019.04.017.

Kolhar, S. and Jagtap, J. (2023) Plant trait estimation and classification studies in plant phenotyping using machine vision – A review. *Information Processing in Agriculture* 10(1), 114–135. DOI: 10.1016/j.inpa.2021.02.006.

Kumar, R.S. and Anitha, Y. (2014) An efficient approach for asymmetric data classification. *International Journal of Innovative Research in Advanced Engineering* 1, 157–161.

Kumar, A., Bewerwal, A., Srivastava, D. and Jain, S. (2024) An intelligent yield prediction of crops in smart agriculture management using enhanced fuzzy based AI framework. In: *2024 2nd International Conference on Disruptive Technologies (ICDT)*, IEEE, Greater Noida, India, pp. 1068–1073. DOI: 10.1109/ICDT61202.2024.10489101.

Kumar, S., Patil, R.R. and Rani, R. (2023) Smart IoT-based pesticide recommendation system for rice diseases. In: *Intelligent Systems and Applications*, Vol. 959. Springer, Singapore, pp. 123–135. DOI: 10.1007/978-981-19-6581-4_2.

Kumawat, G., Kumawat, C.K., Chandra, K., Pandey, S., Chand, S. *et al.* (2020) Insights into marker assisted selection and its applications in plant breeding. In: *Plant Breeding-Current and Future Views*, Vol. 959. pp. 123–135. DOI: 10.5772/intechopen.95004.

Lee, S.H., Liaw, Z.R., Chai, Y.H., Ng, S.L., Bonnet, P. *et al.* (2024) Revolutionizing plant pathogen conservation: The past, present, and future of AI in preserving natural ecosystems. *Biodiversity Information Science and Standards* 8, e133055. DOI: 10.3897/biss.8.133055.

Lee, U., Chang, S., Putra, G.A., Kim, H. and Kim, D.H. (2018) An automated, high-throughput plant phenotyping system using machine learning-based plant segmentation and image analysis. *PLOS ONE* 13(4), e0196615. DOI: 10.1371/journal.pone.0196615.

Liang, Y., Zhang, F., Wang, J., Joshi, T., Wang, Y. *et al.* (2011) Prediction of drought-resistant genes in *Arabidopsis thaliana* using SVM-RFE. *PLOS ONE* 6(7), e21750. DOI: 10.1371/journal.pone.0021750.

Liao, L., Xie, M., Zheng, X., Zhou, Z., Deng, Z. *et al.* (2025) Molecular insights fast-tracked: AI in biosynthetic pathway research. *Natural Product Reports*. DOI: 10.1039/D4NP00003J.

Liu, K., Abdullah, A.A., Huang, M., Nishioka, T., Altaf-Ul-Amin, M. *et al.* (2017) Novel approach to classify plants based on metabolite-content similarity. *BioMed Research International* 5296729. DOI: 10.1155/2017/5296729.

Luo, R., Sedlazeck, F.J., Lam, T.W. and Schatz, M.C. (2019) A multi-task convolutional deep neural network for variant calling in single molecule sequencing. *Nature Communications* 10(1), 998. DOI: 10.1038/s41467-019-09025-z.

Lyu, C., Wang, L. and Zhang, J. (2018) Deep learning for DNase I hypersensitive sites identification. *BMC Genomics* 19(Suppl 10), 905. DOI: 10.1186/s12864-018-5283-8.

Mahlein, A.K. (2016) Plant disease detection by imaging sensors - parallels and specific demands for precision agriculture and plant phenotyping. *Plant Disease* 100(2), 241–251. DOI: 10.1094/PDIS-03-15-0340-FE.

Mahlein, A.K., Arnal Barbedo, J.G., Chiang, K.S., Del Ponte, E.M. and Bock, C.H. (2024) From detection to protection: The role of optical sensors, robots, and artificial intelligence in modern plant disease management. *Phytopathology* 114(8), 1733–1741. DOI: 10.1094/PHYTO-01-24-0009-PER.

Mahood, E.H., Kruse, L.H. and Moghe, G.D. (2020) Machine learning: A powerful tool for gene function prediction in plants. *Applications in Plant Sciences* 8(7), e11376. DOI: 10.1002/aps3.11376.

Mathur, A. and Foody, G.M. (2008) Multiclass and binary SVM classification: Implications for training and classification users. *IEEE Geoscience and Remote Sensing Letters* 5(2), 241–245. DOI: 10.1109/LGRS.2008.915597.

Menden, M.P., Iorio, F., Garnett, M., McDermott, U., Benes, C.H. *et al.* (2013) Machine learning prediction of cancer cell sensitivity to drugs based on genomic and chemical properties. *PloS ONE* 8(4), e61318. DOI: 10.1371/journal.pone.0061318.

Mittu, B., Chaturvedi, A. and Singh, M. (2025) Use of pesticide management using AI. In: *Artificial Intelligence in the Food Industry*. CRC Press, pp. 260–276. DOI: 10.1201/9781032633602-13.

Mochida, K., Koda, S., Inoue, K., Hirayama, T., Tanaka, S. *et al.* (2019) Computer vision-based phenotyping for improvement of plant productivity: A machine learning perspective. *GigaScience* 8(1), giy153. DOI: 10.1093/gigascience/giy153.

Montesinos-López, A., Montesinos-López, O.A., Gianola, D., Crossa, J. and Hernández-Suárez, C.M. (2018) Multi-environment genomic prediction of plant traits using deep learners with dense architecture. *G3: Genes|Genomes|Genetics* 8(12), 3813–3828. DOI: 10.1534/g3.118.200740.

Moreno, M., Vilaça, R. and Ferreira, P.G. (2022) Scalable transcriptomics analysis with dask: Applications in data science and machine learning. *BMC Bioinformatics* 23(1), 514. DOI: 10.1186/s12859-022-05065-3.

Mufassirin, M.M.M., Newton, M.A.H. and Sattar, A. (2023) Artificial intelligence for template-free protein structure prediction: A comprehensive review. *Artificial Intelligence Review* 56(8), 7665–7732. DOI: 10.1007/s10462-022-10350-x.

Murmu, S., Sinha, D., Chaurasia, H., Sharma, S., Das, R. *et al.* (2024) A review of artificial intelligence-assisted omics techniques in plant defense: Current trends and future directions. *Frontiers in Plant Science* 15, 1292054. DOI: 10.3389/fpls.2024.1292054.

Murray, P.G., Stevens, A., Leonibus, C., Koledova, E., Chatelain, P. *et al.* (2018) Transcriptomics and machine learning predict diagnosis and severity of growth hormone deficiency. *JCI Insight* 3(7), e93247. DOI: 10.1172/jci.insight.93247.

Myint, K.Z. and Xie, X.Q. (2014) Ligand biological activity predictions using fingerprint-based artificial neural networks (FANN-QSAR). In: *Artificial Neural Networks.* Springer, New York, NY, pp. 149–164. DOI: 10.1007/978-1-4939-2239-0_9.

Myint, K.Z., Wang, L., Tong, Q. and Xie, X.Q. (2012) Molecular fingerprint-based artificial neural networks QSAR for ligand biological activity predictions. *Molecular Pharmaceutics* 9(10), 2912–2923. DOI: 10.1021/mp300237z.

Nabwire, S., Suh, H.K., Kim, M.S., Baek, I. and Cho, B.K. (2021) Application of artificial intelligence in phenomics. *Sensors* 21(13), 4363. DOI: 10.3390/s21134363.

Nagasubramanian, K., Jones, S., Singh, A.K., Sarkar, S., Singh, A. *et al.* (2019) Plant disease identification using explainable 3D deep learning on hyperspectral images. *Plant Methods* 15(1), 98. DOI: 10.1186/s13007-019-0479-8.

Nagasubramanian, K., Singh, A.K., Singh, A., Sarkar, S. and Ganapathysubramanian, B. (2020) Usefulness of interpretability methods to explain deep learning-based plant stress phenotyping. *arXiv Preprint arXiv:2007.05729.*

Nagila, A. and Mishra, A.K. (2024) Detection and categorization of diseases affecting plant leaves with the use of machine learning. In: *2024 OPJU International Technology Conference (OTCON) on Smart Computing for Innovation and Advancement in Industry 4.0*, IEEE, Raigarh, India, pp. 1–6. DOI: 10.1109/OTCON60325.2024.10688104.

Nayak, J.R., Kumar, D., Bisakha, S., Sneha, O. and Dash, P.K. (2024) Digital diagnosis: A machine learning perspective on plant disease detection. In: *2024 IEEE International Conference on Information Technology, Electronics and Intelligent Communication Systems (ICITEICS)*, IEEE, Bangalore, India, pp. 1–5. DOI: 10.1109/ICITEICS61368.2024.10624976.

Negus, K.L., Li, X., Welch, S.M. and Yu, J. (2024) The role of artificial intelligence in crop improvement. *Advances in Agronomy* 184, 1–66. DOI: 10.1016/bs.agron.2023.11.001.

Novakovsky, G., Dexter, N., Libbrecht, M.W., Wasserman, W.W. and Mostafavi, S. (2023) Obtaining genetics insights from deep learning via explainable artificial intelligence. *Nature Reviews Genetics* 24(2), 125–137. DOI: 10.1038/s41576-022-00532-2.

Omaye, J.D., Ogbuju, E., Ataguba, G., Jaiyeoba, O., Aneke, J. *et al.* (2024) Cross-comparative review of machine learning for plant disease detection: Apple, cassava, cotton and potato plants. *Artificial Intelligence in Agriculture* 12, 127–151. DOI: 10.1016/j.aiia.2024.04.002.

Patil, A. (2023) Use of artificial intelligence to hasten progress in plant genetics. *Biology and Life Sciences.* DOI: 10.20944/preprints202309.1067.v1.

Patil, R.R. and Kumar, S. (2022) Priority selection of agro-meteorological parameters for integrated plant diseases management through analytical hierarchy process. *International Journal of Electrical and Computer Engineering (IJECE)* 12(1), 649. DOI: 10.11591/ijece.v12i1.pp649-659.

Poplin, R., Chang, P.-C., Alexander, D., Schwartz, S., Colthurst, T. *et al.* (2018) A universal SNP and small-indel variant caller using deep neural networks. *Nature Biotechnology* 36(10), 983–987. DOI: 10.1038/nbt.4235.

Rahman, S.M. and Ravi, G. (2022) Role of artificial intelligence in pest management. In: *Current Topics in Agricultural Sciences*, Vol. 7. pp. 64–81. DOI: 10.9734/bpi/ctas/v7/2141B.

Rani, P.J., Satyanarayana, P.V., Chamundeswari, N. and Rani, M.G. (2014) A review on marker assisted selection in crop improvement. *International Journal of Applied Biology and Pharmaceutical Technology* 5, 271–281.

Rani, R., Sahoo, J., Bellamkonda, S., Kumar, S. and Pippal, S.K. (2023) Role of artificial intelligence in agriculture: An analysis and advancements with focus on plant diseases. *IEEE Access* 11, 137999–138019. DOI: 10.1109/ACCESS.2023.3339375.

Rathnayake, I., Abeysekara, I. and Karalliyadda, C. (2024) Economic viability of smart pest management practices. In: *Revolutionizing Pest Management for Sustainable Agriculture*. IGI Global, pp. 85–106. DOI: 10.4018/979-8-3693-3061-6.ch004.

Ravari, S.Z., Dehghani, H. and Naghavi, H. (2015) Evaluation of the relationship between salinity stress tolerance indices and some physiological traits in bread wheat. *Iranian Journal of Field Crop Science* 46(3), 423–432. DOI: 10.22059/ijfcs.2015.55846.

Renuka, G. (2022) Role of AI in plant science research. *African Journal of Biological Sciences* 4(1), 223–237. DOI: 10.48047/AFJBS.4.1.2022.223-237.

Rives, A., Meier, J., Sercu, T., Goyal, S., Lin, Z. *et al.* (2021) Biological structure and function emerge from scaling unsupervised learning to 250 million protein sequences. *Proceedings of the National Academy of Sciences* 118(15), e2016239118. DOI: 10.1073/pnas.2016239118.

Russell, S.J. and Norvig, P. (2016) *Artificial intelligence: A modern approach*, 3rd ed. Pearson Education Limited, Malaysia, pp. 978–1.

Sadashiv, N., Garg, A., Kumar, A., Verma, A. and Adhikari, C. (2021) An artificial intelligence based approach for increasing agricultural yield. *Indian Journal of Science and Technology* 14(1), 8–21. DOI: 10.17485/IJST/v14i1.1977.

Sampath, L., Venna, S., Himakara Datta, M. and Tushar, A.M. (2023) What is artificial intelligence in plant breeding? *Just Agriculture* 3(11), 352–358.

Scheurer, M.S. and Slager, R.J. (2020) Unsupervised machine learning and band topology. *Physical Review Letters* 124(22), 226401. DOI: 10.1103/PhysRevLett.124.226401.

Schwartz, M.P., Hou, Z., Propson, N.E., Zhang, J., Engstrom, C.J. *et al.* (2015) Human pluripotent stem cell-derived neural constructs for predicting neural toxicity. *Proceedings of the National Academy of Sciences USA* 112(40), 12516–12521. DOI: 10.1073/pnas.1516645112.

Shaik, R. and Ramakrishna, W. (2014) Machine learning approaches distinguish multiple stress conditions using stress-responsive genes and identify candidate genes for broad resistance in rice. *Plant Physiology* 164(1), 481–495. DOI: 10.1104/pp.113.225862.

Sharma, N., Kumar, M., Sharma, N. and Singh, N.K. (2025) Integration of artificial intelligence (AI) and machine learning (ML) in plant breeding. In: *Plant Stress Tolerance*. CRC Press, pp. 110–117. DOI: 10.1201/9781003457237-5.

Sharma, S., Walia, G.K., Singh, K., Batra, V., Sekhon, A.K. *et al.* (2024) Comparative analysis on crop yield forecasting using machine learning techniques. *Rural Sustainability Research* 52(347), 63–77. DOI: 10.2478/plua-2024-0015.

Sheikh, M., Iqra, F., Ambreen, H., Pravin, K.A., Ikra, M. *et al.* (2024) Integrating artificial intelligence and high-throughput phenotyping for crop improvement. *Journal of Integrative Agriculture* 23(6), 1787–1802. DOI: 10.1016/j.jia.2023.10.019.

Shikha, M., Kanika, A., Rao, A.R., Mallikarjuna, M.G., Gupta, H.S. *et al.* (2017) Genomic selection for drought tolerance using genome-wide SNPS in maize. *Frontiers in Plant Science* 8, 550. DOI: 10.3389/fpls.2017.00550.

Shrestha, S., Subedi, S. and Jiban, J. (2020) Marker-assisted selection: A smart biotechnological strategy for modern plant breeding. *Peruvian Journal of Agronomy* 4(3), 104–120. DOI: 10.21704/pja.v4i3.1490.

Singh, A.V., Chandrasekar, V., Janapareddy, P., Mathews, D.E., Laux, P. *et al.* (2021) Emerging application of nanorobotics and artificial intelligence to cross the BBB: Advances in design, controlled maneuvering, and targeting of the barriers. *ACS Chemical Neuroscience* 12(11), 1835–1853. DOI: 10.1021/acschemneuro.1c00087.

Stanke, M., Keller, O., Gunduz, I., Hayes, A., Waack, S. *et al.* (2006) AUGUSTUS: Ab initio prediction of alternative transcripts. *Nucleic Acids Research* 34(Web Server issue), W435–W439. DOI: 10.1093/nar/gkl200.

Sun, L., Liu, H., Zhang, L. and Meng, J. (2015) LncRScan-SVM: A tool for predicting long non-coding RNAs using support vector machine. *PLOS ONE* 10(10), e0139654. DOI: 10.1371/journal.pone.0139654.

Sundaram, L., Gao, H., Padigepati, S.R., McRae, J.F., Li, Y. *et al.* (2018) Predicting the clinical impact of human mutation with deep neural networks. *Nature Genetics* 50(8), 1161–1170. DOI: 10.1038/s41588-018-0167-z.

Tang, H. and Thomas, P.D. (2016) Tools for predicting the functional impact of nonsynonymous genetic variation. *Genetics* 203(2), 635–647. DOI: 10.1534/genetics.116.190033.

Tavanaei, A., Anandanadarajah, N., Maida, A.S. and Loganantharaj, R. (2017) A deep learning model for predicting tumor suppressor genes and oncogenes from PDB structure. In: *2017 IEEE International Conference on Bioinformatics and Biomedicine (BIBM)*, Kansas City, MO, pp. 1–10 (Vol. bioRxiv). DOI: 10.1109/BIBM.2017.8217722.

Toneyan, S., Tang, Z. and Koo, P.K. (2022) Evaluating deep learning for predicting epigenomic profiles. *Nature Machine Intelligence* 4(12), 1088–1100. DOI: 10.1038/s42256-022-00570-9.

Trieu, T., Martinez-Fundichely, A. and Khurana, E. (2020) DeepMILO: A deep learning approach to predict the impact of non-coding sequence variants on 3D chromatin structure. *Genome Biology* 21(1), 1–11. DOI: 10.1186/s13059-020-01987-4.

Van Dijk, A.D.J., Kootstra, G., Kruijer, W. and Ridder, D. (2021) Machine learning in plant science and plant breeding. *iScience* 24(1), 101890. DOI: 10.1016/j.isci.2020.101890.

Verma, S., Gupta, A.R.S.S.H., Yalla, S., Shreyaand Patel, P.J. (2024) Integrating marker-assisted (MAS) and genomic selection (GS) for plant functional trait improvement. In: *Plant Functional Traits for Improving Productivity*. Springer Nature, Singapore, pp. 203–215. DOI: 10.1007/978-981-97-1510-7_11.

Wang, Y., Huang, T., Liu, J., Lin, Z., Li, S. *et al.* (2015) Soil pH value, organic matter and macronutrients contents prediction using optical diffuse reflectance spectroscopy. *Computers and Electronics in Agriculture* 111, 69–77. DOI: 10.1016/j.compag.2014.11.019.

Warman, C. and Fowler, J.E. (2021) Deep learning-based high-throughput phenotyping can drive future discoveries in plant reproductive biology. *Plant Reproduction* 34(2), 81–89. DOI: 10.1007/s00497-021-00407-2.

Williamson, H.F., Brettschneider, J., Caccamo, M., Davey, R.P., Goble, C. *et al.* (2021) Data management challenges for artificial intelligence in plant and agricultural research. *F1000Research* 10, 324. DOI: 10.12688/f1000research.52204.2.

Wu, K. and Wei, G.W. (2018) Quantitative toxicity prediction using topology based multitask deep neural networks. *Journal of Chemical Information and Modeling* 58(2), 520–531. DOI: 10.1021/acs.jcim.7b00558.

Yamane, J., Aburatani, S., Imanishi, S., Akanuma, H., Nagano, R. *et al.* (2016) Prediction of developmental chemical toxicity based on gene networks of human embryonic stem cells. *Nucleic Acids Research* 44(12), 5515–5528. DOI: 10.1093/nar/gkw450.

Zhang, X., Ibrahim, Z., Khaskheli, M.B., Raza, H., Zhou, F. *et al.* (2024) Integrative approaches to abiotic stress management in crops: Combining bioinformatics educational tools and artificial intelligence applications. *Sustainability* 16(17), 7651. DOI: 10.3390/su16177651.

Zhou, J. and Troyanskaya, O.G. (2015) Predicting effects of noncoding variants with deep learning-based sequence model. *Nature Methods* 12(10), 931–934. DOI: 10.1038/nmeth.3547.

Zhou, J., Theesfeld, C.L., Yao, K., Chen, K.M., Wong, A.K. *et al.* (2018) Deep learning sequence-based ab initio prediction of variant effects on expression and disease risk. *Nature Genetics* 50(8), 1171–1179. DOI: 10.1038/s41588-018-0160-6.

Zhou, L., Pan, S., Wang, J. and Vasilakos, A.V. (2017) Machine learning on big data: Opportunities and challenges. *Neurocomputing* 237, 350–361. DOI: 10.1016/j.neucom.2017.01.026.

Zhu, W., Li, W., Zhang, H. and Li, L. (2025) Big data and artificial intelligence-aided crop breeding: Progress and prospects. *Journal of Integrative Plant Biology* 67(3), 722–739. DOI: 10.1111/jipb.13791.

Zhu, W., Han, R., Shang, X., Zhou, T., Liang, C. *et al.* (2024) The crop GPT project: Call for a global, coordinated effort in precision design breeding driven by AI using biological big data. *Molecular Plant* 17(2), 215–218. DOI: 10.1016/j.molp.2023.12.015.

Zou, J., Huss, M., Abid, A., Mohammadi, P., Torkamani, A. *et al.* (2019) A primer on deep learning in genomics. *Nature Genetics* 51(1), 12–18. DOI: 10.1038/s41588-018-0295-5.

2 Artificial Intelligence in Functional Genomics of Plants

Varun Kumar Sharma[1*†], Shilpy Singh[1], Biswajit Roy[2] and Manoj Kumar Mishra[3*†]

[1]Department of Biotechnology & Microbiology, Noida International University, Gautam Budh Nagar, India; [2]Information Technology Group CSIR-Centre for Cellular & Molecular Biology, Habsiguda, Hyderabad, India; [3]Department of Biotechnology, SR Institute of Management & Technology, Lucknow, India

Abstract

The convergence of artificial intelligence (AI) and functional genomics is transforming plant research, enabling unprecedented insights into gene functions, regulatory networks, and phenotypic traits. This chapter explores the applications of AI methodologies, including machine learning, deep learning, and natural language processing, in functional genomics. These technologies accelerate data analysis, facilitate genomic predictions, and unravel complex biological processes. We also highlight AI-driven advances in omics data integration, precision phenotyping, and trait prediction, which are crucial for crop improvement and sustainable agriculture. The chapter concludes with challenges, ethical considerations, and future directions at the AI–functional genomics interface.[1]

Keywords: Artificial intelligence, functional genomics, machine learning, deep learning, omics data integration, plant genetics, crop improvement

2.1 Introduction

Functional genomics seeks to understand gene functions and interactions, bridging the gap between genotype and phenotype. With the explosion of high-throughput omics data, traditional analytical approaches face limitations in handling complexity and scale (Caudai *et al.*, 2021; Tyagi *et al.*, 2024). It aims to bridge the gap between sequence data and biological function by studying how genes contribute to the complex web of processes that underpin growth, development, stress responses, and metabolism in plants. The scope of this field extends from identifying the functions of individual genes to examining complex regulatory networks and pathways that control plant phenotypes (Huo *et al.*, 2024; Tyagi *et al.*, 2024). This knowledge contributes to the development of crops with improved characteristics, such as higher yield, enhanced nutritional content, and resilience to biotic and abiotic stresses. For instance, functional genomics has facilitated the identification of genes responsible for drought tolerance, enabling scientists to engineer crops that can

*Corresponding authors: varungenetics@gmail.com and manoj30biotech@gmail.com

†Authors contributed equally

DOI: 10.1079/9781800628649.0002

better withstand changing climate conditions (Sami *et al.*, 2021; Zenda *et al.*, 2021; Abdul Aziz *et al.*, 2022). In plant breeding, functional genomics provides a robust toolkit for precision breeding. Through techniques such as RNA sequencing, gene knockout studies, and CRISPR/Cas9 gene editing, breeders can create varieties with targeted improvements, accelerating the development of crops with desirable traits (Sami *et al.*, 2021; Saini *et al.*, 2023). Furthermore, the integration of functional genomics with other omics technologies, like proteomics and metabolomics, enables a holistic understanding of plant systems biology, making agriculture more sustainable and efficient (Zenda *et al.*, 2021; Jain *et al.*, 2024). Overall, plant functional genomics is vital for addressing global food security challenges by promoting the development of crops that are not only high yielding but also resilient to environmental stressors (Tyagi *et al.*, 2024). This field continues to evolve rapidly with the advent of high-throughput sequencing technologies and computational tools that enhance the ability to analyze and interpret vast amounts of genetic and phenotypic data (Zenda *et al.*, 2021; Jain *et al.*, 2024; Tyagi *et al.*, 2024). Artificial intelligence (AI) emerges as a transformative tool, offering robust analytical capabilities to decipher genomic information and derive actionable insights (Caudai *et al.*, 2021). AI refers to the simulation of human intelligence in machines, and involves the development of algorithms that enable machines to perform tasks such as recognizing patterns, making decisions, solving problems, and understanding natural language (Gupta *et al.*, 2021; Nath *et al.*, 2024). Therefore, it has become a transformative tool in many fields, including healthcare, finance, and agriculture, due to its ability to process and analyze large volumes of data quickly and accurately (Nath *et al.*, 2024). In the context of plant functional genomics, AI plays a pivotal role in advancing the understanding of plant biology. Traditional genomics involves the sequencing of DNA, annotating gene functions, and studying gene expression (Fig. 2.1). However, these methods are often constrained by challenges such as data complexity, scalability, and the difficulty of integrating multiple types of data (e.g. genomic, transcriptomic, and phenotypic data) (Hong *et al.*, 2023; Farooq *et al.*, 2024). AI overcomes these limitations by using machine learning (ML) and deep learning (DL) models to process vast datasets, recognize patterns, and make predictions that would otherwise be too complex for human researchers (Taye, 2023). AI tools have enabled researchers to make ground-breaking advances in gene prediction, gene function annotation, and gene interaction mapping. Through supervised learning, AI models can be trained on annotated datasets to predict gene functions, identify new genes associated with important traits, and discover regulatory networks within plants (Caudai *et al.*, 2021; Farooq *et al.*, 2024). In unsupervised learning, AI can identify hidden patterns in genomic data that were previously unknown, leading to new insights into plant biology and helping to improve crop traits such as yield, disease resistance, and stress tolerance (Khan *et al.*, 2022). Additionally, AI has made significant contributions to areas such as epigenomics, transcriptomics, and gene editing. For example, in transcriptomics, AI can analyze RNA sequencing data to identify gene expression patterns that indicate how plants respond to environmental stressors (Cembrowska-Lech *et al.*, 2023; Murmu *et al.*, 2024). In gene editing, AI tools are used to enhance CRISPR/Cas9 precision by predicting guide RNA sequences and minimizing off-target effects. AI has also become a critical component in predictive modeling, helping researchers design better crops and improve food security by integrating data from various biological disciplines (Ansori *et al.*, 2023). This chapter explores the pivotal role of AI in functional genomics, with a focus on its applications in plant systems.

2.2 AI Methodologies in Functional Genomics

AI methodologies are transforming the area of functional genomics by allowing for the study of large and complicated genetic datasets (Caudai *et al.*, 2021). Researchers can identify detailed patterns and relationships in genetic data that were previously impossible to detect using ML and DL methods (Quazi, 2022; Vilhekar and Rawekar, 2024). These AI-driven technologies facilitate the identification of gene functions, regulatory elements, and genetic variants associated with diseases, enhancing our understanding

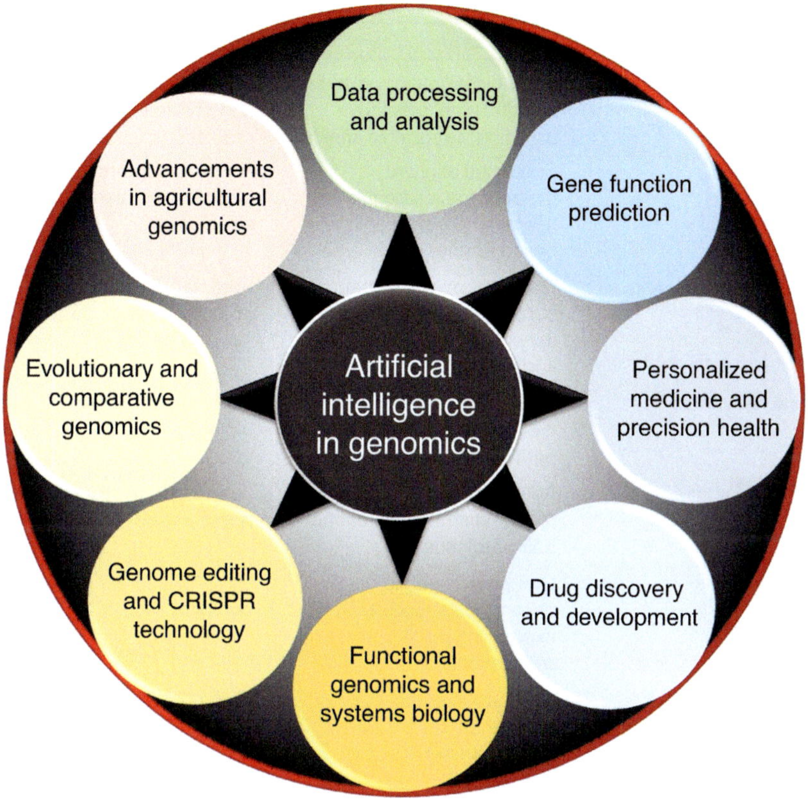

Fig. 2.1. Demonstration of the possible involvement of AI in genomic research and development. (Author's own image.)

of biological systems and speeding up the creation of customized treatment (Caudai *et al.*, 2021; Gupta *et al.*, 2021; Quazi, 2022; Vilhekar and Rawekar, 2024). Table 2.1 summarizes the major AI methodologies and their applications.

2.2.1 Machine learning (ML)

ML algorithms such as support vector machines (SVMs), random forests, and gradient boosting are extensively used for classifying genes, predicting gene functions, and identifying genomic markers (Vanhaeren *et al.*, 2020; Quazi, 2022). Common techniques include supervised learning (training a model on labeled data) and unsupervised learning (finding hidden patterns in unlabeled data). The workflow of ML is illustrated in brief (Fig. 2.2).

Additional aspects of machine learning in functional genomics (Libbrecht and Noble, 2015; Caudai *et al.*, 2021) include:

- **Transfer learning:** Leveraging pre-trained models on similar datasets to reduce computational resources and improve model efficiency.
- **Reinforcement learning:** Utilizing feedback-driven learning for adaptive genomic predictions and optimization of breeding strategies.
- **Automated feature engineering:** Employing AI-driven techniques to automatically extract and refine features from large, unstructured datasets, enhancing prediction accuracy.
- **Integrative analysis:** Combining omics data types (e.g. transcriptomics with

Table 2.1. Use of AI methodologies in plant genomics. (Author's own table.)

AI methodology	Applications	Key benefits
Machine learning (ML)	Gene classification, marker identification, predictive breeding, integrative analysis	Enhanced accuracy, adaptability, and scalability in genomic predictions
Deep learning (DL)	Sequence analysis, transcriptomics, regulatory network modeling, image-based phenotyping	Ability to capture non-linear patterns and hierarchical data structures
Natural language processing (NLP)	Literature mining, database curation, ontology development, text summarization	Automated extraction of insights from vast textual data
Graph neural networks	Analysis of gene–protein–metabolite interactions, pathway discovery	Comprehensive modeling of biological networks
Reinforcement learning	Adaptive optimization of breeding strategies, dynamic trait prediction	Learning from feedback to refine predictions over time
Clustering algorithms	Grouping of genes or samples based on expression profiles or phenotypic traits	Unveiling hidden patterns in complex datasets
Transfer learning	Application of pre-trained models for genomic analysis across species	Reduction in training time and computational costs
Ensemble learning	Integration of multiple models for improved genomic predictions	Enhanced predictive performance and robustness
Automated feature engineering	Extraction of key features from high-dimensional data for gene and trait analysis	Minimizes manual intervention and improves the reproducibility of results
Dimensionality reduction techniques	Simplification of high-throughput data to focus on critical variables	Reduces computational complexity while retaining essential biological information

proteomics) using ensemble learning methods to gain a comprehensive understanding of gene interactions.

- **Clustering and dimensionality reduction:** Implementing algorithms like k-means clustering or principal component analysis (PCA) to identify hidden patterns in high-dimensional data.

2.2.2 Deep learning

A subset of ML, DL involves neural networks with multiple layers that can automatically extract hierarchical features from complex data (Alzubaidi *et al.*, 2021; Sarker, 2021). This approach is particularly effective for tasks like DNA sequence analysis, image-based genomic data (e.g. histopathology slides), and predictive modeling for disease outcomes. DL models, such as convolutional neural networks (CNNs) and recurrent neural networks (RNNs), are used for sequence analysis, transcriptomics, and epigenomics research (Alzubaidi *et al.*, 2021; Caudai *et al.*, 2021; Gupta *et al.*, 2021; Taye, 2023). These models excel in capturing nonlinear and hierarchical patterns in huge datasets. Gene expression prediction is predicting how genes are expressed in different situations, such as environmental pressures or developmental

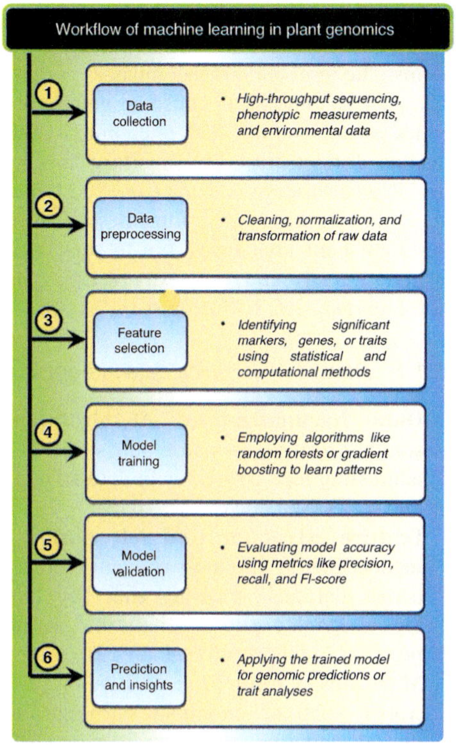

Fig. 2.2. Demonstration of the workflow of machine learning in plant genomics. (1) Data collection → (2) Data preprocessing → (3) Feature selection → (4) Model training → (5) Model validation → (6) Prediction and insights. (Author's own image.)

phases (Libbrecht and Noble, 2015; Vilhekar and Rawekar, 2024). For example, CNNs can be applied to genomic sequences to identify regulatory elements and predict their impact on gene expression. These models can learn intricate patterns within the DNA sequence that influence gene activity. RNNs, in contrast, can be used to model temporal gene expression changes by capturing dependencies between time points (Li *et al.*, 2023; Yue *et al.*, 2023).

The use of DL in gene expression prediction has several advantages. It enables researchers to uncover complex regulatory mechanisms, identify key genes involved in stress responses, and understand the dynamics of gene networks (Li *et al.*, 2023). This knowledge can be applied to enhance crop resilience by engineering plants with optimized gene expression profiles for better growth, yield, and stress tolerance (Sami *et al.*, 2021; Li *et al.*, 2023). By leveraging DL for gene expression prediction, scientists can accelerate the pace of discoveries in plant functional genomics and develop innovative solutions to address agricultural challenges.

2.2.3 Natural language processing

Natural language processing (NLP) is an emerging AI methodology with transformative potential in functional genomics. NLP focuses on enabling machines to interpret, analyze, and generate human language (Oikonomou *et al.*, 2024; Sim *et al.*, 2024). In the context of genomics, NLP aids researchers in extracting valuable insights from vast amounts of textual data, such as scientific publications, patents, and databases (Harmston *et al.*, 2010; Zeng *et al.*, 2019; Oikonomou *et al.*, 2024; Sim *et al.*, 2024).

2.2.3.1 Applications of NLP in functional genomics

Literature mining (Chen *et al.*, 2021; Oikonomou *et al.*, 2024; Sim *et al.*, 2024):

- **Gene annotation:** Automated extraction of gene functions, pathways, and associated phenotypes from scientific literature.
- **Knowledge discovery:** Identification of novel gene–disease or gene–trait associations by analyzing co-occurrence patterns in texts.

Database curation (Harmston *et al.*, 2010; Zeng *et al.*, 2019; Oikonomou *et al.*, 2024):

- **Data integration:** NLP tools facilitate the extraction and structuring of data from disparate sources into unified databases.
- **Automated updates:** Dynamic curation of databases using NLP to incorporate the latest research findings.
- **Hypothesis generation:** NLP models predict potential gene interactions or regulatory mechanisms based on existing knowledge.
- **Ontology development:** Assisting in the creation and refinement of controlled vocabularies and ontologies for plant genomics.

- **Text summarization:** Automated summarization of complex articles and datasets, enabling quicker assimilation of key insights by researchers.

2.2.3.2 Techniques in NLP

Techniques in NLP include (Harmston *et al.*, 2010; Zeng *et al.*, 2019; Oikonomou *et al.*, 2024):

- **Named entity recognition:** Identifying and categorizing entities such as gene names, species, and phenotypic traits in unstructured text.
- **Relation extraction:** Unveiling relationships between entities, such as gene–protein or gene–disease links.
- **Topic modeling:** Using algorithms like Latent Dirichlet allocation to discover themes and topics within large corpora of text.
- **Question-answering systems:** AI-driven systems that respond to specific queries related to genomics using structured and unstructured data sources.

Addressing language ambiguity is crucial in plant functional genomics, as domain-specific terminologies and polysemous words can pose significant challenges. Enhancing the scalability of NLP tools is necessary to process the rapidly growing amount of textual data in this field (Harmston *et al.*, 2010; Zeng *et al.*, 2019). Ensuring model interpretability is essential, as it fosters trust among researchers by providing transparent and explainable outputs. Additionally, integrating NLP with ML and DL creates a comprehensive analysis pipeline that combines text-based and numerical genomic data. By leveraging these advancements in NLP, researchers can accelerate discoveries in plant functional genomics, enabling better data utilization and hypothesis generation (Harmston *et al.*, 2010; Zeng *et al.*, 2019; Oikonomou *et al.*, 2024; Sim *et al.*, 2024).

2.3 Applications in Plant Functional Genomics

2.3.1 Omics data integration

Omics data integration is a pivotal aspect of modern functional genomics, aiming to combine data from diverse biological layers to provide a holistic understanding of plant systems. The term 'omics' encompasses various fields such as genomics, transcriptomics, proteomics, metabolomics, and phenomics, each providing distinct but complementary insights into biological processes (Zenda *et al.*, 2021; Jain *et al.*, 2024). However, integrating these datasets poses significant challenges due to their vast size, heterogeneity, and complexity (Zenda *et al.*, 2021; Jain *et al.*, 2024; Murmu *et al.*, 2024).

2.3.1.1 Role of AI in omics data integration

- **Data harmonization:** AI algorithms standardize and normalize omics datasets, addressing inconsistencies in data formats and scales (Sen and Orešič, 2023).
- **Feature extraction and selection:** ML models identify the most relevant features across datasets, ensuring that key biological signals are not lost in the integration process (Pudjihartono *et al.*, 2022).
- **Multi-omics modeling:** AI-driven frameworks, such as ensemble learning and neural networks, create predictive models that combine inputs from various omics layers to identify gene regulatory networks and pathway interactions (Jain *et al.*, 2024).
- **Network analysis:** Tools like graph neural networks analyze relationships and interactions between genes, proteins, and metabolites, unveiling complex regulatory pathways (Murmu *et al.*, 2024).
- **Visualization:** AI-powered visualization tools render integrated data into interpretable formats, such as heatmaps, network diagrams, or 3D structures, aiding researchers in hypothesis generation (Hohman *et al.*, 2018; Zenda *et al.*, 2021).

2.3.1.2 Application

Omics data integration has various promising applications in plant functional genomics, particularly in agricultural enhancement. Here are a few significant applications (Hohman *et al.*, 2018; Zenda *et al.*, 2021; Jain *et al.*, 2024).

- **Enhanced crop breeding:** By integrating genomics, transcriptomics, proteomics, and metabolomics data, researchers can identify genes and pathways associated with desirable traits such as high yield, disease resistance, and stress tolerance. This helps in developing improved crop varieties.

- **Understanding plant responses:** Multi-omics approaches help in understanding how plants respond to various biotic (e.g. pests and diseases) and abiotic (e.g. drought, salinity) stresses. This knowledge can be used to develop crops that are more resilient to environmental challenges.
- **Precision agriculture:** Omics data can be used to develop precision agriculture techniques, where specific interventions are applied based on the genetic and molecular profile of individual plants or fields. This can lead to more efficient use of resources and better crop management.
- **Nutritional improvement:** By analyzing the metabolic profiles of plants, researchers can identify and enhance the nutritional content of crops, making them healthier for consumption.

2.3.2 Precision phenotyping

Precision phenotyping refers to the accurate and high-throughput measurement of plant traits to establish clear genotype–phenotype correlations. Traditional phenotyping methods are often labor intensive, time consuming, and prone to human error, limiting their scalability (Cembrowska-Lech *et al*., 2023). Precision phenotyping leverages advanced imaging technologies, sensors, and AI to overcome these limitations. High-throughput imaging techniques, such as hyperspectral, thermal, and fluorescence imaging, capture detailed traits, while sensors monitor environmental factors and physiological parameters in real-time (Caudai *et al*., 2021; Gupta *et al*., 2021). AI-powered data analysis identifies subtle trait variations and genetic determinants, and robotic platforms enable automated monitoring. Applications of precision phenotyping include robust genotype–phenotype associations, assessment of plant stress tolerance, enhancement of breeding programs, and early disease diagnostics (Quazi, 2022; Taye, 2023; Yue *et al*., 2023). The approach offers scalability, accuracy, and speed, although challenges such as data integration, cost, and standardization remain. Overall, precision phenotyping drives innovations in crop improvement by providing insights into complex plant traits

2.3.3 Gene function prediction

Gene function prediction is a central task in genomics, as understanding the role of specific genes in biological processes helps to decode complex traits, diseases, and cellular functions. Traditional methods of gene function annotation, based on experimental approaches, are labor intensive and time consuming (Abdul Aziz *et al*., 2022; Li *et al*., 2023; Vilhekar and Rawekar, 2024). AI has emerged as a powerful tool in accelerating gene function prediction and annotation by analyzing large genomic datasets, identifying patterns, and making predictions based on known biological principles (Li *et al*., 2023; Vilhekar and Rawekar, 2024) (Fig. 2.3).

AI, particularly ML and DL, enables the annotation of genes, prediction of their functions, and identification of gene interactions in ways that were not previously possible (Caudai *et al*., 2021; Abdul Aziz *et al*., 2022; Li *et al*., 2023).

2.3.3.1 AI in annotating and predicting gene functions

AI-powered models are used to annotate genes by predicting their biological roles based on sequence information and functional data. Gene annotation involves associating specific functions with genes, such as identifying whether a gene encodes a protein involved in metabolism, signal transduction, or immune response (Caudai *et al*., 2021; Abdul Aziz *et al*., 2022). In functional genomics, AI techniques, such as DL and SVMs, are employed to analyze genomic sequences and predict gene function with greater accuracy and efficiency (Caudai *et al*., 2021; Yue *et al*., 2023; Tyagi *et al*., 2024).

2.3.3.2 Sequence-based prediction

ML models can predict gene function by analyzing sequence motifs, regulatory elements, and known functional domains. For instance, deep CNNs are trained to identify sequence patterns that correspond to specific functional regions, such as promoters, enhancers, and

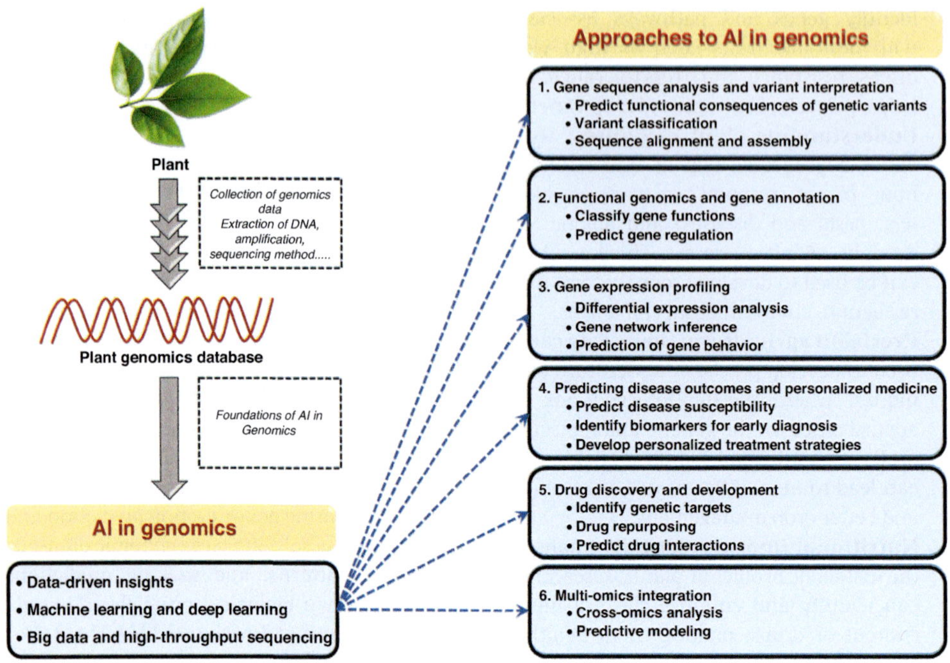

Fig. 2.3. Demonstration of different approaches of AI in genomics. (Author's own image.)

transcription-factor-binding sites (Libbrecht and Noble, 2015). These models can predict whether a given gene may be involved in specific pathways or processes by learning patterns from large datasets of known gene annotations.

2.3.3.3 Gene ontology prediction

Gene ontology (GO) terms categorize gene functions into three domains: biological processes, molecular functions, and cellular components. AI algorithms, including DL and random forests, are used to predict GO terms for genes based on their genomic sequences and expression patterns (Ersoz *et al.*, 2023). These models can predict the biological functions of genes for which the roles are not yet experimentally validated, accelerating the annotation of newly sequenced genomes (Pazos Obregón *et al.*, 2022; Ersoz *et al.*, 2023).

2.3.3.4 Homology-based prediction

AI models can predict gene functions by leveraging homology-based approaches. By comparing the sequences of unknown genes with those of known genes from related species, ML algorithms can infer the function of a gene based on its similarity to genes with well-characterized roles. BLAST and other homology search tools integrated with AI are frequently used to predict gene functions across a wide variety of organisms (Vadapalli *et al.*, 2022; Karlsen *et al.*, 2023).

2.3.3.5 Protein structure and function prediction

AI has revolutionized the prediction of protein structures, which is closely tied to gene function. AlphaFold, an AI system developed by DeepMind, predicts the three-dimensional structure of proteins based on their amino acid sequences. By understanding protein structure, AI models can predict how proteins interact with other molecules and determine the biological role of the gene that encodes them (Caudai *et al.*, 2021; Nussinov *et al.*, 2022).

2.3.3.6 Non-coding region prediction

AI is also used to predict the functions of non-coding genomic regions, which constitute a large portion of the genome. These regions often play critical roles in regulating gene expression. AI tools such as DeepSEA and DeepBind use DL to predict the functional impact of mutations in non-coding regions, identifying potential enhancers, silencers, and other regulatory elements (Caudai *et al.*, 2021; Li *et al.*, 2023; Yousef and Allmer, 2023).

2.3.4 Trait prediction and crop improvement

Plant breeding is the process of selecting and developing new plant varieties with desirable traits, such as higher yield, disease resistance, and better environmental adaptability. Traditional breeding techniques are often slow and labor intensive, requiring multiple generations to achieve significant improvements (Saini *et al.*, 2023; Tyagi *et al.*, 2024). AI is revolutionizing plant breeding by enabling faster, more efficient identification of gene–trait associations and enhancing genomic selection, paving the way for improved crop varieties and more sustainable agriculture (Khan *et al.*, 2022; Farooq *et al.*, 2024).

2.3.4.1 AI for identifying gene–trait associations

A key challenge in plant breeding is identifying the specific genetic factors (genes, alleles, or markers) that are responsible for particular traits, such as drought tolerance, pest resistance, or disease resistance. These complex traits often involve multiple genes interacting with each other and the environment, making it difficult to pinpoint the exact genetic basis (Saini *et al.*, 2023; Tyagi *et al.*, 2024). AI techniques, particularly ML and DL, can analyze vast amounts of genomic and phenotypic data to identify gene–trait associations with high precision. AI models, such as random forests, SVMs, and neural networks, can be trained on large-scale datasets of genotypic and phenotypic information to identify correlations between specific genetic markers and traits of interest (Abdul Aziz *et al.*, 2022; Li *et al.*, 2023; Taye, 2023). These models are particularly valuable when dealing with complex traits that are polygenic (controlled by many genes) and influenced by environmental factors. By processing genomic data along with phenotypic measurements, AI can uncover hidden patterns and relationships that might not be apparent through traditional statistical methods. This allows researchers and breeders to make more accurate predictions about which genes contribute to desirable traits, enabling the selection of plants with the best genetic potential for breeding programs (Khan *et al.*, 2022; Saini *et al.*, 2023; Farooq *et al.*, 2024).

AI also facilitates the identification of quantitative trait loci (QTLs), regions of the genome associated with variation in traits. Through techniques like genome-wide association studies (GWAS), AI algorithms can identify genetic loci linked to complex traits and estimate their effects, accelerating the process of marker-assisted selection (Caudai *et al.*, 2021; Saini *et al.*, 2023; Sahito *et al.*, 2024; Tyagi *et al.*, 2024).

2.3.4.2 Genomic selection using machine learning

Genomic selection is a powerful tool that enables breeders to select plants based on their genetic makeup rather than solely relying on phenotypic observations. This approach has been particularly transformative in crops with long breeding cycles or when phenotypic data is difficult to obtain (Harmston *et al.*, 2010; Alemu *et al.*, 2024; Tyagi *et al.*, 2024). AI, particularly ML algorithms, has greatly enhanced the accuracy and efficiency of genomic selection by enabling the analysis of large, high-dimensional datasets that include both genomic and phenotypic information (Libbrecht and Noble, 2015; Caudai *et al.*, 2021; Quazi, 2022; Yue *et al.*, 2023). In genomic selection, ML models are trained on historical data, where both the genetic information (genotype) and the observed traits (phenotype) of plants are known. These models learn to associate specific genomic patterns with phenotypic traits, enabling them to predict the performance of new plants based on their genetic data (Danilevicz *et al.*, 2022; Alemu *et al.*, 2024). AI tools, such as random forests, neural networks, and deep learning, are particularly effective at handling the large, complex datasets generated by genomic research. These models can consider interactions between multiple genes and environmental factors, improving predictions for traits that involve complex genetic networks

(Quazi, 2022; Farooq *et al.*, 2024). The main advantage of using AI in genomic selection is its ability to handle large datasets and provide predictions at a much faster rate compared to traditional breeding methods. This significantly shortens the time required to develop new plant varieties with desirable traits. For example, AI-driven genomic selection can be used to predict the performance of hybrid plants or identify elite parent lines in breeding programs, allowing breeders to focus on the most promising genetic combinations without the need for extensive field trials (Quazi, 2022; Alemu *et al.*, 2024; Farooq *et al.*, 2024).

AI also helps in predicting the genetic value of individuals within a population, enabling the selection of plants that are most likely to yield desirable traits, even before they are phenotypically assessed (Khan *et al.*, 2022; Feng *et al.*, 2024). This makes genomic selection particularly useful for breeding programs aimed at improving crops for specific traits, such as yield, pest resistance, or drought tolerance, where traditional phenotypic screening may be slow or impractical (Khan *et al.*, 2022; Farooq *et al.*, 2024; Feng *et al.*, 2024).

2.3.4.3 Benefits

The integration of AI into plant breeding and trait improvement offers several significant benefits. First, it accelerates the breeding process by enabling more precise identification of desirable traits and genetic markers. This allows breeders to make better-informed decisions and select superior plants more quickly, leading to the development of improved crop varieties in shorter timeframes (Khan *et al.*, 2022; Yoosefzadeh Najafabadi *et al.*, 2023; Farooq *et al.*, 2024). AI also helps breeders handle the complexity of polygenic traits and environmental interactions, which are often difficult to manage using traditional breeding methods. Moreover, AI-enhanced genomic selection reduces the reliance on labor-intensive and time-consuming phenotypic evaluations, making it possible to conduct breeding programs at a larger scale. The ability to predict traits based on genetic information alone improves efficiency and reduces the costs associated with field trials (Quazi, 2022; Cembrowska-Lech *et al.*, 2023; Murmu *et al.*, 2024; Vilhekar and Rawekar, 2024).

2.4 Case Studies

2.4.1 Case study on rice (*Oryza sativa*): AI in stress tolerance and yield improvement

Rice (*Oryza sativa*) is a staple crop and a critical source of food for billions of people worldwide. With climate change, however, rice productivity is increasingly challenged by environmental stressors like drought, salinity, and temperature fluctuations (Muthayya *et al.*, 2014; Mohidem *et al.*, 2022). AI has emerged as a powerful tool for improving rice stress tolerance and yield (Muthayya *et al.*, 2014; Liu *et al.*, 2021; Mohidem *et al.*, 2022; Shubhika *et al.*, 2024).

- **Stress tolerance in rice:** AI-based approaches, particularly ML models, have been used to identify genetic markers linked to stress-tolerance traits. By integrating genomic, transcriptomic, and phenotypic data, AI tools can predict how rice varieties will respond to various stressors. ML models trained on large datasets of rice gene expression profiles under different stress conditions can identify stress-responsive genes and regulatory networks (Liu *et al.*, 2021; Mohidem *et al.*, 2022). These models can predict how specific genetic variations influence stress tolerance, helping to guide breeding programs that select for drought and salinity-resistant varieties. For example, AI algorithms have been used to uncover key genes involved in regulating osmotic stress responses and to predict how these genes interact with environmental factors (Mohidem *et al.*, 2022; Shubhika *et al.*, 2024).
- **Yield improvement in rice:** AI applications are also playing a key role in improving rice yield through genomic selection and predictive modeling. By analyzing genomic data alongside environmental variables and phenotypic traits, AI tools can identify genetic markers associated with high yield potential. Predictive models can simulate how changes in specific genes or regulatory networks may affect rice growth and productivity under different environmental conditions (Shubhika *et al.*, 2024). These models also incorporate data from field

trials, sensor networks, and environmental variables, providing a more accurate prediction of rice yield and optimizing breeding decisions for higher-yielding varieties (Liu *et al.*, 2021; Mohidem *et al.*, 2022). By selecting for genes related to photosynthesis efficiency, root architecture, and nutrient uptake, AI-driven insights are leading to the development of rice varieties that are better suited for global food security (Liu *et al.*, 2021; Mohidem *et al.*, 2022; Shubhika *et al.*, 2024).

2.4.2 Case study on maize (*Zea mays*): AI in genetic diversity and trait discovery

Maize (*Zea mays*) is another important crop, known for its high genetic diversity and adaptability to different growing conditions. AI has become an essential tool in understanding maize genetics, enabling the discovery of new traits and improving the efficiency of breeding programs aimed at enhancing productivity and resilience (Chekole and Ahmed, 2023; Baio *et al.*, 2023; Han *et al.*, 2024).

- **Genetic diversity in maize:** AI-driven techniques are being used to analyze the vast genetic diversity of maize, which is essential for developing new hybrids and varieties with desirable traits. By applying ML algorithms to genomic data, AI can help map the genetic architecture of maize and identify genetic variants that contribute to important traits such as disease resistance, nutrient content, and abiotic stress tolerance (Croci *et al.*, 2023: Patel *et al.*, 2024). For instance, AI models that integrate GWAS data with phenotypic data allow researchers to identify loci associated with key traits and predict their effects on the overall genetic diversity of maize. These insights help breeders choose parent lines for hybridization, ensuring the retention of beneficial genetic variants and the development of new maize varieties with improved performance (Croci *et al.*, 2023; Sahito *et al.*, 2024; Han *et al.*, 2024).
- **Trait discovery in maize:** AI is playing a pivotal role in trait discovery in maize by analyzing complex datasets that combine genomic, transcriptomic, and phenotypic information (Patel *et al.*, 2024; Sahito *et al.*, 2024). ML models are able to identify novel genetic loci associated with traits like drought tolerance, pest resistance, and improved grain quality. For example, AI algorithms have been used to pinpoint genes associated with plant height, ear size, and flowering time—traits that are important for increasing maize yield (Baio *et al.*, 2023). Predictive models that integrate multi-omics data are able to identify not only individual genes but also the interactions between genes that contribute to complex traits. This leads to the discovery of new genetic pathways and potential targets for genetic modification or breeding (Baio *et al.*, 2023; Croci *et al.*, 2023; Han *et al.*, 2024).

2.5 AI in Stress Response Genomics

Plants are constantly exposed to a variety of stressors, including both abiotic stresses (such as drought, salinity, temperature extremes, and nutrient deficiencies) and biotic stresses (caused by pathogens like bacteria, fungi, and viruses). These stresses can significantly reduce crop yield and quality, making stress tolerance and resistance crucial traits in plant breeding (Alemu *et al.*, 2024; Tyagi *et al.*, 2024). AI is revolutionizing the way we understand and improve stress tolerance and resistance in plants by enabling more accurate identification of genes involved in stress responses and by predicting plant resistance to pathogens. The ability of AI to analyze large-scale genomic, transcriptomic, and phenotypic data makes it an invaluable tool for improving plant resilience to environmental challenges (Zenda *et al.*, 2021; Cembrowska-Lech *et al.*, 2023; Jain *et al.*, 2024; Shubhika *et al.*, 2024).

2.5.1 AI in identifying genes responsible for abiotic and biotic stress tolerance

One of the key applications of AI in plant stress tolerance is the identification of genes responsible for resisting or tolerating various stress conditions (Shubhika *et al.*, 2024). Abiotic stresses like drought and salinity can adversely affect plant growth and development, while biotic stresses, such as fungal or bacterial infections, can devastate crops. Understanding the genetic basis of plant responses to these stresses is essential for developing resilient crops (Liu *et al.*, 2021; Shubhika *et al.*, 2024).

AI models, particularly ML and DL algorithms, can analyze genomic, transcriptomic, and phenotypic data to identify stress-responsive genes and regulatory pathways that control plant tolerance to stress (Nath *et al.*, 2024; Shubhika *et al.*, 2024). These models are trained on large datasets that combine genetic information with observable traits (such as drought resistance, salt tolerance, or pathogen resistance). AI tools can identify key genes and biomarkers associated with specific stress tolerances, even in complex, polygenic traits where multiple genes and environmental factors interact (Khan *et al.*, 2022; Farooq *et al.*, 2024; Nath *et al.*, 2024).

- **Abiotic stress tolerance:** For abiotic stresses like drought, heat, and salinity, AI models can analyze how gene expression changes under stress conditions and predict which genes are involved in stress response mechanisms (Alemu *et al.*, 2024; Tyagi *et al.*, 2024). Deep neural networks and random forests have been used to identify genes that regulate processes such as osmotic adjustment, ion homeostasis, and antioxidative responses, all of which are critical for stress tolerance (Kong and Yu, 2018; Trono and Pecchioni, 2022; Tyagi *et al.*, 2024). These models can also predict the potential genetic pathways that plants might use to mitigate damage from extreme environmental conditions, helping breeders target specific genes for enhancement in crops (Trono and Pecchioni, 2022; Tyagi *et al.*, 2024).
- **Biotic stress resistance:** AI models are also being used to identify genes that contribute to resistance against pathogens. ML techniques, including SVMs and decision trees, are used to identify gene expression patterns that correlate with pathogen resistance (Abdul Aziz *et al.*, 2022; Li *et al.*, 2023; Taye, 2023; Shubhika *et al.*, 2024). For example, AI can identify plant genes involved in pathogen recognition, immune signaling, and defense responses, which are crucial for plant resistance to bacterial, fungal, and viral infections. AI can analyze data from various omics platforms, such as transcriptomics (RNA-seq) and metabolomics, to pinpoint genes that play critical roles in immune responses and resistance mechanisms (Cembrowska-Lech *et al.*, 2023; Feng *et al.*, 2024; Murmu *et al.*, 2024).

2.5.2 Predicting resistance to pathogens using AI models

Predicting a plant's resistance to pathogens is one of the most valuable applications of AI in plant breeding, as it can help breeders select plants that are less susceptible to diseases, reducing the need for chemical pesticides and improving crop yields (Jafar *et al.*, 2024). AI models can be used to predict which plants will be resistant to specific pathogens based on their genetic makeup, environmental conditions, and historical performance (Trono and Pecchioni, 2022; Tyagi *et al.*, 2024).

AI tools can integrate large datasets of genomic, phenotypic, and environmental data to train predictive models that estimate the likelihood of a plant developing resistance to a given pathogen. These models are trained to recognize genetic patterns and interactions that correlate with pathogen resistance, allowing breeders to identify resistant varieties more efficiently (Caudai *et al.*, 2021; Saini *et al.*, 2023; Tyagi *et al.*, 2024; Farooq *et al.*, 2024).

2.5.2.1 Machine learning for pathogen resistance prediction

Random forests, deep neural networks, and gradient boosting machines are commonly used in plant pathology and breeding to predict pathogen resistance (Kong and Yu, 2018; Han *et al.*, 2024). These models can analyze the genetic sequences of plants to identify markers associated with resistance genes. By combining genomic data with environmental data, such as soil conditions and climate factors, AI can predict how plants will respond to diseases in different growing conditions (Kong and Yu, 2018; Liu *et al.*, 2021; Han *et al.*, 2024).

2.5.2.2 *Disease resistance modeling*

AI can also be used to model plant–pathogen interactions at a deeper level, helping to predict how different pathogens will affect plants under varying environmental conditions (Farooq *et al.*, 2024; Jafar *et al.*, 2024). For example, AI models can simulate how pathogens spread and how plant immune responses are activated, providing insights into which genetic traits might provide stronger resistance in different climates or farming systems (Khan *et al.*, 2022; Jafar *et al.*, 2024). This predictive capability allows breeders to prioritize resistance traits that are likely to perform well in specific regions or under particular stress conditions, facilitating more targeted and efficient breeding programs (Saini *et al.*, 2023; Farooq *et al.*, 2024; Tyagi *et al.*, 2024).

Additionally, AI models can be used to develop disease susceptibility indices, which combine genetic and environmental data to predict the likelihood of disease outbreaks in different plant populations. These predictive models are invaluable for managing diseases in agriculture, allowing farmers to adopt preventative measures before an outbreak occurs, thus reducing the need for costly and potentially harmful pesticides (Farooq *et al.*, 2024; Jafar *et al.*, 2024; Tyagi *et al.*, 2024).

2.6 Challenges and Ethical Considerations

As AI continues to be integrated into genomic research and applications, particularly in areas like genetic modification and crop improvement, it is crucial to address the ethical and regulatory concerns that arise (Abdul Aziz *et al.*, 2022). These concerns span issues related to the responsible use of AI in genetic engineering, the impact on biodiversity and ecosystems, and the governance of AI-driven biotechnologies. Ensuring that AI is used ethically and in compliance with appropriate regulatory frameworks is essential for fostering public trust and ensuring the sustainability of agricultural and biotechnological advancements (Abdul Aziz *et al.*, 2022; Khan *et al.*, 2022; Li *et al.*, 2023; Farooq *et al.*, 2024).

2.6.1 Data challenges

AI systems rely heavily on large-scale datasets to make accurate predictions and generate meaningful insights. However, the quality of these datasets is crucial to the success of AI models (Danilevicz *et al.*, 2022; Pudjihartono *et al.*, 2022; Feng *et al.*, 2024). In the context of plant genomics, several data-related challenges need to be addressed:

- **Data quality:** Genomic, transcriptomic, proteomic, and phenotypic data used in AI models can be noisy, incomplete, or of varying quality. Inaccurate or low-quality data can lead to poor model performance, limiting the effectiveness of AI tools in genomic studies (Caudai *et al.*, 2021; Feng *et al.*, 2024). For example, errors in sequencing data or inconsistencies in phenotypic measurements can introduce significant noise into the analysis, making it difficult to accurately identify gene–environment interactions or predict plant traits (Alemu *et al.*, 2024; Feng *et al.*, 2024; Tyagi *et al.*, 2024).
- **Data bias:** Many AI models are trained on datasets that may not fully represent the genetic diversity of plant species or the range of environmental conditions under which crops grow. This bias can lead to overfitting of models to specific populations, geographical regions, or environmental conditions, thus limiting their ability to generalize across different scenarios (Baio *et al.*, 2023; Chekole and Mohammed Ahmed, 2023; Han *et al.*, 2024). For instance, genomic datasets often focus on a few model crops or specific varieties, which may not capture the full genetic diversity of global crop species. Consequently, AI models trained on biased data may fail to accurately predict traits in less studied varieties or environments, potentially hindering the development of universally resilient

crops (Liu *et al.*, 2021; Baio *et al.*, 2023; Han *et al.*, 2024; Shubhika *et al.*, 2024).

- **Generalization issues:** AI models, especially those based on deep learning, may perform well on data similar to that on which they were trained, but struggle to generalize to new, unseen datasets. For example, a model trained on one set of genetic data might not be applicable to another crop species or environmental condition (Baio *et al.*, 2023; Han *et al.*, 2024; Shubhika *et al.*, 2024). Generalizing across diverse genetic backgrounds, environmental variables, and complex trait interactions remains a significant challenge for AI applications in plant genomics. Overcoming these issues requires the development of more robust models that can handle variability and ensure that predictions are accurate across different crop types and environmental scenarios (Khan *et al.*, 2022; Li *et al.*, 2023; Saini *et al.*, 2023; Farooq *et al.*, 2024; Tyagi *et al.*, 2024).

2.6.2 Ethical implications

As AI-enabled genomic technologies advance, appropriate regulatory frameworks must be developed to ethical implications. These frameworks should aim to balance innovation with safety, ensuring that AI applications in genomics are deployed responsibly and ethically (Coghlan *et al.*, 2024; Harishbhai Tilala *et al.*, 2024). Several key considerations for regulatory oversight include:

- **International regulations and standards:** Currently, regulatory oversight of AI in genomics varies widely between countries and regions. For example, the European Union and the United States have different approaches to the regulation of genetically modified organisms (GMOs) (Abdul Aziz *et al.*, 2022; Tachikawa and Matsuo, 2023). In some regions, there is a regulatory focus on the process by which genetic modifications are made (such as CRISPR gene editing), while in others, the focus may be on the product (i.e. the genetically modified organism itself) (Sami *et al.*, 2021; Saini *et al.*, 2023). Given the global nature of agricultural trade, international regulations and standards for AI-driven genetic modification are necessary to ensure consistency and fairness in the regulation of GMOs across borders (Sami *et al.*, 2021; Abdul Aziz *et al.*, 2022; Tachikawa and Matsuo, 2023; Saini *et al.*, 2023).
- **Safety and risk assessment:** Regulatory bodies must ensure that AI-enabled genomic technologies are safe for both human consumption and the environment. This involves comprehensive risk assessments that consider not only the direct effects of genetic modifications but also potential long-term ecological impacts (Liu *et al.*, 2021; Quazi, 2022). For example, the release of genetically modified crops into the wild should be carefully monitored to assess their impact on surrounding ecosystems and biodiversity. Regulators may also need to address the specific risks posed by AI, such as unintended biases in AI models that could result in unforeseen negative outcomes (Liu *et al.*, 2021; Abdul Aziz *et al.*, 2022; Quazi, 2022; Harishbhai Tilala *et al.*, 2024).
- **Transparency and accountability:** AI models used in genomics need to be transparent in their decision-making processes. Regulatory frameworks must require clear documentation of how AI models are developed, trained, and validated, especially when used for genetic modification purposes (Kiseleva *et al.*, 2022; Sami *et al.*, 2021). This ensures that stakeholders, including farmers, consumers, and environmental groups, understand how decisions are made, what data are used, and how potential risks are mitigated. Additionally, clear accountability mechanisms must be in place in the event that AI-driven modifications lead to harmful consequences (Coghlan *et al.*, 2024; Huo *et al.*, 2024; Nath *et al.*, 2024).
- **Ethical review committees:** To ensure that AI applications in genetic modification are ethically sound, regulatory frameworks should require the establishment of independent ethical review committees (Tachikawa and Matsuo, 2023). These committees would assess the potential

societal, environmental, and ethical impacts of AI-driven genomic technologies. This review process would allow for a more balanced and inclusive approach, taking into account diverse perspectives from different stakeholders, including the public, environmental organizations, and indigenous communities (Coghlan *et al.*, 2024; Harishbhai Tilala *et al.*, 2024).

- **Public engagement and education:** Public trust in AI-driven genetic modification technologies is critical. Regulatory frameworks should include mechanisms for public engagement, such as transparency in decision-making processes and educational campaigns that explain the benefits and risks of AI in plant genomics (Coghlan *et al.*, 2024; Huo *et al.*, 2024; Nath *et al.*, 2024). Involving the public in discussions about the ethical implications and societal impacts of AI-driven genetic modification will foster greater understanding and acceptance of these technologies (Gerke *et al.*, 2020; Tachikawa and Matsuo, 2023; Coghlan *et al.*, 2024).

2.7 Future Directions in AI-Enabled Plant Genomics

As AI continues to reshape the field of plant functional genomics, exciting new developments are on the horizon (Libbrecht and Noble, 2015; Caudai *et al.*, 2021). These advancements promise to not only enhance our understanding of plant biology but also revolutionize agricultural practices, making them more efficient, sustainable, and adaptable to the challenges posed by climate change. Key future directions for AI in plant genomics include advances in AI algorithms and the integration of quantum computing (Caudai *et al.*, 2021; Li *et al.*, 2023; Yue *et al.*, 2023).

2.7.1 Advances in AI algorithms for functional genomics

The field of AI in plant genomics is evolving rapidly, with several exciting advancements in AI algorithms on the horizon. New and improved algorithms are expected to enhance the power of AI models, enabling deeper insights into plant biology (Caudai *et al.*, 2021; Pazos Obregón *et al.*, 2022; Li *et al.*, 2023). Key areas of development include the following examples.

2.7.1.1 Deep learning and neural networks

Continued progress in deep learning architectures will allow AI to model even more complex relationships within genomic data. Advanced neural networks, such as CNNs and RNNs, may become more capable of detecting intricate patterns in genomic sequences and predicting functional gene interactions across diverse plant species (Alzubaidi *et al.*, 2021; Li *et al.*, 2023; Yue *et al.*, 2023). These algorithms could significantly improve gene function prediction, facilitate the identification of regulatory networks, and aid in the design of more resilient crop varieties (Alzubaidi *et al.*, 2021; Yue *et al.*, 2023).

2.7.1.2 Explainable AI

As AI models grow in complexity, one of the major challenges has been their "black-box" nature, which makes it difficult to understand how predictions are made. The future of AI in genomics lies in the development of explainable AI (XAI) models that provide transparency, allowing scientists to interpret the reasoning behind predictions (Budhkar *et al.*, 2025). This will help improve the trust and applicability of AI-driven genomic research, particularly in critical areas such as gene editing, trait prediction, and crop breeding (Caudai *et al.*, 2021; Danilevicz *et al.*, 2022; Feng *et al.*, 2024; Budhkar *et al.*, 2025).

2.7.1.3 Transfer learning and multi-omics integration

As genomic datasets grow, there is a need for more effective ways to integrate different types of omics data—genomic, transcriptomic, proteomic, and metabolomics—into a unified AI framework. Advances in transfer learning will allow models trained on one dataset to be applied to others with fewer data requirements, enhancing the generalization ability of AI tools

(Caudai *et al.*, 2021; Feng *et al.*, 2024). This will be crucial in integrating multi-omics data to provide a comprehensive understanding of plant biology, especially for complex traits like disease resistance or drought tolerance (Caudai *et al.*, 2021; Feng *et al.*, 2024; Shubhika *et al.*, 2024).

2.7.2 Integration of quantum computing and AI in genomics

One of the most exciting possibilities for the future of AI in genomics is the integration of quantum computing. Quantum computing has the potential to revolutionize the way AI models process and analyze large genomic datasets, accelerating research in ways that classical computers cannot (Chae *et al.*, 2024; Wójcik-Gront *et al.*, 2024; Sunki *et al.*, 2025).

2.7.2.1 Quantum machine learning

Quantum computers can solve certain types of problems much faster than traditional computers. Integrating quantum computing with machine learning algorithms (quantum machine learning) could drastically improve the ability of AI to handle the massive datasets typical in genomics (Chae *et al.*, 2024). With quantum computing, AI could run highly complex simulations of gene–environment interactions, predict phenotypic outcomes with greater accuracy, and optimize plant breeding strategies in a fraction of the time (Maraveas *et al.*, 2024; Wójcik-Gront *et al.*, 2024).

2.7.2.2 Improving computational power

The future of plant genomics relies on the ability to process increasingly large datasets from whole-genome sequencing, transcriptomics, phenotyping, and environmental data (Cembrowska-Lech *et al.*, 2023; Feng *et al.*, 2024). The ability of quantum computing to solve optimization and computational problems more efficiently than classical computing systems could dramatically speed up genomic analysis, making AI models faster and more scalable. This enhanced computational power will enable AI to better predict complex traits and interactions that are critical for crop improvement (Chae *et al.*, 2024; Feng *et al.*, 2024; Maraveas *et al.*, 2024; Wójcik-Gront *et al.*, 2024).

2.7.3 Democratizing AI tools

Democratizing AI tools in plant genomics involves making advanced AI technologies accessible to a broader range of researchers, including those without extensive computational expertise. This can be achieved through user-friendly platforms, open-source software, and cloud-based services that simplify complex analyses (Hong *et al.*, 2023; Tyagi *et al.*, 2024). For example, tools like BioAgents use multi-agent systems to streamline bioinformatics workflows, making it easier for researchers to handle large datasets and perform sophisticated analyses. Additionally, AI-powered platforms can assist in tasks such as plant species identification, disease diagnosis, and crop yield prediction, enhancing research efficiency and accelerating the development of resilient crop varieties. By democratizing AI tools, we can foster innovation and collaboration, ultimately contributing to sustainable agriculture and food security (Hong *et al.*, 2023; Saini *et al.*, 2023; Jain *et al.*, 2024; Tyagi *et al.*, 2024).

2.7.4 AI and the future of precision agriculture

The integration of AI into precision agriculture will be a game-changer for the future of farming, enhancing both crop productivity and sustainability. Precision agriculture involves using data-driven approaches to optimize every aspect of farming, from seed selection to irrigation and pest management (Liu *et al.*, 2021; Jafar *et al.*, 2024). AI is set to play a pivotal role in this transformation:

2.7.4.1 Data-driven decision making

AI will be increasingly used to analyze large volumes of data from various sources, such as sensors, drones, satellites, and genomic data, to optimize farming practices. These AI-driven systems will provide real-time recommendations to farmers on planting schedules, irrigation management, fertilizer application, and pest control. This will increase efficiency while reducing costs and environmental impacts (Coghlan *et al.*, 2024; Huo *et al.*, 2024; Nath *et al.*, 2024).

2.7.4.2 Autonomous systems

The future of precision agriculture will likely include the widespread adoption of autonomous systems, powered by AI, that can monitor crops, detect early signs of disease or stress, and apply treatments with high precision (Jafar *et al.*, 2024; Maraveas *et al.*, 2024; Nath *et al.*, 2024). Drones, automated harvesters, and AI-driven robots will transform how farms are managed, allowing farmers to respond to issues in real time and reducing the need for large labor forces (Mana *et al.*, 2024).

2.7.4.3 Optimized crop breeding

AI will also continue to play a key role in the development of new crop varieties tailored to specific environmental conditions. With the aid of AI, breeders can better select for traits like drought resistance, disease resistance, and higher nutritional content, while also ensuring that crops are optimized for yield in a sustainable manner (Yoosefzadeh Najafabadi *et al.*, 2023; Feng *et al.*, 2024; Tyagi *et al.*, 2024).

2.8 Conclusion

AI has already made important contributions to the field of plant functional genomics, allowing researchers to evaluate complicated genetic data at previously unreachable scales and speeds. AI-powered technologies have aided in the discovery of genes connected to desired features such as drought tolerance, disease resistance, and increased nutritional content. AI has improved crop improvement efficiency and precision through machine learning, deep learning, and data integration by better predicting gene functions, regulatory networks, and phenotypic consequences. AI has also expedited our understanding of plant biology by facilitating the integration of multi-omics data, allowing researchers to investigate gene–environment relationships and anticipate plant responses to environmental challenges. These advances are expected to change crop breeding and assist in meeting the challenges of climate change, food security, and sustainable agriculture.

References

Abdul Aziz, M., Brini, F., Rouached, H. and Masmoudi, K. (2022) Genetically engineered crops for sustainably enhanced food production systems. *Frontiers in Plant Science* 13, 1027828. DOI: 10.3389/fpls.2022.1027828.

Alemu, A., Åstrand, J., Montesinos-López, O.A., Isidro Y Sánchez, J., Fernández-Gónzalez, J. *et al.* (2024) Genomic selection in plant breeding: Key factors shaping two decades of progress. *Molecular Plant* 17(4), 552–578. DOI: 10.1016/j.molp.2024.03.007.

Alzubaidi, L., Zhang, J., Humaidi, A.J., Al-Dujaili, A., Duan, Y. *et al.* (2021) Review of deep learning: Concepts, CNN architectures, challenges, applications, future directions. *Journal of Big Data* 8(1), 53. DOI: 10.1186/s40537-021-00444-8.

Ansori, A.N., Antonius, Y., Susilo, R.J., Hayaza, S., Kharisma, V.D. *et al.* (2023) Application of CRISPR-cas9 genome editing technology in various fields: A review. *Narra J* 3(2), e184. DOI: 10.52225/narra.v3i2.184.

Baio, F.H.R., Santana, D.C., Teodoro, L.P.R., Oliveira, I.C. d, Gava, R. *et al.* (2023) Maize yield prediction with machine learning, spectral variables and irrigation management. *Remote Sensing* 15(1), 79. DOI: 10.3390/rs15010079.

Budhkar, A., Song, Q., Su, J. and Zhang, X. (2025) Demystifying the black box: A survey on explainable artificial intelligence (XAI) in bioinformatics. *Computational and Structural Biotechnology Journal* 27, 346–359. DOI: 10.1016/j.csbj.2024.12.027.

Caudai, C., Galizia, A., Geraci, F., Le Pera, L., Morea, V. *et al.* (2021) AI applications in functional genomics. *Computational and Structural Biotechnology Journal* 19, 5762–5790. DOI: 10.1016/j.csbj.2021.10.009.

Cembrowska-Lech, D., Krzemińska, A., Miller, T., Nowakowska, A., Adamski, C. *et al.* (2023) An integrated multi-omics and artificial intelligence framework for advance plant phenotyping in horticulture. *Biology* 12(10), 1298. DOI: 10.3390/biology12101298.

Chae, E., Choi, J. and Kim, J. (2024) An elementary review on basic principles and developments of qubits for quantum computing. *Nano Convergence* 11(1), 11. DOI: 10.1186/s40580-024-00418-5.

Chekole, F.C. and Mohammed Ahmed, A. (2023) Future climate implication on maize (*Zea mays*) productivity with adaptive options at Harbu district, Ethiopia. *Journal of Agriculture and Food Research* 11, 100480. DOI: 10.1016/j.jafr.2022.100480.

Chen, H.O., Lin, P.C., Liu, C.R., Wang, C.S. and Chiang, J.H. (2021) Contextualizing genes by using text-mined co-occurrence features for cancer gene panel discovery. *Frontiers in Genetics* 12, 771435. DOI: 10.3389/fgene.2021.771435.

Coghlan, S., Gyngell, C. and Vears, D.F. (2024) Ethics of artificial intelligence in prenatal and pediatric genomic medicine. *Journal of Community Genetics* 15(1), 13–24. DOI: 10.1007/s12687-023-00678-4.

Croci, M., Impollonia, G., Meroni, M. and Amaducci, S. (2023) Dynamic maize yield predictions using machine learning on multi-source data. *Remote Sensing* 15(1), 100. DOI: 10.3390/rs15010100.

Danilevicz, M.F., Gill, M., Anderson, R., Batley, J., Bennamoun, M. *et al.* (2022) Plant genotype to phenotype prediction using machine learning. *Frontiers in Genetics* 13, 822173. DOI: 10.3389/fgene.2022.822173.

Ersoz, N.S., Bakir-Gungor, B. and Yousef, M. (2023) GeNetOntology: Identifying affected gene ontology terms via grouping, scoring, and modeling of gene expression data utilizing biological knowledge-based machine learning. *Frontiers in Genetics* 14, 1139082. DOI: 10.3389/fgene.2023.1139082.

Farooq, M.A., Gao, S., Hassan, M.A., Huang, Z., Rasheed, A. *et al.* (2024) Artificial intelligence in plant breeding. *Trends in Genetics* 40(10), 891–908. DOI: 10.1016/j.tig.2024.07.001.

Feng, W., Gao, P. and Wang, X. (2024) AI breeder: Genomic predictions for crop breeding. *New Crops* 1, 100010. DOI: 10.1016/j.ncrops.2023.12.005.

Gerke, S., Minssen, T. and Cohen, G. (2020) Ethical and legal challenges of artificial intelligence-driven healthcare. *Artificial Intelligence in Healthcare* 295–336. DOI: 10.1016/B978-0-12-818438-7.00012-5.

Gupta, R., Srivastava, D., Sahu, M., Tiwari, S., Ambasta, R.K. *et al.* (2021) Artificial intelligence to deep learning: Machine intelligence approach for drug discovery. *Molecular Diversity* 25(3), 1315–1360. DOI: 10.1007/s11030-021-10217-3.

Han, Y., Wang, K., Yang, F., Pan, S., Liu, Z. *et al.* (2024) Prediction of maize cultivar yield based on machine learning algorithms for precise promotion and planting. *Agricultural and Forest Meteorology* 355, 110123. DOI: 10.1016/j.agrformet.2024.110123.

Harishbhai Tilala, M., Kumar Chenchala, P., Choppadandi, A., Kaur, J., Naguri, S. *et al.* (2024) Ethical considerations in the use of artificial intelligence and machine learning in health care: A comprehensive review. *Cureus* 16(6), e62443. DOI: 10.7759/cureus.62443.

Harmston, N., Filsell, W. and Stumpf, M.P.H. (2010) What the papers say: Text mining for genomics and systems biology. *Human Genomics* 5(1), 17–29. DOI: 10.1186/1479-7364-5-1-17.

Hohman, F.M., Kahng, M., Pienta, R. and Chau, D.H. (2018) Visual analytics in deep learning: An interrogative survey for the next frontiers. *IEEE Transactions on Visualization and Computer Graphics*. DOI: 10.1109/TVCG.2018.2843369.

Hong, K., Radian, Y., Manda, T., Xu, H. and Luo, Y. (2023) The development of plant genome sequencing technology and its conservation and application in endangered gymnosperms. *Plants (Basel, Switzerland)* 12(23), 4006. DOI: 10.3390/plants12234006.

Huo, Q., Song, R. and Ma, Z. (2024) Recent advances in exploring transcriptional regulatory landscape of crops. *Frontiers in Plant Science* 15, 1421503. DOI: 10.3389/fpls.2024.1421503.

Jafar, A., Bibi, N., Naqvi, R.A., Sadeghi-Niaraki, A. and Jeong, D. (2024) Revolutionizing agriculture with artificial intelligence: Plant disease detection methods, applications, and their limitations. *Frontiers in Plant Science* 15, 1356260. DOI: 10.3389/fpls.2024.1356260.

Jain, A., Sarsaiya, S., Singh, R., Gong, Q., Wu, Q. *et al.* (2024) Omics approaches in understanding the benefits of plant-microbe interactions. *Frontiers in Microbiology* 15, 1391059. DOI: 10.3389/fmicb.2024.1391059.

Karlsen, S.T., Rau, M.H., Sánchez, B.J., Jensen, K. and Zeidan, A.A. (2023) From genotype to phenotype: Computational approaches for inferring microbial traits relevant to the food industry. *FEMS Microbiology Reviews* 47(4), fuad030. DOI: 10.1093/femsre/fuad030.

Khan, M.H.U., Wang, S., Wang, J., Ahmar, S., Saeed, S. *et al.* (2022) Applications of artificial intelligence in climate-resilient smart-crop breeding. *International Journal of Molecular Sciences* 23(19), 11156. DOI: 10.3390/ijms231911156.

Kiseleva, A., Kotzinos, D. and De Hert, P. (2022) Transparency of AI in healthcare as a multilayered system of accountabilities: Between legal requirements and technical limitations. *Frontiers in Artificial Intelligence* 5, 879603. DOI: 10.3389/frai.2022.879603.

Kong, Y. and Yu, T. (2018) A deep neural network model using random forest to extract feature representation for gene expression data classification. *Scientific Reports* 8(1), 16477. DOI: 10.1038/s41598-018-34833-6.

Li, Z., Gao, E., Zhou, J., Han, W., Xu, X. *et al.* (2023) Applications of deep learning in understanding gene regulation. *Cell Reports Methods* 3(1), 100384. DOI: 10.1016/j.crmeth.2022.100384.

Libbrecht, M.W. and Noble, W.S. (2015) Machine learning applications in genetics and genomics. *Nature Reviews Genetics* 16(6), 321–332. DOI: 10.1038/nrg3920.

Liu, L.W., Ma, X., Wang, Y.M., Lu, C.T. and Lin, W.S. (2021) Using artificial intelligence algorithms to predict rice (*Oryza sativa* L.) growth rate for precision agriculture. *Computers and Electronics in Agriculture* 187, 106286. DOI: 10.1016/j.compag.2021.106286.

Mana, A.A., Allouhi, A., Hamrani, A., Rehman, S., el Jamaoui, I. *et al.* (2024) Sustainable AI-based production agriculture: Exploring AI applications and implications in agricultural practices. *Smart Agricultural Technology* 7, 100416. DOI: 10.1016/j.atech.2024.100416.

Maraveas, C., Konar, D., Michopoulos, D.K., Arvanitis, K.G. and Peppas, K.P. (2024) Harnessing quantum computing for smart agriculture: Empowering sustainable crop management and yield optimization. *Computers and Electronics in Agriculture* 218, 108680. DOI: 10.1016/j.compag.2024.108680.

Mohidem, N.A., Hashim, N., Shamsudin, R. and Che Man, H. (2022) Rice for food security: Revisiting its production, diversity, rice milling process and nutrient content. *Agriculture* 12(6), 741. DOI: 10.3390/agriculture12060741.

Murmu, S., Sinha, D., Chaurasia, H., Sharma, S., Das, R. *et al.* (2024) A review of artificial intelligence-assisted omics techniques in plant defense: Current trends and future directions. *Frontiers in Plant Science* 15, 1292054. DOI: 10.3389/fpls.2024.1292054.

Muthayya, S., Sugimoto, J.D., Montgomery, S. and Maberly, G.F. (2014) An overview of global rice production, supply, trade, and consumption. *Annals of the New York Academy of Sciences* 1324(1), 7–14. DOI: 10.1111/nyas.12540.

Nath, P.C., Mishra, A.K., Sharma, R., Bhunia, B., Mishra, B. *et al.* (2024) Recent advances in artificial intelligence towards the sustainable future of agri-food industry. *Food Chemistry* 447, 138945. DOI: 10.1016/j.foodchem.2024.138945.

Nussinov, R., Zhang, M., Liu, Y. and Jang, H. (2022) AlphaFold, artificial intelligence (AI), and allostery. *The Journal of Physical Chemistry. B* 126(34), 6372–6383. DOI: 10.1021/acs.jpcb.2c04346.

Oikonomou, E.D., Karvelis, P., Giannakeas, N., Vrachatis, A. and Glavas, E. (eds) (2024) How natural language processing derived techniques are used on biological data: A systematic review. *Network Modeling Analysis in Health Informatics and Bioinformatics* 13(1), 23. DOI: 10.1007/s13721-024-00458-1.

Patel, R., Memon, J., Kumar, S., Patel, D.A., Sakure, A.A. *et al.* (2024) Genetic diversity and population structure of maize (*Zea may* L.) inbred lines in association with phenotypic and grain qualitative traits using SSR genotyping. *Plants (Basel, Switzerland)* 13(6), 823. DOI: 10.3390/plants13060823.

Pazos Obregón, F., Silvera, D., Soto, P., Yankilevich, P., Guerberoff, G. *et al.* (2022) Gene function prediction in five model eukaryotes exclusively based on gene relative location through machine learning. *Scientific Reports* 12(1), 11655. DOI: 10.1038/s41598-022-15329-w.

Pudjihartono, N., Fadason, T., Kempa-Liehr, A.W. and O'Sullivan, J.M. (2022) A review of feature selection methods for machine learning-based disease risk prediction. *Frontiers in Bioinformatics* 2, 927312. DOI: 10.3389/fbinf.2022.927312.

Quazi, S. (2022) Artificial intelligence and machine learning in precision and genomic medicine. *Medical Oncology* 39(8), 120. DOI: 10.1007/s12032-022-01711-1.

Sahito, J.H., Zhang, H., Gishkori, Z.G.N., Ma, C., Wang, Z. *et al.* (2024) Advancements and prospects of genome-wide association studies (GWAS) in maize. *International Journal of Molecular Sciences* 25(3), 1918. DOI: 10.3390/ijms25031918.

Saini, H., Thakur, R., Gill, R., Tyagi, K. and Goswami, M. (2023) CRISPR/Cas9-gene editing approaches in plant breeding. *GM Crops & Food* 14(1), 1–17. DOI: 10.1080/21645698.2023.2256930.

Sami, A., Xue, Z., Tazein, S., Arshad, A., He Zhu, Z. *et al.* (2021) CRISPR-Cas9-based genetic engineering for crop improvement under drought stress. *Bioengineered* 12(1), 5814–5829. DOI: 10.1080/21655979.2021.1969831.

Sarker, I.H. (2021) Deep learning: A comprehensive overview on techniques, taxonomy, applications and research directions. *SN Computer Science* 2(6), 420. DOI: 10.1007/s42979-021-00815-1.

Sen, P. and Orešič, M. (2023) Integrating omics data in genome-scale metabolic modeling: A methodological perspective for precision medicine. *Metabolites* 13(7), 855. DOI: 10.3390/metabo13070855.

Shubhika, S., Patel, P., Singh, R., Tripathi, A., Prajapati, S. *et al.* (2024) Application of artificial intelligence techniques to addressing and mitigating biotic stress in paddy crop: A review. *Plant Stress* 14, 100592. DOI: 10.1016/j.stress.2024.100592.

Sim, J.A., Huang, X., Horan, M.R., Baker, J.N. and Huang, I.C. (2024) Using natural language processing to analyze unstructured patient-reported outcomes data derived from electronic health records for cancer populations: A systematic review. *Expert Review of Pharmacoeconomics & Outcomes Research* 24(4), 467–475. DOI: 10.1080/14737167.2024.2322664.

Sunki, K., Reddy, C.K.K., Reddy, D.M.K. and Doss, S. (2025) The role of quantum artificial intelligence in healthcare advancements. In: Gunjan, V.K., Senatore, S. and Kumar, A. (eds) *Cybernetics, Human Cognition, and Machine Learning in Communicative Applications*. Cognitive Science and Technology. Springer, Singapore, pp. 981–997. DOI: 10.1007/978-981-97-8533-9_10.

Tachikawa, M. and Matsuo, M. (2023) Divergence and convergence in international regulatory policies regarding genome-edited food: How to find a middle ground. *Frontiers in Plant Science* 14, 1105426. DOI: 10.3389/fpls.2023.1105426.

Taye, M.M. (2023) Understanding of machine learning with deep learning: Architectures, workflow, applications and future directions. *Computers* 12(5), 91. DOI: 10.3390/computers12050091.

Trono, D. and Pecchioni, N. (2022) Candidate genes associated with abiotic stress response in plants as tools to engineer tolerance to drought, salinity and extreme temperatures in wheat: An overview 11(23), 3358. DOI: 10.3390/plants11233358.

Tyagi, A., Mir, Z.A., Almalki, M.A., Deshmukh, R. and Ali, S. (2024) Genomics-assisted breeding: A powerful breeding approach for improving plant growth and stress resilience. *Agronomy* 14(6), 1128. DOI: 10.3390/agronomy14061128.

Vadapalli, S., Abdelhalim, H., Zeeshan, S. and Ahmed, Z. (2022) Artificial intelligence and machine learning approaches using gene expression and variant data for personalized medicine. *Briefings in Bioinformatics* 23(5), bbac191. DOI: 10.1093/bib/bbac191.

Vanhaeren, T., Divina, F., García-Torres, M., Gómez-Vela, F., Vanhoof, W. *et al.* (2020) A comparative study of supervised machine learning algorithms for the prediction of long-range chromatin interactions. *Genes* 11(9), 985. DOI: 10.3390/genes11090985.

Vilhekar, R.S. and Rawekar, A. (2024) Artificial intelligence in genetics. *Cureus* 16(1), e52035. DOI: 10.7759/cureus.52035.

Wójcik-Gront, E., Zieniuk, B. and Pawełkowicz, M. (2024) Harnessing AI-powered genomic research for sustainable crop improvement. *Agriculture* 14(12), 2299. DOI: 10.3390/agriculture14122299.

Yoosefzadeh Najafabadi, M., Hesami, M. and Eskandari, M. (2023) Machine learning-assisted approaches in modernized plant breeding programs. *Genes* 14(4), 777. DOI: 10.3390/genes14040777.

Yousef, M. and Allmer, J. (2023) Deep learning in bioinformatics. *Turkish Journal of Biology = Turk Biyoloji Dergisi* 47(6), 366–382. DOI: 10.55730/1300-0152.2671.

Yue, T., Wang, Y., Zhang, L., Gu, C., Xue, H. *et al.* (2023) Deep learning for genomics: From early neural nets to modern large language models. *International Journal of Molecular Sciences* 24(21), 15858. DOI: 10.3390/ijms242115858.

Zenda, T., Liu, S., Dong, A., Li, J., Wang, Y. *et al.* (2021) Omics-facilitated crop improvement for climate resilience and superior nutritive value. *Frontiers in Plant Science* 12, 774994. DOI: 10.3389/fpls.2021.774994.

Zeng, Z., Deng, Y., Li, X., Naumann, T. and Luo, Y. (2019) Natural language processing for EHR-based computational phenotyping. *IEEE/ACM Transactions on Computational Biology and Bioinformatics* 16(1), 139–153. DOI: 10.1109/TCBB.2018.2849968.

3 Artificial Intelligence Approaches for Constructing Digital Plant Proteomics

Pradip Kumar Averi Dubey[1], Kannan Vijayarani[2], Mani Manoj[1], Shanmugam Gavaskar[3], Asirvatham Alwin Robert[4], Jeyabal Philomenathan Antony Prabhu[1], Arumugam Vijaya Anand[1] and Alagarsamy Sumithra[5]*

[1]Department of Human Genetics and Molecular Biology, Bharathiar University, Coimbatore, India; [2]Department of Education, Bharathiar University, Bharathiar University, Coimbatore, India; [3]Department of Computer Application, Bharathiar University, Coimbatore, India; [4]Department of Endocrinology and Diabetes, Prince Sultan Military Medical City, Ministry of Defense, Riyadh, Kingdom of Saudi Arabia; [5]Department of Computer Science and Engineering, SNS College of Technology, Coimbatore, India

Abstract

Plant biology has been one of the many scientific disciplines artificial intelligence (AI) has quickly revolutionized in recent years. Plant proteomics, the large-scale study of proteins within plant systems, stands to benefit immensely from AI-driven innovations, offering new pathways to understand plant physiology, adaptation, and growth at a molecular level. This chapter focuses on a comprehensive exploration of the role of AI in advancing plant proteomics. It delves into the digital transformation of proteomics, presenting state-of-the-art machine learning and deep learning approaches for protein identification, quantitative analysis, functional annotation, and protein–protein interaction prediction. The chapter further explores integrative AI approaches, personalized plant proteomics for crop improvement, and multi-omics data integration, all aimed at pushing the boundaries of plant science.

Keywords: protein–protein interaction, artificial intelligence, proteomics, digital plant, data acquisition

3.1 Introduction to Plant Proteomics

Plant proteomics is a rapidly evolving area that integrates proteomic techniques to enhance our knowledge of plant biology, particularly in response to environmental stresses. This discipline focuses on identifying and describing proteins, which are important for plant growth, development, and stress tolerance. Proteogenomics combines proteomics with genomics and transcriptomics, refining gene models and enhancing our knowledge of plant protein functions under diverse conditions (Song *et al.*, 2023). Proteomics is vital in studying plant responses to abiotic stresses, like flooding and drought, which are critical for crop yield and resilience (Chen and Harmon, 2006). The insights gained from proteomic studies are

*Corresponding author: sumithra.a.cse@snsct.org

DOI: 10.1079/9781800628649.0003

instrumental in crop breeding, allowing for the recognition of stress-responsive proteins that can be targeted for developing resilient crop varieties (Riyazuddin *et al.*, 2022).

3.1.1 Importance of proteomics in plant biology

Proteomics plays a significant role in advancing plant biology, particularly in crop improvement and stress response analysis. By studying the protein composition of plants, investigators can acquire insights into genetic functions, stress responses, and overall plant health, which are essential for sustainable agriculture. Proteomics aids in identifying proteins that influence agronomic traits, thereby enhancing crop yields and stress toleration (Alvares *et al.*, 2024). Proteomic studies reveal how plants adapt to abiotic and biotic stresses through changes in protein expression and modifications (Chauhan *et al.*, 2023). These data are vital for developing crops that can withstand environmental challenges, ultimately contributing to food security. High-throughput proteomic techniques, particularly mass spectrometry (MS), enable comprehensive protein analysis, complementing genomic studies (Mergner and Kuster, 2022). The ongoing advancements in omics technologies are expected to revolutionize crop improvement strategies (Chauhan *et al.*, 2023).

3.1.2 Historical perspective and evolution of plant proteomics

The historical perspective of plant proteomics reveals a significant evolution in methodologies and applications since the term 'proteome' was first familiarized in 1994. Initially, plant proteomics relied heavily on gel electrophoresis techniques, which, despite being considered somewhat outdated, remain valuable for investigating protein species and post-translational modifications (PTMs) (Jorrin-Novo *et al.*, 2019). The field gained momentum with the sequencing of the *Arabidopsis thaliana* genome in the year 2000, establishing it as a model organism and paving the way for extensive research on various crops (Jamet and Santoni, 2018). Advances in MS and computational tools have further refined proteomic analyses, enabling detailed insights into protein dynamics and interactions under various environmental stresses (Desa *et al.*, 2023). Proteomics has become crucial in understanding plant–pathogen interactions, providing insights into resistance mechanisms. The incorporation of proteomics with other "omics" techniques has significantly enhanced our knowledge of plant biology and its responses to environmental challenges (Lin, 2020).

The development of plant proteomics has been characterized by significant advancements in methodologies and technologies since the early 2000s. Initially reliant on gel electrophoresis, the field has transitioned through various generations of techniques, including isotopic labeling and shotgun proteomics, enhancing the ability to analyze complex protein mixtures and PTMs (Jorrin-Novo *et al.*, 2019). The release of the *A. thaliana* genome catalyzed research, establishing it as a model organism and leading to over 650 publications in plant proteomics (Jamet and Santoni, 2018). The facility of the International Plant Proteomics Organization in 2011 has fostered collaboration and knowledge exchange among researchers, promoting the application of proteomics in plant biology (Heazlewood *et al.*, 2016). Despite these advancements, the full potential of proteomics in understanding plant adaptation and stress responses remains to be fully realized, indicating a need for further exploration and integration of transcriptomic and proteomic data (Voelckel *et al.*, 2017).

3.1.3 Challenges in plant proteomics

Plant proteomics faces several significant challenges that hinder its advancement. One major issue is the complexity of plant tissues, which possess various cell types and structures, making protein extraction difficult due to the presence of robust cell walls and interfering compounds like phenolic substances and pigments (Balotf *et al.*, 2022). The over-representation of high-abundance proteins in proteomic studies often masks the identification of less abundant but crucial proteins involved in plant–pathogen interactions and stress responses (Zhou *et al.*, 2022). The lack of comprehensive genomic databases for the vast number of plant species further complicates proteomic analyses, particularly for those not deemed economically relevant (Musavi *et al.*, 2022). Moreover, while proteomics has made significant strides, it is still considered less mature compared to genomics and transcriptomics, requiring specialized expertise and technology to implement effectively (Mustafa *et al.*, 2022).

3.1.4 Digital transformation in proteomics

Digital transformation in plant proteomics is significantly enhancing research capabilities and applications in the field. The incorporation of big data and advanced bioinformatics tools has revolutionized the analysis of plant proteomes, allowing for the recognition of thousands of proteins through high-throughput MS (Subba *et al.*, 2019). This progress is complemented by translational plant proteomics, which aims to apply basic scientific knowledge to practical challenges like food safety and crop improvement (Agrawal *et al.*, 2012). Additionally, the use of digital technologies, including smart sensors and Internet of Things (IoT) tools, facilitates real-time observation of plant growth factors, thereby optimizing agricultural practices (Nezamova *et al.*, 2022). While traditional methods like two-dimensional gel electrophoresis (2-DE) remain valuable, the shift towards more modern techniques, such as shotgun proteomics, reflects the ongoing evolution in the field (Rabilloud, 2014). Collectively, these advancements underscore the importance of digital transformation in enhancing the efficiency and effectiveness of plant proteomics research and applications (Stewart *et al.*, 2011).

3.2 Fundamentals of Artificial Intelligence in Proteomics

Artificial intelligence (AI) is an important tool in advancing plant proteomics by enhancing the analysis and interpretation of complex biological data. Machine learning (ML) and deep learning algorithms are employed to decipher intricate patterns within proteomic datasets, facilitating a deeper understanding of plant defense mechanisms against abiotic and biotic stresses (Kaminuma, 2022; Murmu *et al.*, 2024). The integration of AI in digital phenomics allows for the recognition of structural features of proteins, such as those found in ubiquitin regulatory X (UBX) proteins, which are crucial for cellular proteostasis (Zhang *et al.*, 2022a; Harfouche *et al.*, 2023). AI-driven technologies enable real-time monitoring and precise control of protein functionalities, thereby improving the design of functional plant-based foods (Alasi *et al.*, 2024). As AI continues to evolve, it offers robust tools for researchers to explore the molecular intricacies of plant proteomes, ultimately paving the path for innovative agricultural solutions and sustainable food production.

3.2.1 Introduction to AI and ML

AI and ML are revolutionizing digital plant proteomics by enabling the investigation of vast and complex datasets generated from omics technologies. These advanced computational methods facilitate the extraction of significant insights from large-scale proteomics data, which is crucial for understanding protein functions and interactions in plants (Khïarak *et al.*, 2018). AI algorithms, especially deep learning are practiced at recognizing intricate patterns within these datasets, thereby enhancing the exploration of plant defense mechanisms and various stresses (Murmu *et al.*, 2024). Moreover, ML techniques, such as transfer learning, allow researchers to leverage existing data to predict protein behavior and optimize engineering processes (Shukla, 2024). The integration of AI in digital phenomics further supports the analysis of phenotypic data, providing a comprehensive framework for interpreting the relationships between genotype and phenotype (Harfouche *et al.*, 2023). Overall, the synergy of AI and ML in plant proteomics is paving the way for innovative approaches in plant science and biotechnology (Kaminuma, 2022).

Current applications of AI and ML in digital plant proteomics include the use of algorithms for predicting protein localization, analyzing complex omics data, and enhancing protein engineering through sequence–structure–function relationships. ML algorithms, such as TargetP and Localizer, are employed to predict sub-cellular localization of proteins, although their reliability varies significantly across species, with accuracy ranging from 75% to as low as 2% in some cases (Gould *et al.*, 2023). AI-assisted omics techniques facilitate the exploration of plant defense mechanisms by analyzing genomics, transcriptomics, and proteomics data, providing insights into plant responses to stressors (Kaminuma, 2022; Murmu *et al.*, 2024). However, limitations persist, including the scarcity of comprehensive sequence and structural data for plant proteins, which hampers the effectiveness of these methods (Shukla, 2024). The performance of prediction algorithms is often compromised by the evolutionary stretch between training and query species, highlighting the need for improved models (Gould *et al.*, 2023).

3.2.2 AI techniques in biological sciences

AI techniques are increasingly being applied in plant biological sciences to enhance various aspects of agriculture and plant research. ML and deep learning algorithms are pivotal in analyzing complex biological data, like genomics, phenomics, and transcriptomics, facilitating crop enhancement and disease resistance (Kaminuma, 2022; Maknickiene, 2022). AI aids in early disease diagnosis by classifying plant conditions through image recognition, achieving high accuracy rates in identifying plant leaf diseases (Kanade and Prasad, 2021). AI models utilize environmental data to predict crop yields and optimize agricultural practices, contributing to sustainable farming (Gupta *et al.*, 2024). The integration of AI in plant genomics allows for the prediction of plant genomes and the reduction of chemical fertilizer usage, promoting eco-friendly practices (Osama *et al.*, 2015).

Recent progress in AI has greatly improved the analysis of protein structures in plant biological sciences. Notably, AlphaFold2 has emerged as a leading tool, providing highly accurate predictions of three-dimensional protein structures, which has been instrumental in functional annotation and understanding of cysteine disulfides and metal binding sites in plants (Willems *et al.*, 2023). Additionally, ML techniques are increasingly being integrated into computational structural biology (Fig. 3.1), addressing challenges specific to plant sciences, and enabling deeper insights into protein functions (Chen and Shukla, 2022). The combination of AI with traditional structural biology approaches, like nuclear magnetic resonance and cryo-electron microscopy, offers unique opportunities for characterizing biomolecules, thereby enriching our understanding of plant proteomes (Karamanos and Matthews, 2024). The accessibility of extensive structural data through repositories like the RCSB Protein Data Bank enhances the potential for novel drug design and protein engineering in plant research (Nam, 2023).

3.3 Data Acquisition and Preprocessing in Plant Proteomics

Data acquisition and preprocessing in plant proteomics involve several sophisticated techniques and tools to ensure accurate protein identification and quantification. Proteomics can utilize both targeted and untargeted data acquisition strategies, such as the targeted data acquisition/data-dependent acquisition (TDA/DDA) approach, which enhances the quantification of proteins by combining both methods for improved results compared to conventional DDA (Hart-Smith, 2020). Data-independent acquisition (DIA) has appeared as a strong method, providing in-depth insights into protein diversity and PTMs through advanced MS techniques (Fahrner *et al.*, 2022). The integration of open-source software tools within frameworks like Galaxy facilitates reproducible data processing and analysis, addressing the complexities associated with large datasets (Yang and Qiao, 2023). The development of various software devices for data preprocessing, statistical investigation, and visualization has significantly advanced the field, enabling researchers to navigate the challenges of high-throughput proteomics (Misra, 2018).

3.3.1 Proteomics data acquisition methods

Plant proteomics data acquisition methods have evolved significantly, incorporating various

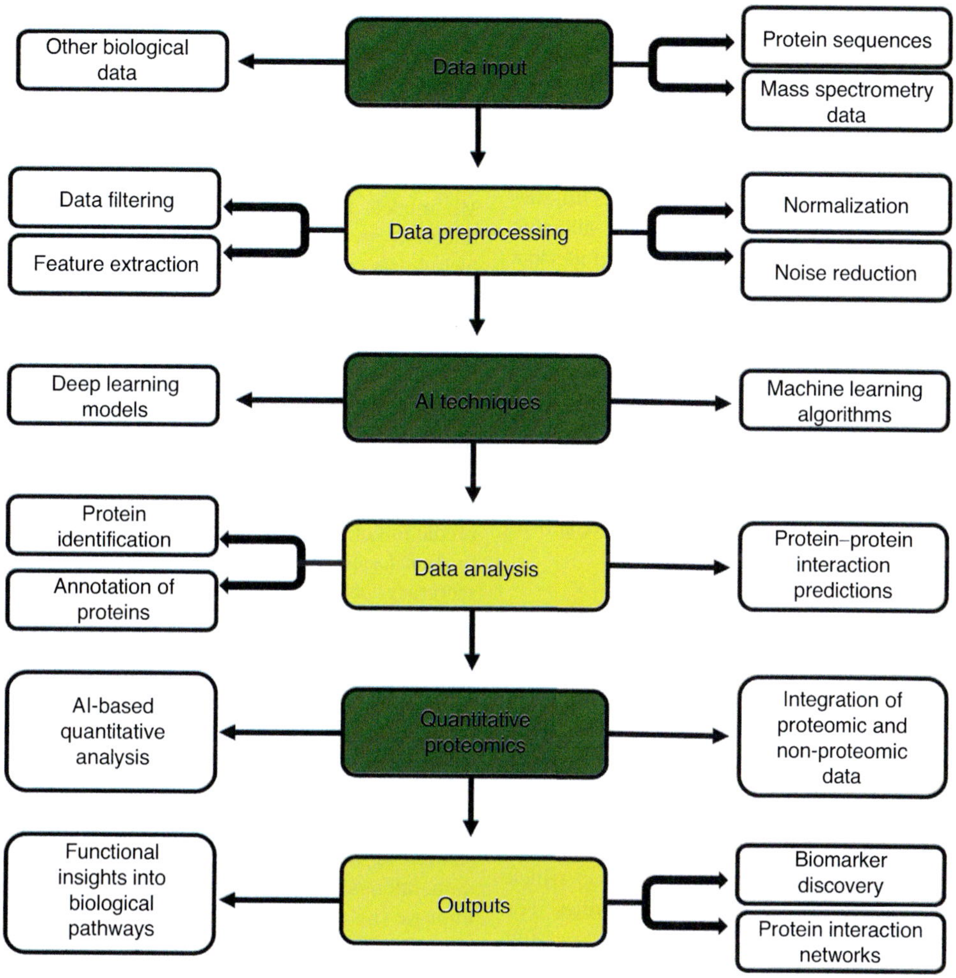

Fig. 3.1. Application of AI in biological sciences. (Author's own image.)

strategies to enhance protein quantification and analysis. Traditional approaches often relied on DDA techniques; however, recent advancements have introduced direct DIA, which offers improved quantitative accuracy and reduced sample interference, making it preferable for plant tissue analysis (Mehta *et al.*, 2020). Additionally, combining targeted and untargeted data acquisition methods, such as TDA/DDA, has been shown to enhance the quantification of both specific and broad protein sets (Hart-Smith, 2020). The integration of labeling techniques, such as 15N labeling, further refines quantitative proteomics by providing reliable data extraction workflows (Chen and Harmon, 2006). Moreover, the continuous evolution of bioinformatics tools and protocols for protein extraction and analysis has facilitated the application of these methods across various plant species, thereby advancing our understanding of plant biology (Basu *et al.*, 2022).

3.3.2 Data preprocessing techniques

Data preprocessing in plant proteomics is crucial for ensuring the accuracy and reliability of results. Techniques typically involve several steps, like data transformation, missing value imputation, and outlier detection, which are essential for preparing raw data for analysis. The use of 2-DE remains prevalent for protein separation, but it requires careful experimental design and statistical analysis to validate findings (Valledor *et al.*, 2014). Additionally, the integration of high-throughput MS techniques has revolutionized data acquisition, allowing for the recognition of thousands of proteins in a single analysis (Subba *et al.*, 2019). The outcome of various software tools and databases has facilitated the annotation and visualization of large proteomic datasets, addressing challenges in data analysis (Misra, 2018). Overall, these preprocessing techniques are vital for extracting meaningful insights from complex proteomic data, ultimately enhancing our knowledge of plant responses to diverse conditions (Cánovas *et al.*, 2004).

3.3.3 AI in data preprocessing

AI plays an important role in data preprocessing within plant proteomics, enhancing the accuracy and efficiency of data analysis. The integration of ML algorithms facilitates the handling of large, complex datasets, which are characteristic of proteomics studies. For instance, AI can assist in the preprocessing stages, including transformation, missing value imputation, and outlier detection, ensuring that the data are clean and reliable for subsequent analysis (Valledor *et al.*, 2014). Moreover, AI-driven methods can optimize data acquisition in MS, improving the quantification and identification of proteins and peptides, which is vital for proteomic research (Coon *et al.*, 2015). The application of ML not only facilitates the data management strategy but also enables researchers to extract meaningful understandings from vast amounts of heterogeneous data, thereby addressing the challenges posed by the increasing volume of omics data in plant biology (Khïarak *et al.*, 2018). Overall, AI enhances the robustness of proteomic analyses, paving the way for more accurate interpretations of plant responses to various stimuli (Murmu *et al.*, 2024).

3.4 AI-Based Techniques for Protein Identification

3.4.1 ML models for protein identification

ML models have revolutionized protein identification in plants by leveraging advanced techniques and feature extraction strategies. For instance, the GraphIdn model utilizes deep representation learning and graph convolutional networks to analyze protein topology, achieving high accuracy in identifying plant vacuole proteins (Sui *et al.*, 2023). Similarly, the ESMAraPPI model employs a pre-trained protein language model to correct sequences into high-dimensional vectors, enhancing the prediction of protein–protein interactions (PPIs) in *Arabidopsis* (Zhou *et al.*, 2023). The SE-small secretory peptides (SSPs) model incorporates transformer encoders and a feature modification module to enhance the robustness of small secretory peptide recognition (Wang *et al.*, 2023a). The RBPLight model combines deep and shallow learning algorithms with evolutionary features to accurately identify RNA-binding proteins (Pradhan *et al.*, 2023). These models demonstrate the importance of integrating structural, sequence-derived, and evolutionary data, alongside innovative ML techniques, to improve the accuracy and efficiency of protein identification in plant systems (Shukla, 2024).

ML-based approaches for protein identification in plant species offer significant advantages over traditional methods, which are often time-consuming and require extensive expertise. For instance, the usage of convolutional neural networks has shown high accuracy in identifying medicinal plants, achieving up to 98% recognition rates, which surpasses the capabilities of manual identification methods (Udaya and Usharani, 2022). ML techniques facilitate rapid investigation of sequence–structure–function relationships in proteins, enabling more efficient protein engineering despite challenges like limited sequence data (Shukla, 2024). Advanced models such as the extended protein feature method enhance predictive capabilities by combining various feature extraction techniques, leading to improved performance in protein–protein interaction predictions (Emmanuel *et al.* 2025). Lastly, ML models for predicting SSPs in plants demonstrate robustness through innovative feature correction strategies, streamlining the identification process compared to traditional experimental methods (Wang *et al.*, 2023a).

3.4.2 Deep learning in proteomics

Deep-learning-based approaches have shown significant advantages over traditional ML methods in the analysis of plant proteomics data. Traditional methods, such as linear regression and simpler algorithms, often struggle to capture complex relationships within the data, limiting their effectiveness in tasks like genomic prediction and protein interaction modeling. In contrast, deep learning models, such as the deep neural network genomic prediction (DNNGP) and advanced architectures like AlphaFold2, excel in handling multi-omics data and predicting intricate patterns, leading to improved efficiency and accuracy (Lee *et al.*, 2022; Wang *et al.*, 2023b). Deep learning approaches can dynamically learn characteristics from raw data, reducing the risk of overfitting and enhancing convergence rates, which is particularly beneficial in high-dimensional datasets typical in plant proteomics (Kaminuma, 2022). The integration of deep learning techniques in plant bioinformatics represents a transformative shift, enabling more robust and insightful analyses compared to traditional ML approaches (Shoaib *et al.*, 2023).

Deep learning algorithms can significantly enhance the accuracy of protein identification in plant proteomics. Approaches like MonoMS1 leverage deep learning to utilize MS1 features alongside traditional MS/MS data, improving identification rates in complex samples such as human serum and single-cell proteomes (Dai *et al.*, 2024). SpecEncoder employs deep metric learning to transform MS/MS spectra into robust embedding vectors, facilitating more accurate peptide identification through a hybrid search approach (Liu *et al.*, 2024). In the realm of plant proteomics, deep representation learning methods, as demonstrated by prPred-DRLF, effectively encode amino acids into numerical vectors, achieving high accuracy and performance metrics for predicting plant resistance proteins (Le, 2022). These advancements illustrate that deep learning not only facilitates the identification procedure but also improves the reliability of predictions, making it a valuable tool in plant omics research (Wang *et al.*, 2022).

3.4.3 Case studies and applications

Recent advancements in plant protein identification have been significantly enhanced through different proteomic methods and molecular methods. For instance, the approach of liquid chromatography–tandem mass spectrometry (LC-MS/MS) has been pivotal in characterizing proteins from therapeutic plants utilized in diabetes therapy, revealing their biological functions related to metabolism and cellular respiration (Pedrete *et al.*, 2019). The use of targeted PCR and next-generation sequencing has proven effective in authenticating plant-based protein powders, addressing issues of adulteration and ingredient identity (Faller *et al.*, 2019). The exploration of PPIs through affinity purification and proximity labeling techniques has expanded our understanding of protein networks in plants, which is crucial for elucidating their biological roles (Kerbler *et al.*, 2021). Lastly, computational approaches have been conceived to predict PPIs on a genome scale, providing a comprehensive resource for understanding protein interactions in key crops like soybean and maize (Ding and Kihara, 2019).

3.5 Quantitative Proteomics and AI

Quantitative proteomics, combined with AI, is revolutionizing plant research by enabling a deeper knowledge of plant responses to various stresses and environmental circumstances. For instance, studies have utilized tandem mass tag labeling and MS to identify thousands of proteins in plants, revealing significant changes in protein expression under different irrigation methods, such as ebb-and-flow subirrigation in tomatoes (Ding and Kihara, 2019). The incorporation of multi-omics data, including proteomics, with AI approaches facilitates high-throughput plant phenotyping, allowing researchers to manage the complexity of biological data effectively (Cembrowska-Lech *et al.*, 2023). AI algorithms can analyze vast datasets generated from proteomic studies, such as those identifying redox-modified proteins in response to fluctuating light conditions (Chen *et al.*, 2022), and assist in deciphering the intricate molecular mechanisms underlying plant defense and adaptation (Murmu *et al.*, 2024). This synergy between quantitative proteomics and AI holds immense prospects for advancing sustainable agriculture and enhancing crop resilience.

3.5.1 Overview of quantitative proteomics

Quantitative proteomics in plants has emerged as a vital tool for understanding protein dynamics and their roles in diverse biological functions, particularly in response to environmental pressures. Recent advancements have enabled the identification of a wide array of proteins, facilitating crop breeding by linking protein profiles to nutritional value and yield under adverse conditions (Thiellement *et al.*, 2007). Techniques such as 15N labeling and parallel reaction monitoring enhance the reliability of protein quantification, allowing for detailed analysis of protein modifications (Heazlewood *et al.*, 2016). Investigations have revealed the application of quantitative proteomics in mapping the redox regulation of photosynthesis, revealing how fluctuating light affects protein stability and activity (Chen *et al.*, 2022). The development of optimized protein extraction protocols has significantly improved the quality of proteomic analyses in crop plants, enabling the recognition of thousands of proteins in a short timeframe (Brajkovic *et al.*, 2023). This overarching strategy is vital for boosting crop resilience and enhancing our knowledge of plant biology.

The field of quantitative proteomics presents considerable promise for improving crop yields and comprehending how plants react to environmental stresses. By analyzing changes in protein abundances, PTMs, and protein interactions under stress conditions, researchers can identify key biomarker proteins that contribute to stress toleration (Chauhan *et al.*, 2023). This information is crucial for developing stress-tolerant crops through targeted breeding strategies, as it bridges the gap between genomics and functional biology (Alugoju and Tencomnao, 2023). Proteomic studies on crops like wheat and barley have revealed differentially abundant proteins that correlate with stress resilience, providing insights into the genetic basis of adaptability (Kosová *et al.*, 2023). A thorough understanding of plant proteomics is made possible by the integration of cutting-edge proteomic techniques like MS and protein microarrays, which could eventually be used to enhance agricultural yield when plants experience abiotic stress (Yashaswini *et al.*, 2024).

3.5.2 AI models for quantitative analysis

AI models play a vital part in quantitative analysis across different domains. They enhance traditional methods by integrating ML and deep learning techniques, which facilitate tasks such as prediction, classification, and optimization in complex systems (Zhao, 2024). For instance, AI algorithms can analyze vast datasets, including unstructured data, to improve business valuation accuracy and efficiency (Farahani, 2024). AI-driven mathematical modeling allows for the optimization of processes, such as route planning, demonstrating significant improvements in operational costs (Meng, 2023). Explainable AI methods are essential for interpreting the decision-making processes of AI models, particularly in applications like remote sensing image classification, where understanding model behavior is critical. The integration of AI in quantitative analysis not only enhances data processing capabilities but also provides deeper insights and more reliable outcomes in various fields (Torrentira Jr, M.C, 2024).

The development of deep learning algorithms, such as d: pPop, facilitates the prediction of peptide detectability, which is crucial for designing targeted proteomics assays (Zimmer *et al.*, 2018). These improvements in AI not only enhance the selection of proteotypic peptides but also address the complexities of omics data, enabling researchers to unravel plant defense mechanisms against various stresses (Murmu *et al.*, 2024). Moreover, the incorporation of AI with proteomics has directed significant insights into the molecular responses of crops to abiotic stresses, thereby aiding in the identification of candidate genes for marker-aided selection. However, the successful application of these AI models necessitates rigorous experimental design and statistical analysis to ensure reliable results (Valledor *et al.*, 2014). Overall, AI-driven approaches are transforming plant proteomics, offering robust tools for advancing agricultural research and crop protection strategies (Mirzaei *et al.*, 2016).

3.5.3 Integrative AI approaches

Integrative AI and proteomics techniques can significantly enrich our knowledge of plant–microbe interactions by providing a complete analysis of the molecular mechanisms involved. Proteomics identifies essential proteins that play critical roles in these interactions, while AI can analyze complex datasets, revealing patterns and insights that may not be immediately apparent through conventional techniques (Jain *et al.*, 2024). The incorporation of proteomics with other omics technologies, like metabolomics and transcriptomics, allows for a holistic perspective of the dynamic relations between plants and microbes, elucidating pathways involved in beneficial, pathogenic, and commensal relationships (Kimotho and Maina, 2024). Modern bioinformatics tools and ML can optimize data processing, improving proteomic data interpretation, and making it easier to identify important proteins and their roles in the context of interactions between microbes and plants (Xu and Fan, 2024).

Integrative AI approaches greatly improve the identification of novel plant proteins and their functions by leveraging multi-omics data and advanced analytical techniques. The incorporation of proteomics, genomics, metabolomics, and transcriptomics allows for a comprehensive knowledge of plant biology, facilitating the elucidation of complex traits and regulatory networks essential for genetic modification (Zhang *et al.*, 2022b; Cembrowska-Lech *et al.*, 2023). AI algorithms, particularly ML and deep learning, are adept at managing the vast scale of omics data, enabling researchers to derive meaningful insights into plant defense mechanisms and other functional roles of proteins (Murmu *et al.*, 2024). Furthermore, AI tools like AlphaFold provide accurate predictions of protein structures, which can guide experimental research and hypothesis generation regarding plant UBX proteins and their functions (Zhang *et al.*, 2022a). This synergy between AI and proteomics not only accelerates the discovery of novel proteins but also enhances our knowledge of their biological roles, paving the way for creative applications in crop improvement and sustainable agriculture (Alasi *et al.*, 2024).

AI can enhance the investigation of large-scale proteomic data, enabling the recognition of key proteins related to desirable traits, thus facilitating the outcome of genetically modified crops with improved yields and stress tolerance (Farooq *et al.*, 2024). The capabilities of AI in big data analysis can bridge the genotype–phenotype gap, allowing for more precise trait selection and optimization of breeding programs (Sheikh *et al.*, 2024). The integration of high-throughput phenotyping with AI can streamline the assessment of phenotypic traits, while ML can refine genome editing techniques, enhancing their efficiency and specificity (Chen *et al.*, 2024).

3.6 Functional Annotation and Protein–protein Interaction Prediction

Functional annotation and protein–protein interaction prediction are crucial in digital plant proteomics, especially given the prompt advances in high-throughput sequencing techniques. Various computational approaches have been developed to predict protein functions by incorporating heterogeneous data sources, such as protein motifs, sequence similarities, and interaction data, which enhance the accuracy of functional annotations (Benso *et al.*, 2013). For instance, the GOstruct framework effectively combines disparate data sources to improve functional predictions, demonstrating superior performance compared to traditional sequence-based models (Sokolov *et al.*, 2013). The incorporation of functional interrelationships, such as those derived from gene ontology (GO), can significantly enhance prediction accuracy by recognizing co-occurring functional terms (Zhang and Dai, 2011). Tools like Function Annotator further facilitate the annotation process by providing comprehensive outputs, including GO term assignments and enzyme annotations, which are particularly beneficial for non-model organisms (Chen *et al.*, 2017).

The functional annotation of plant proteomes faces several challenges and limitations, primarily due to the rapid increase in uncharacterized proteins resulting from large-scale genome sequencing projects. Many protein sequences in public databases lack accompanying experimental data, making it difficult to assign accurate functions without relying on homology searches, which can be problematic when mutations affect critical functional residues (George *et al.*, 2005). Precise predictions require the incorporation of heterogeneous data sources, but this is made more difficult by species-specific data that may be difficult to compare (Sokolov *et al.*, 2013). Current computational methods, while improving, still struggle with the sheer volume of new proteins and the need for high-throughput functional predictions, as traditional experimental approaches are too slow to keep pace (Benso *et al.*, 2013). Furthermore, existing frameworks often require enhancements to effectively utilize diverse data types, which is crucial for comprehensive functional annotation (McDermott and Samudrala, 2003).

3.6.1 AI for functional annotation

AI-assisted functional annotation of digital plant proteomics data significantly enhances our understanding of plant–microbe interactions by leveraging deep learning ideals to predict protein functions from unaligned amino acid sequences (Fig. 3.2). This approach addresses the limitations of traditional alignment-based techniques, which fail to annotate a substantial portion of microbial proteins, thereby expanding the functional coverage of databases like Pfam by more than 9.5% (Bileschi *et al.*, 2022). Proteomics methods, like 2-DE and MS, allow for the recognition of pathogenicity and defense-associated proteins expressed during plant–pathogen relations, revealing critical understandings of the molecular mechanisms underlying these relations (Kav *et al.*, 2007; Afroz *et al.*, 2011). Furthermore, understanding the signaling pathways involved in plant immune responses, including PAMP-triggered immunity and effector-triggered immunity, is important for developing approaches to enrich plant resistance against pathogens (Pandey and Chatterjee, 2013).

AI-powered functional annotation of digital plant proteomics data holds significant potential for developing novel crops with enhanced stress tolerance. By leveraging large-scale proteomics, researchers can recognize and characterize proteins concerned with plant responses to various abiotic pressures, like salinity, drought, and thermal pressure (Hashiguchi *et al.*, 2010). This approach enables the elucidation of complex protein networks and signaling pathways that govern strain responses, facilitating the discovery of key genes and proteins that can be targeted for genetic modification (Kamal *et al.*, 2010; Afroz *et al.*, 2011). The incorporation of AI with proteomics can enhance the understanding of plant–microbial interactions and hormone signaling, which are crucial for stress tolerance (Barkla *et al.*, 2013). The consequence of transgenic plants expressing specific stress-related proteins, such as NF-YB proteins, exemplifies how these insights can be translated into practical applications for crop improvement (Nelson *et al.*, 2008).

3.6.2 Predicting PPIs

In digital plant proteomics, various computational methods and algorithms are employed to predict PPIs. Sequence-based approaches, such as the multivariate mutual information method, utilize feature representation to contrast interaction pairs from non-interaction pairs, achieving high accuracy rates in datasets like *Saccharomyces cerevisiae* and *Helicobacter pylori* (Ding *et al.*, 2016). Multimodal deep polynomial networks integrate multiple protein features, enhancing prediction performance with accuracies exceeding 99% across different datasets (Lei *et al.*, 2018). The applicability vector machine

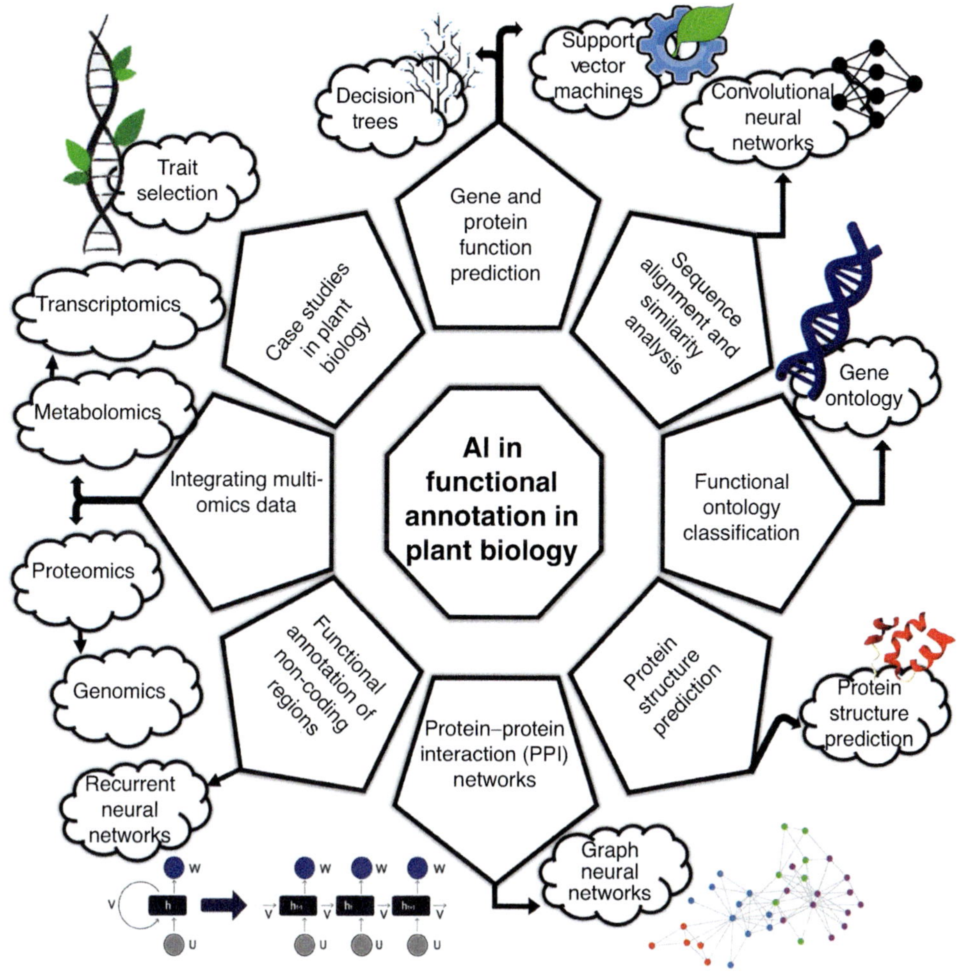

Fig. 3.2. AI in functional annotation in plant biology. (Author's own image.)

model combined with bi-gram probabilities also shows promise, achieving accuracies of 94.57 and 90.57% on yeast and *H. pylori* datasets (An *et al.*, 2016). Optimization techniques like differential evolution and stochastic learning automata are used to maximize individual objectives related to PPI networks, outperforming traditional methods in sensitivity and specificity (Chowdhury *et al.*, 2016). Overall, these computational strategies significantly advance the understanding of PPIs in plant proteomics (Xia *et al.*, 2010).

Predicting PPIs in digital plant proteomics has significant applications for understanding plant evolution and stress reactions. By elucidating the complex networks of PPIs, researchers can gain an understanding of the molecular mechanisms reinforcing plant responses to abiotic stresses, which are mediated by various protein kinases and signaling pathways (Chen *et al.*, 2021). The recognition of specific proteins concerned with plant–pathogen interactions through proteomic studies can improve our knowledge of defense mechanisms and the role of stress-associated proteins (Afroz *et al.*, 2011). The integration of high-throughput technologies for mapping protein interactions allows for the construction of comprehensive interactome networks, which can reveal functional specificities and regulatory components essential for plant growth and adaptation (Uhrig, 2006). Ultimately, these insights can inform breeding strategies for climate-resilient crops, contributing to improved agricultural sustainability (Hirt, 2009).

3.7 Proteome-Wide Analysis and Biomarker Discovery

Proteome-wide analysis is important in biomarker discovery, leveraging advanced techniques in proteomics to identify potential diagnostic markers across various diseases. The integration of high-throughput omic techniques, such as genomics and proteomics, enhances the feasibility of discovering novel biomarkers, particularly in complex fields like organ transplantation. However, challenges remain, including the need for rigorous study design, sample selection, and data normalization to ensure reliable results (Sigdel and Sarwal, 2008). Recent advancements in nanoproteomics and novel sensor technologies have improved the sensitivity and selectivity of biomarker detection, facilitating the recognition of clinically relevant biomarkers (Dasilva *et al.*, 2012). Additionally, the application of functional and chemical proteomics aids in understanding disease pathways and discovering new druggable targets, further enriching the biomarker landscape (Savino *et al.*, 2012). Despite the potential, the field has faced setbacks due to over-optimism and methodological limitations, necessitating a more critical approach to biomarker validation (Lescuyer *et al.*, 2007; Tambor *et al.*, 2010).

3.7.1 AI in proteome-wide analysis

AI-assisted proteome-wide analysis significantly enhances the accuracy of digital plant proteomics by leveraging advanced ML algorithms to improve data acquisition and analysis. These algorithms can dynamically adapt MS methods during data collection, thereby increasing the recognition accuracy of proteins and peptides in complex plant samples (Coon *et al.*, 2015). The integration of AI in proteomics facilitates the recognition of low-abundance proteins, which are often masked by high-abundance counterparts, thus providing a more extensive view of the plant proteome (Fröhlich *et al.*, 2012). AI technologies make it possible to predict experimental peptide measurements based on amino acid sequences. This increases the dependability of analytical workflows by guaranteeing that experimental results match theoretical expectations (Mann *et al.*, 2021). This combination of improved sensitivity, dynamic range, and predictive accuracy allows for a deeper understanding of plant responses to different stimuli, ultimately advancing the field of plant proteomics (Heazlewood, 2011).

AI-driven proteome-wide analysis holds excellent potential for discovering new biomarkers linked to plant stress response and disease resistance. Understanding plant adaptation mechanisms depends on the ability of proteomics to identify variations in protein expression in response to abiotic triggers like temperature swings and chemical exposure (Rocco *et al.*, 2013). Additionally, research has demonstrated that proteomic methods can clarify the functions of certain proteins concerned with interactions between plants and pathogens, improving our knowledge of disease-defense mechanisms (Hashiguchi *et al.*, 2010). The recognition of novel biomarkers that connect with stress responses and resistance qualities can be facilitated by the incorporation of AI technology, which can streamline the analysis of complex proteome data (Nesatyy and Suter, 2008). Researchers can gain a new understanding of the molecular mechanisms of plant resilience by utilizing these sophisticated analytical tools, which will finally lead to the development of more effective crop protection approaches in dealing with environmental adversities (Mayrose *et al.*, 2006).

3.7.2 Biomarker discovery with AI

Biomarker discovery in plant proteomics is significantly enhanced by AI techniques, which facilitate the analysis of complex biological data. ML, particularly deep learning, has appeared as a strong tool for predicting peptide measurements from amino acid sequences, thereby improving the reliability of proteomic analyses (Mann *et al.*, 2021). The periodic analysis of differentially expressed proteins through proteomics allows for the identification of key proteins concerned with metabolic pathways, which is crucial for comprehending plant biology (Jacobs *et al.*, 2000). The incorporation of large-scale omics, like transcriptomes and metabolomes, into a comprehensive digital framework, enables deeper insights into gene expression networks across various plants (Ohyanagi *et al.*, 2015). This multi-omics approach, combined with AI-driven data interpretation, not only accelerates biomarker discovery but also enhances the functional analysis of proteins, paving the way for advancements in plant biotechnology (Narayanan *et al.*, 2002).

AI-assisted biomarker discovery holds significant potential in precision agriculture and crop improvement by enhancing the efficiency of crop breeding and management practices. By integrating AI with metabolomics, researchers can identify metabolic biomarkers that indicate plant stress responses, which is crucial for growing resilient crops capable of resisting environmental challenges (Agarrwal and Nair, 2020). AI technologies can analyze large datasets generated through precision agriculture, allowing for better modeling of agricultural systems and informed decision making regarding irrigation, fertilization, and pest management (Sudduth *et al.*, 2020). The combination of AI with genomics and phenomics can facilitate the recognition of genetic markers related to desirable traits, thereby streamlining the selection of elite breeding lines (Varshney and Tuberosa, 2007). Ultimately, these advancements can be directed to improved crop yields and nutritional quality, addressing future food security challenges while minimizing environmental impacts (Peters *et al.*, 2020).

3.7.3 Integration with genomic and transcriptomic data

ML algorithms can be used to integrate genomic and transcriptomic data to construct digital plant proteomics by enhancing gene prediction and functional annotation processes. For instance, the Seqping pipeline utilizes self-training Hidden Markov Models alongside transcriptomic data to improve gene predictions in plant genomes, achieving high accuracy in species like *Oryza sativa* and *A. thaliana* (Chan *et al.*, 2017). Additionally, the Plant Omics Data Center (PODC) employs natural language processing (NLP) methods for the curation of gene expression networks, facilitating a deeper understanding of molecular biology across various plant species (Ohyanagi *et al.*, 2015). e.g. ENES integrates expressed sequence tags with metabolic pathway information, allowing for comprehensive analysis and functional annotation of plant genes (Masoudi-Nejad *et al.*, 2007). Collectively, these approaches leverage ML to enrich the accuracy and efficiency of genomic data analysis, eventually contributing to the development of robust digital proteomics frameworks in plants (Pertea and Salzberg, 2002).

Integrating genomic and transcriptomic data for constructing digital plant proteomics presents several key challenges. One major issue is the complexity of data analysis, as ultra-high-throughput sequencing generates vast amounts of data that require computationally intensive processing and standardized alignment procedures for effective interpretation (Wang *et al.*, 2010). The incorporation of diverse data types, including gene expression profiles and PPIs, often reveals limited overlap, complicating the identification of functional gene modules (Heyndrickx and Vandepoele, 2012). The need for comprehensive functional annotations and the development of high-throughput techniques to map proteomes are critical for understanding metabolic pathways across different plant tissues (Whitelegge, 2002). Addressing these challenges involves enhancing computational tools, improving data curation through NLP, and fostering collaboration among researchers to share resources and methodologies, as exemplified by initiatives like the PODC (Ohyanagi *et al.*, 2015).

3.8 Data Integration and Multi-Omics Approaches

3.8.1 Importance of data integration

Data integration is crucial for constructing digital plant proteomics as it allows for a complete knowledge of complex biological systems. By integrating various data sources, such as genomic, transcriptomic, and proteomic information, researchers can uncover non-obvious relationships between genes and proteins, enhancing the reliability of functional predictions (Jansen *et al.*, 2002). High-throughput technologies, including protein microarrays and MS, facilitate the analysis of large-scale proteomic data, enabling the identification of protein interactions and functions (Kersten *et al.*, 2002). Integrating metabolomics with proteomics provides insights into the dynamic molecular phenotype of plants, revealing biochemical regulation through metabolite network dynamics (Weckwerth, 2008). The PODC exemplifies this integration by offering a repository of gene expression networks across multiple species, enhancing functional annotation and evolutionary understanding (Ohyanagi *et al.*, 2015).

Integrating diverse data sources for constructing digital plant proteomics presents several key challenges. One major issue is the representational heterogeneity of data across various biological sources, which complicates the integration process due to differing data formats and structures (Hernandez and Kambhampati, 2004). Additionally, the vast volume and complexity of data generated from high-throughput proteomic experiments necessitate advanced computational methods for effective integration and functional interpretation (Huang *et al.*, 2007). The incorporation of multiple omics data types further complicates the analysis, as researchers must navigate the myriad of options and techniques available for data processing and storage (Heazlewood and Millar, 2003). Moreover, the lack of overlap among different experimental datasets can hinder the identification of functional gene modules, emphasizing the need for comprehensive approaches that combine various data types to enhance biological understanding (Heyndrickx and Vandepoele, 2012).

3.8.2 AI techniques for data integration

In digital plant proteomics, several AI techniques can enhance data integration and analysis. Symbolic ML approaches, such as nearest neighbor and identification tree techniques, can be utilized for classifying and interpreting complex proteomic data. Artificial neural networks can model intricate biological relationships and predict outcomes based on proteomic profiles, facilitating the integration of diverse datasets. Genetic algorithms may also be employed to optimize the selection of relevant features from large-scale proteomic data, improving the accuracy of analyses (Narayanan *et al.*, 2002). Integrating metabolomics and proteomics through high-throughput shotgun proteomics and multivariate statistics can reveal correlative biomarkers, enhancing the interpretation of molecular dynamics in plant systems (Weckwerth, 2008). The development of comprehensive omics resources, such as gene expression networks, can further support the incorporation of proteomic data with genomic and transcriptomic facts, providing a holistic view of plant biology (Ohyanagi *et al.*, 2015).

ML algorithms significantly enhance the accuracy of data integration in digital plant proteomics by enabling the analysis of diverse omics data types, through sophisticated integration strategies. For instance, multi-omics integration methods, as discussed in the literature, allow for the combination of complementary biological layers, leading to improved diagnostic and classification biomarkers (Picard *et al.*, 2021). Frameworks like FIDDLE facilitate the learning of unified representations from multiple genomic datasets, which can predict functional outcomes without extensive manual feature selection (Eser and Churchman, 2016). Specific ML models, like support vector machines, have been designed to classify plastid-type proteins based on sequence attributes, demonstrating high accuracy in distinguishing between plastid and non-plastid proteins (Kaundal *et al.*, 2013). Altogether, these advancements underscore the critical role of ML in refining data integration processes, ultimately enhancing the understanding of plant proteomics (Lye *et al.*, 2021).

3.9 AI in Personalized Plant Proteomics

AI is revolutionizing personalized plant proteomics by enhancing the recognition and quantification of proteins relevant to specific traits and conditions. ML techniques, such as those developed for predicting peroxisome-targeted proteins in plants, have shown high accuracy in identifying novel protein targeting signals, which is essential for understanding plant responses to strain and environmental changes (Lingner *et al.*, 2011). AI-driven approaches in MS are improving the reliability of peptide measurements, facilitating biomarker discovery that outperforms traditional assays (Mann *et al.*, 2021). Targeted proteomics, illustrated by selected reaction monitoring, leverages AI to develop predictive marker panels for traits like disease resistance and yield in crops, thus enabling precision breeding (Chawade *et al.*, 2016). Innovative AI algorithms enhance data acquisition in proteomics, increasing sample coverage and reproducibility, which is essential for comprehensive proteomic investigations (Coon *et al.*, 2015).

3.9.1 Concept of personalized plant proteomics

Personalized plant proteomics refers to the tailored application of proteomic technologies to understand and enhance plant responses to various environmental and biological factors, ultimately improving agricultural practices and

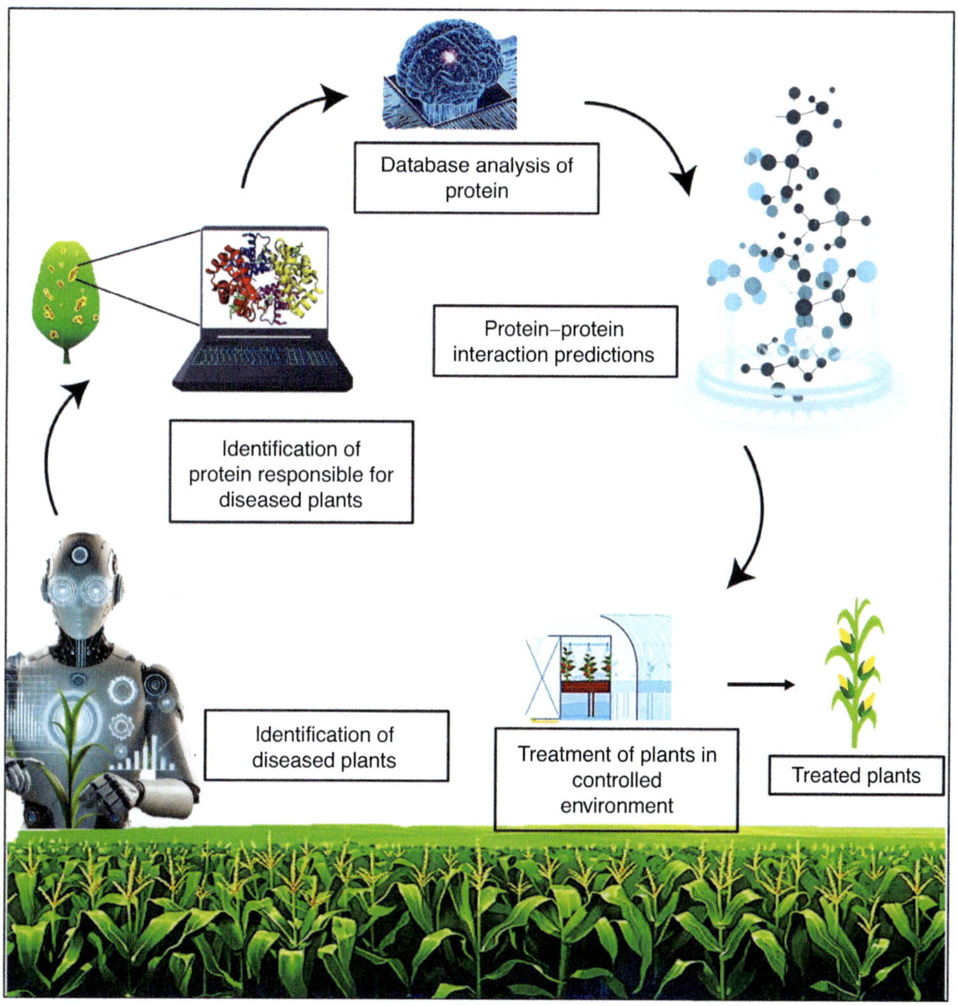

Fig. 3.3. Personalized proteomics for plants. (Author's own image.)

crop resilience (Fig. 3.3). This concept leverages high-throughput techniques such as 2-DE and MS to analyze protein expression and PTMs, which are crucial for regulating plant functions and interactions with microbes (Kwon *et al.*, 2006). By identifying specific proteins involved in metabolic pathways and plant–microbe interactions, researchers can develop strategies for personalized crop management that cater to the unique genetic and environmental contexts of different plant varieties (Kav *et al.*, 2007). The integration of pharmacoproteomics principles can enhance the knowledge of plant responses to stressors, leading to the consequence of targeted interventions that optimize growth and yield (Jain, 2004).

3.9.2 AI-driven approaches for crop improvement

AI-driven approaches have significant potential for crop improvement, which is necessary for addressing global food safety concerns, especially with the projected population increase to 9 billion by 2050 demanding a 70–80% rise in food production (Chauhan *et al.*, 2022). These technologies can enhance agricultural productivity through intelligent techniques such as ML and deep learning, which analyze complex data related to phenomes, genomes, and environmental conditions to optimize crop yields and predict diseases (Singh and Jain, 2022). AI applications like robotics and remote sensing can facilitate precision agriculture, enabling efficient resource management and pest control (Torero, 2021). By integrating AI with existing agricultural practices, farmers can achieve better environmental sustainability while minimizing the use of harmful chemicals (Sanyal, 2011).

Implementing AI-driven approaches for crop improvement in developing countries presents both significant challenges and opportunities. One major challenge is the need for robust data managing techniques to handle the extensive amounts of data generated by AI technologies, which is crucial for effective decision making in agriculture. Additionally, the economic costs associated with adopting these technologies can be prohibitive, particularly for smallholder farmers (Chaterji *et al.*, 2020). However, the incorporation of AI with other techniques, like IoT and robotics, offers opportunities to enhance productivity and efficiency in agricultural practices, enabling better soil management and pest control (Hasan *et al.*, 2022). AI can encourage the evolution of predictive models that help farmers adapt to climate variability and optimize resource use, ultimately contributing to food safety in the face of a growing population (Peters *et al.*, 2020).

3.9.3 Case studies

Personalized plant proteomics is increasingly leveraging AI to enhance the understanding of plant biology and improve applications in agriculture and medicine. High-throughput technologies, such as shotgun proteomics, have enabled the recognition of thousands of proteins, as seen in studies of Arabidopsis pollen, which revealed over 3500 proteins, including many previously uncharacterized ones (Grobei *et al.*, 2009). The integration of AI and ML in genomic computing allows for more sophisticated analyses of expression profiling data, facilitating the recognition of protein interactions and modifications (Kell *et al.*, 2001). Bioinformatics can play an important part in therapeutic plant research, streamlining the discovery of plant-based medicines by analyzing large datasets generated from proteomic studies (Sharma and Sarkar, 2013). As plant proteomics evolves, the combination of these advanced technologies will likely lead to significant breakthroughs in understanding plant functions and developing personalized approaches to plant-based remedies (Kersten *et al.*, 2002).

3.10 Future Directions and Challenges

The integration of AI in plant proteomics presents both promising future directions and significant challenges. AI technologies, particularly ML and deep learning, are enhancing the predictive capabilities of MS-based proteomics, allowing for improved accuracy in peptide measurements from amino acid sequences independently, which is crucial for biomarker discovery (Mann *et al.*, 2021). The field faces hurdles, however, such as the complexity of plant proteomes, which are influenced by various developmental and environmental factors, making comprehensive proteomic profiling difficult (Heazlewood, 2011). The need for advanced computational models that can simulate biological processes *in silico* is essential for effective metabolic engineering (Whitelegge, 2002). The development of plant-specific proteomic techniques and improved sample extraction methods is also necessary to overcome the limitations posed by recalcitrant tissues and secondary metabolites (Dubey and Grover, 2001).

3.11 Conclusion

The integration of AI into plant proteomics marks a pivotal transformation in the capabilities of biological research. AI has revolutionized protein identification, functional annotation, and quantitative analysis, significantly revving the pace at which plant molecular information can be processed and understood. By combining proteomics with ML models and deep learning algorithms, researchers are now able to extract noteworthy understandings from complex datasets and identify potential biomarkers for plant health, growth, and disease resistance. The future of plant proteomics lies in personalized approaches, AI-driven crop improvement, and comprehensive multi-omics data integration. As AI continues to develop, however, challenges such as data standardization, interpretability, and resource limitations need to be managed to fully harness the potential of AI. This chapter has provided a roadmap for AI-enabled advancements in plant proteomics, offering valuable insights for future research and practical applications in agricultural science.

Acknowledgments

All the authors are thankful to their respective Universities and Institutions for their valuable support.

Data availability

The authors confirm that the data supporting the findings of this study are available within the book/chapter.

References

Afroz, A., Ali, G.M., Mir, A. and Komatsu, S. (2011) Application of proteomics to investigate stress-induced proteins for improvement in crop protection. *Plant Cell Reports* 30(5), 745–763. DOI: 10.1007/s00299-010-0982-x.

Agarrwal, R. and Nair, S. (2020) In: *Metabolomics-Assisted Crop Improvement*. Woodhead Publishing, pp. 263–274.

Agrawal, G.K., Pedreschi, R., Barkla, B.J., Bindschedler, L.V., Cramer, R. *et al.* (2012) Translational plant proteomics: A perspective. *Journal of Proteomics* 75(15), 4588–4601. DOI: 10.1016/j.jprot.2012.03.055.

Alasi, S.O., Sanusi, M.S., Sunmonu, M.O., Odewole, M.M. and Adepoju, A.L. (2024) Exploring recent developments in novel technologies and AI integration for plant-based protein functionality: A review. *Journal of Agriculture and Food Research* 15, 101036. DOI: 10.1016/j.jafr.2024.101036.

Alugoju, P. and Tencomnao, T. (2023) In: *Medicinal Plants Proteomics in Response to Abiotic Stresses*. Springer Nature Singapore, Singapore, pp. 35–107.

Alvares, K.V., Souza, J.A.L. and Silva Freitas, G. (2024) Agricultural future: Proteomics as a tool in crop breeding. *Seven Editora* 80–96.

An, J.Y., Meng, F.R., You, Z.H., Chen, X., Yan, G.Y. *et al.* (2016) Improving protein–protein interactions prediction accuracy using protein evolutionary information and relevance vector machine model. *Protein Science* 25(10), 1825–1833. DOI: 10.1002/pro.2991.

Balotf, S., Wilson, R., Tegg, R.S., Nichols, D.S. and Wilson, C.R. (2022) Shotgun proteomics as a powerful tool for the study of the proteomes of plants, their pathogens, and plant–pathogen interactions. *Proteomes* 10(1), 5. DOI: 10.3390/proteomes10010005.

Barkla, B.J., Castellanos-Cervantes, T., León, J.L.D., Matros, A., Mock, H.-P. *et al.* (2013) Elucidation of salt stress defense and tolerance mechanisms of crop plants using proteomics--current achievements and perspectives. *Proteomics* 13(12–13), 1885–1900. DOI: 10.1002/pmic.201200399.

Basu, P., Kruse, C.P., Luesse, D.R. and Wyatt, S.E. (2022) Plant proteomic data acquisition and data analyses: Lessons from spaceflight. *Plant Gravitropism: Methods and Protocols* 199–214. DOI: 10.1007/978-1-0716-1677-2_13.

Benso, A., Carlo, S., Ur Rehman, H., Politano, G., Savino, A. *et al.* (2013) A combined approach for genome wide protein function annotation/prediction. *Proteome Science* 11(1), 1–12. DOI: 10.1186/1477-5956-11-S1-S1.

Bileschi, M.L., Belanger, D., Bryant, D.H., Sanderson, T., Carter, B. *et al.* (2022) Using deep learning to annotate the protein universe. *Nature Biotechnology* 40(6), 932–937. DOI: 10.1038/s41587-021-01179-w.

Brajkovic, S., Rugen, N., Agius, C., Berner, N., Eckert, S. *et al.* (2023) Getting ready for large-scale proteomics in crop plants. *Nutrients* 15(3), 783. DOI: 10.3390/nu15030783.

Cánovas, F.M., Dumas-Gaudot, E., Recorbet, G., Jorrin, J., Mock, H.-P. *et al.* (2004) Plant proteome analysis. *Proteomics* 4(2), 285–298. DOI: 10.1002/pmic.200300602.

Cembrowska-Lech, D., Krzemińska, A., Miller, T., Nowakowska, A., Adamski, C. *et al.* (2023) An integrated multi-omics and artificial intelligence framework for advance plant phenotyping in horticulture. *Biology* 12(10), 1298. DOI: 10.3390/biology12101298.

Chan, K.L., Rosli, R., Tatarinova, T.V., Hogan, M., Firdaus-Raih, M. *et al.* (2017) Seqping: Gene prediction pipeline for plant genomes using self-training gene models and transcriptomic data. *BMC Bioinformatics* 18(1), 1–7. DOI: 10.1186/s12859-016-1426-6.

Chaterji, S., DeLay, N., Evans, J., Mosier, N., Engel, B. *et al.* (2020) Artificial intelligence for digital agriculture at scale: Techniques, policies, and challenges. DOI: 10.48550/arXiv.2001.09786.

Chauhan, C., Thakur, A., Sharma, V., Yadav, R.R., Sivakumar, K.B. *et al.* (2023) In: *Proteomics as a Tool for Analyzing Plant Responses to Abiotic and Biotic Stresses*. Academic Press, pp. 35–68.

Chauhan, U., Sharma, D., Saleem, S., Kumar, M. and Singh, S.P. (2022) In: *Artificial Intelligence-Based Sustainable Agricultural Practices*. pp. 1–16.

Chawade, A., Alexandersson, E., Bengtsson, T., Andreasson, E. and Levander, F. (2016) Targeted proteomics approach for precision plant breeding. *Journal of Proteome Research* 15(2), 638–646. DOI: 10.1021/acs.jproteome.5b01061.

Chen, J. and Shukla, D. (2022) Integration of machine learning with computational structural biology of plants. *The Biochemical Journal* 479(8), 921–928. DOI: 10.1042/BCJ20200942.

Chen, Q., Xiao, Y., Ming, Y., Peng, R., Hu, J. *et al.* (2022) Quantitative proteomics reveals redox-based functional regulation of photosynthesis under fluctuating light in plants. *Journal of Integrative Plant Biology* 64(11), 2168–2186. DOI: 10.1111/jipb.13348.

Chen, S. and Harmon, A.C. (2006) Advances in plant proteomics. *Proteomics* 6(20), 5504–5516. DOI: 10.1002/pmic.200600143.

Chen, T.W., Gan, R.C., Fang, Y.K., Chien, K.Y., Liao, W.C. *et al.* (2017) Function annotator, a versatile and efficient web tool for non-model organism annotation. *Scientific Reports* 7(1), 10430. DOI: 10.1038/s41598-017-10952-4.

Chen, X., Ding, Y., Yang, Y., Song, C., Wang, B. *et al.* (2021) Protein kinases in plant responses to drought, salt, and cold stress. *Journal of Integrative Plant Biology* 63(1), 53–78. DOI: 10.1111/jipb.13061.

Chen, L., Liu, G. and Zhang, T. (2024) Integrating machine learning and genome editing for crop improvement. *aBIOTECH* 5(2), 262–277. DOI: 10.1007/s42994-023-00133-5.

Chowdhury, A., Rakshit, P. and Konar, A. (2016) Protein-protein interaction network prediction using stochastic learning automata induced differential evolution. *Applied Soft Computing* 49, 699–724. DOI: 10.1016/j.asoc.2016.08.053.

Coon, J.J., Westphall, M.S., McAlister, G. and Bailey, D. (2015) U.S [Patent no.9,040,903]. Washington, DC, U.S. Patent and Trademark Office.

Dai, Y., Yang, Y., Wu, E., Shen, C. and Qiao, L. (2024) Deep learning powers protein identification from precursor MS information. *Journal of Proteome Research* 23(9), 3837–3846. DOI: 10.1021/acs.jproteome.4c00118.

Dasilva, N., Díez, P., Matarraz, S., González-González, M., Paradinas, S. *et al.* (2012) Biomarker discovery by novel sensors based on nanoproteomics approaches 12(2), 2284–2308. DOI: 10.3390/s120202284.

Desa, S., Saidin, S., Azeyanty, J.A., Alobaidi, K.H., Al-Taie, B.S. *et al.* (2023) In: *Plant-Environment Interactions: Proteomics, Metabolomics and Genetic Engineering Perspective*. Springer Nature Switzerland, Cham, pp. 15–51.

Ding, Z. and Kihara, D. (2019) Computational identification of protein-protein interactions in model plant proteomes. *Scientific Reports* 9(1), 8740. DOI: 10.1038/s41598-019-45072-8.

Ding, Y., Tang, J. and Guo, F. (2016) Predicting protein-protein interactions via multivariate mutual information of protein sequences. *BMC Bioinformatics* 17, 1–13. DOI: 10.1186/s12859-016-1253-9.

Dubey, H. and Grover, A. (2001) Current initiatives in proteomics research: The plant perspective. *Current Science* 262–269.

Emmanuel, J., Isewon, I., Olasehinde, G. and Oyelade, J. (2025) An extended feature representation technique for predicting sequenced-based host-pathogen protein-protein interaction. *Current Bioinformatics* 20(3), 229–245. DOI: 10.2174/0115748936286848240108074303.

Eser, U. and Churchman, L.S. (2016) FIDDLE: An integrative deep learning framework for functional genomic data inference. *Bioinformatics*. DOI: 10.1101/081380.

Fahrner, M., Föll, M.C., Grüning, B.A., Bernt, M., Röst, H. *et al.* (2022) Democratizing data-independent acquisition proteomics analysis on public cloud infrastructures via the galaxy framework. *GigaScience* 11, giac005. DOI: 10.1093/gigascience/giac005.

Faller, A.C., Arunachalam, T., Shanmughanandhan, D., Kesanakurti, P., Shehata, H.R. *et al.* (2019) Investigating appropriate molecular and chemical methods for ingredient identity testing of plant-based protein powder dietary supplements. *Scientific Reports* 9(1), 12130. DOI: 10.1038/s41598-019-48467-9.

Farahani, M.S. (2024) Analysis of business valuation models with AI emphasis. *Sustainable Economies* 2(3), 132. DOI: 10.62617/se.v2i3.132.

Farooq, M.A., Gao, S., Hassan, M.A., Huang, Z., Rasheed, A. *et al.* (2024) Artificial intelligence in plant breeding. *Trends in Genetics* 40(10), 891–908. DOI: 10.1016/j.tig.2024.07.001.

Fröhlich, A., Gaupels, F., Sarioglu, H., Holzmeister, C., Spannagl, M. *et al.* (2012) Looking deep inside: Detection of low-abundance proteins in leaf extracts of Arabidopsis and phloem exudates of pumpkin. *Plant Physiology* 159(3), 902–914. DOI: 10.1104/pp.112.198077.

George, R.A., Spriggs, R.V., Bartlett, G.J., Gutteridge, A., MacArthur, M.W. *et al.* (2005) Effective function annotation through catalytic residue conservation. *Proceedings of the National Academy of Sciences* 102(35), 12299–12304. DOI: 10.1073/pnas.0504833102.

Gould, S.B., Magiera, J., García García, C. and Raval, P.K. (2023) A cautionary tale on organelle proteome prediction algorithms: Limits and opportunities. *bioRxiv*. DOI: 10.1101/2023.12.18.572050.

Grobei, M.A., Qeli, E., Brunner, E., Rehrauer, H., Zhang, R. *et al.* (2009) Deterministic protein inference for shotgun proteomics data provides new insights into Arabidopsis pollen development and function. *Genome Research* 19(10), 1786–1800. DOI: 10.1101/gr.089060.108.

Gupta, D.K., Pagani, A., Zamboni, P. and Singh, A.K. (2024) AI-powered revolution in plant sciences: Advancements, applications, and challenges for sustainable agriculture and food security. *Exploration of Foods and Foodomics* 2(5), 443–459. DOI: 10.37349/eff.2024.00045.

Harfouche, A.L., Nakhle, F., Harfouche, A.H., Sardella, O.G., Dart, E. *et al.* (2023) A primer on artificial intelligence in plant digital phenomics: Embarking on the data to insights journey. *Trends in Plant Science* 28(2), 154–184. DOI: 10.1016/j.tplants.2022.08.021.

Hart-Smith, G. (2020) Combining targeted and untargeted data acquisition to enhance quantitative plant proteomics experiments. *Plant Proteomics: Methods and Protocols* 169–178. DOI: 10.1007/978-1-0716-0528-8_13.

Hasan, M.M., Islam, M.U. and Sadeq, M.J. (2022) Towards the technological adaptation of advanced farming through artificial intelligence, the internet of things, and robotics: A comprehensive overview. *Artificial Intelligence and Smart Agriculture Technology* 21–42. DOI: 10.1201/9781003299059-2.

Hashiguchi, A., Ahsan, N. and Komatsu, S. (2010) Proteomics application of crops in the context of climatic changes. *Food Research International* 43(7), 1803–1813. DOI: 10.1016/j.foodres.2009.07.033.

Heazlewood, J.L. (2011) The green proteome: Challenges in plant proteomics. *Frontiers in Plant Science* 2, 6. DOI: 10.3389/fpls.2011.00006.

Heazlewood, J.L. and Millar, A.H. (2003) Integrated plant proteomics – putting the green genomes to work. *Functional Plant Biology* 30(5), 471–482. DOI: 10.1071/FP03036.

Heazlewood, J.L., Jorrín-Novo, J.V., Agrawal, G.K., Mazzuca, S. and Lüthje, S. (2016) Editorial: International plant proteomics organization (INPPO) World Congress 2014. *Frontiers in Plant Science* 7, 1190. DOI: 10.3389/fpls.2016.01190.

Hernandez, T. and Kambhampati, S. (2004) Integration of biological sources: Current systems and challenges ahead. *ACM SIgmod Record* 33(3), 51–60. DOI: 10.1145/1031570.1031583.

Heyndrickx, K.S. and Vandepoele, K. (2012) Systematic identification of functional plant modules through the integration of complementary data sources. *Plant Physiology* 159(3), 884–901. DOI: 10.1104/pp.112.196725.

Hirt, H. (ed.) (2009) *Plant Stress Biology: From Genomics to Systems Biology*. John Wiley & Sons.

Huang, H., Hu, Z.Z., Arighi, C.N. and Wu, C.H. (2007) Integration of bioinformatics resources for functional analysis of gene expression and proteomic data. *Frontiers in Bioscience* 12, 5071–5088. DOI: 10.2741/2449.

Jacobs, D.I., Heijden, R. and Verpoorte, R. (2000) Proteomics in plant biotechnology and secondary metabolism research. *Phytochemical Analysis: An International Journal of Plant Chemical and Biochemical Techniques* 11(5), 277–287. DOI: 10.1002/1099-1565(200009/10)11:53.0.CO;2-S.

Jain, K.K. (2004) Role of pharmacoproteomics in the development of personalized medicine. *Pharmacogenomics* 5(3), 331–336. DOI: 10.1517/phgs.5.3.331.29830.

Jain, A., Sarsaiya, S., Singh, R., Gong, Q., Wu, Q. *et al.* (2024) Omics approaches in understanding the benefits of plant-microbe interactions. *Frontiers in Microbiology* 15, 1391059. DOI: 10.3389/fmicb.2024.1391059.

Jamet, E. and Santoni, V. (2018) Editorial for special issue: 2017 plant proteomics. *Proteomes* 6(3), 28. DOI: 10.3390/proteomes6030028.

Jansen, R., Lan, N., Qian, J. and Gerstein, M. (2002) Integration of genomic datasets to predict protein complexes in yeast. *Journal of Structural and Functional Genomics* 2(2), 71–81. DOI: 10.1023/a:1020495201615.

Jorrin-Novo, J.V., Komatsu, S., Sanchez-Lucas, R. and Francisco, L.E.R.. (2019) Gel electrophoresis-based plant proteomics: Past, present, and future. happy 10th anniversary. *Journal of Proteomics* 198, 1–10. DOI: 10.1016/j.jprot.2018.08.016.

Kamal, A.H.M., Kim, K.H., Shin, K.H., Choi, J.S., Baik, B.K. *et al.* (2010) Abiotic stress responsive proteins of wheat grain determined using proteomics technique. *Australian Journal of Crop Science* 4(3), 196–208.

Kaminuma, E. (2022) In: *Introduction of Deep Learning Approaches in Plant Omics Research*. CABI, Wallingford UK, pp. 217–223.

Kanade, P. and Prasad, J.P. (2021) Machine learning techniques in plant conditions classification and observation. In: *5th International Conference on Computing Methodologies and Communication (ICCMC)*, IEEE, pp. 729–734.

Karamanos, T.K. and Matthews, S. (2024) Biomolecular NMR in the AI-assisted structural biology era: Old tricks and new opportunities. *Biochimica et Biophysica Acta (BBA) - Proteins and Proteomics* 1872(2), 140949. DOI: 10.1016/j.bbapap.2023.140949.

Kaundal, R., Sahu, S.S., Verma, R. and Weirick, T. (2013) Identification and characterization of plastid-type proteins from sequence-attributed features using machine learning. In: *BMC Bioinformatics*, Vol. 14. BioMed Central, pp. 1–14. DOI: 10.1186/1471-2105-14-S14-S7.

Kav, N.N., Srivastava, S., Yajima, W. and Sharma, N. (2007) Application of proteomics to investigate plant-microbe interactions. *Current Proteomics* 4(1), 28–43. DOI: 10.2174/157016407781387357.

Kell, D.B., Darby, R.M. and Draper, J. (2001) Genomic computing. Explanatory analysis of plant expression profiling data using machine learning. *Plant Physiology* 126(3), 943–951. DOI: 10.1104/pp.126.3.943.

Kerbler, S.M., Natale, R., Fernie, A.R. and Zhang, Y. (2021) From affinity to proximity techniques to investigate protein complexes in plants. *International Journal of Molecular Sciences* 22(13), 7101. DOI: 10.3390/ijms22137101.

Kersten, B., Bürkle, L., Kuhn, E.J., Giavalisco, P., Konthur, Z. *et al.* (2002) Large-scale plant proteomics. In: Town, C. (ed.) *Functional Genomics*. Springer, Dordrecht, pp. 133–141. DOI: 10.1007/978-94-010-0448-0_9.

Khïarak, J.N., Valïzadeh-kamran, R., Heydarïyan, A. and Damghanï, N. (2018) Big data analysis in plant science and machine learning tool applications in genomics and proteomics. *International Journal of Computational and Experimental Science and Engineering* 4(2), 23–31. DOI: 10.22399/ijcesen.414984.

Kimotho, R.N. and Maina, S. (2024) Unraveling plant-microbe interactions: Can integrated omics approaches offer concrete answers. *Journal of Experimental Botany* 75(5), 1289–1313. DOI: 10.1093/jxb/erad448.

Kosová, K., Prášil, I.T., Klíma, M., Nesvadba, Z., Vítámvás, P. *et al.* (2023) Proteomics of wheat and barley cereals in response to environmental stresses: Current state and future challenges. *Journal of Proteomics* 282, 104923. DOI: 10.1016/j.jprot.2023.104923.

Kwon, S.J., Choi, E.Y., Choi, Y.J., Ahn, J.H. and Park, O.K. (2006) Proteomics studies of post-translational modifications in plants. *Journal of Experimental Botany* 57(7), 1547–1551. DOI: 10.1093/jxb/erj137.

Le, N.Q.K. (2022) Potential of deep representative learning features to interpret the sequence information in proteomics. *Proteomics* 22(1–2), e2100232. DOI: 10.1002/pmic.202100232.

Lee, D., Xiong, D., Wierbowski, S., Li, L., Liang, S. *et al.* (2022) Deep learning methods for 3D structural proteome and interactome modeling. *Current Opinion in Structural Biology* 73, 102329. DOI: 10.1016/j.sbi.2022.102329.

Lei, H., Wen, Y., You, Z., Elazab, A., Tan, E.L. *et al.* (2018) Protein–protein interactions prediction via multimodal deep polynomial network and regularized extreme learning machine. *IEEE Journal of Biomedical and Health Informatics* 23(3), 1290–1303. DOI: 10.1109/JBHI.2018.2845866.

Lescuyer, P., Hochstrasser, D. and Rabilloud, T. (2007) How shall we use the proteomics toolbox for biomarker discovery. *Journal of Proteome Research* 6(9), 3371–3376. DOI: 10.1021/pr0702060.

Lin, S. (2020) Recent insights into proteomics in plant pathology. In: *Proceedings of the 2020 7th International Conference on Biomedical and Bioinformatics Engineering*, pp. 119–124. DOI: 10.1145/3444884.3444907.

Lingner, T., Kataya, A.R., Antonicelli, G.E., Benichou, A., Nilssen, K. *et al.* (2011) Identification of novel plant peroxisomal targeting signals by a combination of machine learning methods and *in vivo* subcellular targeting analyses. *The Plant Cell* 23(4), 1556–1572. DOI: 10.1105/tpc.111.084095.

Liu, K., Tao, C., Ye, Y. and Tang, H. (2024) SpecEncoder: Deep metric learning for accurate peptide identification in proteomics 40(1), i257–i265. DOI: 10.1093/bioinformatics/btae220.

Lye, K.O., Mishra, S. and Molinaro, R. (2021) A multi-level procedure for enhancing accuracy of machine learning algorithms. *European Journal of Applied Mathematics* 32(3), 436–469. DOI: 10.1017/S0956792520000224.

Maknickiene, N. (2022) Application of evolutionary artificial intelligence. An exploratory literature review. *Applied Business* 1, 22–31. DOI: 10.57005/ab.2022.1.4.

Mann, M., Kumar, C., Zeng, W.F. and Strauss, M.T. (2021) Artificial intelligence for proteomics and biomarker discovery. *Cell Systems* 12(8), 759–770. DOI: 10.1016/j.cels.2021.06.006.

Masoudi-Nejad, A., Goto, S., Jauregui, R., Ito, M., Kawashima, S. *et al.* (2007) EGENES: Transcriptome-based plant database of genes with metabolic pathway information and expressed sequence tag indices in KEGG. *Plant Physiology* 144(2), 857–866. DOI: 10.1104/pp.106.095059.

Mayrose, M., Ekengren, S.K., Melech-Bonfil, S., Martin, G.B. and Sessa, G. (2006) A novel link between tomato GRAS genes, plant disease resistance and mechanical stress response. *Molecular Plant Pathology* 7(6), 593–604. DOI: 10.1111/j.1364-3703.2006.00364.x.

McDermott, J. and Samudrala, R. (2003) Bioverse: Functional, structural and contextual annotation of proteins and proteomes. *Nucleic Acids Research* 31(13), 3736–3737. DOI: 10.1093/nar/gkg550.

Mehta, D., Scandola, S. and Uhrig, R.G. (2020) Direct data-independent acquisition (direct DIA) enables substantially improved label-free quantitative proteomics in Arabidopsis. *bioRxiv*. DOI: 10.1101/2020.11.07.372276.

Meng, Y. (2023) Mathematical modeling and simulation analysis based on artificial intelligence algorithm. In: *2023 2nd International Conference on 3D Immersion, Interaction and Multi-Sensory Experiences (ICDIIME)*, IEEE, pp. 347–351.

Mergner, J. and Kuster, B. (2022) Plant proteome dynamics. *Annual Review of Plant Biology* 73(1), 67–92. DOI: 10.1146/annurev-arplant-102620-031308.

Mirzaei, M., Wu, Y., Handler, D., Maher, T., Pascovici, D. *et al.* (2016) Applications of quantitative proteomics in plant research. *Agricultural Proteomics: Crops, Horticulture, Farm Animals, Food, Insect and Microorganisms* 1, 1–29. DOI: 10.1007/978-3-319-43275-5_1.

Misra, B.B. (2018) Updates on resources, software tools, and databases for plant proteomics in 2016-2017. *Electrophoresis* 39(13), 1543–1557. DOI: 10.1002/elps.201700401.

Murmu, S., Sinha, D., Chaurasia, H., Sharma, S., Das, R. *et al.* (2024) A review of artificial intelligence-assisted omics techniques in plant defense: Current trends and future directions. *Frontiers in Plant Science* 15, 1292054. DOI: 10.3389/fpls.2024.1292054.

Musavi, S.A.M., Ahmad, S., Abbas, S., Athar, H.U.R. and Ahmed, I. (2022) Omics technology: Revolution in plant biology. In: *Principles and Practices of OMICS and Genome Editing for Crop Improvement*. Springer International Publishing, Cham, pp. 197–212.

Mustafa, G., Farooq, A., Riaz, Z., Hasan, M. and Amna, K. (2022) Nano-proteomics of stress tolerance in crop plants. *Sustainable Agriculture Reviews* 53, 373–397. DOI: 10.1007/978-3-030-86876-5_15.

Nam, K.H. (2023) AI-based protein models enhance the accuracy of experimentally determined protein crystal structures. *Frontiers in Molecular Biosciences* 10, 1208810. DOI: 10.3389/fmolb.2023.1208810.

Narayanan, A., Keedwell, E.C. and Olsson, B. (2002) Artificial intelligence techniques for bioinformatics. *Applied Bioinformatics* 1(4), 191–222.

Nelson, D., Heard, J., Cherian, S., Jayaprakash, T., Dasgupta, S. *et al.* (2008). U.S. Patent Application.

Nesatyy, V.J. and Suter, M.J. -F (2008) Analysis of environmental stress response on the proteome level. *Mass Spectrometry Reviews* 27(6), 556–574. DOI: 10.1002/mas.20177.

Nezamova, O.A., Stupina, A.A. and Shagaeva, O.A. (2022) The main directions of agro-industrial complex development in the near future. In: *IOP Conference Series: Earth and Environmental Science*, IOP Publishing, p. 012053 (Vol. 1076).

Ohyanagi, H., Takano, T., Terashima, S., Kobayashi, M., Kanno, M. *et al.* (2015) Plant omics data center: An integrated web repository for interspecies gene expression networks with NLP-based curation. *Plant & Cell Physiology* 56(1), e9. DOI: 10.1093/pcp/pcu188.

Osama, K., Mishra, B.N. and Somvanshi, P. (2015) *Machine Learning Techniques in Plant Biology*. The Omics of Plant Science, PlantOmics, pp. 731–754.

Pandey, A. and Chatterjee, S. (2013) Signaling in plant-microbe interactions. *Plant Stress* 7, 52–59.

Pedrete, T.A., Hauser-Davis, R.A. and Moreira, J.C. (2019) Proteomic characterization of medicinal plants used in the treatment of diabetes. *International Journal of Biological Macromolecules* 140, 294–302. DOI: 10.1016/j.ijbiomac.2019.08.035.

Pertea, M. and Salzberg, S.L. (2002) Computational gene finding in plants. *Plant Molecular Biology* 48(1–2), 39–48. DOI: 10.1023/A:1013770123580.

Peters, D.P.C., Rivers, A., Hatfield, J.L., Lemay, D.G., Liu, S. *et al.* (2020) Harnessing AI to transform agriculture and inform agricultural research. *IT Professional* 22(3), 16–21. DOI: 10.1109/MITP.2020.2986124.

Picard, M., Scott-Boyer, M.P., Bodein, A., Périn, O. and Droit, A. (2021) Integration strategies of multi-omics data for machine learning analysis. *Computational and Structural Biotechnology Journal* 19, 3735–3746. DOI: 10.1016/j.csbj.2021.06.030.

Pradhan, U.K., Meher, P.K., Naha, S., Pal, S., Gupta, S. *et al.* (2023) RBPLight: A computational tool for discovery of plant-specific RNA-binding proteins using light gradient boosting machine and ensemble of evolutionary features. *Briefings in Functional Genomics* 22(5), 401–410. DOI: 10.1093/bfgp/elad016.

Rabilloud, T. (2014) How to use 2D gel electrophoresis in plant proteomics. *Plant Proteomics: Methods and Protocols* 43–50. DOI: 10.1007/978-1-62703-631-3_4.

Riyazuddin, R., Choudhary, A.K., Khatri, N., Sarkar, A., Agrawal, G.K. *et al.* (2022) Proteomics as a tool to understand the biology of agricultural crops. In: *Bioinformatics in Agriculture*. Academic Press, pp. 107–122.

Rocco, M., Arena, S., Renzone, G., Scippa, G.S., Lomaglio, T. *et al.* (2013) Proteomic analysis of temperature stress-responsive proteins in *Arabidopsis thaliana* rosette leaves. *Molecular bioSystems* 9(6), 1257–1267. DOI: 10.1039/c3mb70137a.

Sanyal, P. (2011) *Application of Intelligent Techniques: Towards Improvement of Crop Productivity*. LAP Lambert Academic Publishing.

Savino, R., Paduano, S., Preianò, M. and Terracciano, R. (2012) The proteomics big challenge for biomarkers and new drug-targets discovery. *International Journal of Molecular Sciences* 13(11), 13926–13948. DOI: 10.3390/ijms131113926.

Sharma, V. and Sarkar, I.N. (2013) Bioinformatics opportunities for identification and study of medicinal plants. *Briefings in Bioinformatics* 14(2), 238–250. DOI: 10.1093/bib/bbs021.

Sheikh, M., Iqra, F., Ambreen, H., Pravin, K.A., Ikra, M. *et al.* (2024) Integrating artificial intelligence and high-throughput phenotyping for crop improvement. *Journal of Integrative Agriculture* 23(6), 1787–1802. DOI: 10.1016/j.jia.2023.10.019.

Shoaib, M., Shah, B., Sayed, N., Ali, F., Ullah, R. *et al.* (2023) Deep learning for plant bioinformatics: An explainable gradient-based approach for disease detection. *Frontiers in Plant Science* 14, 1283235. DOI: 10.3389/fpls.2023.1283235.

Shukla, D. (2024) Challenges and opportunities in machine learning-guided plant protein engineering. *Biophysical Journal* 123(3), 455a. DOI: 10.1016/j.bpj.2023.11.2768.

Sigdel, T.K. and Sarwal, M.M. (2008) The proteogenomic path towards biomarker discovery. *Pediatric Transplantation* 12(7), 737–747. DOI: 10.1111/j.1399-3046.2008.01018.x.

Singh, S. and Jain, P. (2022) Applications of artificial intelligence for the development of sustainable agriculture. In: *Agro-Biodiversity and Agri-Ecosystem Management. Springer Nature Singapore.* Springer Nature Singapore, Singapore, pp. 303–322. DOI: 10.1007/978-981-19-0928-3_16.

Sokolov, A., Funk, C., Graim, K., Verspoor, K. and Ben-Hur, A. (2013) Combining heterogeneous data sources for accurate functional annotation of proteins 14, 1–13.

Song, Y.C., Das, D., Zhang, Y., Chen, M.X., Fernie, A.R. *et al.* (2023) Proteogenomics-based functional genome research: Approaches, applications, and perspectives in plants. *Trends in Biotechnology* 41(12), 1532–1548. DOI: 10.1016/j.tibtech.2023.05.010.

Stewart, C.N., Touraev, A., Citovsky, V. and Tzfira, T. (eds) (2011) *Plant Transformation Technologies*. John Wiley & Sons.

Subba, P., Narayana Kotimoole, C. and Prasad, T.S.K. (2019) Plant proteome databases and bioinformatic tools: An expert review and comparative insights. *Omics: A Journal of Integrative Biology* 23(4), 190–206. DOI: 10.1089/omi.2019.0024.

Sudduth, K.A., Woodward-Greene, M.J., Penning, B.W., Locke, M.A., Rivers, A.R. *et al.* (2020) AI down on the farm. *IT Professional* 22(3), 22–26. DOI: 10.1109/MITP.2020.2986104.

Sui, J., Chen, J., Chen, Y., Iwamori, N. and Sun, J. (2023) Identification of plant vacuole proteins by using graph neural network and contact maps. *BMC Bioinformatics* 24(1), 357. DOI: 10.1186/s12859-023-05475-x.

Tambor, V., Fučíková, A., Lenco, J., Kacerovský, M., Řeháček, V. *et al.* (2010) Application of proteomics in biomarker discovery: A primer for the clinician. *Physiological Research* 59(4), 471–497. DOI: 10.33549/physiolres.931758.

Thiellement, H., Zivy, M., Damerval, C. and Méchin, V. (2007) *Plant Proteomics*. Human Press, Totowa, NJ, USA.

Torero, M. (2021) Robotics and AI in food security and innovation: Why they matter and how to harness their power. *Science, Ethics, and Policy, Robotics, AI, and Humanity* 99–107.

Torrentira Jr, M.C. (2024) Capabilities and application of artificial intelligence (AI) models in qualitative and quantitative data mining, data processing and data analysis. *European Journal of Education Studies* 11(9), 365–371. DOI: 10.46827/ejes.v11i9.5521.

Udaya, C.S. and Usharani, M. (2022) Medicinal plant identification using machine learning techniques: Automatic recognition of medicinal plants. In: *Handbook of Research on Advances in Data Analytics and Complex Communication Networks*. IGI Global, pp. 120–130.

Uhrig, J.F. (2006) Protein interaction networks in plants. *Planta* 224(4), 771–781. DOI: 10.1007/s00425-006-0260-x.

Valledor, L., Romero-Rodríguez, M.C. and Jorrin-Novo, J.V. (2014) Standardization of data processing and statistical analysis in comparative plant proteomics experiment. *Plant Proteomics: Methods and Protocols* 51–60. DOI: 10.1007/978-1-62703-631-3_5.

Varshney, R.K. and Tuberosa, R. (2007) *Genomics-Assisted Crop Improvement, Vol 2: Genomics Applications in Crops*. Springer.

Voelckel, C., Gruenheit, N. and Lockhart, P. (2017) Evolutionary transcriptomics and proteomics: Insight into plant adaptation. *Trends in Plant Science* 22(6), 462–471. DOI: 10.1016/j.tplants.2017.03.001.

Wang, R., Zhou, Z., Wu, X., Jiang, X., Zhuo, L. *et al.* (2023a) An effective plant small secretory peptide recognition model based on feature correction strategy. *Journal of Chemical Information and Modeling* 64(7), 2798–2806. DOI: 10.1021/acs.jcim.3c00868.

Wang, K., Abid, M.A., Rasheed, A., Crossa, J., Hearne, S. *et al.* (2023b) DNNGP, a deep neural network-based method for genomic prediction using multi-omics data in plants. *Molecular Plant* 16(1), 279–293. DOI: 10.1016/j.molp.2022.11.004.

Wang, L., Li, P. and Brutnell, T.P. (2010) Exploring plant transcriptomes using ultra high-throughput sequencing. *Briefings in Functional Genomics* 9(2), 118–128. DOI: 10.1093/bfgp/elp057.

Wang, Y., Xu, L., Zou, Q. and Lin, C. (2022) PrPred-DRLF: Plant R protein predictor using deep representation learning features. *Proteomics* 22(1–2), 2100161. DOI: 10.1002/pmic.202100161.

Weckwerth, W. (2008) Integration of metabolomics and proteomics in molecular plant physiology--coping with the complexity by data-dimensionality reduction. *Physiologia Plantarum* 132(2), 176–189. DOI: 10.1111/j.1399-3054.2007.01011.x.

Whitelegge, J.P. (2002) Plant proteomics: Blasting out of a mudpit. *Proceedings of the National Academy of Sciences* 99(18), 11564–11566. DOI: 10.1073/pnas.192449199.

Willems, P., Huang, J., Messens, J. and Breusegem, F. (2023) Functionally annotating cysteine disulfides and metal binding sites in the plant kingdom using AlphaFold2 predicted structures. *Free Radical Biology and Medicine* 194, 220–229. DOI: 10.1016/j.freeradbiomed.2022.12.001.

Xia, J.F., Wang, S.L. and Lei, Y.K. (2010) Computational methods for the prediction of protein-protein interactions. *Protein and Peptide Letters* 17(9), 1069–1078. DOI: 10.2174/092986610791760405.

Xu, Y. and Fan, K.T. (2024) *1MSU-DOE Plant Research Laboratory*. Michigan State University Agricultural Biotechnology Research Center, Academia Sinica.

Yang, Y. and Qiao, L. (2023) Data-independent acquisition proteomics methods for analyzing post-translational modifications. *Proteomics* 23(7–8), 2200046. DOI: 10.1002/pmic.202200046.

Yashaswini, R., Sagar, S.P., Kuchanur, P.H., Raghavendra, V.C., Prasad, B.K. *et al.* (2024) *Plant Proteome in Response to Abiotic Stresses*. Intechopen.

Zhang, X.F. and Dai, D.Q. (2011) A framework for incorporating functional interrelationships into protein function prediction algorithms. *IEEE/ACM Transactions on Computational Biology and Bioinformatics* 9(3), 740–753. DOI: 10.1109/TCBB.2011.148.

Zhang, J., Vancea, A.I. and Arold, S.T. (2022a) Targeting plant UBX proteins: AI-enhanced lessons from distant cousins. *Trends in Plant Science* 27(11), 1099–1108. DOI: 10.1016/j.tplants.2022.05.012.

Zhang, R., Zhang, C., Yu, C., Dong, J. and Hu, J. (2022b) Integration of multi-omics technologies for crop improvement: Status and prospects. *Frontiers in Bioinformatics* 2, 1027457. DOI: 10.3389/fbinf.2022.1027457.

Zhao, T. (2024) Artificial intelligence in mathematical modeling of complex systems. *EAI Endorsed Transactions on E-Learning* 10. DOI: 10.4108/eetel.5256.

Zhou, K., Lei, C., Zheng, J., Huang, Y. and Zhang, Z. (2023) Pre-trained protein language model sheds new light on the prediction of Arabidopsis protein-protein interactions. *Plant Methods* 19(1), 141. DOI: 10.1186/s13007-023-01119-6.

Zhou, Q., Ma, K., Hu, H., Xing, X., Huang, X. *et al.* (2022) Extracellular vesicles: Their functions in plant-pathogen interactions. *Molecular Plant Pathology* 23(6), 760–771. DOI: 10.1111/mpp.13170.

Zimmer, D., Schneider, K., Sommer, F., Schroda, M. and Mühlhaus, T. (2018) Artificial intelligence understands peptide observability and assists with absolute protein quantification. *Frontiers in Plant Science* 9, 1559. DOI: 10.3389/fpls.2018.01559.

4 Artificial Intelligence in Plant Systems Biology

Anurag Yadav[1]* and Kusum Yadav[2]

[1]Department of Microbiology, C.P. College of Agriculture, Sardarkrushinagar Dantiwada Agricultural University, Sardarkrushinagar, Dist. Banaskantha, India; [2]Department of Biochemistry, University of Lucknow, Lucknow, India

Abstract

Artificial intelligence (AI) is reshaping plant systems biology by unifying multiscale data into predictive frameworks that decode genotype-to-phenotype relationships with new resolution. This chapter synthesizes advances across the pipeline: deep learning architectures that annotate genomes, predict gene functions, and reconstruct dynamic gene regulatory networks; integrative algorithms that fuse genomics, transcriptomics, metabolomics, and phenomics into unified models; and computer-vision platforms that automate high-throughput phenotyping under biotic and abiotic stress. We show how AI-driven pathway optimization accelerates metabolic engineering, while analytics forecast plant–environment interactions, guiding climate-resilient crop design and yield optimization. Case studies demonstrate AI-enabled precision breeding, synthetic pathway design and CRISPR target selection, illustrating translational impact from model species to orphan crops. Performance benchmarks and interpretability strategies highlight limitations and opportunities for next-generation, explainable models. We outline data governance and ethical and infrastructural imperatives for responsible deployment. Collectively, AI propels plant biology toward a predictive, design-driven paradigm essential for sustainable agriculture.

Keywords: CRISPR, deep learning, genotype-to-phenotype, plant–environment interactions, precision breeding

4.1 Introduction

Integrating computational intelligence, particularly artificial intelligence (AI) and machine learning (ML), with plant systems biology marks a transformative paradigm shift, fundamentally enhancing our ability to decode the intricate mechanisms governing plant life. As complex multicellular organisms, plants exhibit many biological processes that span multiple scales—from molecular interactions within cells to ecosystem-level dynamics. Traditional experimental approaches, while invaluable, often struggle to capture the full scope of this complexity due to limitations in scalability, resolution, and the integration of heterogeneous datasets. In contrast, AI and ML provide a powerful toolkit that enables the analysis, modeling, and prediction of biological systems with unprecedented accuracy and efficiency (Wang *et al.*, 2020a). Unlike traditional ways, AI can analyze simultaneously the large and noisy datasets generated by genomics, transcriptomics, proteomics, and metabolomics. To illustrate, using deep learning

*Corresponding author: anuragyadav123@gmail.com

DOI: 10.1079/9781800628649.0004

involves convolutional neural networks (CNNs), which help to analyze pictures of plants for traits, predict functions of different genes, identify regulating elements and understand processes within the plant—all tasks that are difficult to achieve with usual approaches (Bai *et al.*, 2024).

A defining strength of AI in plant systems biology lies in its ability to uncover patterns and relationships within large datasets that are often imperceptible to human researchers. Applying unsupervised techniques such as clustering and dimensionality reduction has found new co-expression networks and metabolic pathways in plants, explaining how development and stress are regulated (Yan and Wang, 2022). Similarly, supervised learning models leverage genotypic data to predict phenotypic outcomes, accelerating the discovery of key genetic markers for desirable traits (Montesinos-López *et al.*, 2019). This predictive power extends to forecasting plant behavior under environmental stressors like drought, salinity, or pathogen attack, optimizing resource allocation, and enhancing crop resilience (Montesinos-López *et al.*, 2019). In addition, AI makes it possible to integrate diverse multi-omics data by creating unified models representing the interacting relationships across the various biological processes (Guo *et al.*, 2023), leading to the discovery of novel gene networks and regulatory mechanisms. Fig. 4.1 illustrates the integrative framework of AI and ML in plant systems biology, highlighting how genomic, transcriptomic, and phenomic data streams converge into AI/ML models to yield comprehensive biological insights and practical applications. The convergence of computational intelligence and biological complexity also revolutionizes plant breeding and genetic engineering.

Advanced AI algorithms speed up the process of producing desired crop traits by providing highly reliable predictions for genetic modifications, which reduces both the time and expense required by conventional breeding techniques. Nevertheless, the process of combining these two approaches presents many complications. Some main topics of ongoing research remain connected to challenges like ensuring data quality, interpreting models accurately and making domain-specific modifications (Farooq *et al.*, 2024). Nonetheless, the rapid advancements in AI algorithms and the increasing availability of high-quality biological data pave the way for transformative breakthroughs. By bridging the computational and biological sciences gap, AI can change how we view plant

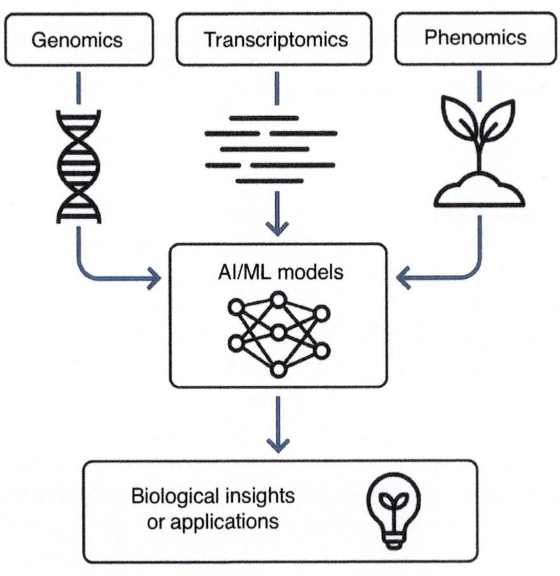

Fig. 4.1. Integrative conceptual framework for AI-driven plant systems biology. (Author's own image.)

life and lead to new advancements in farming, biotechnology, and caring for the environment.

In this chapter, we will explore the state of AI in plant systems biology to showcase the ways in which AI enhances the exploration of raw data and to identify the obstacles and potentialities associated with advancing AI in this field. By synthesizing recent advancements and case studies, we aim to illustrate how AI is revolutionizing our approach to plant research, from basic science to applied biotechnology.

The chapter seeks to offer a detailed overview of AI applications, covering methodologies, tools, and their biological implications. We aim to demonstrate how AI enhances data interpretation by predicting phenotypic outcomes from genotypic data and accelerating the discovery of genetic markers for desirable traits. This transformative impact is crucial for advancing plant science. The chapter will address challenges like data quality, model interpretability, and the need for domain-specific adaptations. Opportunities include leveraging AI for sustainable agriculture and addressing global food security, aligning with the goals of recent reviews like the one on AI in plant sciences for sustainable agriculture (Gupta *et al.*, 2024).

We illustrate practical applications by compiling case studies, such as the role of AI in genomic selection and CRISPR-based genome editing, enhancing the understanding of the potential of AI in plant breeding and genetic engineering. We seek to identify gaps in current methodologies, such as the need for plant-specific AI models, and propose research directions to enhance the synergy between computational intelligence and biological complexity. This work includes exploring interdisciplinary collaboration. Through this comprehensive analysis, we aspire to foster a deeper appreciation of AI's potential in advancing plant systems biology and to inspire innovative solutions to pressing agricultural and environmental challenges, such as climate change and food security.

4.2 AI for Decoding Plant Complexity

The complexity of plants comes from the various interactions between genetic, molecular, and environmental aspects at all levels and over different periods. Traditional approaches to studying plant biology often focus on isolated components, such as individual genes or pathways, which can limit our ability to understand the system as a whole. AI is a powerful tool for bridging these gaps, enabling multiscale modeling that integrates data from genomes to phenomes. By leveraging AI-driven approaches, researchers can now construct comprehensive models that capture the dynamic relationships between genetic information, molecular processes, and observable traits (Wang *et al.*, 2020a).

4.2.1 Genomic analysis, gene function prediction, and network modeling

AI has revolutionized plant genomics by transforming how genomes are analyzed, annotated, and interpreted. Leveraging high-throughput sequencing data, AI models reveal hidden patterns that elucidate gene functions and regulatory mechanisms essential for growth, development, and stress responses. Deep neural networks such as DeepGO (Kulmanov *et al.*, 2018) and its improved variant DeepGOPlus (Kulmanov and Hoehndorf, 2020) predict gene ontology terms directly from protein sequences, efficiently annotating thousands of genes in crops like rice, wheat, and maize without extensive manual feature engineering.

Beyond protein-coding regions, AI tools now address the critical roles of non-coding regulatory elements. For example, DeepSEA uses CNNs to relate changes in DNA sequence patterns with corresponding epigenetic marks, revealing how specific variations in non-coding DNA affect the traits of drought tolerance and resistance to diseases (Zhou and Troyanskaya, 2015). Integrating high-resolution epigenomic data with these approaches offers a comprehensive view of genome regulation, crucial for understanding the interplay between multiple genes and regulatory regions.

Moreover, the advent of multi-omics integration, combining transcriptomics, proteomics, metabolomics, and phenomics, has deepened insights into complex regulatory networks governing plant development and stress adaptation.

AI-driven genomic selection pipelines, which predict breeding values from genome-wide markers, rapidly replace traditional selection methods and facilitate marker-assisted breeding and gene editing via CRISPR/Cas9 (Crossa *et al.*, 2017). In parallel, advancements in protein structure prediction, as exemplified by AlphaFold (Jumper *et al.*, 2021a), are refining functional annotations by providing accurate structural models of plant proteins.

AI allows researchers to understand and alter plant molecular networks, resulting in the discovery of novel techniques and enhancing the yield of agricultural crops. Researchers can visualize and predict how metabolic pathways respond to diverse environmental cues or genetic perturbations by integrating high-resolution omics data with computational modeling frameworks. A noteworthy demonstration of this approach involves the application of constraint-based modeling in conjunction with ML to optimize metabolic networks for enhanced production of secondary metabolites in *Arabidopsis thaliana* (Sweetlove and Fernie, 2018). In such workflows, genome-scale metabolic models of *Arabidopsis* can be constructed and validated using high-throughput datasets that capture transcriptional and biochemical changes. ML techniques are then employed to refine flux predictions, enabling the identification of rate-limiting steps, regulatory bottlenecks, or alternative pathways that might otherwise go unnoticed. This integrative modeling pipeline offers a blueprint for targeted metabolic engineering, wherein researchers can precisely modify key genes or regulatory elements, thereby boosting the capacity of a plant for valuable phytochemicals or improving stress resilience. Beyond Arabidopsis, these methods have become increasingly relevant for crop species where genomic resources are rapidly expanding, thus laying the groundwork for rational strategies to enhance agronomic traits such as drought tolerance, nutrient-use efficiency, and yield (De Oliveira Dal'Molin *et al.*, 2010; Cheung *et al.*, 2014). Significant progress in AI techniques and data integration has paved the way for using AI-guided manipulation of metabolic pathways to advance our understanding of plant physiology as well as create new crops better adapted to a changing world (Wurtzel *et al.*, 2019).

Several AI and ML models have been applied for tasks such as gene function prediction, variant prioritization, and genomic selection. Table 4.1 provides a comparative overview of commonly used AI models in plant genomic analysis, summarizing their performance, key features, and application domains.

Table 4.1. Comparisons of AI models for plant genomic analysis. (Author's own table.)

Primary purpose	Accuracy	Method/algorithm	Training data	Applications	References
Predict gene expression profiles	More than 80% accuracy	Interpretable deep learning models	Gene flanking regions and expression profiles from *A. thaliana*, *Solanum lycopersicum*, Sorghum *bicolor*, *Zea mays*	Understanding gene regulation, genetic variation, cross-species regulatory features	Peleke *et al.*, 2024
SNP variant calling from low-depth ONT data	Competitive with state-of-the-art (e.g. Clair)	Generative adversarial networks (GANs)	ONT sequencing data (applicable to plants)	Variant calling in plants for breeding	Yang *et al.*, 2022
De novo genome assembly (path identification)	Similar or superior contiguity to state-of-the-art	Geometric deep learning (likely GNNs)	Assembly graphs from various species, including plants	Improving genome assembly quality for plants	Vrček *et al.*, 2025
Genomic prediction of phenotypic traits	Varies by data set, often competitive with linear models	Deep neural networks (e.g. DNNGP)	Multi-omics data (genomic, transcriptomic) from plants	Accelerating plant breeding, predicting agronomic traits	Wang *et al.*, 2023

DNNGP, deep neural network genomic prediction; GNN, graph neural networks; ONT, Oxford Nanopore Technology; SNP, single nucleotide polymorphism.

4.2.2 Gene regulatory networks and dynamic modeling

Gene regulatory networks (GRNs) are fundamental to understanding how genes interact to control cellular processes, development, and responses to environmental stimuli in plants. These networks consist of complex interactions between transcription factors (TFs), target genes, and regulatory elements, collectively determining gene expression patterns. Traditional methods for reconstructing GRNs, such as experimental validation and statistical inference, are often labor intensive and limited in scalability. ML is proving to be essential for understanding and dissecting GRNs, allowing researchers to forecast relationships between genes, discover important regulators, and reconstruct network structures (Wang *et al.*, 2020a).

ML algorithms are particularly well suited for GRN inference because they can handle large-scale omics data, such as transcriptomics, epigenomics, and proteomics. A range of ML models including support vector machines (SVMs) and random forests (RFs) have been used to predict the associations between transcription factors and enhancers based on data deriving from TF binding sites, gene expression levels, and chromatin files (Huynh-Thu *et al.*, 2010). Clustering and dimensionality reduction are common unsupervised methods employed to find patterns of coordinated gene behavior and discover candidate regulatory mechanisms (Langfelder and Horvath, 2008).

GENIE3 (Gene Network Inference with Ensemble of Trees) is a widely used ML-based tool for inferring GRNs. It employs a tree-based ensemble method to predict regulatory interactions from gene expression data. In a landmark study, Huynh-Thu *et al.* (2010) applied GENIE3 to reconstruct GRNs in *A. thaliana* using RNA-seq data from more than 1000 samples. The study revealed the involvement of novel TFs in regulating flowering time and stress responses, and uncovering some of the molecular processes associated with plant responses to changing environments. This work demonstrated the potential of ML to uncover hidden regulatory relationships and generate testable hypotheses for experimental validation. Another notable tool in this field is DeepGRN, which utilizes deep learning to integrate multi-omics data and predict regulatory interactions with high precision (Wang *et al.*, 2020a).

4.2.3 Integrating multi-omics data and gene regulatory networks

Integrating multi-omics data has significantly improved the accuracy and robustness of GRN inference. For example, combining transcriptomics with chromatin accessibility data (e.g. ATAC-Seq) allows researchers to identify TF-binding sites and predict regulatory interactions more accurately. A study in maize used ML to integrate RNA-seq and ATAC-seq data, reconstructing a GRN that governs root development and nutrient uptake (Wang *et al.*, 2020a). The strategy revealed essential regulators of root architecture and established how genetic and epigenetic elements interplay to regulate gene expression.

In crop plants, ML has been instrumental in identifying regulatory networks associated with agronomically important traits. For instance, a study on rice used ML to analyze transcriptomic data from plants exposed to drought stress, identifying a GRN involved in drought tolerance. The study revealed several TFs that regulate the expression of stress-responsive genes, providing potential targets for genetic engineering (Wang *et al.*, 2020a). Similarly, ML-based GRN inference has been used in wheat to identify grain quality and yield regulators, offering insights into the genetic basis of these traits (Chen *et al.*, 2023b).

4.2.3.1 Integrating multiscale data for predictive modeling

One of the most significant contributions of AI to plant systems biology is its ability to integrate data across multiple scales. As an example, scientists have applied AI to bring together genomic, transcriptomic, and phenomic data to help predict traits such as crop yield and tolerance to stresses. A notable case study involved using RF algorithms to predict maize yield based on genomic and environmental data, achieving high accuracy and providing actionable insights for crop improvement (Montesinos-López *et al.*,

2019). A separate study applied deep learning to investigate how genes work together to control growth and development in *Arabidopsis* (Takou *et al.*, 2024). A compelling example of AI-driven multiscale modeling is its application in understanding and improving crop drought tolerance. Researchers have used ML algorithms to integrate genomic, transcriptomic, and phenomic data to identify key genes and pathways associated with drought responses. For instance, a study on rice employed deep learning to analyze transcriptomic data and predict the expression of drought-responsive genes under varying water availability conditions (Wang *et al.*, 2020a). This approach identified potential targets for genetic engineering and provided a framework for predicting crop performance under drought stress.

4.2.3.2 Dynamic modelling of GRNs

Beyond static network inference, ML has enabled the dynamic modeling of GRNs, capturing how regulatory interactions change over time or in response to environmental stimuli. For example, time-series RNA-seq data have been used to train ML models that predict the temporal dynamics of gene expression during plant development. A study in Arabidopsis used dynamic Bayesian networks to model the GRN controlling leaf development, revealing how gene expression patterns shift as leaves mature (Wang *et al.*, 2020a). In this way, plant scientists have relied on CNNs to assess leaf area, the overall height of the tree/plant, and signs of diseases from plant pictures.

4.2.4 AI for plant phenotyping and trait prediction

Rapid, accurate, and scalable measurement of plant traits—often described as *phenomics*—is now the principal bottleneck that limits the pace of crop improvement. AI methods, particularly computervision pipelines, deep learning and multimodal datafusion models, are overcoming this constraint by transforming heterogeneous sensor streams (RGB, hyperspectral, LiDAR, thermal, and fluorescence) into biologically meaningful traits at unprecedented speed and resolution. By automating the detection of morphology, physiology, and stress signatures, AI-driven phenotyping closes the long-standing genotype-to-phenotype gap, feeding trait predictions directly into genomic selection, CRISPR editing and decision-support systems for breeders and growers. Foundational studies that married machine learning with highthroughput imaging have matured into integrated, field-scale platforms capable of predicting complex agronomic traits and stress tolerance across thousands of genotypes (Furbank and Tester, 2011). The subsections that follow survey these advances, beginning with high-throughput image-based pipelines and progressing to stress-specific applications and multimodal data integration.

4.2.4.1 Highthroughput phenotyping and trait prediction

At the phenome level, AI has transformed the field of plant phenotyping by enabling high-throughput, non-destructive analysis of plant traits. For example, using CNNs, it is possible to examine plant images and record leaf area, height, and whether a plant is suffering from disease. A case study involving wheat demonstrated the use of CNNs to detect and classify diseases from leaf images with over 95% accuracy, significantly reducing the time and labor required for manual phenotyping (Mohanty *et al.*, 2016). Similarly, AI-powered platforms like PlantCV have been developed to automate the extraction of phenotypic data from imaging datasets, facilitating large-scale studies of plant growth and development (Fahlgren *et al.*, 2015b).

High-throughput phenotyping increasingly relies on diverse imaging platforms, ranging from RGB and hyperspectral to LiDAR and thermal cameras, that capture morphological, physiological, and stress-response traits under both biotic and abiotic conditions. These platforms differ in their spatial resolution, spectral depth, and suitability for specific traits or crop species. Table 4.2 summarizes selected imaging systems commonly used in trait phenotyping, highlighting their technical specifications, target traits, and stress detection capabilities across major crops. This overview provides a

Table 4.2. Details of some selected imaging platforms used for trait phenotyping for biotic and abiotic stress and other morphological traits in crops. (Author's own table.)

A. Biotic and abiotic stresses

Platform	Traits recorded	Crop	References
Automated video tracking platform	Resistance to aphids and other piercing-sucking insects	Lettuce (*Lactuca sativa*) and Arabidopsis (*Arabidopsis thaliana*)	Kloth *et al*., 2015
BreedVision	Lodging, plant moisture content, biomass yield or tiller density	Triticale (*x*Triticosecale Wittmack L.)	Busemeyer *et al*., 2013
DroughtSpotter® precision-irrigation & weight platform	Automated pot weighing and irrigation; high-resolution transpiration and biomass dynamics for drought or combined heat/salinity stress	Wheat, barley, sorghum (greenhouse and growth-chamber trials)	Tovignan *et al*., 2023
Field monitoring support system	Occurrence of the rice bug in the field	Rice (*Oryza sativa*)	Fukatsu *et al*., 2012
Growscreenfluoro	Leaf growth and chlorophyll fluorescence, which allows detection of stress tolerance	Arabidopsis (*A. thaliana*)	Jansen *et al*., 2009
HyperART	Leaf traits such as disease severity or leaf chlorophyll content	Barley (*Hordeum vulgare*), maize (*Zea mays*), tomato (*Lycopersicon esculentum*) and rapeseed (*Brassica rapa*)	Bergsträsser *et al*., 2015
LemnaTec 3D scanalyzer system	Salinity tolerance traits	Rice (*O. sativa*)	Hairmansis *et al*., 2014
Liaphen	Leaf expansion or transpiration rate in response to water deficit	Sunflower (*Helianthus annuus*)	Gosseau *et al*., 2019
PhenoBox	Detection of head smut fungus and corn smut on Brachypodium and maize, respectively, and salt stress response in tobacco	Brachypodium (*Brachypodium distachyon*), maize (*Z. mays*), tobacco (*Nicotiana benthamiana*)	Czedik-Eysenberg *et al*., 2018
PhénoField	Characterization of different abiotic stresses	Wheat (*Triticum aestivum*)	Beauchêne *et al*., 2019
PhenoImage	Plant responses to water stress	Wheat (*T. aestivum*), sorghum (Sorghum bicolor)	Zhu *et al*., 2021
Phenopsis	Plant responses to water stress	Arabidopsis (*A. thaliana*)	Granier *et al*., 2006
Phenotic	Evaluation of plant resistance to pathogens, evaluation of virulence of pathogens	Horticultural crops	Boureau, 2020
PhenoMobile ground-based multi-sensor buggy	LiDAR canopy height, and biomass, hyperspectral indices, canopy temperature for rust infection, heat and water-deficit screening	Wheat (*T. aestivum*) breeding nurseries	Deery *et al*., 2014

Continued

Table 4.2. Continued

A. Biotic and abiotic stresses			
Phenovision	Detection of drought stress and recovery	Maize (*Z. mays*)	Asaari *et al.*, 2019
Phnodyn	Soil water status (drought scenarios), leaf elongation rate, and micrometeorological variables	Rice (*O. sativa*) and maize (*Z. mays*)	Sadok *et al.*, 2007
PlantArray functional-physiology (lysimetric) platform	Continuous gravimetric transpiration, canopy conductance, real-time growth and water-use efficiency during drought/heat/salinity cycles	Barley (*Hordeum vulgare*), cowpea, pepper, rice, maize	Halperin *et al.*, 2017
PlantEye F500/F600 multispectral 3-D scanner	3-D leaf-area, plant height, multispectral-derived stress indices enabling early diagnosis of salt and drought injury	Wheat (*T. aestivum*); tomato, sorghum and other model crops	Maphosa *et al.*, 2016
PlantScreen™ robotic XYZ system	Drought tolerance traits	Rice (*O. sativa*)	Kim *et al.*, 2020c
Radix	Root- and shoot-related traits under control and stress conditions	Maize (*Z. mays*)	Le Marié *et al.*, 2016
RhizoTubes	Root-related traits under non-stressed and stressed conditions	Medicago (*Medicago truncatula*), pea (*Pisum sativum*), rapeseed (*Brassica napus*), grapes (*Vitis vinifera*), wheat (*T. aestivum*)	Jeudy *et al.*, 2016
TerraSentia autonomous under-canopy robot	In-row RGB + LiDAR measures of plant and ear height, stem diameter, LAI across nitrogen, drought and disease gradients	Maize (*Z. mays*) across 142 field sites	DeBruin *et al.*, 2025
UAV hyperspectral imaging platform	Flight-based hyperspectral indicators (e.g. red-edge NDVI, PRI) to dissect heat- and drought-yield relationships	Durum wheat (*T. turgidum* ssp. *durum*)	Mérida-García *et al.*, 2024
Morphological and physiological traits (recorded under unstressed conditions)			
3DPhenoMVS pipeline	SfM/MVS-derived point clouds for plant height, leaf length/width, relative leaf area, volume throughout growth	*Solanum lycopersicum*	Wang *et al.*, 2022b
Bellwether Phenotyping Platform (controlled-environment conveyor facility)	Daily RGB, NIR, and fluorescence imagery for plant height, projected leaf area, biomass accumulation, color, water-use efficiency	*Setaria viridis* and *S. italica*	Fahlgren *et al.*, 2015a
CropQuant	Performance-related traits	Wheat (*T. aestivum*)	Zhou *et al.*, 2017
DIRT/3D root phenotyping	Image-based 3-D quantification of 18 root-architecture traits (nodal root angle, diameter, whorl spacing, etc.)	*Z. mays*	Liu *et al.*, 2021

Continued

Table 4.2. Continued

A. Biotic and abiotic stresses			
ETH Field Phenotyping Platform (FIP) (cable-suspended system)	High-resolution canopy height, biomass indices, multispectral reflectance, and thermal traits across 1 ha plots	*T. aestivum*	Kirchgessner *et al.*, 2016
Field Scanalyzer	Morphological traits	Wheat (*T. aestivum*)	Virlet *et al.*, 2016
High-throughput rice phenotyping facility (HRPF)	Agronomic traits	Rice (*O. sativa*)	Yang *et al.*, 2014
MVS-Pheno	Plant height and leaf traits	Maize (*Z. mays*)	Wu *et al.*, 2020
Phenoarch	Tracking maize ear and silks growth	Maize (*Z. mays*)	Brichet *et al.*, 2017
PhenoArch phenotyping platform	Light interception and radiation-use efficiency	Maize (*Z. mays*)	Cabrera-Bosquet *et al.*, 2016
Phenobot 1.0	Biomass-related traits	Sorghum (*S. bicolor*)	Salas Fernandez *et al.*, 2017
Phenocart	Morphological traits	Wheat (*T. aestivum*)	Crain *et al.*, 2016
PhenoRoots	Root-related traits	Cotton (*Gossypium hirsutum L.*)	Martins *et al.*, 2019
Phenoscope	Complex traits that are involved in growth responses to the environment	Arabidopsis (*A. thaliana*)	Tisné *et al.*, 2013
Phenovator	Growth, photosynthesis, and spectral reflectance	Arabidopsis (*A. thaliana*)	Flood *et al.*, 2016
Plant root monitoring platform (PlaRoM)	Root-related traits	Arabidopsis (*A. thaliana*)	Yazdanbakhsh and Fisahn, 2009
Rhizoponics	Root-related traits	Arabidopsis (*A. thaliana*)	Mathieu *et al.*, 2015
Self-propelled electric HTPP platform	Plant height	Wheat (*T. aestivum*)	Pérez-Ruiz *et al.*, 2020
TERRA-REF Field Scanner (3-axis gantry, Arizona)	Multi-sensor data for canopy height, LAI, hyperspectral VIs (NDVI, PRI), and canopy temperature	*S. bicolor*, *T. aestivum*	Lebauer *et al.*, 2020
UAV RGB-SfM platform (field plots)	Time-series plant height and growth rate from crop-surface models	*T. aestivum*	Holman *et al.*, 2016
Zeppelin NT aircraft	Leaf area index, leaf biomass, early vigor, plant height	Maize (*Z. mays*)	Liebisch *et al.*, 2015

LiDAR, light detection and ranging; RGB, red, green, blue; LAI, leaf area index.

practical reference for selecting appropriate tools in phenomics-driven breeding programs.

4.2.4.2 Highthroughput phenotyping: automating plant trait analysis

High-throughput phenotyping (HTP) leverages advanced sensors and AI to measure plant traits rapidly and non-destructively on a large scale. This approach transforms crop breeding and stress biology by overcoming the "phenotyping bottleneck,"' enabling the evaluation of hundreds or thousands of genotypes under real-world conditions (Furbank and Tester, 2011). Modern HTP employs tools like hyperspectral imaging, unmanned aerial vehicles (UAVs) carrying multispectral or thermal cameras, LiDAR scanning for 3D structures, and deep learning models for image analysis. These AI-driven technologies can objectively quantify subtle plant responses that human observers might miss, especially under stress conditions. Below, we examine case studies across major crops (wheat, maize, rice, soybean, etc.) and stressors (drought, salinity, temperature extremes, pathogens) to illustrate the power of AI-powered HTP, citing recent peer-reviewed findings throughout (Gill *et al.*, 2022).

4.2.4.3 Phenotyping under stress

Highthroughput phenotyping (HTP) allied with artificial intelligence is transforming how breeders and physiologists unravel plant responses to environmental and biotic stresses. Multisensor platforms, from hyperspectral and thermal cameras to LiDAR and RGB drones, collect rich temporal data that capture both subtle biochemical shifts and gross structural changes before symptoms are apparent. Machine and deeplearning models convert these massive image streams into quantitative "itraits" such as canopy temperature, water status indices, 3D volume, or pixel-level disease severity, enabling early, objective, and scalable discrimination of tolerant genotypes. By integrating multimodal imagery with genomic information, researchers map stressresilience loci and predict yield under drought, salinity, temperature extremes, or pathogen attack in days rather than weeks, thereby closing the phenotype-to-gene loop with unprecedented resolution. AI-powered HTP collectively provides a unifying framework that accelerates stress breeding pipelines and offers actionable, real-time diagnostics for farmers managing climate-induced variability (Fig. 4.2).

4.2.4.3.1 PHENOTYPING UNDER DROUGHT STRESS. Drought causes significant reductions in crop production around the globe (Kim *et al.*, 2021). AI-based HTP methods have been extensively applied to detect drought stress early, quantify its impact on growth, and identify tolerant genotypes. Recent advances in sensor-based HTP have made it possible to quantify crop water status non-destructively and at scale.

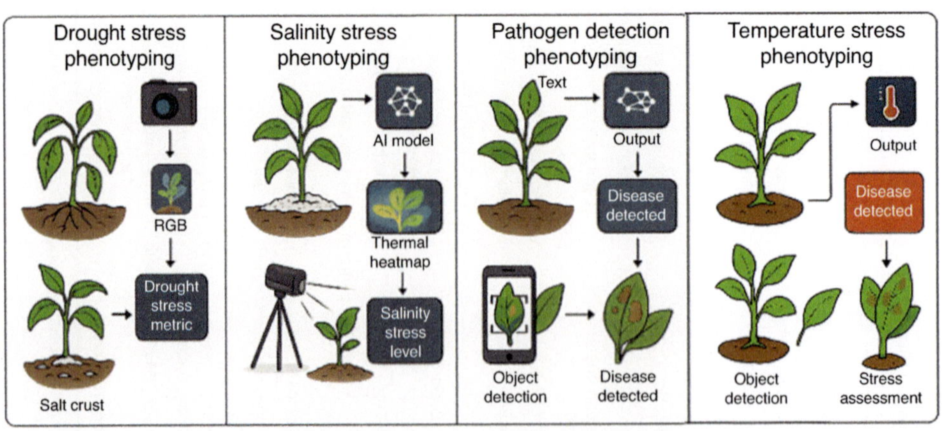

Fig. 4.2. AI-driven plant phenotyping under abiotic and biotic stress conditions. (Author's own image.)

Hyperspectral imaging for water status: Hyperspectral sensors capture reflectance in hundreds of narrow wavelengths, which can reveal water content and physiological status. For example, Okyere *et al.* (2024) used hyperspectral imaging with machine learning to detect drought stress in wheat. They developed new spectral indices that, when combined with known vegetation indices, identified drought-treated wheat (with low stomatal conductance and photosynthesis) with > 94% accuracy (Okyere *et al.*, 2024). The models could even predict physiological measures like stomatal conductance from spectra, showing how hyperspectral data plus AI can *monitor and predict* crop drought status. Similarly, hyperspectral reflectance was used to estimate relative water content in drought-stressed maize, achieving over 87% accuracy (Kim *et al.*, 2021)—a strong indicator that hyperspectral imaging can quantify plant water status under field drought.

UAV remote sensing for field drought trials: Drones equipped with RGB and multispectral cameras allow rapid field scanning of plots to assess drought effects. A recent field study in maize demonstrated this at a breeding trial scale. Pereira *et al.* (2025) grew 21 transgenic maize lines (plus controls) under well-watered vs water-limited regimes and flew drones weekly to capture multi-temporal images. Canopy indices derived from cheap RGB images were sufficient to distinguish drought-stressed plants, enabling early identification of the most drought-tolerant lines. Continuous monitoring over the season provided insight into how each genotype's growth trajectory changed under water deficit. The image data is fed into predictive models that accurately project final performance (e.g. yield) under drought, dramatically accelerating the selection of resilient varieties. This case illustrates that UAV imagery and AI analysis can replace laborious manual measurements, reducing evaluation time from weeks to hours.

Thermal imaging for canopy temperature: Canopy temperature is a proxy for plant water status (well-watered plants transpire and stay cooler). UAV-mounted thermal cameras have been used to screen genotypes for drought/heat tolerance by mapping canopy temperature. For instance, Qin *et al.* (2022) evaluated wheat under deficit irrigation using drone thermal imaging. They found that thermal indicators of drought resistance were most informative after the grain-filling stage. This multi-temporal thermal approach helped identify wheat lines that maintained cooler canopies (thus higher transpiration) under limited water, which is desirable for drought avoidance. The study highlights that the timing of phenotyping is crucial—early-season thermal data was less valuable, whereas late-season imagery differentiated drought-resilient varieties.

3D LiDAR scanning: Water deficit often stunts plant growth—something 3D scanning can quantify. LiDAR (light detection and ranging) sensors measure plant height and volume at high resolution by generating point clouds. In chickpeas, 3D laser scanning data combined with ML accurately predicted plot-level evapotranspiration (water use) under drought stress (Su *et al.*, 2019). The ability to remotely estimate water loss via canopy structure is valuable, as evapotranspiration reflects how a genotype copes with drought. In this case, an SVM model using LiDAR-derived traits outperformed other methods for gauging drought response. Similarly, structure-from-motion photogrammetry (using drone images to reconstruct 3D crop models) has been used to monitor height reductions or wilting due to drought. When tracked over time, these structural phenotypes allow breeders to pick drought-tolerant plants that sustain growth and canopy development under water scarcity.

Deep learning for drought classification: AI models can also directly classify or detect stress from images. For example, Niu *et al.* (2024) developed an improved YOLOv8 deep-learning pipeline to pinpoint drought-stressed regions in aerial pictures of maize. Their UAV-based system detected early drought symptoms with ~ 90% precision and ~ 89% mAP (mean average precision), significantly outperforming earlier models. An explainable AI approach on UAV images has also been shown for potatoes, where a lightweight CNN classified drought stress in field plots with high accuracy (Patra and Sahoo, 2024). These deep-learning models can process complex image features (leaf color, texture, posture) that correlate with stress, enabling *automated stress scouting*. The upshot is that breeders or farmers could use a drone and a trained model to survey vast fields for real-time drought stress hotspots.

4.2.4.3.2 PHENOTYPING UNDER SALINITY STRESS. Soil salinity is another significant abiotic stress that impairs plant growth by causing ion toxicity and osmotic stress (Shrivastava and Kumar, 2015). Salinity primarily reduces plant growth and vigor. UAV imaging can track this by measuring canopy size over time. Johansen *et al.* (2019) showed that wild tomato genotypes exposed to salt stress could be distinguished by differences in canopy coverage dynamics measured from drone-based RGB and multispectral imagery (Johansen *et al.*, 2019). By comparing time-series growth curves of stressed vs control plants, they identified salt-sensitive lines (with stunted canopy development) vs tolerant ones (Ninomiya, 2022). Notably, this detection was possible before severe visual symptoms appeared, underscoring how multi-temporal imaging and AI analysis enable *early diagnostics* of salinity effects.

Vegetation indices derived from hyperspectral or multispectral data can be salinity stress markers. Researchers used a hyperspectral-derived photochemical reflectance index (PRI) in quinoa to successfully identify salt-stressed plants (Ninomiya, 2022). The PRI, related to photosynthetic light-use efficiency, shifted in salt-treated quinoa compared to controls. This shift indicates that hyperspectral sensors captured stress-induced changes in leaf biochemistry and interpreted them via an index. More broadly, salinity causes changes in leaf pigment, water content, and structure that alter reflectance; AI algorithms can combine multiple narrow-band signals to classify salt stress severity. For instance, ML models analyzing full-spectrum data (400–1000 nm) have been used to predict salinity tolerance in rice and other cereals (Gill *et al.*, 2022).

A combination of sensor data can improve salinity assessments. In one study on chickpeas under greenhouse conditions, RGB imaging was used alongside chlorophyll fluorescence measurements to evaluate salt stress, with data integrated by partial least squares regression (Gill *et al.*, 2022. The models could quantify stress impacts on growth and photosynthesis. In field settings, deploying multispectral cameras (to gauge vegetation indices) and LiDAR (to measure salt-induced stunting) via UAV provides a richer picture. Such multimodal HTP was suggested to capture both "color" and "structure" aspects of salinity damage (Kim *et al.*, 2020a). For example, a recent cotton study highlighted the potential of HTP to combat salinity by correlating spectral indices and 3D traits with yield under salt stress (Khalid *et al.*, 2023).

Rice is often grown in paddy soils prone to salinity build-up. Al-Tamimi *et al.* (2016) demonstrated HTP of rice under salt stress in field plots using imaging. They found that certain growth and greenness metrics (extracted from RGB images) were strongly associated with genetic loci for salinity tolerance. This method helped map salt-tolerance quantitative trait loci by linking image-derived phenotypes with genotype data. It is an early example of integrating AI phenotyping with genomics for stress breeding.

4.2.4.3.3 PHENOTYPING UNDER TEMPERATURE EXTREMES. Heat and cold stress significantly impact crop development, and AI-driven phenotyping offers ways to assess plant responses to temperature in high throughput. High temperatures often lead to increased canopy temperatures and faster wilting. Thermal infrared (TIR) cameras on drones or ground robots measure canopy temperature as a direct readout of plant water status and transpirational cooling. In wheat, UAV-based thermal imaging has been used to evaluate genotype performance under hot, drought-like conditions. As mentioned, one study found that thermal differences between varieties emerged clearly during grain filling under heat/drought stress (Qin *et al.*, 2022). Genotypes that kept cooler canopies (indicating better cooling or deeper roots accessing water) were deemed more heat/drought resistant. Thermal imagery thus serves as an *instantaneous indicator* of heat stress: AI algorithms can threshold or classify images to flag "hot" (stressed) plants vs "cool" (unstressed) ones. Zhou *et al.* (2021) employed multispectral and thermal UAV data to monitor soybean plots, successfully detecting heat-induced water stress (and even transient flooding stress) by changes in canopy temperature combined with vegetation indices.

Controlled-environment phenotyping platforms (indoor HTP) use conveyor systems and imaging to expose diverse genotypes to heat or cold treatments and quantify their responses. AI models process images of these plants to extract stress symptoms like leaf rolling, senescence, or growth rate reduction. For example, a phenomics study on wheat seedlings under heat stress used machine vision to measure traits like leaf wilting and color change, achieving objective stress ratings correlating with classical heat tolerance scores (this is often referenced in phenotyping literature, though specific recent examples are in development). Deep learning can enhance this by tracking subtle changes over time; e.g. time-series image analysis via recurrent neural networks could detect if a plant recovered at night or continued to deteriorate under repeated hot days.

While less frequently reported than drought/heat, HTP has also been applied to cold or frost tolerance. Imaging can capture leaf discoloration or necrosis due to freezing, and 3D scanning can measure reduced growth. In winter wheat and rapeseed breeding, drone surveys after frost events have been used to quantify survival and damage in plots, with AI segmenting live vs dead tissue. Multispectral indices like normalized difference vegetation index (NDVI) drop sharply in cold-damaged plants, which ML models can use to single out the most cold-tolerant lines. These high-throughput measures align with yield losses due to frost, helping validate the selections (Gill *et al.*, 2022).

Importantly, phenotyping under temperature stress is being combined with genome-wide association studies. Recent research integrated drone-based phenotyping of wheat under heat and drought with genetic data to predict yield and identify loci for stress resilience (Zhang *et al.*, 2024c). The model used 24 image-derived traits ("i-traits", such as canopy temperature and greenness over time) and ML to successfully predict yield under stress (Okyere *et al.*, 2024).

4.2.4.3.4 PHENOTYPING UNDER PATHOGEN STRESS (DISEASE DETECTION). Biotic stresses like fungal, bacterial, or viral diseases can devastate crop yields. Traditional disease scoring is labor intensive and subjective, but AI-powered phenotyping is revolutionizing pathogen stress detection through imaging and pattern recognition:

Modern approaches use deep CNNs to analyze images of crop canopies or leaves and automatically recognize disease symptoms. A prime example is wheat yellow rust (stripe rust) detection via UAV images. Pan *et al.* (2021) trained a segmentation CNN (PSPNet) to classify each pixel in drone images as healthy wheat, rust-infected wheat, or soil. The model achieved ~ 98% accuracy in identifying yellow rust infection in wheat fields. It was even transferable to new field conditions with minimal loss of accuracy, demonstrating robust generalization (Pan *et al.*, 2021). This AI system can survey large trial plots or farms, flagging disease outbreaks far faster than scouts on foot. Similarly, researchers have developed deep-learning classifiers for *multiple* wheat diseases (e.g. powdery mildew, leaf rust) with high accuracies of ~ 97% using curated image datasets (Long *et al.*, 2023).

AI methods can detect diseases even before visible symptoms appear by using hyperspectral cameras that sense subtle biochemical changes. For instance, an innovative study targeted bacterial leaf blight (BLB) in rice with an asymptomatic incubation period. Hyperspectral images of rice leaves were collected before lesions were visible, and a 3D-CNN was trained to distinguish BLB-infected leaves from healthy ones. The model reached ~ 95.4% accuracy in detecting early, asymptomatic infections (Cao *et al.*, 2022). Key spectral wavelengths related to infection (around 530–570 nm) were identified via the network's saliency analysis. This result shows the promise of combining hyperspectral data with deep learning to catch diseases very early, enabling proactive management.

Beyond classification, HTP can quantify disease severity and its effect on plant performance. For example, UAV multispectral imagery has been used to create disease stress indices—areas of reduced chlorophyll or changed canopy temperature correlate with infection foci. In one study on maize, drone images captured the progression of northern leaf blight, and ML models predicted yield loss based on the fractional area of the canopy affected (with decent accuracy, linking stress to outcome). Likewise, thermal imaging can reveal "hot spots" on leaves where pathogen-induced stomatal closure leads to heat-up. AI algorithms can map these to quantify how much of the canopy is under pathogen attack. Such quantitative phenotypes can be used in genetics studies to find disease-resistance genes (Pignon *et al.*, 2021).

The same deep learning techniques are being extended to other crops and pathogens—from soybean (e.g. foliar diseases like rust and blight) to fruit trees (identifying pest or fungal infestations from aerial images) (Upadhyay *et al.*, 2025). Many researchers are now deploying drones or ground robots in breeding nurseries to score disease resistance automatically. Integrating these tools into breeding programs means selections for disease resistance can be made with unprecedented precision and at scale, even in early generations (Kamilaris and Prenafeta-Boldú, 2018).

4.2.4.4 Integrating multimodal phenotyping and future perspectives

The case studies above underscore that no single sensor suffices for all stress phenotyping—instead, integrating multiple AI-driven tools yields the best insights. For example, drought-tolerance selection might combine RGB/multispectral indices (for biomass and greenness), thermal imaging (for transpiration cooling), and LiDAR (for structural robustness) to capture the multifaceted plant responses (Kim *et al.*, 2021). ML models can fuse these data streams to produce composite stress metrics or predict yield under stress more reliably than any trait (Fig. 4.3). In one study, an ensemble XGBoost model was used to combine wheat spike counts from deep learning, spectral features, and texture data into a yield prediction under field conditions (Chen *et al.*, 2023a), illustrating how multi-source data integration via AI can directly inform breeding decisions.

Another emerging trend is the use of time-series data and longitudinal analysis. Rather than a single snapshot, repeated measurements allow AI models to understand stress dynamics (onset, peak, recovery). In the maize drought study, continuous UAV monitoring was "essential in understanding how plants respond to water stress" over time (De Castro, 2025). Such temporal phenomics, enabled by drones and automated analysis, opens the door to identifying whether a genotype is tolerant and when and how its tolerance manifests (e.g. escaping stress

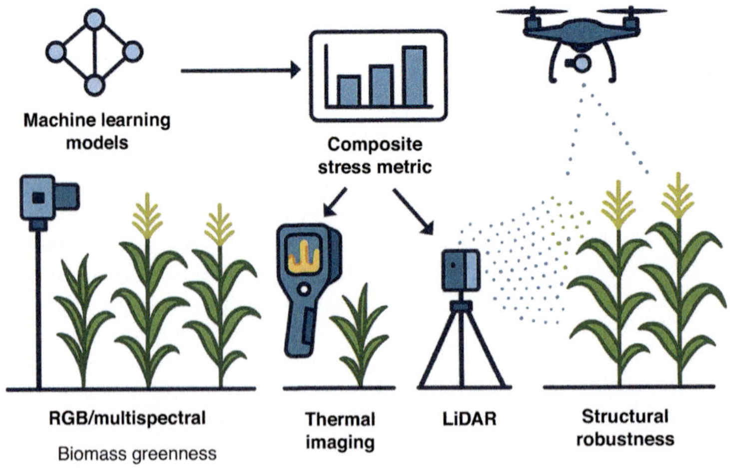

Fig. 4.3. Integrating multimodal phenotyping and future perspectives in crop stress assessment using AI. (Author's own image.)

by early flowering vs enduring stress via deep roots). For high-impact research, the integration of HTP with genetics is key. The ultimate goal is to link phenotypic data to genetic markers, accelerating marker-assisted or genomic selection. We now see many studies where AI-extracted phenotypic traits under stress are used in genome-wide association studies or genomic prediction models (Zhang *et al.* 2024a, b). This approach has successfully identified genomic regions associated with drought-adaptive traits in wheat using hyperspectral indices (Mérida-García *et al.*, 2024). As AI phenotyping data become ever richer, the accuracy of predicting breeding values for stress tolerance will further improve.

4.2.5 Integrating multi-omics data for comprehensive models

Combining data from different omics levels (genomics, transcriptomics, proteomics, and metabolomics) has led to marked improvements in the performance of predictive models. For instance, a study in maize used ML to integrate transcriptomic and metabolomic data, reconstructing a metabolic network that governs nitrogen utilization. The model identified key enzymes and regulatory genes that control nitrogen assimilation, providing targets for improving crop nitrogen use efficiency (Chowdhury *et al.*, 2021).

Predictive modeling has also been applied to study metabolic responses to environmental stress. In rice, researchers used ML to analyze metabolomic data from plants exposed to drought and salinity. The model predicted changes in metabolic fluxes and identified key metabolites involved in stress tolerance, such as proline and trehalose. These insights have been used to develop crops with enhanced resilience to abiotic stress (Wang *et al.*, 2020a).

4.3 AI for Modeling Plant Metabolism and Pathways

Plant metabolism is a set of interconnected chemical reactions that drive a range of fundamental functions within a plant, including growth, development and responses to environmental cues. Understanding these metabolic pathways is critical for improving crop yield, nutritional quality, and resilience to environmental challenges. However, the sheer complexity of plant metabolism, involving thousands of metabolites and enzymes, makes it challenging to study using traditional experimental approaches alone. Powered by AI and ML, predictive modeling has emerged as a transformative tool for unraveling plant metabolic networks and pathways. By integrating multi-omics data and leveraging computational algorithms, researchers can predict metabolic fluxes, identify key regulatory nodes, and engineer pathways for desired outcomes (Sweetlove and Fernie, 2018).

4.3.1 Predicting and optimizing metabolic pathways

AI and ML have transformed the study of plant metabolic pathways, offering powerful tools to dissect complex, multi-omics datasets and predict metabolic fluxes with precision. These techniques excel at navigating large, heterogeneous data, revealing hidden patterns and nonlinear relationships (Min *et al.*, 2017; Zampieri *et al.*, 2017) that accelerate discoveries in plant physiology and pinpoint targets for crop enhancement. Table 4.3 summarizes key studies highlighting AI-driven optimization of metabolic pathways for crop enhancement. It details specific AI models, targeted metabolic pathways, plant species studied, and the resulting improvements in crop traits. ML techniques—deep neural networks, RFs, and SVMs—effectively link genotype and phenotype data, shedding light on plant metabolic networks. In *Arabidopsis thaliana*, deep learning has analyzed gene expression to uncover co-regulated gene modules tied to secondary metabolism, enabling hypotheses about regulatory nodes to boost phytochemical yields) (De Oliveira Dal'Molin *et al.*, 2010; Sweetlove and Fernie, 2018). Constraint-based models, like flux balance analysis (FBA), predict fluxes and gain refined accuracy when paired with ML. For instance, ML-enhanced FBA models in Arabidopsis accurately forecast growth and nutrient use (Sweetlove and Fernie,

Table 4.3. Optimized metabolic pathways for crop enhancement with AI predictions and models (Author's own table)

Optimized metabolic pathways for crop enhancement	AI predictions	Name of models used	Name of algorithms used	References
Photosynthesis Efficiency (C3 to C4/ CAM conversion)	AI models predict key enzymatic bottlenecks (e.g. RuBisCO) and suggest gene edits to enhance carbon fixation efficiency in C3 crops like rice, potentially increasing yield by 20–50%.	CatBoost, LightGBM, RF, XGBoost	Gradient boosting, RF	Wang *et al.*, 2022a
Fatty acid biosynthesis (seed oil content)	Machine learning identifies regulatory genes (e.g. ATP:citrate lyase) to increase fatty acid synthesis, predicting a 15–30% rise in oil content in crops like Arabidopsis and oilseed rape.	SVM, deep learning	Support vector machines, neural networks	Mahood *et al.*, 2020
Nitrogen use efficiency (alanine metabolism)	AI-driven flux analysis predicts overexpression of glutamate-pyruvate aminotransferase (GPT) to enhance nitrogen assimilation, improving yield by up to 12% in crops like canola and rice.	SVM, deep learning	Support vector machines, neural networks	Mahood *et al.*, 2020
Terpenoid production (stress resilience)	Graph Transformer and CNN models (GTC) predict metabolic pathways for terpenoid synthesis, enhancing stress tolerance in tomato by 25% through targeted gene cluster activation.	GTC (graph transformer and CNN)	Deep learning (graph transformer, CNN)	Bao *et al.*, 2023
Starch accumulation (storage poot development)	Transcriptome-integrated AI models (e.g. rMeCBMx-HPCOF) predict carbon flux optimization in cassava, increasing starch yield by 10–20% in high-yield cultivars.	Machine learning model for image analysis	Likely CNN or other image processing ML algorithms	Earles *et al.*, 2018
Lignin modification (biomass quality)	Network analysis predicts enzyme correlations (e.g. arogenate dehydratase) to reduce lignin by up to 70%, improving digestibility in Arabidopsis and bioenergy crops.	Graph neural network	Deep learning (graph neural network)	König-Mattern *et al.*, 2024
Isoprenoid biosynthesis (nutritional quality)	AI-guided pathway engineering predicts enhanced carotenoid production (e.g. via isoprenoid pathways), boosting vitamin A content in maize by 15–25%.	GTC (graph transformer and CNN)	Deep learning (graph transformer, CNN)	Bao *et al.*, 2023
Phenylpropanoid pathways (antioxidant content)	Supervised learning predicts flux redirection to increase flavonoid synthesis, enhancing antioxidant levels in *Brassica napus* by 20–30%.	GTC (graph transformer and CNN)	Deep learning (graph transformer, CNN)	Bao *et al.*, 2023

2018). Deep learning frameworks, such as DeepMetabolism, reconstruct pathways from genomic and transcriptomic data (Wang *et al.*, 2020a), while RFs and Bayesian networks infer networks and key regulatory points (Escorcia-Rodríguez *et al.*, 2023).

Constraint-based modeling constructs genome-scale metabolic networks using FBA to

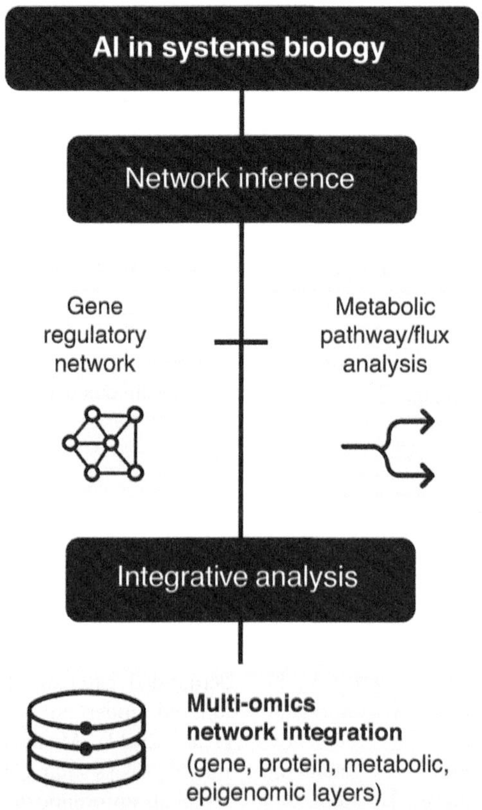

Fig. 4.4. AI-driven network inference and integrative analysis in systems biology. (Author's own image.)

estimate fluxes under specific conditions. ML integration enhances these predictions via adaptive flux sampling or optimized constraints. A key case in *Arabidopsis* combined these methods with AI to optimize specialized metabolite production, identifying modifications that increase yields without harming plant fitness (Sweetlove and Fernie, 2018). This integration yields a detailed metabolic roadmap for targeting enzymes in crop engineering.

Fusing transcript, protein, metabolite, and epigenetic data via ML reveals master regulators and stress-responsive control points (Fig. 4.4). In tomato (*Solanum lycopersicum*), integrative models blending gene expression and flux analyses boosted anthocyanin production, creating healthier varieties with added benefits like pathogen resistance and longer shelf life (Jian *et al*., 2023).

In rice and maize breeding, ML models integrating genomic and metabolite data predict nitrogen-use efficiency, surpassing traditional methods (Crossa *et al*., 2017). This integration narrows the focus to elite lines, speeding up breeding and cutting costs. In *Arabidopsis*, predictive modeling of carbon metabolism under varying conditions uncovered metabolic rewiring for energy and biomass optimization under stress (Sweetlove and Fernie, 2018). Figure 4.5 illustrates an integrated framework demonstrating how AI and ML techniques, plant metabolic pathway analysis, and constraint-based modeling collaboratively contribute to crop enhancement. AI/ML techniques such as

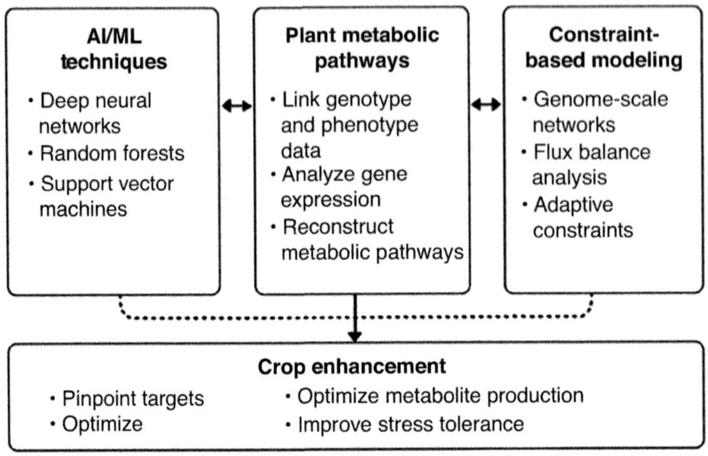

Fig. 4.5. Integration of AI/ML techniques, plant metabolic pathway analysis, and constraint-based modeling for crop enhancement. (Author's own image.)

deep neural networks, RFs, and support vector machines process complex datasets, facilitating the analysis of plant metabolic pathways by linking genotype and phenotype data, gene expression analysis, and pathway reconstruction. Concurrently, constraint-based modeling, including genome-scale networks and flux balance analysis, provides adaptive constraints. Together, these components synergistically pinpoint metabolic targets, optimize metabolite production, and improve stress tolerance, significantly advancing precision breeding strategies. AI also drives secondary metabolite engineering. In *Nicotiana benthamiana*, ML optimized gene expression for artemisinin biosynthesis, markedly increasing production of this antimalarial compound (Wang *et al.*, 2020a). This result underscores AI's potential in biotechnological innovation.

4.3.2 Image recognition and segmentation: Extracting biological features and real-time monitoring of plant growth and stress responses

AI-driven image recognition and segmentation have transformed plant phenotyping by enabling rapid, precise extraction of biological features from plant images, surpassing traditional methods that relied on manual segmentation and classical ML algorithms like SVMs, which demanded extensive feature engineering and faltered under environmental variability (Gill *et al.*, 2022). Modern deep learning models, notably CNNs such as U-Net, Mask R-CNN, and DeepLab, automatically learn robust feature representations, adeptly segmenting complex plant scenes into components like leaves, stems, and fruits, even in challenging field conditions (Gill *et al.*, 2022). Transfer learning enhances this adaptability by achieving an F1-score of approximately 0.97 in segmenting Arabidopsis and tobacco leaves (Meraj *et al.*, 2024). These models excel in applications like plant disease detection, achieving up to 96% pixel-wise accuracy in segmenting diseased leaves and over 95% accuracy in classifying crop pathologies (Gill *et al.*, 2022), while in yield estimation, Mask R-CNN has delivered around 97% accuracy in fruit counting in apple orchards (Maraveas, 2024). Real-time analysis further bolsters agricultural decision making, with ML applied to UAV imagery accurately predicting crop nitrogen status for swift, non-destructive diagnostics at scale (Kou *et al.*, 2022). Figure 4.6 demonstrates the integration of AI models with environmental inputs for plant growth and stress monitoring. Artificial neural networks, RF, and hybrid models utilize data from remote sensing, weather conditions,

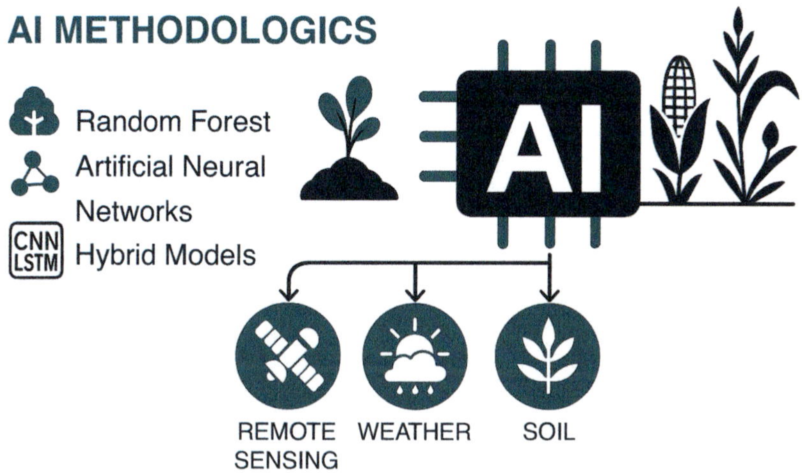

Fig. 4.6. AI methodologies and multisource data integration for precision crop yield prediction. (Author's own image.)

and soil parameters to enhance the precision of agricultural phenotyping.

This technology proves invaluable in plant breeding and stress detection, offering detailed, non-destructive trait analysis and early identification of biotic and abiotic stresses. Automated segmentation of growth traits like leaf count and size informs the selection of high-yielding varieties (Maraveas, 2024), while precise yield component analysis elucidates metabolic processes such as carbon allocation (Maraveas, 2024). AI models detect subtle stress indicators, facilitating timely interventions (Gill *et al.*, 2022; Maraveas, 2024), and by linking image-derived features to metabolic models, they offer the potential for inferring pathway activities from phenotypic data, such as correlating leaf coloration changes with pigment concentrations (Kisiel *et al.*, 2023). Though still emerging, this integration could enable predictions of internal metabolic states, bridging visible traits with plant physiology.

Despite these advances, AI in plant phenotyping faces significant challenges. The demand for large, diverse, well-annotated datasets hinders model generalization across varied species, genotypes, and environments (Walsh *et al.*, 2024), while the opacity of many models undermines trust, though explainable AI efforts are progressing (Mostafa *et al.*, 2023). Standardization of imaging protocols remains essential for reproducibility across sensors and conditions (Walsh *et al.*, 2024), and computational costs, alongside the complexity of integrating multi-modal data like genomic and environmental inputs, pose ongoing obstacles to fully realizing the potential of AI in this field.

4.4 Predictive Analytics in Plant Systems Biology

Advancements in predictive analytics have reshaped plant systems biology by enabling precise forecasts of plant responses under varying environmental scenarios. Researchers can integrate diverse datasets—including genomics, phenomics, environmental sensors, and remote sensing—to unravel complex plant–environment interactions by employing sophisticated AI methods, particularly ML and deep learning. These insights not only facilitate early detection and management of environmental stressors but also guide the optimization of crop yields and resilience, thus paving the path toward more robust and sustainable agricultural systems amid escalating climatic challenges.

4.4.1 Forecasting plant behavior under environmental stress

In an era of rapid climate change, ensuring global food security demands advanced strategies to predict and manage plant responses to environmental stress. Drought, temperature extremes, and pest outbreaks threaten crop yields, often eluding traditional methods that struggle to identify early stress signals, thus delaying critical interventions. Machine learning and deep learning empower AI to undertake advanced analysis of multi-domain datasets such as satellites, genomics and in-field sensors for early detection of stress signs. Techniques such as CNNs excel in image-based stress diagnosis, while recurrent neural networks (RNNs) analyze sequential data like weather patterns, enabling early detection of physiological shifts and supporting timely actions like optimized irrigation (Khaki *et al.*, 2020a). A significant leap forward lies in integrating phenomic data with real-time climate and soil metrics, essential for decoding the interplay of genotypic, phenotypic, and environmental factors. Deep neural networks (DNNs) reveal hidden relationships within multi-modal data—fusing satellite imagery, soil sensors, and genomic profiles—to enhance yield predictions under adverse conditions, fostering climate-resilient cultivars (Boukhris *et al.*, 2024).

The interpretability of AI models, vital for translating predictions into practical insights, remains challenging yet critical. Attention-based networks and explainable AI frameworks illuminate key diagnostic features, such as leaf reflectance or evapotranspiration patterns, enriching our grasp of plant–environment dynamics (Wang *et al.*, 2024). Complementing this, synergy with robotics and IoT sensors bolsters data collection—autonomous systems and networks monitoring field conditions in real-time refine models for adaptive responses like precise pesticide application. Recent evidence highlights AI's prowess: SVMs achieve high precision in stress prediction ($R^2 = 0.96$), recursive SVMs classify rice stress responses with perfect accuracy, and RF models forecast plant–pathogen interactions at 95% accuracy (Murmu *et al.*, 2024). The field's growth, propelled by deep learning and datasets like PlantVillage, is evident in surging publications (Walsh *et al.*, 2024). Looking forward, unified platforms integrating diverse data sources could tackle compound stressors, though issues like data quality and model opacity demand rigorous validation. The fusion of AI with advanced analytics revolutionizes plant stress management, securing sustainable agriculture amid climatic uncertainty.

4.4.2 AI-driven crop yield prediction and optimization

Recent studies showcase how AI enhances crop yield prediction and optimization through advanced algorithms. For example, a study in the Midwestern United States compared a DNN with traditional models like RF and SVM for predicting maize and soybean yields. The DNN, fine-tuned with specific parameters, delivered a root mean square error (RMSE) of 0.765 ton/ha for maize and 0.285 ton/ha for soybeans, outperforming RF and SVM with a 21–33% improvement in mean absolute error (MAE) for maize and 17–22% for soybeans. This result highlights the ability of AI to detect complex patterns in datasets from sources like MODIS, PRISM, and GLDAS, improving prediction accuracy significantly (Kim *et al.*, 2019). Similarly, a large-scale European study by Athanasiadis *et al.* (2022) applied ML to forecast yields for multiple crops across the Netherlands, Germany, and France. Using data from crop simulations, weather, remote sensing, and soil characteristics, the model achieved normalized RMSE (NRMSE) values close to those of the MARS Crop Yield Forecasting System (MCYFS). Soft wheat in the Netherlands recorded an NRMSE of 7.87, compared to 6.32 from MCYFS, showing AI's adaptability across diverse agricultural settings and its utility for early-season predictions.

Various AI methodologies contribute uniquely to this field. RF, as used by Prasad *et al.* (2021) for cotton yield prediction in India, offers robustness in handling regional data. Artificial neural networks (ANN), applied by Singh Boori *et al.* (2023) in the Fergana Valley, excel at modeling non-linear relationships. Meanwhile, deep learning techniques like CNNs and long-short term memory (LSTM) networks shine with complex data. Khaki *et al.* (2020c) combined CNN and RNN in a framework that processed spatial and temporal data, such as satellite imagery and time series, for accurate yield predictions. A review by Yan *et al.* (2024) of papers from 1965 to 2023 further emphasized the rise of hybrid models like CNN–LSTM, noting the importance of vegetation indices like NDVI and the enhanced vegetative index (EVI) in boosting accuracy. Table 4.4 summarizes various AI methodologies employed for crop yield prediction, highlighting their key applications, performance metrics, and advantages across diverse crops and geographical regions.

Emerging trends point to exciting possibilities. Integrating multimodal data—remote sensing, IoT sensors, and meteorological inputs—could enrich predictions further. Hybrid approaches blending crop simulation models with machine learning, as seen in Cao *et al.* (2021a) for winter wheat in northern China, combine mechanistic insights with data-driven precision. Techniques like transfer learning and ensemble methods also promise to enhance model performance across different crops and regions, supporting broader applications in agriculture.

AI-driven crop yield prediction and optimization empower farmers with precise tools for decision making, resource management, and risk mitigation, paving the way for smarter, more sustainable farming practices through ongoing advancements in ML and deep learning.

4.4.3 Simulating plant–environment interactions for climate resilience

Simulating plant–environment interactions is a cornerstone of modern agricultural research, particularly as climate change introduces stressors such as drought, extreme temperatures, and altered precipitation patterns. These simulations enable researchers to predict plant responses to environmental changes, informing the development of crop varieties with enhanced climate resilience. By integrating data from genomics, phenomics, and environmental monitoring, AI facilitates precise modeling of plant responses, offering insights critical for sustainable agriculture.

A notable application is the work of Dussarrat *et al.* (2022), who studied 24 plant species in the Atacama Desert, one of the most arid regions on Earth. Using ML-based generalized linear models, they analyzed multispecies ecological metabolomics, identifying 39 common metabolites that predicted environmental fluctuations with 79% accuracy. This study established the plant metabolome as a robust predictor of environmental conditions, validated across independent datasets. The research shows that AI technology can help identify metabolic processes which could potentially form the basis for creating more durable crop varieties that are better able to withstand different environmental challenges.

Further advancing this field, Harfouche *et al.* (2019) underscore the role of AI in integrating multi-omics data—genomics, phenomics, and environmental metrics—to predict plant performance under diverse scenarios. By linking genotype to phenotype, AI enables the selection of traits conferring resilience to abiotic stresses like water scarcity, accelerating the development of climate-adapted cultivars. This approach reduces the time and resources required compared to traditional breeding methods, transforming experimental design to anticipate plant responses under projected climate scenarios.

The interpretability of AI models is essential for translating predictions into practical agricultural applications. Explainable AI frameworks, such as SHAP (SHapley Additive exPlanations), enhance transparency by identifying which features—genetic markers, metabolic compounds, or environmental factors—drive predictions. For instance, Jafar *et al.* (2024) applied SHAP to quantify feature contributions in crop yield predictions, revealing key drivers like rainfall. Similarly, Cartolano *et al.* (2024) used SHAP and LIME (Local Interpretable

Table 4.4. AI models for crop yield prediction.

Model type	Specific algorithm	Input data	Accuracy metric	Reported accuracy	Crop	Region	Reference
Deep Learning	CNN-RNN	Yield performance, management, weather, soil	RMSE	8% of average yield	Soybean	US Corn Belt	Khaki *et al.*, 2020b
Deep Learning	CNN-DNN	Weather, soil, farm management	R^2	R^2=0.87	Corn	US Corn Belt	Jabed and Azmi Murad, 2024
Deep Learning	CNN-RNN	Yield performance, management, weather, soil	RMSE	9% of average yield	Corn	US Corn Belt	Jabed and Azmi Murad, 2024
Deep Learning	LSTM	Satellite vegetation indexes, meteorological indexes, soil properties	R^2, RMSE	R^2=0.77–0.87, RMSE = 298.11–724 kg/ha	Rice	China	Cao *et al.*, 2021b
Deep Learning	LSTM	Remote sensing imagery	R^2	R^2 = 0.93	Winter wheat	China	Jabed and Azmi Murad, 2024
Deep Learning	LSTM	Time-series satellite data, meteorological data	R^2	R^2 = 0.859	Rice	Korea	Jabed and Azmi Murad, 2024
Deep Learning	LSTM + CNN	Meteorological, remote sensing, soil data	R^2, RMSE	R^2=0.77, RMSE = 721 kg/ha	Winter wheat	Main wheat-producing regions of China	Wang *et al.*, 2020b
Deep Learning	LSTM + 1D-CNN	Satellite and meteorological data	R^2, RMSE	R^2 = 0.86, RMSE = 610 kg/ha	Rice	South and North Korea	Jeong *et al.*, 2022
Machine Learning	RF	Crop planting areas, climate, satellite, soil properties	R^2	R^2=0.66	Winter wheat	China	Jabed and Azmi Murad, 2024
Machine Learning	RF	Weather, soil, etc. (assumed)	R^2	R^2 = 0.963	Various crops	Rajasthan, India	Jhajharia *et al.*, 2023
Machine Learning	RF	Satellite vegetation indexes, meteorological indexes, soil properties	R^2, RMSE	R^2 = 0.76–0.82, RMSE = 366–723.3 kg/ha	Rice	China	Cao *et al.*, 2021b

Model-agnostic Explanations) in smart agriculture to understand feature importance for crop recommendations, such as humidity for apples. These techniques foster trust among researchers and farmers by elucidating the biological mechanisms underlying resilience.

4.5 AI in Plant Breeding and Genetic Engineering

Advances in AI have sparked transformative changes in agricultural biotechnology, significantly accelerating plant breeding and genetic engineering. By harnessing computational tools and ML algorithms, researchers have moved beyond conventional methods to predict complex phenotypes, optimize genetic pathways, and perform precise genome edits at unprecedented speed and accuracy. These AI-driven approaches facilitate the rapid development of superior crop varieties with enhanced yield, stress resilience, nutritional quality, and adaptability to shifting climatic conditions, marking a new era in precision agriculture and biotechnology.

4.5.1 Accelerating precision breeding with machine learning

Integrating AI, particularly ML, into plant breeding has transformed agriculture by accelerating precision breeding. As global populations surge and climate change escalates, the demand for crops with superior yield, resilience, and adaptability intensifies. Precision breeding leverages advanced genetic and phenotypic analyses, enhanced by the ability of ML to process vast omics datasets—encompassing genomics, phenomics, and environmental sources. This data-driven approach shifts breeding from trial-and-error to optimized, rapid development of improved varieties. Figure 4.7 illustrates the integration of omics data, ML methodologies, environmental information, and traditional plant breeding approaches into a cohesive framework for precision breeding. This holistic model highlights how data-driven methods streamline trait prediction, candidate selection, and the development of resilient crop varieties.

ML excels at revealing patterns in complex datasets, enabling trait prediction and candidate

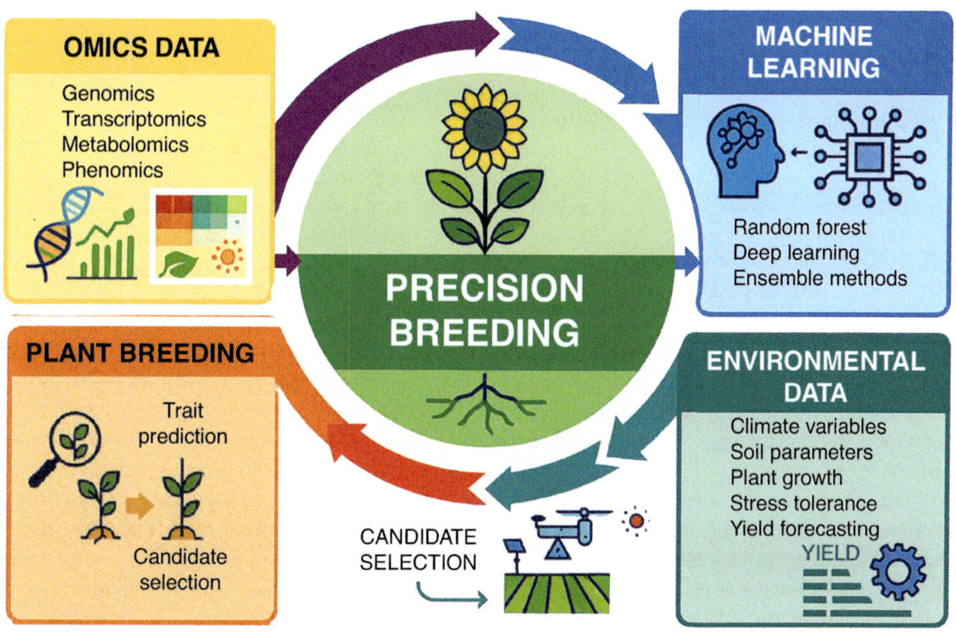

Fig. 4.7. Precision-breeding framework integrating omics, environment, and machine learning. (Author's own image.)

selection without extensive field trials. For example, a RF algorithm in soybean breeding predicted seed yield using hyperspectral reflectance data from 292 accessions across six environments, identifying key phenomic traits early and reducing end-of-season phenotyping (Parmley *et al.*, 2019). In maize, genomic selection with ML predicted traits of 34,000 untested hybrids from 9000 tested ones using multi-omics data, boosting precision with DeepGS and cutting costs (Ma *et al.*, 2018). Beyond yield, ML enhances stress tolerance amid climate change. In wheat breeding, CNNs analyzed drone imagery and meteorological data to predict growth stages under drought, with the growth stage prediction (GSP)-AI platform selecting resilient varieties (Shen *et al.*, 2024). In rice, ensemble learning—integrating multi-layer perceptrons, radial basis function networks, and RFs—forecasted yield for 250 genotypes across four environments, speeding the identification of top cultivars (Yu *et al.*, 2023). Institutions like CIMMYT (the International Maize and Wheat Improvement Center) have shown that ML in genomic selection slashes field trial time and cost, improving yield (Crossa *et al.*, 2017). Computer vision and deep neural networks further refine genotype-to-phenotype detection (Choi *et al.*, 2024), while large-scale public data on crops of strawberry, pepper, grape, tomato, and paprika integrate environmental variables to enhance disease resistance predictions (Lee and Yun, 2023).

4.5.2 Designing synthetic pathways for enhanced traits

Integrating AI into synthetic biology has fundamentally transformed the engineering of enhanced crop traits, enabling unprecedented precision in improving yield, nutritional content, stress resilience, and disease resistance. By harnessing extensive multi-omics data and ML algorithms, researchers can efficiently navigate the design–build–test–learn cycle, decoding the intricate networks of plant metabolism to pinpoint and optimize critical genetic and enzymatic components. AI-driven tools have been pivotal in projects such as introducing the C4 photosynthetic pathway into C3 crops like rice, where they identify key genes and regulatory motifs to construct synthetic gene circuits that boost carbon fixation under high temperatures and light intensity (South *et al.*, 2019). Similarly, biofortification initiatives, exemplified by the development of Golden Rice with elevated provitamin A levels, have leveraged AI to analyze metabolic flux and target rate-limiting steps (Paine *et al.*, 2005). In enhancing stress tolerance, AI models integrate transcriptomic data to engineer soybean varieties resilient to drought and waterlogging by optimizing osmoprotectant production (Valliyodan *et al.*, 2017). Disease resistance has also benefited from AI-optimized synthetic antimicrobial peptides and engineered immune receptors, which reduce pathogen load while minimizing off-target effects (Jones *et al.*, 2016).

Central to these advancements is the meticulous design of synthetic pathways, beginning with identifying and characterizing desired traits through a profound understanding of plant physiology, genetics, and metabolic networks, often informed by comprehensive omics data. Pathway design follows, involving the selection of suitable enzymes or genes—potentially from other organisms or engineered for enhanced function—using computational tools like optimization-based methods and retrobiosynthesis to ensure feasibility and efficiency (Islam *et al.*, 2024). Computational modeling, including flux balance analysis and metabolic modeling, is indispensable for simulating the synthetic pathway's behavior, addressing gaps in plant metabolism knowledge, and optimizing flux (Gao and Zhao, 2024). Implementation employs cutting-edge genetic engineering techniques, notably CRISPR-Cas9 for precise genome editing, complemented by Agrobacterium-mediated transformation and transient expression systems to achieve correct expression within the plant's cellular milieu. Rigorous testing and optimization ensue, encompassing phenotyping, metabolite level assessments, and performance evaluations under diverse conditions, with iterative refinements addressing challenges like unintended metabolic interactions or suboptimal expression.

Successful applications underscore the potential of plant synthetic biology: enhancing photosynthesis has increased biomass production through modifications to the Calvin cycle or photorespiratory pathways (Batista-Silva *et al.*, 2020); engineering metabolic pathways has yielded novel compounds, such as antifungals with activity rivaling commercial fungicides (Calgaro-Kozina *et al.*, 2020); and synthetic directed evolution techniques have bolstered traits like drought resistance and yield using CRISPR-Cas-dependent and -independent technologies (Rao *et al.*, 2021). Advancements in computational approaches are driving the concept of designing synthetic pathways, which can help accelerate global efforts to improve crop productivity while addressing growing challenges to food security and the environment.

4.5.3 AI-guided CRISPR applications for targeted plant genome editing

CRISPR-Cas9, a revolutionary gene-editing tool, enables precise modifications to plant genomes by introducing targeted double-strand breaks, which are repaired through mechanisms like non-homologous end joining (NHEJ) or homology-directed repair (HDR). However, challenges such as variable editing efficiency, off-target effects, and the complexity of plant genomes necessitate advanced optimization strategies. AI, particularly ML, addresses these challenges by analyzing large-scale omics data, genomics, transcriptomics, and epigenomics, to predict and enhance editing outcomes. This integration is critical as global populations surge and climate change intensifies, demanding crops with superior yield, stress tolerance, and nutritional content.

One of the primary applications of AI in CRISPR is the design of highly efficient guide RNAs (gRNAs), which are essential for directing the Cas9 enzyme to the target DNA sequence. ML models leverage nucleotide composition, GC content, RNA secondary structure, and epigenetic states to predict gRNA activity (Fig. 4.8). For instance, the Rule Set 3 model incorporates the identity of the trans-activating CRISPR RNA (tracrRNA) and uses gradient boosting to achieve superior predictive accuracy, improving on-target cleavage while minimizing off-target effects (DeWeirdt *et al.*, 2022). Specialized tools like FORECasT-BE for base editors and DeepPE for prime editors further refine predictions for advanced editing techniques, ensuring precision in applications such as single nucleotide variant corrections in plants.

AI is also crucial in optimizing Cas protein variants, which are crucial for expanding the range of targetable sequences and improving specificity. The DeepSpCas9variants model, developed by Kim *et al.* (2020b), evaluates 13 different Cas9 variants across 26,891 target sequences to assess their specificity and compatibility with protospacer adjacent motifs (PAMs), enabling researchers to select the most suitable variant for a given application. Advanced structural prediction tools like AlphaFold2 provide accurate three-dimensional models of Cas proteins, facilitating the engineering of variants with enhanced properties, such as reduced off-target activity, which is vital for safe plant genome editing (Jumper *et al.*, 2021b).

Ensuring the specificity of genome editing is paramount to avoid unintended genetic

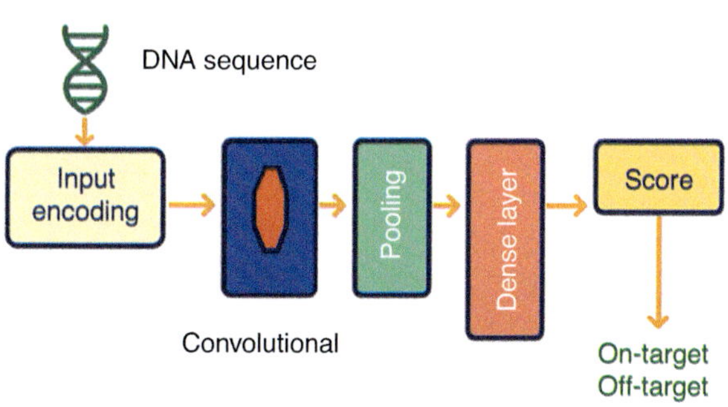

Fig. 4.8. Deep learning framework for predicting on-target and off-target effects of CRISPR gRNA. (Author's own image.)

changes, and AI-driven models are instrumental in this regard. ML approaches, such as Elevation, DeepCRISPR, and CnnCrispr, predict potential off-target sites by integrating sequence data and chromatin accessibility information (Gupta *et al.*, 2024). These models identify off-target genes using techniques such as GUIDE-seq and CIRCLE-seq, helping choose gRNAs capable of limiting off-target effects, thereby improving the safety and effectiveness of CRISPR technology in plants. This ability is crucial for regulatory approval and public acceptance of edited crops (Gupta *et al.*, 2024).

AI contributes significantly to forecasting the outcomes of genome editing events, predicting the types of insertions, deletions, or repairs that may occur post-editing. Models like FORECasT, SPROUT, inDelphi, and CROTON analyze DNA sequence features, such as GC content and microhomology length, to predict repair outcomes via NHEJ or microhomology-mediated end joining (MMEJ) pathways. The CROTON model, employing deep multi-task convolutional neural networks, achieves high accuracy in predicting editing results directly from raw sequences, which is essential for designing experiments aimed at specific genetic modifications, such as gene knockouts or precise insertions in crops (Dixit *et al.*, 2024).

The practical application of AI-guided CRISPR in crop breeding has accelerated the development of improved plant varieties. ML facilitates gene discovery by identifying genes associated with desirable agronomic traits. For example, the GRAiN model has been used to identify genes linked to drought resistance in rice, enabling targeted editing to enhance this trait, which is crucial for climate-resilient agriculture. Similarly, QTG-Finder2, developed by Lin *et al.* (2020), identified 13 candidate genes for plant height in rice, including the SD1 gene, known for its role in the Green Revolution, facilitating precise editing for improved stature and yield. Additionally, AI-driven analysis of cis-regulatory elements has led to tools like the CNN model for predicting tomato fruit ripening regulators and iCREPCP for identifying core promoter sequences, as demonstrated by Peleke *et al.* (2024) and Akagi *et al.* (2022), enhancing the ability to fine-tune gene expression for traits like fruit quality and stress tolerance (Zhao *et al.*, 2024).

Epigenome editing, which involves modifying DNA methylation or histone marks without altering the genetic sequence, can benefit from AI-optimized gRNA design, offering a non-invasive approach to enhance gene regulation. Fusing ML with single-cell CRISPR screening approaches like Perturb-Seq and CROP-seq may allow for a deeper understanding of the genetics underlying plant phenotypes that are determined by many genes working together (Chen *et al.*, 2024a). Recent breakthroughs in computational tools allow us to develop synthetic biological pathways. These can dramatically speed up actions towards boosting crop yields and meeting rising demands for food and nutrition, as well as managing agricultural impacts (Li *et al.*, 2021). These advancements will likely contribute significantly to sustainable agriculture and global food security (Chen *et al.*, 2024b).

The integration of AI with CRISPR technology is poised to unlock new frontiers in plant genome editing. Figure 4.9 shows how AI tools enhance every step of CRISPR-Cas9 genome editing in crops. AI-assisted gRNA design, Cas9 protein engineering, off-target and repair-outcome prediction, AI-driven gene discovery, and regulatoryelement mapping are all feeding into future applications such as epigenome editing and single-cell CRISPR screens.

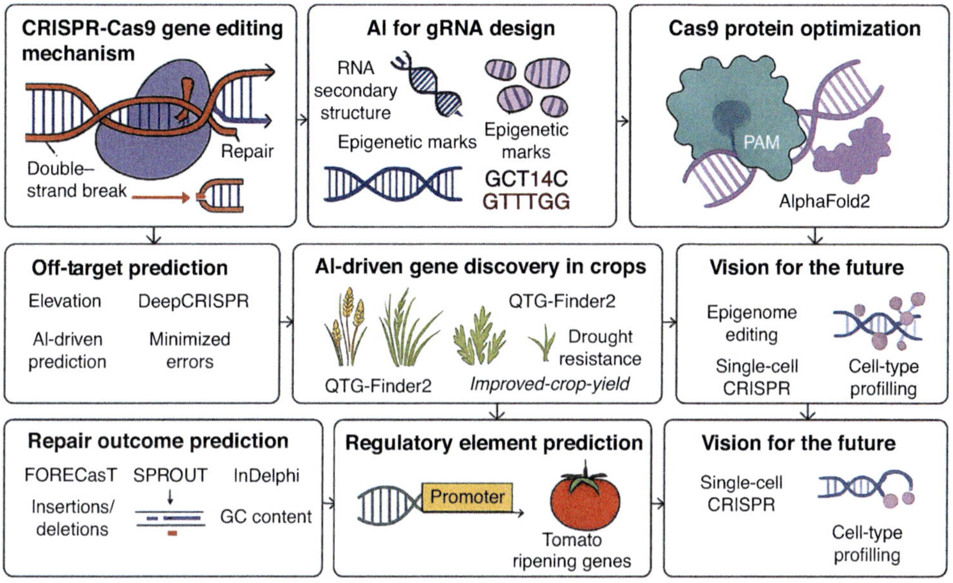

Fig. 4.9. AI-guided CRISPR applications for targeted plant genome engineering. (Author's own image.)

References

Akagi, T., Masuda, K., Kuwada, E., Takeshita, K., Kawakatsu, T. *et al.* (2022) Genome-wide cis-decoding for expression design in tomato using cistrome data and explainable deep learning. *The Plant Cell* 34(6), 2174–2187. DOI: 10.1093/plcell/koac079.

Al-Tamimi, N., Brien, C., Oakey, H., Berger, B., Saade, S. *et al.* (2016) Salinity tolerance loci revealed in rice using high-throughput non-invasive phenotyping. *Nature Communications* 7, 13342. DOI: 10.1038/ncomms13342.

Asaari, M.S.M., Mertens, S., Dhondt, S., Inzé, D., Wuyts, N. *et al.* (2019) Analysis of hyperspectral images for detection of drought stress and recovery in maize plants in a high-throughput phenotyping platform. *Computers and Electronics in Agriculture* 162, 749–758. DOI: 10.1016/j.compag.2019.05.018.

Athanasiadis, G., Meijsen, J.J., Helenius, D., Schork, A.J., Ingason, A. *et al.* (2022) A comprehensive map of genetic relationships among diagnostic categories based on 48.6 million relative pairs from the Danish genealogy. *Proceedings of the National Academy of Sciences* 119(6), E2118688119. DOI: 10.1073/pnas.2118688119.

Bai, W., Li, C., Li, W., Wang, H., Han, X. *et al.* (2024) Machine learning assists prediction of genes responsible for plant specialized metabolite biosynthesis by integrating multi-omics data. *BMC Genomics* 25(1), 418. DOI: 10.1186/s12864-024-10258-6.

Bao, H., Zhao, J., Zhao, X., Zhao, C., Lu, X. *et al.* (2023) Prediction of plant secondary metabolic pathways using deep transfer learning. *BMC Bioinformatics* 24(1), 348. DOI: 10.1186/s12859-023-05485-9.

Batista-Silva, W., Fonseca-Pereira, P., Martins, A.O., Zsögön, A., Nunes-Nesi, A. *et al.* (2020) Engineering improved photosynthesis in the era of synthetic biology. *Plant Communications* 1(2), 100032. DOI: 10.1016/j.xplc.2020.100032.

Beauchêne, K., Leroy, F., Fournier, A., Huet, C., Bonnefoy, M. *et al.* (2019) Management and characterization of abiotic stress via phénofield®, a high-throughput field phenotyping platform. *Frontiers in Plant Science* 10, 904. DOI: 10.3389/fpls.2019.00904.

Bergsträsser, S., Fanourakis, D., Schmittgen, S., Cendrero-Mateo, M.P., Jansen, M. *et al.* (2015) HyperART: Non-invasive quantification of leaf traits using hyperspectral absorption-reflectance-transmittance imaging. *Plant Methods* 11(1), 1. DOI: 10.1186/s13007-015-0043-0.

Boukhris, A., Jilali, A. and Sadiq, A. (2024) Satellite imagery, big data, iot and deep learning techniques for wheat yield prediction in morocco. *Results in Control and Optimization* 17, 100489. DOI: 10.1016/j.rico.2024.100489.

Boureau, T. (2020) *PHENOTIC Platform*. Portail Data INRAE.

Brichet, N., Fournier, C., Turc, O., Strauss, O., Artzet, S. *et al.* (2017) A robot-assisted imaging pipeline for tracking the growths of maize ear and silks in a high-throughput phenotyping platform. *Plant Methods* 13(1), 1–12. DOI: 10.1186/s13007-017-0246-7.

Busemeyer, L., Mentrup, D., Möller, K., Wunder, E., Alheit, K. *et al.* (2013) BreedVision—a multi-sensor platform for non-destructive field-based phenotyping in plant breeding. *Sensors* 13(3), 2830–2847. DOI: 10.3390/s130302830.

Cabrera-Bosquet, L., Fournier, C., Brichet, N., Welcker, C., Suard, B. *et al.* (2016) High-throughput estimation of incident light, light interception and radiation-use efficiency of thousands of plants in a phenotyping platform. *New Phytologist* 212(1), 269–281. DOI: 10.1111/nph.14027.

Calgaro-Kozina, A., Vuu, K.M., Keasling, J.D., Loqué, D., Sattely, E.S. *et al.* (2020) Engineering plant synthetic pathways for the biosynthesis of novel antifungals. *ACS Central Science* 6(8), 1394–1400. DOI: 10.1021/acscentsci.0c00241.

Cao, J., Zhang, Z., Luo, Y., Zhang, L., Zhang, J. *et al.* (2021a) Wheat yield predictions at a county and field scale with deep learning, machine learning, and google earth engine. *European Journal of Agronomy* 123, 126204. DOI: 10.1016/j.eja.2020.126204.

Cao, J., Zhang, Z., Tao, F., Zhang, L., Luo, Y. *et al.* (2021b) Integrating multi-source data for rice yield prediction across China using machine learning and deep learning approaches. *Agricultural and Forest Meteorology* 297, 108275. DOI: 10.1016/j.agrformet.2020.108275.

Cao, Y., Yuan, P., Xu, H., Martínez-Ortega, J.F., Feng, J. *et al.* (2022) Detecting asymptomatic infections of rice bacterial leaf blight using hyperspectral imaging and 3-dimensional convolutional neural network with spectral dilated convolution. *Frontiers in Plant Science* 13, 963170. DOI: 10.3389/fpls.2022.963170.

Cartolano, A., Cuzzocrea, A. and Pilato, G. (2024) Analyzing and assessing explainable AI models for smart agriculture environments. *Multimedia Tools and Applications* 83(12), 37225–37246. DOI: 10.1007/s11042-023-17978-z.

Chen, F., Chen, L., Yan, Z., Xu, J., Feng, L. *et al.* (2024a) Recent advances of CRISPR-based genome editing for enhancing staple crops. *Frontiers in Plant Science* 15. DOI: 10.3389/fpls.2024.1478398.

Chen, J., Zhou, J., Li, Q., Li, H., Xia, Y. *et al.* (2023a) CropQuant-air: An AI-powered system to enable phenotypic analysis of yield- and performance-related traits using wheat canopy imagery collected by low-cost drones. *Frontiers in Plant Science* 14, 1219983. DOI: 10.3389/fpls.2023.1219983.

Chen, L., Liu, G. and Zhang, T. (2024b) Integrating machine learning and genome editing for crop improvement. *aBIOTECH* 5(2), 262–277. DOI: 10.1007/s42994-023-00133-5.

Chen, Y., Guo, Y., Guan, P., Wang, Y., Wang, X. *et al.* (2023b) A wheat integrative regulatory network from large-scale complementary functional datasets enables trait-associated gene discovery for crop improvement. *Molecular Plant* 16(2), 393–414. DOI: 10.1016/j.molp.2022.12.019.

Cheung, C.Y.M., Poolman, M.G., Fell, David., Ratcliffe, R.G. and Sweetlove, L.J. (2014) A diel flux balance model captures interactions between light and dark metabolism during day-night cycles in C3 and crassulacean acid metabolism leaves. *Plant Physiology* 165(2), 917–929. DOI: 10.1104/pp.113.234468.

Choi, J., Byun, S.-W., Chae, N., Lim, J.H., Lim, T. *et al.* (2024) A study on phenotype prediction using an artificial intelligence-based data augmentation approach. In: *2024 IEEE International Conference on Big Data (BigData)*, IEEE, pp. 4679–4684. DOI: 10.1109/BigData62323.2024.10825288.

Chowdhury, N.B., Schroeder, W.L., Sarkar, D., Amiour, N., Quilleré, I. *et al.* (2021) Dissecting the metabolic reprogramming of maize root under nitrogen-deficient stress conditions. *Journal of Experimental Botany* 73(1), 275–291. DOI: 10.1093/jxb/erab435.

Crain, J.L., Wei, Y., Barker III, J., Thompson, S.M. and Alderman, P.D. (2016) Development and deployment of a portable field phenotyping platform. *Crop Science* 56(3), 965–975. DOI: 10.2135/cropsci2015.05.0290.

Crossa, J., Pérez-Rodríguez, P., Cuevas, J., Montesinos-López, O., Jarquín, D. *et al.* (2017) Genomic selection in plant breeding: Methods, models, and perspectives. *Trends in Plant Science* 22(11), 961–975. DOI: 10.1016/j.tplants.2017.08.011.

Czedik-Eysenberg, A., Seitner, S., Güldener, U., Koemeda, S., Jez, J. *et al.* (2018) The 'phenobox', a flexible, automated, open-source plant phenotyping solution. *The New Phytologist* 219(2), 808–823. DOI: 10.1111/nph.15129.

De Castro, P.D. (2025) Use of drones enhances genetic improvement of drought-tolerant corn plants [Online]. Available at: https://Jornal.Unicamp.Br/En/Noticias/2025/01/06/Uso-de-drones-potencializa-melhoramento-genetico-de-plantas-de-milho-tolerantes-a-seca/

De Oliveira Dal'Molin, C.G., Quek, L.-E., Palfreyman, R.W., Brumbley, S.M. and Nielsen, L.K. (2010) AraGEM, a genome-scale reconstruction of the primary metabolic network in arabidopsis. *Plant Physiology* 152(2), 579–589. DOI: 10.1104/pp.109.148817.

DeBruin, J., Aref, T., Tirado Tolosa, S., Hensley, R., Underwood, H. *et al.* (2025) Breaking the field phenotyping bottleneck in maize with autonomous robots. *Communications Biology* 8(1), 467. DOI: 10.1038/s42003-025-07890-7.

Deery, D., Jimenez-Berni, J., Jones, H., Sirault, X. and Furbank, R. (2014) Proximal remote sensing buggies and potential applications for field-based phenotyping. *Agronomy* 4(3), 349–379. DOI: 10.3390/agronomy4030349.

DeWeirdt, P.C., McGee, A.V., Zheng, F., Nwolah, I., Hegde, M. *et al.* (2022) Accounting for small variations in the tracrRNA sequence improves sgRNA activity predictions for CRISPR screening. *Nature Communications* 13(1), 5255. DOI: 10.1038/s41467-022-33024-2.

Dixit, S., Kumar, A., Srinivasan, K., Vincent, P.M.D.R. and Ramu Krishnan, N. (2024) Advancing genome editing with artificial intelligence: opportunities, challenges, and future directions. *Frontiers in Bioengineering and Biotechnology* 11, 1335901. DOI: 10.3389/fbioe.2023.1335901.

Dussarrat, T., Prigent, S., Latorre, C., Bernillon, S., Flandin, A. *et al.* (2022) Predictive metabolomics of multiple Atacama plant species unveils a core set of generic metabolites for extreme climate resilience. *The New Phytologist* 234(5), 1614–1628. DOI: 10.1111/nph.18095.

Earles, J.M., Knipfer, T., Tixier, A., Orozco, J., Reyes, C. *et al.* (2018) *In vivo* quantification of plant starch reserves at micrometer resolution using x-ray microct imaging and machine learning. *The New Phytologist* 218, 1260–1269. DOI: 10.1111/nph.15068.

Escorcia-Rodríguez, J.M., Gaytan-Nuñez, E., Hernandez-Benitez, E.M., Zorro-Aranda, A., Tello-Palencia, M.A. *et al.* (2023) Improving gene regulatory network inference and assessment: the importance of using network structure. *Frontiers in Genetics* 14, 1143382. DOI: 10.3389/fgene.2023.1143382.

Fahlgren, N., Feldman, M., Gehan, M.A., Wilson, M.S., Shyu, C. *et al.* (2015a) A versatile phenotyping system and analytics platform reveals diverse temporal responses to water availability in setaria. *Molecular Plant* 8(10), 1520–1535. DOI: 10.1016/j.molp.2015.06.005.

Fahlgren, N., Gehan, M.A. and Baxter, I. (2015b) Lights, camera, action: High-throughput plant phenotyping is ready for a close-up. *Current Opinion in Plant Biology* 24, 93–99. DOI: 10.1016/j.pbi.2015.02.006.

Farooq, M.A., Gao, S., Hassan, M.A., Huang, Z., Rasheed, A. *et al.* (2024) Artificial intelligence in plant breeding. *Trends in Genetics* 40(10), 891–908. DOI: 10.1016/j.tig.2024.07.001.

Flood, P.J., Kruijer, W., Schnabel, S.K., Schoor, R., Jalink, H. *et al.* (2016) Phenomics for photosynthesis, growth and reflectance in *Arabidopsis thaliana* reveals circadian and long-term fluctuations in heritability. *Plant Methods* 12, 14. DOI: 10.1186/s13007-016-0113-y.

Fukatsu, T., Watanabe, T., Hu, H., Yoichi, H. and Hirafuji, M. (2012) Field monitoring support system for the occurrence of *Leptocorisa chinensis* dallas (Hemiptera: Alydidae) using synthetic attractants, field servers, and image analysis. *Computers and Electronics in Agriculture* 80, 8–16. DOI: 10.1016/j.compag.2011.10.005.

Furbank, R.T. and Tester, M. (2011) Phenomics--technologies to relieve the phenotyping bottleneck. *Trends in Plant Science* 16(12), 635–644. DOI: 10.1016/j.tplants.2011.09.005.

Gao, Y. and Zhao, C. (2024) Development and applications of metabolic models in plant multi-omics research. *Frontiers in Plant Science* 15, 1361183. DOI: 10.3389/fpls.2024.1361183.

Gill, T., Gill, S.K., Saini, D.K., Chopra, Y., Koff, J.P. *et al.* (2022) A comprehensive review of high throughput phenotyping and machine learning for plant stress phenotyping (3), 156–183. DOI: 10.1007/s43657-022-00048-z.

Gosseau, F., Blanchet, N., Varès, D., Burger, P., Campergue, D. *et al.* (2019) Heliaphen, an outdoor high-throughput phenotyping platform for genetic studies and crop modeling. *Frontiers in Plant Science* 9. DOI: 10.3389/fpls.2018.01908.

Granier, C., Aguirrezabal, L., Chenu, K., Cookson, S.J., Dauzat, M. *et al.* (2006) PHENOPSIS, an automated platform for reproducible phenotyping of plant responses to soil water deficit in *Arabidopsis thaliana* permitted the identification of an accession with low sensitivity to soil water deficit. *New Phytologist* 169(3), 623–635. DOI: 10.1111/j.1469-8137.2005.01609.x.

Guo, H., Lv, X., Li, Y. and Li, M. (2023) Attention-based GCN integrates multi-omics data for breast cancer subtype classification and patient-specific gene marker identification. *Briefings in Functional Genomics* 22(5), 463–474. DOI: 10.1093/bfgp/elad013.

Gupta, D.K., Pagani, A., Zamboni, P. and Singh, A.K. (2024) AI-powered revolution in plant sciences: Advancements, applications, and challenges for sustainable agriculture and food security. *Exploration of Foods and Foodomics* 2(5), 443–459. DOI: 10.37349/eff.2024.00045.

Hairmansis, A., Berger, B., Tester, M. and Roy, S.J. (2014) Image-based phenotyping for non-destructive screening of different salinity tolerance traits in rice. *Rice* 7(1), 16. DOI: 10.1186/s12284-014-0016-3.

Halperin, O., Gebremedhin, A., Wallach, R. and Moshelion, M. (2017) High-throughput physiological phenotyping and screening system for the characterization of plant-environment interactions. *The Plant Journal* 89(4), 839–850. DOI: 10.1111/tpj.13425.

Harfouche, A.L., Jacobson, D.A., Kainer, D., Romero, J.C., Harfouche, A.H. *et al.* (2019) Accelerating climate resilient plant breeding by applying next-generation artificial intelligence. *Trends in Biotechnology* 37(11), 1217–1235. DOI: 10.1016/j.tibtech.2019.05.007.

Holman, F., Riche, A., Michalski, A., Castle, M., Wooster, M. *et al.* (2016) High throughput field phenotyping of wheat plant height and growth rate in field plot trials using UAV based remote sensing. *Remote Sensing* 8(12), 1031. DOI: 10.3390/rs8121031.

Huynh-Thu, V.A., Irrthum, A., Wehenkel, L. and Geurts, P. (2010) Inferring regulatory networks from expression data using tree-based methods. *PLoS ONE* 5(9), e12776. DOI: 10.1371/journal.pone.0012776.

Islam, M.T., Liu, Y., Hassan, M.M., Abraham, P.E., Merlet, J. *et al.* (2024) Advances in the application of single-cell transcriptomics in plant systems and synthetic biology. *Biodesign Research* 6, 0029. DOI: 10.34133/bdr.0029.

Jabed, M.A. and Azmi Murad, M.A. (2024) Crop yield prediction in agriculture: A comprehensive review of machine learning and deep learning approaches. *Biodesign Research* 10, E40836. DOI: 10.1016/j.heliyon.2024.e40836.

Jafar, A., Bibi, N., Naqvi, R.A., Sadeghi-Niaraki, A. and Jeong, D. (2024) Revolutionizing agriculture with artificial intelligence: plant disease detection methods, applications, and their limitations. *Frontiers in Plant Science* 15, 1356260. DOI: 10.3389/fpls.2024.1356260.

Jansen, M., Gilmer, F., Biskup, B., Nagel, K.A., Rascher, U. *et al.* (2009) Simultaneous phenotyping of leaf growth and chlorophyll fluorescence via GROWSCREEN FLUORO allows detection of stress tolerance in *Arabidopsis thaliana* and other rosette plants. *Functional Plant Biology* 36(11), 902–914. DOI: 10.1071/FP09095.

Jeong, S., Ko, J. and Yeom, J.-M. (2022) Predicting rice yield at pixel scale through synthetic use of crop and deep learning models with satellite data in south and North Korea. *The Science of the Total Environment* 802, 149726. DOI: 10.1016/j.scitotenv.2021.149726.

Jeudy, C., Adrian, M., Baussard, C., Bernard, C., Bernaud, E. *et al.* (2016) RhizoTubes as a new tool for high throughput imaging of plant root development and architecture: test, comparison with pot grown plants and validation. *Plant Methods* 12, 1–18. DOI: 10.1186/s13007-016-0131-9.

Jhajharia, K., Mathur, P., Jain, S. and Nijhawan, S. (2023) Crop yield prediction using machine learning and deep learning techniques. *Procedia Computer Science* 218, 406–417. DOI: 10.1016/j.procs.2023.01.023.

Jian, W., Ou, X., Sun, L., Chen, Y., Liu, S. *et al.* (2023) Characterization of anthocyanin accumulation, nutritional properties, and postharvest attributes of transgenic purple tomato. *Food Chemistry* 408, 135181. DOI: 10.1016/j.foodchem.2022.135181.

Johansen, K., Morton, M.J.L., Malbeteau, Y.M., Aragon, B., Al-Mashharawi, S.K. *et al.* (2019) Unmanned aerial vehicle-based phenotyping using morphometric and spectral analysis can quantify responses of wild tomato plants to salinity stress. *Frontiers in Plant Science* 10, 370. DOI: 10.3389/fpls.2019.00370.

Jones, J.D.G., Vance, R.E. and Dangl, J.L. (2016) Intracellular innate immune surveillance devices in plants and animals. *Science* 354(6316), aaf6395. DOI: 10.1126/science.aaf6395.

Jumper, J., Evans, R., Pritzel, A., Green, T., Figurnov, M. *et al.* (2021a) Highly accurate protein structure prediction with alphafold. *Nature* 596(7873), 583–589. DOI: 10.1038/s41586-021-03819-2.

Jumper, J., Evans, R., Pritzel, A., Green, T., Figurnov, M. *et al.* (2021b) Highly accurate protein structure prediction with alphafold. *Nature* 596(7873), 583–589. DOI: 10.1038/s41586-021-03819-2.

Kamilaris, A. and Prenafeta-Boldú, F.X. (2018) Deep learning in agriculture: A survey. *Computers and Electronics in Agriculture* 147, 70–90. DOI: 10.1016/j.compag.2018.02.016.

Khaki, S., Wang, L. and Archontoulis, S.V. (2020a) A CNN-RNN framework for crop yield prediction. *Frontiers in Plant Science* 10, 1750. DOI: 10.3389/fpls.2019.01750.

Khaki, S., Wang, L. and Archontoulis, S.V. (2020b) A CNN-RNN framework for crop yield prediction. *Frontiers in Plant Science* 10. DOI: 10.3389/fpls.2019.01750.

Khaki, S., Wang, L. and Archontoulis, S.V. (2020c) A CNN-RNN framework for crop yield prediction. *Frontiers in Plant Science* 10, 2019. DOI: 10.3389/fpls.2019.01750.

Khalid, M., Amjad, I., Fatima, I., Ammar, A. and Rasheed, Z. (2023) HIGH throughput phenotyping: A revolutionary approach to combat salinity stress in cotton. *Trends in Biotechnology and Plant Sciences* 1, 7–15. DOI: 10.62460/TBPS/2023.008.

Kim, J., Kim, K.-S., Kim, Y. and Chung, Y.S. (2020a) A short review: Comparisons of high-throughput phenotyping methods for detecting drought tolerance. *Scientia Agricola* 78(4), E20190300300. DOI: 10.1590/1678-992x-2019-0300.

Kim, M., Lee, C., Hong, S., Kim, S.L., Baek, J.-H. *et al.* (2021) High-throughput phenotyping methods for breeding drought-tolerant crops. *International Journal of Molecular Sciences* 22(15), 8266. DOI: 10.3390/ijms22158266.

Kim, N., Ha, K.-J., Park, N.-W., Cho, J., Hong, S. *et al.* (2019) A comparison between major artificial intelligence models for crop yield prediction: Case study of the Midwestern United States, 2006–2015. *ISPRS International Journal of Geo-Information* 8(5), 240. DOI: 10.3390/ijgi8050240.

Kim, N., Kim, H.K., Lee, S., Seo, J.H., Choi, J.W. *et al.* (2020b) Prediction of the sequence-specific cleavage activity of cas9 variants. *Nature Biotechnology* 38(11), 1328–1336. DOI: 10.1038/s41587-020-0537-9.

Kim, S.L., Kim, N., Lee, H., Lee, E., Cheon, K.-S. *et al.* (2020c) High-throughput phenotyping platform for analyzing drought tolerance in rice. *Planta* 252(3), 38. DOI: 10.1007/s00425-020-03436-9.

Kirchgessner, N., Liebisch, F., Yu, K., Pfeifer, J., Friedli, M. *et al.* (2016) The ETH field phenotyping platform FIP: A cable-suspended multi-sensor system. *Functional Plant Biology* 44(1), 154–168. DOI: 10.1071/FP16165.

Kisiel, A., Krzemińska, A., Cembrowska-Lech, D. and Miller, T. (2023) Data science and plant metabolomics. *Metabolites* 13(3), 454. DOI: 10.3390/metabo13030454.

Kloth, K.J., Ten Broeke, C.J., Thoen, M.P., Hanhart-Van Den Brink, M., Wiegers, G.L. *et al.* (2015) High-throughput phenotyping of plant resistance to aphids by automated video tracking. *Plant Methods* 11, 4. DOI: 10.1186/s13007-015-0044-z.

König-Mattern, L., Sanchez Medina, E.I., Komarova, A.O., Linke, S., Rihko-Struckmann, L. *et al.* (2024) Machine learning-supported solvent design for lignin-first biorefineries and lignin upgrading. *Chemical Engineering Journal* 495, 153524. DOI: 10.1016/j.cej.2024.153524.

Kou, J., Duan, L., Yin, C., Ma, L., Chen, X. *et al.* (2022) Predicting leaf nitrogen content in cotton with UAV RGB images. *Sustainability* 14(15), 9259. DOI: 10.3390/su14159259.

Kulmanov, M. and Hoehndorf, R. (2020) DeepGOPlus: Improved protein function prediction from sequence. *Bioinformatics* 36(2), 422–429. DOI: 10.1093/bioinformatics/btz595.

Kulmanov, M., Khan, M.A. and Hoehndorf, R. (2018) DeepGO: Predicting protein functions from sequence and interactions using a deep ontology-aware classifier. *Bioinformatics* 34(4), 660–668. DOI: 10.1093/bioinformatics/btx624.

Langfelder, P. and Horvath, S. (2008) WGCNA: An R package for weighted correlation network analysis. *BMC Bioinformatics* 9, 1–13. DOI: 10.1186/1471-2105-9-559.

Le Marié, C., Kirchgessner, N., Flütsch, P., Pfeifer, J., Walter, A. *et al.* (2016) RADIX: Rhizoslide platform allowing high throughput digital image analysis of root system expansion. *Plant Methods* 12(1), 40. DOI: 10.1186/s13007-016-0140-8.

Lebauer, D., Burnette, M.A., Demieville, J., Fahlgren, N., French, A.N. *et al.* (2020) Terra-ref, an open reference data set from high resolution genomics, phenomics, and imaging sensors. *Dryad.* DOI: 10.5061/dryad.4b8gtht99.

Lee, S. and Yun, C.M. (2023) A deep learning model for predicting risks of crop pests and diseases from sequential environmental data. *Plant Methods* 19(1), 145. DOI: 10.1186/s13007-023-01122-x.

Li, Y., Chen, J., Tsai, S.Q. and Cheng, Y. (2021) Easy-prime: A machine learning–based prime editor design tool. *Genome Biology* 22(1), 235. DOI: 10.1186/s13059-021-02458-0.

Liebisch, F., Kirchgessner, N., Schneider, D., Walter, A. and Hund, A. (2015) Remote, aerial phenotyping of maize traits with a mobile multi-sensor approach. *Plant Methods* 11, 9. DOI: 10.1186/s13007-015-0048-8.

Lin, F., Lazarus, E.Z. and Rhee, S.Y. (2020) QTG-finder2: A generalized machine-learning algorithm for prioritizing QTL causal genes in plants. *G3: Genes|Genomes|Genetics* 10(7), 2411–2421. DOI: 10.1534/g3.120.401122.

Liu, S., Barrow, C.S., Hanlon, M., Lynch, J.P. and Bucksch, A. (2021) DIRT/3D: 3D root phenotyping for field-grown maize (*Zea mays*). *Plant Physiology* 187(2), 739–757. DOI: 10.1093/plphys/kiab311.

Long, M., Hartley, M., Morris, R.J. and Brown, J.K.M. (2023) Classification of wheat diseases using deep learning networks with field and glasshouse images. *Plant Pathology* 72(3), 536–547. DOI: 10.1111/ppa.13684.

Ma, W., Qiu, Z., Song, J., Li, J., Cheng, Q. *et al.* (2018) A deep convolutional neural network approach for predicting phenotypes from genotypes. *Planta* 248(5), 1307–1318. DOI: 10.1007/s00425-018-2976-9.

Mahood, E.H., Kruse, L.H. and Moghe, G.D. (2020) Machine learning: A powerful tool for gene function prediction in plants. *Applications in Plant Sciences* 8(7), e11376. DOI: 10.1002/aps3.11376.

Maphosa, L., Thoday-Kennedy, E., Vakani, J., Phelan, A., Badenhorst, P. *et al.* (2016) Phenotyping wheat under salt stress conditions using a 3D laser scanner. *Israel Journal of Plant Sciences* 1–8. DOI: 10.1080/07929978.2016.1243405.

Maraveas, C. (2024) Image analysis artificial intelligence technologies for plant phenotyping: Current state of the art. *AgriEngineering* 6(3), 3375–3407. DOI: 10.3390/agriengineering6030193.

Martins, S.M., Brito, G.G.D., Gonçalves, W.D.C., Tripode, B.M.D., Lartaud, M. *et al.* (2019) PhenoRoots: An inexpensive non-invasive phenotyping system to assess the variability of the root system architecture. *Scientia Agricola* 77(5), E20180420. DOI: 10.1590/1678-992x-2018-0420.

Mathieu, L., Lobet, G., Tocquin, P. and Périlleux, C. (2015) Rhizoponics": A novel hydroponic rhizotron for root system analyses on mature *Arabidopsis thaliana* plants. *Plant Methods* 11, 3. DOI: 10.1186/s13007-015-0046-x.

Meraj, T., Sharif, M.I., Raza, M., Alabrah, A., Kadry, S. *et al.* (2024) Computer vision-based plants phenotyping: A comprehensive survey. *iScience* 27(1), 108709. DOI: 10.1016/j.isci.2023.108709.

Mérida-García, R., Gálvez, S., Solís, I., Martínez-Moreno, F., Camino, C. *et al.* (2024) High-throughput phenotyping using hyperspectral indicators supports the genetic dissection of yield in durum wheat grown under heat and drought stress. *Frontiers in Plant Science* 15, 1470520. DOI: 10.3389/fpls.2024.1470520.

Min, S., Lee, B. and Yoon, S. (2017) Deep learning in bioinformatics. *Briefings in Bioinformatics* 18, 851–869. DOI: 10.1093/bib/bbw068.

Mohanty, S.P., Hughes, D.P. and Salathé, M. (2016) Using deep learning for image-based plant disease detection. *Frontiers in Plant Science* 7, 1419. DOI: 10.3389/fpls.2016.01419.

Montesinos-López, O.A., Martín-Vallejo, J., Crossa, J., Gianola, D., Hernández-Suárez, C.M. *et al.* (2019) A benchmarking between deep learning, support vector machine and Bayesian threshold best linear unbiased prediction for predicting ordinal traits in plant breeding. *G3: Genes|Genomes|Genetics* 9(2), 601–618. DOI: 10.1534/g3.118.200998.

Mostafa, S., Mondal, D., Panjvani, K., Kochian, L. and Stavness, I. (2023) Explainable deep learning in plant phenotyping. *Frontiers in Artificial Intelligence* 6, 1203546. DOI: 10.3389/frai.2023.1203546.

Murmu, S., Sinha, D., Chaurasia, H., Sharma, S., Das, R. *et al.* (2024) A review of artificial intelligence-assisted omics techniques in plant defense: Current trends and future directions. *Frontiers in Plant Science* 15, 1292054. DOI: 10.3389/fpls.2024.1292054.

Ninomiya, S. (2022) High-throughput field crop phenotyping: Current status and challenges. *Breeding Science* 72(1), 3–18. DOI: 10.1270/jsbbs.21069.

Niu, S., Nie, Z., Li, G. and Zhu, W. (2024) Early drought detection in maize using UAV images and YOLOv8+. *Drones* 8(5), 170. DOI: 10.3390/drones8050170.

Okyere, F.G., Cudjoe, D.K., Virlet, N., Castle, M., Riche, A.B. *et al.* (2024) Hyperspectral imaging for phenotyping plant drought stress and nitrogen interactions using multivariate modeling and machine learning techniques in wheat. *Remote Sensing* 16(18), 3446. DOI: 10.3390/rs16183446.

Paine, J.A., Shipton, C.A., Chaggar, S., Howells, R.M., Kennedy, M.J. *et al.* (2005) Improving the nutritional value of golden rice through increased pro-vitamin a content. *Nature Biotechnology* 23(4), 482–487. DOI: 10.1038/nbt1082.

Pan, Q., Gao, M., Wu, P., Yan, J. and Li, S. (2021) A deep-learning-based approach for wheat yellow rust disease recognition from unmanned aerial vehicle images. *Sensors (Basel)* 21(19), 6540. DOI: 10.3390/s21196540.

Parmley, K., Nagasubramanian, K., Sarkar, S., Ganapathysubramanian, B. and Singh, A.K. (2019) Development of optimized phenomic predictors for efficient plant breeding decisions using phenomic-assisted selection in soybean. *Plant Phenomics* 5809404. DOI: 10.34133/2019/5809404.

Patra, A.K. and Sahoo, L. (2024) Explainable light-weight deep learning pipeline for improved drought stress identification. *Frontiers in Plant Science* 15, 1476130. DOI: 10.3389/fpls.2024.1476130.

Peleke, F.F., Zumkeller, S.M., Gültas, M., Schmitt, A. and Szymański, J. (2024) Deep learning the cis-regulatory code for gene expression in selected model plants. *Nature Communications* 15(1), 3488. DOI: 10.1038/s41467-024-47744-0.

Pereira, H.D., Nonato, J.V.A., Duarte, R.C.R.M., Gerhardt, I.R., Dante, R.A. *et al.* (2025) Temporal field phenomics of transgenic maize events subjected to drought stress: Cross-validation scenarios and machine learning models. *The Plant Phenome Journal* 8(1), E70015. DOI: 10.1002/ppj2.70015.

Pérez-Ruiz, M., Prior, A., Martinez-Guanter, J., Apolo-Apolo, O.E., Andrade-Sanchez, P. *et al.* (2020) Development and evaluation of a self-propelled electric platform for high-throughput field phenotyping in wheat breeding trials. *Computers and Electronics in Agriculture* 169, 105237. DOI: 10.1016/j.compag.2020.105237.

Pignon, C.P., Fernandes, S.B., Valluru, R., Bandillo, N., Lozano, R. *et al.* (2021) Phenotyping stomatal closure by thermal imaging for GWAS and TWAS of water use efficiency-related genes. *Plant Physiology* 187(4), 2544–2562. DOI: 10.1093/plphys/kiab395.

Prasad, N.R., Patel, N.R. and Danodia, A. (2021) Crop yield prediction in cotton for regional level using random forest approach. *Spatial Information Research* 29(2), 195–206. DOI: 10.1007/s41324-020-00346-6.

Qin, W., Wang, J., Ma, L., Wang, F., Hu, N. *et al.* (2022) UAV-based multi-temporal thermal imaging to evaluate wheat drought resistance in different deficit irrigation regimes. *Remote Sensing* 14(21), 5608. DOI: 10.3390/rs14215608.

Rao, G.S., Jiang, W. and Mahfouz, M. (2021) Synthetic directed evolution in plants: Unlocking trait engineering and improvement. *Synthetic Biology* 6(1), ysab025. DOI: 10.1093/synbio/ysab025.

Sadok, W., Naudin, P., Boussuge, B., Muller, B., Welcker, C. *et al.* (2007) Leaf growth rate per unit thermal time follows QTA-dependent daily patterns in hundreds of maize lines under naturally fluctuating conditions. *Plant, Cell & Environment* 30(2), 135–146. DOI: 10.1111/j.1365-3040.2006.01611.x.

Salas Fernandez, M.G., Bao, Y., Tang, L. and Schnable, P.S. (2017) A high-throughput, field-based phenotyping technology for tall biomass crops. *Plant Physiology* 174(4), 2008–2022. DOI: 10.1104/pp.17.00707.

Shen, L., Ding, G., Jackson, R., Ali, M., Liu, S. *et al.* (2024) GSP-AI: An AI-powered platform for identifying key growth stages and the vegetative-to-reproductive transition in wheat using trilateral drone imagery and meteorological data. *Plant Phenomics* 6, 0255. DOI: 10.34133/plantphenomics.0255.

Shrivastava, P. and Kumar, R. (2015) Soil salinity: A serious environmental issue and plant growth promoting bacteria as one of the tools for its alleviation. *Saudi Journal of Biological Sciences* 22(2), 123–131. DOI: 10.1016/j.sjbs.2014.12.001.

Singh Boori, M., Choudhary, K., Paringer, R. and Kupriyanov, A. (2023) Machine learning for yield prediction in fergana valley, central Asia. *Journal of the Saudi Society of Agricultural Sciences* 22(2), 107–120. DOI: 10.1016/j.jssas.2022.07.006.

South, P.F., Cavanagh, A.P., Liu, H.W. and Ort, D.R. (2019) Synthetic glycolate metabolism pathways stimulate crop growth and productivity in the field. *Science* 363(6422), eaat9077. DOI: 10.1126/science.aat9077.

Su, Y., Wu, F., Ao, Z., Jin, S., Qin, F. *et al.* (2019) Evaluating maize phenotype dynamics under drought stress using terrestrial lidar. *Plant Methods* 15, 11. DOI: 10.1186/s13007-019-0396-x.

Sweetlove, L.J. and Fernie, A.R. (2018) The role of dynamic enzyme assemblies and substrate channelling in metabolic regulation. *Nature Communications* 9(1), 2136. DOI: 10.1038/s41467-018-04543-8.

Takou, M., Bellis, E.S. and Lasky, J.R. (2024) Predicting gene expression responses to environment in arabidopsis thaliana using natural variation in DNA sequence. *Evolutionary Biology*. DOI: 10.1101/2024.04.25.591174.

Tisné, S., Serrand, Y., Bach, L., Gilbault, E., Ben Ameur, R. *et al.* (2013) Phenoscope: An automated large-scale phenotyping platform offering high spatial homogeneity. *The Plant Journal* 74(3), 534–544. DOI: 10.1111/tpj.12131.

Tovignan, T.K., Basha, Y., Windpassinger, S., Augustine, S.M., Snowdon, R. *et al.* (2023) Precision phenotyping of agro-physiological responses and water use of sorghum under different drought scenarios. *Agronomy* 13(3), 722. DOI: 10.3390/agronomy13030722.

Upadhyay, A., Chandel, N.S., Singh, K.P., Chakraborty, S.K., Nandede, B.M. *et al.* (2025) Deep learning and computer vision in plant disease detection: A comprehensive review of techniques, models, and trends in precision agriculture. *Artificial Intelligence Review* 58(3), 92. DOI: 10.1007/s10462-024-11100-x.

Valliyodan, B., Ye, H., Song, L., Murphy, M., Shannon, J.G. *et al.* (2017) Genetic diversity and genomic strategies for improving drought and waterlogging tolerance in soybeans. *Journal of Experimental Botany* 68(8), 1835–1849. DOI: 10.1093/jxb/erw433.

Virlet, N., Sabermanesh, K., Sadeghi-Tehran, P. and Hawkesford, M.J. (2016) Field scanalyzer: An automated robotic field phenotyping platform for detailed crop monitoring. *Functional Plant Biology* 44(1), 143–153. DOI: 10.1071/FP16163.

Vrček, L., Bresson, X., Laurent, T., Schmitz, M., Kawaguchi, K. *et al.* (2025) Geometric deep learning framework for de novo genome assembly. *Genome Research* 35(4), 839–849. DOI: 10.1101/gr.279307.124.

Walsh, J.J., Mangina, E. and Negrão, S. (2024) Advancements in imaging sensors and AI for plant stress detection: a systematic literature review. *Plant Phenomics* 6, 0153. DOI: 10.34133/plantphenomics.0153.

Wang, J., Ma, A., Ma, Q., Xu, D. and Joshi, T. (2020a) Inductive inference of gene regulatory network using supervised and semi-supervised graph neural networks. *Computational and Structural Biotechnology Journal* 18, 3335–3343. DOI: 10.1016/j.csbj.2020.10.022.

Wang, K., Abid, M.A., Rasheed, A., Crossa, J., Hearne, S. *et al.* (2023) DNNGP, a deep neural network-based method for genomic prediction using multi-omics data in plants. *Molecular Plant* 16(1), 279–293. DOI: 10.1016/j.molp.2022.11.004.

Wang, Q., Li, C., Huang, L., Chen, L., Zheng, Q. *et al.* (2024) Research on rapeseed seedling counting based on an improved density estimation method. *Agriculture* 14(5), 783. DOI: 10.3390/agriculture14050783.

Wang, X., Huang, J., Feng, Q. and Yin, D. (2020b) Winter wheat yield prediction at county level and uncertainty analysis in main wheat-producing regions of china with deep learning approaches. *Remote Sensing* 12(11), 1744. DOI: 10.3390/rs12111744.

Wang, Y., Dai, X., Fu, D., Li, P. and Du, B. (2022a) PGD: A machine learning-based photosynthetic-related gene detection approach. *BMC Bioinformatics* 23(1), 183. DOI: 10.1186/s12859-022-04722-x.

Wang, Y., Hu, S., Ren, H., Yang, W. and Zhai, R. (2022b) 3DPhenoMVS: A low-cost 3D tomato phenotyping pipeline using 3D reconstruction point cloud based on multiview images. *Agronomy* 12(8), 1865. DOI: 10.3390/agronomy12081865.

Wu, S., Wen, W., Wang, Y., Fan, J., Wang, C. *et al.* (2020) MVS-pheno: A portable and low-cost phenotyping platform for maize shoots using multiview stereo 3D reconstruction. *Plant Phenomics* 1848437. DOI: 10.34133/2020/1848437.

Wurtzel, E.T., Vickers, C.E., Hanson, A.D., Millar, A.H., Cooper, M. *et al.* (2019) Revolutionizing agriculture with synthetic biology. *Nature Plants* 5(12), 1207–1210. DOI: 10.1038/s41477-019-0539-0.

Yan, J. and Wang, X. (2022) Unsupervised and semi-supervised learning: The next frontier in machine learning for plant systems biology. *The Plant Journal* 111(6), 1527–1538. DOI: 10.1111/tpj.15905.

Yan, T., Rashid, J., Saleem, M.S., Ahmad, S. and Faheem, M. (2024) A hybrid deep learning approach for green energy forecasting in Asian countries. *Computers, Materials & Continua* 81(2), 2685–2708. DOI: 10.32604/cmc.2024.058186.

Yang, H., Gu, F., Zhang, L. and Hua, X.-S. (2022) Using generative adversarial networks for genome variant calling from low depth ONT sequencing data. *Scientific Reports* 12(1), 8725. DOI: 10.1038/s41598-022-12346-7.

Yang, W., Guo, Z., Huang, C., Duan, L., Chen, G. *et al.* (2014) Combining high-throughput phenotyping and genome-wide association studies to reveal natural genetic variation in rice. *Nature Communications* 5, 5087. DOI: 10.1038/ncomms6087.

Yazdanbakhsh, N. and Fisahn, J. (2009) High throughput phenotyping of root growth dynamics, lateral root formation, root architecture and root hair development enabled by plarom. *Functional Plant Biology* 36(11), 938–946. DOI: 10.1071/FP09167.

Yu, W., Yang, G., Li, D., Zheng, H., Yao, X. *et al.* (2023) Improved prediction of rice yield at field and county levels by synergistic use of SAR, optical and meteorological data. *Agricultural and Forest Meteorology* 342, 109729. DOI: 10.1016/j.agrformet.2023.109729.

Zampieri, M., Sekar, K., Zamboni, N. and Sauer, U. (2017) Frontiers of high-throughput metabolomics. *Current Opinion in Chemical Biology* 36, 15–23. DOI: 10.1016/j.cbpa.2016.12.006.

Zhang, Z., Qu, Y., Ma, F., Lv, Q., Zhu, X. *et al.* (2024a) Integrating high-throughput phenotyping and genome-wide association studies for enhanced drought resistance and yield prediction in wheat. *New Phytologist* 243(5), 1758–1775. DOI: 10.1111/nph.19942.

Zhang, Z., Qu, Y., Ma, F., Lv, Q., Zhu, X. *et al.* (2024b) Integrating high-throughput phenotyping and genome-wide association studies for enhanced drought resistance and yield prediction in wheat. *New Phytologist* 243(5), 1758–1775. DOI: 10.1111/nph.19942.

Zhang, Z., Qu, Y., Ma, F., Lv, Q., Zhu, X. *et al.* (2024c) Integrating high-throughput phenotyping and genome-wide association studies for enhanced drought resistance and yield prediction in wheat. *New Phytologist* 243(5), 1758–1775. DOI: 10.1111/nph.19942.

Zhao, P., Feng, Y., Jiang, P. and Zhang, W. (2024) Identification and functional characterization of i-motifs in plants. *Trends in Plant Science* 29(4), 495–496. DOI: 10.1016/j.tplants.2024.02.002.

Zhou, J. and Troyanskaya, O.G. (2015) Predicting effects of noncoding variants with deep learning–based sequence model. *Nature Methods* 12(10), 931–934. DOI: 10.1038/nmeth.3547.

Zhou, J., Reynolds, D., Cornu, T.L., Websdale, D., Orford, S. *et al.* (2017) CropQuant: An automated and scalable field phenotyping platform for crop monitoring and trait measurements to facilitate breeding and digital agriculture. *Plant Biology*. DOI: 10.1101/161547.

Zhou, J., Mou, H., Zhou, J., Ali, M.L., Ye, H. *et al.* (2021) Qualification of soybean responses to flooding stress using UAV-based imagery and deep learning. *Plant Phenomics* 9892570. DOI: 10.34133/2021/9892570.

Zhu, F., Saluja, M., Dharni, J.S., Paul, P., Sattler, S.E. *et al.* (2021) Phenoimage : An open-source graphical user interface for plant image analysis. *The Plant Phenome Journal* 4(1). DOI: 10.1002/ppj2.20015.

5 Artificial Intelligence-Enabled Single-Cell Resolution and Spatial Plant Multi-omics

Hannah Elizabeth Stephen Babu[1], Mani Manoj[1], Pragya Pallavi[2], Agnishwar Girigoswami[3], Koyeli Girigoswami[3], Pemula Gowtham[3], Asirvatham Alwin Robert[4] and Arumugam Vijaya Anand[1]*

[1]Department of Human Genetics and Molecular Biology, Bharathiar University, Coimbatore, India; [2]Center for Global Health Research, Saveetha Institute of Medical and Technical Sciences, Chennai, India; [3]Faculty of Allied Health Sciences, Chettinad Academy of Research and Education, Chennai, India; [4]Department of Endocrinology and Diabetes, Prince Sultan Military Medical City, Ministry of Defense, Riyadh, Kingdom of Saudi Arabia

Abstract

The ability to investigate the complex cellular landscapes and regulatory systems within plants is made possible by single-cell and spatial omics methods, which are transforming the field of plant biology. The study of cell types and interactions in their natural surroundings is made easier by the preservation of geographical information provided by spatial transcriptomics, while single-cell RNA sequencing has exposed transcriptional heterogeneities. Modern developments, including spatially improved-resolution omics-sequencing, combines these techniques to get over the drawbacks of conventional methods and offer high-resolution insights into plant transcriptomes. Notwithstanding the potential of these technologies, there are still obstacles to overcome, such as difficulties with tissue dissection and data integration. Multi-omics applications in medicinal plants demonstrate the extensive implications of these technologies in basic and applied plant research, as well as the possibility of bio-manufacturing natural products. This chapter provides insights into the influence of artificial intelligence in single-cell and spatial plant multi-omics in providing the detailed output of data interpretation.

Keywords: Artificial intelligence, single cell, spatial omics, multi-omics, RNA sequencing

5.1 Introduction

Single-cell and spatial omics methods are revolutionizing plant biology by enabling researchers to explore the intricate cellular landscapes and regulatory mechanisms within plants. Single-cell RNA sequencing (scRNA-seq) has indicated transcriptional heterogeneities, while spatial transcriptomics allows for the preservation of spatial information, facilitating the study of cell types and interactions in their native environments (Chen *et al.*,

*Corresponding author: avahgmb@buc.edu.in

DOI: 10.1079/9781800628649.0005

2023a; Rhaman *et al.*, 2024). Recent advancements, such as spatially improved-resolution omics-sequencing (Stereo-seq), combine these approaches to overcome the limitations of traditional methods, providing high-resolution insights into plant transcriptomes (Bawa *et al.*, 2024). Despite the promise of these technologies, challenges remain, including difficulties in tissue dissection and data integration (Yin *et al.*, 2023). The benefit of multi-omics in medicinal plants further highlights the potential for bio manufacturing natural products, showcasing the broad implications of these technologies in both basic and applied plant studies (Chen *et al.*, 2024).

Single-cell and spatial omics techniques have revolutionized our knowledge of plant developmental biology by enabling detailed analysis of cell heterogeneity and regulatory mechanisms. These approaches allow researchers to identify distinct cell types and their specific gene expression patterns, as demonstrated in a spatially determined single-cell atlas of soybean, which uncovered 103 cell types and their regulatory networks (Zhang *et al.*, 2024). Single-cell multi-omics approaches facilitate the integration of transcriptomic, proteomic, and epigenomic data, providing a complete picture of cellular functions and interactions (Lv *et al.*, 2024). The application of spatial transcriptomics has further enhanced the ability to study cell-to-cell interactions and reconstruct cell-fate lineages, addressing challenges in plant tissue dissection. Overall, these technologies are crucial for elucidating the complexities of plant development and hold promise for future horticultural research (Zhang *et al.*, 2023).

5.1.1 Overview of single-cell omics

Single-cell omics has appeared as a transformative process in plant biology, enabling researchers to dissect the complex cellular heterogeneity and responses of plants to environmental stimuli. This technology encompasses various methodologies, including single-cell metabolomics, proteomics, and transcriptomics, which give high-dimensional insights into cellular functions and regulatory mechanisms at an unprecedented resolution. For instance, single-cell transcriptomics has revealed distinct cellular responses to abiotic and biotic factors, highlighting the variability in how different cell types react to environmental changes (Tenorio-Berrío and Dubois, 2024). Additionally, single-cell proteomics (SCP) has identified stress-related proteins in Arabidopsis, showcasing the dynamic nature of protein interactions under stress conditions (Fulcher *et al.*, 2024). Advances in single-cell mass spectrometry have allowed for the quantification of natural products at the cellular level, revealing their diverse distribution across plant tissues (Vu *et al.*, 2024). Collectively, these innovations underscore the prospect of single-cell omics to enhance our knowledge of plant biology and its intricate cellular networks (Redmond *et al.*, 2024).

Single-cell omics methods, such as SCP and single-cell assay for transposase-accessible chromatin with high-throughput sequencing (scATAC-seq), are pivotal in uncovering novel plant cell types and their functions. SCP allows for the resolution of cellular heterogeneity by quantifying thousands of proteins from single cells, enabling the differentiation of closely related cell types, as demonstrated in Arabidopsis root cells where distinct proteins were identified in cortex and endodermis cells (Montes *et al.*, 2024). This differentiation aids in identifying biomarkers and candidate proteins for functional genomics. scATAC-seq facilitates the characterization of chromatin availability, revealing cell-type-specific regulatory elements and enabling the recognition of unique cell types based on chromatin features (Lu *et al.*, 2024). Together, these methods provide a comprehensive understanding of the complex regulatory mechanisms and developmental trajectories of plant cells, advancing our knowledge of plant biology (Rhaman *et al.*, 2024).

5.1.2 Spatial omics and its significance

Spatial omics techniques, particularly spatial transcriptomics, illustrate a transformative approach in plant biology, enabling investigators to examine gene expression at a single-cell level within the spatial context of tissues. This technology has overcome previous challenges in plant research, such as tissue dissection and data integration, allowing for the recognition of cell types, reconstruction of cell-fate lineages, and exploration of cell-to-cell interactions (Yin *et al.*, 2023). The benefit of spatial multi-omics in herbal plants enhances our understanding of biosynthesis pathways and the production of natural products, which is crucial for industrial applications (Chen *et al.*, 2024). Integrating omics technologies aids in elucidating plant–microbe interactions and improving agricultural practices through biological control and bioremediation (Bastas and Kumar, 2023). Overall, spatial omics not only provide insights into the complex architecture of plant tissues but also facilitate the exploration of ecological dynamics and disease spread, thereby significantly advancing our knowledge of plant biology (Rao *et al.*, 2021).

Spatial omics technologies facilitate the recognition of novel regulatory networks and signaling pathways in plants by enabling the systematic measurement of gene expression across different tissue regions, revealing spatial heterogeneity in gene activity and cellular interactions (Chen *et al.*, 2023a). For instance, studies have shown that spatial transcriptomics can uncover regulatory modules that govern functional specialization in plant leaves, highlighting the roles of specific transcription factor families in regulating genes associated with critical biological processes like photosynthesis and environmental sensing (Robertson and Wilkins, 2023). Advancements in spatial epigenome–transcriptome co-mapping allow researchers to explore how epigenetic modifications influence gene regulation at nearly single-cell resolution, providing insights into cell dynamics and transcriptional responses (Jin and Jin, 2023). These insights have consequential involvement for plant breeding and biotechnology, as they can characterize the progress of crops with enhanced traits, such as improved stress resistance and yield, by targeting specific regulatory pathways (Bressan *et al.*, 2023).

5.1.3 Challenges in single-cell and spatial omics

Single-cell and spatial omics technologies face several challenges that hinder their full potential. One significant issue is the complexity of data integration, as these technologies generate large, high-dimensional datasets that require sophisticated computational methods for analysis. The lack of standardized computational pipelines complicates the integration of diverse data types, necessitating the application of automation and expert systems strategies to visualize and interpret the artificial intelligence (AI)-driven data effectively (Dezem *et al.*, 2024). The reliance on reference datasets can limit the discovery of novel cell populations, as existing methods may overlook biologically relevant variations (Zhu *et al.*, 2024). The inherent ambiguity in cell annotations, due to noisy and sparse data, poses challenges in accurately identifying cell types and states (Karin *et al.*, 2024). In plant systems, the structural complexity of cells adds another layer of difficulty, limiting the application of single-cell technologies (Rhaman *et al.*, 2024).

5.2 Fundamentals of AI in Single-Cell and Spatial Omics

5.2.1 Introduction to AI and machine learning

AI and machine learning (ML) are revolutionizing the analysis of single-cell and spatial omics by addressing the complexities of high-dimensional datasets. Techniques such as the Nicheformer model leverage large-scale datasets to learn unified cellular representations, enhancing the understanding of cellular interactions within their microenvironments (Schaar *et al.*, 2024). Methods like the dual-path graph attention auto-encoder (SSGATE) facilitate the incorporation of multi-omics data, improving the exploration of spatial heterogeneity (Lv *et al.*, 2024). AI-driven approaches also tackle challenges in cell annotation, enabling the identification of stem-like cell populations in tumors without prior cell classification (Dezem *et al.*, 2023). The quick development of AI tools has led to innovative cell annotation algorithms that effectively manage issues like data sparsity and batch effects (Hou *et al.*, 2024).

Integrating AI and ML in single-cell and spatial omics significantly enhances our understanding of biological systems by enabling the analysis of complex, high-dimensional datasets. Techniques such as the SSGATE facilitate the consolidation of single-cell and spatial data, improving performance in downstream analyses like transcriptomics and proteomics (Lv *et al.*, 2024). Models like Spatial Scope leverage deep generative approaches to achieve single-cell resolution from spatial transcriptomics data, allowing for detailed characterization of tissue structures and cellular interactions (Wan *et al.*, 2023). The concept of annotation trainability helps identify meaningful biological signals within datasets, aiding in the interpretation of cellular diversity and dynamics (Karin *et al.*, 2024). Lastly, Nicheformer, a transformer-based model, effectively learns cellular representations from vast datasets, enabling predictions of spatial contexts and enhancing the understanding of cellular microenvironments (Schaar *et al.*, 2024).

5.2.2 AI applications in omics data analysis

AI plays an important part in the analysis of omics data in plants, enhancing our comprehension of complex biological systems. Integrating multi-omics data and AI techniques, particularly ML and deep learning, facilitates the extraction of meaningful insights from wide datasets, which traditional methods struggle to manage (Cembrowska-Lech *et al.*, 2023; Murmu *et al.*, 2024). These AI-driven approaches enable high-throughput plant phenotyping, predictive breeding, and improved management of plant stress responses, thereby addressing challenges posed by climate change and biotic stresses (Isewon *et al.*, 2022). Moreover, the implementation of automated algorithms allows for the identification of intricate patterns within omics data, optimizing the selection of algorithms based on specific research tasks. The development of explainable AI systems further enhances the interpretability of results, making data analytics more accessible to researchers (Harfouche *et al.*, 2023).

5.3 Data Acquisition Techniques for Single-Cell and Spatial Omics

Data acquisition techniques for single-cell and spatial omics in plants have advanced significantly, primarily through the development of scRNA-seq and spatial transcriptomics. scRNA-seq allows researchers to detect transcriptional heterogeneities at the single-cell level, although it lacks high-resolution spatial information (Bawa *et al.*, 2024). To address this limitation, spatial transcriptomics has emerged, enabling the analysis of cell types and interactions within their native tissue context, thus preserving spatial information (Yin *et al.*, 2023). Recent innovations, such as Stereo-seq, combine the strengths of single-cell transcriptomics with spatial data, enhancing our understanding of plant biology (Bawa *et al.*, 2024). Integrating these technologies with live imaging and time-resolved datasets facilitates the exploration of cell plasticity and developmental responses, providing a comprehensive view of plant resilience (Nolan and Shahan, 2023). Despite these advancements, challenges remain in sample processing and data analysis, necessitating further research for broader applicability in plant studies (Chen *et al.*, 2023b). The AI-driven single-cell and spatial omics analysis workflow is explained in Fig. 5.1.

5.3.1 Single-cell omics techniques

Single-cell omics techniques hold significant potential for advancing plant breeding and crop improvement by providing detailed insights into cellular heterogeneity and regulatory mechanisms. These techniques, including single-cell transcriptomics, proteomics, and epigenomics, enable researchers to dissect complex biological function at the individual cell level, revealing how individual cells respond to environmental stresses and developmental cues (Rhaman *et al.*, 2024). For instance, SCP can identify stress-related proteins, facilitating the selection of resilient plant varieties (Fulcher *et al.*, 2024). Single-plant-omics can track transcriptional changes during critical developmental transitions, allowing breeders to select desirable traits based on intrinsic biological age (Redmond *et al.*, 2024). Integrating multi-omics approaches supplements our interpretation of gene regulatory networks (GRNs) and chromatin accessibility, which can inform targeted breeding strategies (Zhang *et al.*, 2023).

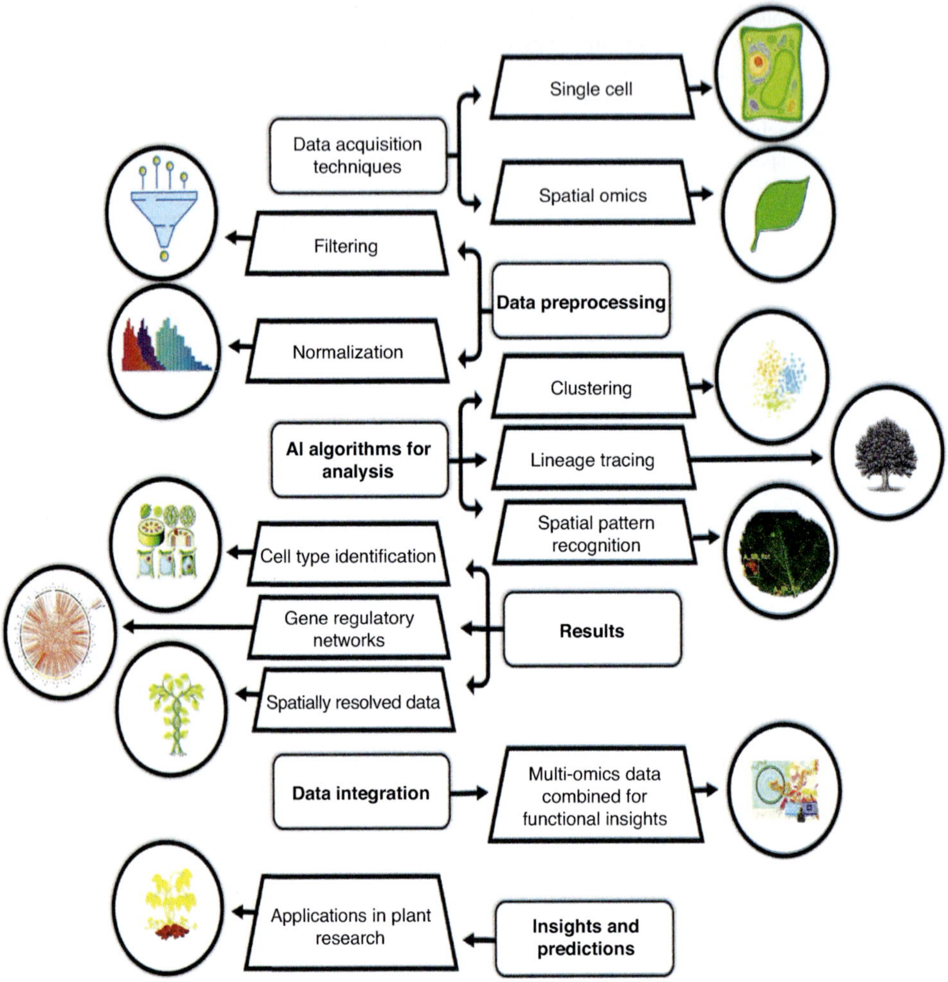

Fig. 5.1. AI-driven single-cell and spatial omics analysis workflow. (Author's own image.)

Applying single-cell omics techniques in plant research faces several technical challenges and limitations. One significant hurdle is the inherent complexity of plant cell structures, which complicates the isolation and analysis of individual cells, leading to difficulties in obtaining high-quality data (Rhaman *et al.*, 2024). Methodology like scRNA-seq often results in the loss of spatial data, which is crucial for understanding the context of cellular functions within tissues (Adema *et al.*, 2024). Single-cell assays, such as scATAC-seq, show high sparsity and noise in the data, making it challenging to accurately infer GRNs (Lu *et al.*, 2024). The quantification of metabolites at the single-cell level also has difficulties owing to the uneven distribution of natural products across different cell types (Vu *et al.*, 2024).

5.3.2 Spatial transcriptomics and proteomics

Spatial transcriptomics and proteomics are emerging as transformative tools in plant research, enabling the perusal of gene expression and protein distribution at unprecedented spatial resolutions. Recent advancements in spatial transcriptomics allow researchers to study single-cell transcriptomes while preserving spatial data, which is crucial for comprehending cell-to-cell interactions and tissue organization in plants (Yin *et al.*, 2023). Coupled with proteogenomics, these technologies can distil gene models and reveal dynamic changes in protein-coding genes during plant development and in response to environmental stressors (Song *et al.*, 2023). The consolidation of spatial transcriptomics with proteomics enhances our interpretation of the functional genome, providing an understanding of the molecular mechanisms underlying plant growth stress responses (Bartels *et al.*, 2024; Wang *et al.*, 2024). Despite their potential, the benefit of these technologies in plant investigation is still developing, focusing attention on the need for further exploration and methodological advancements to fully leverage their capabilities (Chen *et al.*, 2023b).

Integrating spatial transcriptomics and proteomics data to understand plant developmental processes presents several challenges and limitations. One significant issue is the current inability of scRNA-seq to provide high-resolution spatial information, which is pivotal for the interpretation of the specific functions of cells in their respective spatial contexts (Bawa *et al.*, 2024). While spatial transcriptomics preserves spatial information, it often lacks the resolution needed to discern subcellular structures, limiting insights into cellular differentiation and interactions (Choe *et al.*, 2023). Integrating these omics technologies is hindered by methodological limitations, such as the need for standardized protocols and the complexity of data analysis (Chen *et al.*, 2023b). The spatiotemporal aspects of plant biology are often overlooked, necessitating a more comprehensive approach that combines molecular, biochemical, and ecological perspectives to fully capture the dynamics of plant development (Munné-Bosch, 2022).

5.3.3 Preprocessing and data integration

Data integration from single-cell and spatial omics approaches significantly enhances our knowledge of plant development and responses to environmental stimuli by providing a complete view of cellular behavior in context; scRNA-seq allows for the discernment of transcriptional heterogeneities, while spatial transcriptomics (ST) offers understandings into tissue architecture and cellular. Integrating these technologies, like Stereo-seq and SpatialScope, enables researchers to achieve single-cell resolution and infer transcriptome-extensive assertion levels, facilitating the recognition of spatially differentially expressed genes and cellular communication (Wan *et al.*, 2023; Bawa *et al.*, 2024). Moreover, the combination of live imaging with spatial transcriptomics allows for time-resolved datasets that map cellular responses to stimuli, thereby elucidating developmental plasticity and resilience mechanisms in plants (Nolan and Shahan, 2023). This integrated approach ultimately fosters a clear knowledge of the intricate regulatory mechanisms that control plant adaptation to their environment (Rhaman *et al.*, 2024).

Integrating data from various spatial and single-cell omics platforms in plant research presents several key challenges, primarily due to the loss of spatial data during scRNA-seq, which complicates biological interpretation (Adema *et al.*, 2024). Existing computational methods often fail to utilize all available data domains, such as molecular expression, spatial positioning, and tissue imaging, limiting the discovery of biologically relevant cell populations (Zhu *et al.*, 2024). The inherent ambiguity in cell annotations further complicates data integration, as it can lead to misinterpretations of cellular states and relationships. To address these challenges, innovative approaches like CellSNAP can enhance cell population delineation by integrating multiple data types, while models like Liam can facilitate the simultaneous integration of multimodal data, accounting for batch consequence and keeping hold of biological distinction (Rautenstrauch and Ohler, 2024).

5.4 AI-Driven Analysis of Single-Cell Omics Data

AI-driven analysis of single-cell omics data in plants is revolutionizing our understanding of cellular heterogeneity and developmental processes. Recent advances in single-cell transcriptomics have enabled the detailed characterization of plant cell populations, revealing distinct identities and metabolic activities within tissues such as leaves (Tenorio Berrío *et al.*, 2022). Tools like scDEC utilize deep generative neural networks to analyze single-cell ATAC-seq data, effectively addressing challenges posed by high sparsity and dimensionality, while also inferring cell trajectories during differentiation (Liu *et al.*, 2021). The deep cross-omics cycle attention model further enhances this analysis by integrating transcriptomics and epigenomics data, allowing for the dissection of cellular heterogeneity and the construction of regulatory networks (Zuo *et al.*, 2021). The VIA algorithm facilitates scalable trajectory inference across various omics datasets, capturing complex cellular pathways in large-scale studies (Stassen *et al.*, 2021).

5.4.1 Clustering and cell type identification

Key molecular mechanisms underlying plant cell type identification and clustering involve advanced techniques for analysing gene expression and epigenetic profiles. Laser microdissection (LM) permits the precise isolation of specific cell types, enabling global analyses of gene expression and metabolite profiling, which are crucial for understanding cellular responses to stimuli and developmental processes (Ohtsu *et al.*, 2007). scRNA-seq and scATAC-seq provide insights into transcriptional and epigenetic heterogeneity, respectively, facilitating the classification of diverse cell populations (Zamanighomi *et al.*, 2018; Zhang *et al.*, 2018). The integration of cell type-specific transcriptome profiles with metabolite data enhances the understanding of complex regulatory networks, revealing interactions that are often obscured in whole-organ analyses (Rogers *et al.*, 2012). Computational tools like the Genome Cluster Database support the clustering of protein sequences, aiding in the identification of functional gene families and their roles in plant biology (Horan *et al.*, 2005).

The recognition and clustering of specific plant cell types have significant implications for crop improvement and plant breeding. By utilizing advanced techniques such as LM combined with high-throughput technologies, researchers can analyze gene expression and metabolite profiles at the cellular level, leading to a clear knowledge of the biological processes that govern plant development and response to environmental stimuli (Ohtsu *et al.*, 2007). This cell type-specific profiling can facilitate the identification of superior alleles through direct allele selection, enhancing the efficiency of marker-assisted selection in breeding programs (Sorrells and Wilson, 1997). In-field phenotyping using automated 3D imaging allows for the correlation of observable plant characteristics with their underlying genetics, streamlining the selection process and improving crop yields (Sodhi *et al.*, 2017). Overall, these methodologies enable more precise breeding strategies by integrating genetic information with phenotypic data, ultimately directing the development of improved crop varieties (Wenzel, 2006).

5.4.2 Trajectory inference and lineage tracing

Trajectory inference in plant development is influenced by several key factors, including physiological determinants, environmental responses, and genetic mechanisms. The physiological components of growth, such as leaf growth and flowering, are essential for understanding overall plant performance and are modeled mathematically to quantify their relationships (Charles-Edwards *et al.*, 1986). Mechanistic phonological models also play a crucial role, as they account for temperature responses and can be applied to various developmental processes, including frost hardiness and tree growth (Chuine *et al.*, 2013). The natural coordinate systems formed by growth trajectories in simplistically growing organs provide insights into the alignment of cellular structures, which is vital for understanding growth patterns (Hejnowicz, 1984). Integrating longitudinal data with time-to-event analyses allows for the identification of genetic factors that influence both growth trajectories and developmental events, enhancing our understanding of plant biology (Lin and Wu, 2006). Finally, live imaging techniques reveal the dynamic nature of plant development, capturing rapid changes in organ formation (Reddy *et al.*, 2007).

Lineage tracing techniques play a crucial role in improving our knowledge of plant morphogenesis by providing understandings into the dynamics of cell behavior and gene expression during development. For instance, the computerized trail of stem cell derivation in Arabidopsis shoot apical meristems allows researchers to analyze spatio-temporal patterns of cell growth and division, revealing how these processes contribute to organ differentiation and stem-cell homeostasis (Liu *et al.*, 2010). Computational models of cellular morphogenesis simulate the interactions between genetic regulation and mechanical factors, offering a framework to explore how these elements influence plant form and structure (Rudge and Haseloff, 2005). Genetic analyses *in vitro*, such as those examining somatic embryogenesis, provide valuable data on the genetic basis of morphogenetic processes, allowing for a clear knowledge of the mechanisms underlying plant development (Sugiyama, 2000). Together, these techniques facilitate a comprehensive exploration of the intricate relationships between cellular behavior and morphogenetic outcomes in plants (Ferrús and Garcia-Bellido, 1976).

5.4.3 GRN reconstruction

GRN reconstruction in plants is a complex yet vital area of research that leverages large-scale gene expression data to uncover the interactions among genes that govern biological processes. Various methodologies have been developed, including Bayesian networks, linear time-variant models, and consensus-driven approaches like TopkNet, which integrates multiple algorithms for improved accuracy in GRN inference (Kabir *et al.*, 2010; Vignes *et al.*, 2011). Statistical approaches such as interconnection analysis and probabilistic graphical models are employed to analyze gene expression data effectively (Li *et al.*, 2015). The challenge lies not only in the reconversion of these meshworks but also in validating the inferred interactions and translating them into biological functions (Gaudinier and Brady, 2016). By harnessing diverse data and employing robust computational strategies, researchers aim to elucidate the regulatory mechanisms that underlie important agronomic traits in plants, ultimately enhancing our knowledge of plant biology and enhancing crop management practices (Hase *et al.*, 2013).

Different plant species exhibit varied responses to GRN reconstruction, which is crucial for understanding their developmental and adaptive mechanisms. For instance, the reconstruction of gene networks in Arabidopsis has revealed significant insights into the ERECTA signaling pathway, highlighting how genetic variants influence gene expression and regulatory interactions (Terpstra *et al.*, 2010). The identification of specific mRNA sets in organs like anthers and roots underscores the complexity of gene regulation in response to developmental cues (Goldberg, 1986). The integration of multifactorial approaches, such as Bayesian networks and meta-analysis, has proven effective in reconstructing these networks, as demonstrated in the DREAM5 challenges. Comprehending these networks not only enhances our understanding of plant biology but also has profound implications for plant breeding and biotechnology, enabling the growth of crops with improved traits and resilience to abiotic stresses (Hong-bo, 2006).

5.5 AI in Spatial Omics Data Analysis

AI is increasingly being integrated into spatial omics data analysis in plants, enhancing the understanding of complex biological systems. Techniques such as automation and structured learning are applied to high-throughput spatial transcriptome profiling, allowing for the functional characterization of specialized tissue domains and enhancing the efficiency of data collection and examination (Giacomello *et al.*, 2017). AI facilitates the integration of large-scale omics data, like gene expression networks, by employing natural language processing for better gene annotation and curation (Ohyanagi *et al.*, 2015). The application of spatial statistical analysis in plant breeding, supported by AI, has been shown to improve genotype selection by accounting for spatial dependencies in experimental data, leading to more accurate assessments of genetic materials (Duarte and Vencovsky, 2005). Overall, the convergence of AI and spatial omics is paving the way for innovative approaches in plant phenomics, ultimately contributing to advancements in plant biology and breeding programs (Nabwire *et al.*, 2021).

5.5.1 Spatial pattern recognition and feature extraction

Spatial pattern recognition and feature extraction in plants are critical for understanding plant ecology and community dynamics. Various methods have been devised to inspect spatial patterns, such as Hill's two-term local quadrat variance method and spectral analysis, which help identify the clustering and distribution of plant species and associated organisms like nematodes (Noe and Campbell, 1985). The use of neural network classifiers in automated systems for plant species recognition leverages shape-based features derived from leaf images, enhancing accuracy in classification (Chaki and Parekh, 2011). The analysis of spatial patterns also involves understanding the aggregation of individuals and the stochastic events that influence these patterns, which are essential for comprehending how plant communities are structured and function (Herben and Hara, 2003). The selection of apportion demographic methods, such as the density-corrected neighborhood density function, is crucial for accurately characterizing spatial patterns at various scales (Perry *et al.*, 2006).

Key spatial patterns and features extracted from plant morphology can enhance crop monitoring and management. Morphological image processing techniques can simplify and interpret the shapes of plant structures, facilitating applications such as leaf identification and texture analysis, which are vital for assessing crop health and quality (McDonald and Chen, 1990). Mathematical descriptors like morphological covariance and circular covariance histograms can capture essential characteristics of leaf venation, aiding in plant recognition and classification (Aptoula and Yanikoglu, 2013). The morphological scale-space approach further allows for effective feature extraction by measuring the persistence of image structures, which can reveal important attributes such as size and shape (Vachier, 2001). Understanding the morphological variations among different maize varieties is also essential for optimizing management practices, as these variations can significantly influence growth and yield (Kllng, 1991). Collectively, these methodologies provide a comprehensive framework for improving agricultural practices through enhanced monitoring and management of crop morphology (Bucksch *et al.*, 2017).

5.5.2 Spatially resolved transcriptomics and proteomics

Spatially resolved transcriptomics and proteomics in plants represent a significant advancement in understanding tissue organization and gene expression at high resolution. This approach allows researchers to map gene expression profiles and cellular positions within plant tissues, providing insights into specialized tissue domains and their functions (Rao *et al.*, 2021). Recent methodologies have emerged that integrate transcriptome-wide sequencing with imaging techniques, enabling the recognition and spatial mapping of cell types in complex plant structures (Larsson *et al.*, 2021). The development of computational tools has enhanced the analysis of spatial transcriptomic data, allowing for the quantification of spatial organization and cell–cell communications, which are crucial for understanding biological functions in health and disease. These advancements not only facilitate hypothesis generation but also support experimental designs that can test specific conditions or perturbations, thereby enriching our knowledge of plant biology (Dries *et al.*, 2021).

5.5.3 Case studies in spatial plant-omics

Case studies in spatial plant-omics highlight the intricate relationships between genetic variation, tissue architecture, and microbiome interactions. Research indicates that spatial genetic structure in plant populations is impacted by limited seed and pollen dispersal, leading to local genetic subdivisions due to environmental heterogeneity and isolation by distance (Heywood,

1991). Newfangled spatial transcriptomics has enabled the systematic assessment of gene expression across tissue spaces, providing insights into cellular arrangements and functional characterization of specialized tissue domains (Epperson, 2000). The integration of holo-omics approaches, which combine host and microbial datasets, offers a comprehensive understanding of plant–microbiome interactions, enhancing our ability to predict plant fitness and manipulate these relationships for improved health. Together, these studies highlight the importance of spatial analysis in unraveling the complexities of plant biology and their interactions with the environment and associated microbiomes (Xu *et al.*, 2021).

5.6 Integrative Multi-Omics Approaches Using AI

Multi-omics data incorporation in plants involves the comprehensive analysis of various biological layers, including metabolomics, genomics, transcriptomics, and proteomics, to enhance our knowledge of plant biology and function. The current advancements in high-throughput technologies have led to the generation of extensive omics data, necessitating the development of integration strategies that combine these diverse datasets effectively (Picard *et al.*, 2021). Approaches such as forward, assorted, intermediary, deceased, and stratified integration have been proposed to leverage the complementary information from each omics layer, and ease the pinpointing of diagnostic biomarkers and functional gene modules (Graw *et al.*, 2021). For instance, integrating metabolomics and proteomics data through high-throughput techniques allows for the identification of correlative biomarkers and insights into biochemical regulation (Weckwerth, 2008). Additionally, repositories like the Plant Omics Data Center provide valuable resources for gene expression networks across multiple species, enhancing functional annotation and evolutionary understanding. This holistic approach ultimately aids in unraveling complex biological processes in plants (Heyndrickx and Vandepoele, 2012).

5.6.1 AI techniques for data integration

AI techniques can integrate diverse data types in plant research, particularly through high-throughput phenotyping and multi-omics approaches. AI algorithms, including expert systems and automation, enable the analysis of large datasets, facilitating the extraction of significant patterns and correlations between phenotypic traits and genetic factors, which is essential for crop enhancement and breeding programs (Farooq *et al.*, 2024). AI enhances the competence of data assembling and exploration in plant phenomics by integrating non-invasive imaging techniques, thereby improving the accuracy of trait assessments (Nabwire *et al.*, 2021). Integrating AI with experimental setups, such as controlled plant experimentation chambers, allows for real-time monitoring of environmental parameters, ensuring optimal conditions for data collection (Solano *et al.*, 2022). Overall, the combination of AI and advanced phenotyping technologies is transforming plant biology, enabling more precise trait selection and sustainable agricultural practices (Cembrowska-Lech *et al.*, 2023).

5.6.2 Applications in plant research

Integrative multi-omics approaches have revolutionized plant research by enabling comprehensive analyses of transcriptomics, genomics, metabolomics, and proteomics, thereby enhancing our knowledge of plant biology and improving crop traits. These methods facilitate the identification of biomarkers and integration of compound biological networks, which are crucial for understanding plant responses to environmental stresses and optimizing growth and yield (Paczkowska *et al.*, 2020). For instance, integrating high-throughput techniques, including liquid chromatography–mass spectrometry, allows for the concurrent study on metabolites and proteins, providing insights into biochemical regulation and metabolic network dynamics (Weckwerth, 2008). The development of repositories like the Plant Omics Data Center enhances data accessibility and promotes collaborative research by linking gene expression networks across species (Ohyanagi *et al.*, 2015).

Overall, these integrative strategies are essential for advancing crop improvement and addressing challenges in agricultural sustainability (Yang *et al.*, 2021).

5.7 AI in Single-Cell and Spatial Epigenomics

AI is increasingly being integrated into single-cell and spatial epigenomics, enhancing our understanding of complex biological systems. Recent advancements in single-cell technologies, such as sci-Space and sciMAP-ATAC, allow for high-resolution spatial mapping of gene expression and chromatin accessibility, respectively, facilitating the identification of cell types and their spatial organization within tissues (Srivatsan *et al.*, 2021; Thornton *et al.*, 2021). By leveraging AI, researchers can analyze vast datasets generated from these technologies, enabling the discovery of GRN and enhancers through spatial integration of transcriptomic and epigenomic data (Bravo González-Blas *et al.*, 2020). AI can assist in interpreting single-cell molecular profiles, linking them to clinical phenotypes, and improving precision medicine approaches. This integration of AI with single-cell and spatial epigenomics not only enhances data analysis but also paves the way for intelligent systems capable of making informed decisions in disease diagnosis and treatment (Zeng *et al.*, 2018).

5.7.1 Overview of plant epigenomics

Plant epigenomics encompasses the study of congenital conversion in gene expression and genome stability that occur in the absence of modification to the DNA sequence, primarily through mechanisms such as DNA methoxide and histone modifications. These epigenetic modifications play crucial roles in plant development, stress responses, and reproductive processes, influencing traits such as growth and resilience to environmental changes (Chen *et al.*, 2010; Gutierrez-Marcos and Dickinson, 2012; Kumar and Mohapatra, 2021). Recent advancements in sequencing technologies have enabled detailed mapping of epigenetic landscapes, revealing the vigorous nature of the epigenome in response to internal and external cues (Lloyd and Lister, 2022). Understanding these epigenetic mechanisms is vital for crop improvement, as they can provide insights into adaptive traits and enhance breeding strategies for climate resilience (Kakoulidou *et al.*, 2021).

Plant epigenomics has significant applications in enhancing crop resilience and productivity, particularly in the face of climate change and environmental stresses. By understanding heritable epigenetic changes, researchers can optimize crop traits without altering the DNA sequence, thereby improving gene expression and phenotypic variation (Kakoulidou *et al.*, 2021). The plasticity of the plant epigenome allows for memory of past stress events, enabling better responses to future challenges, a phenomenon known as 'plant stress memory' (Sun *et al.*, 2021). Moreover, transposable elements (TEs) play a crucial role in epigenetic modifications, contributing to genetic variability and stress adaptation (Ramakrishnan *et al.*, 2021). Advances in epigenomic technologies facilitate the characterization of these modifications, which can be harnessed for breeding programs aimed at developing resilient crop varieties (Lloyd and Lister, 2022). Overall, integrating epigenetic insights into breeding strategies can significantly enhance agricultural sustainability and productivity in changing climatic conditions (Chen *et al.*, 2010).

Plant epigenomics faces several challenges that hinder its application in crop improvement and understanding plant responses to environmental changes. One significant challenge is the complexity of epigenetic modifications, such as DNA methoxide and histone modifications, which can vary widely among individuals and species, complicating the mapping of these changes to specific phenotypic traits (Chen *et al.*, 2010; Lloyd and Lister, 2022). The heritable nature of epigenetic exchange can lead to unpredictable outcomes in plant breeding, as these modifications may not always correlate with desired traits (Kakoulidou *et al.*, 2021). The role of TEs in mediating epigenetic changes adds another layer of complexity, as their activation under stress can introduce genetic variability that is difficult to control (Ramakrishnan *et al.*, 2021). The integration of advanced technologies for studying epigenomic patterns remains a challenge, as current methodologies may not fully capture the dynamic nature of the epigenome in response to abiotic stresses (Akhter *et al.*, 2021).

5.7.2 AI approaches in epigenomic analysis

AI approaches in the epigenomic analysis of plants have gained significant traction, leveraging deep learning techniques to enhance the understanding of epigenetic modifications. Tools like DeepSignal-plant utilize deep learning to detect genome-wide cytosine methylations (5mCs) across various contexts (CpG, CHG, CHH) with high accuracy from Nanopore sequencing data, surpassing traditional methods that only focus on CpG contexts (Ni *et al.*, 2021). Similarly, the Smart Model for Epigenetics in Plants (SMEP) predicts multiple epigenomic modifications, achieving over 80% accuracy for 5mC and other modifications, thus facilitating functional research (Wang *et al.*, 2021). PlantDeepSEA offers a web service to predict the regulatory effects of genomic variants, enhancing our understanding of chromatin accessibility and regulatory elements (Zhao *et al.*, 2021). These advancements highlight the transformative role of AI in analyzing complex epigenomic datasets, ultimately contributing to improved genome engineering efforts in plant biology (Caudai *et al.*, 2021).

AI approaches, particularly ML, have significantly advanced epigenomic analysis in plants. For instance, the SMEP utilizes deep learning to predict various epigenomic modifications, achieving high accuracy in identifying DNA and RNA methylation as well as histone modifications across different plant species, including rice and maize (Wang *et al.*, 2021). ML techniques are essential for analysing large-scale genomic data, enabling the identification of biomarkers and understanding complex biological mechanisms (Liu *et al.*, 2013). Integrating these computational methods with genomic data allows for a comprehensive analysis of DNA methylation patterns, which are crucial for gene regulation and phenotypic variation (Cortijo *et al.*, 2014).

AI approaches in the epigenomic analysis of plants face several challenges, primarily due to the complexity of epigenetic regulation and the dynamic nature of chromatin modifications (Fig. 5.2.). The intricate interplay of various epigenetic factors, such as histone modifications and DNA methylation, complicates the development of robust AI models capable of accurately predicting plant responses to environmental stresses (Kim, 2021). The heritable nature of epigenetic changes, which can influence gene expression and phenotypic variation, requires AI systems to account for both current and transgenerational effects (Sun *et al.*, 2021). The limited transferability of knowledge from model plants to crops poses a significant hurdle in applying AI effectively for crop improvement (Kakoulidou *et al.*, 2021). Lastly, the need for high-quality, comprehensive datasets to train AI models is critical, yet often lacking due to the challenges in capturing the multifaceted nature of epigenetic modifications (Dalakouras and Vlachostergios, 2021).

5.8 Future Directions and Challenges in AI-Enabled Single-Cell and Spatial Omics

Future directions in AI-enabled single-cell and spatial omics in plants are poised to considerably enhance our knowledge of plant biology and resilience to climate change. Integrating ML with single-cell transcriptomics is expected to improve prediction tasks related to cellular responses to perturbations, enabling the construction of informative perturbation latent spaces that can model complex biological systems more effectively (Ji *et al.*, 2021). Advancements in GRNs through single-cell sequencing will facilitate the identification of stress resilience mechanisms at the individual cell level, which is essential for food safety in the face of global warming (Tripathi and Wilkins, 2021). The development of spatially reconstructed single-cell transcriptome atlases will allow researchers to link gene expression patterns with cellular identity and

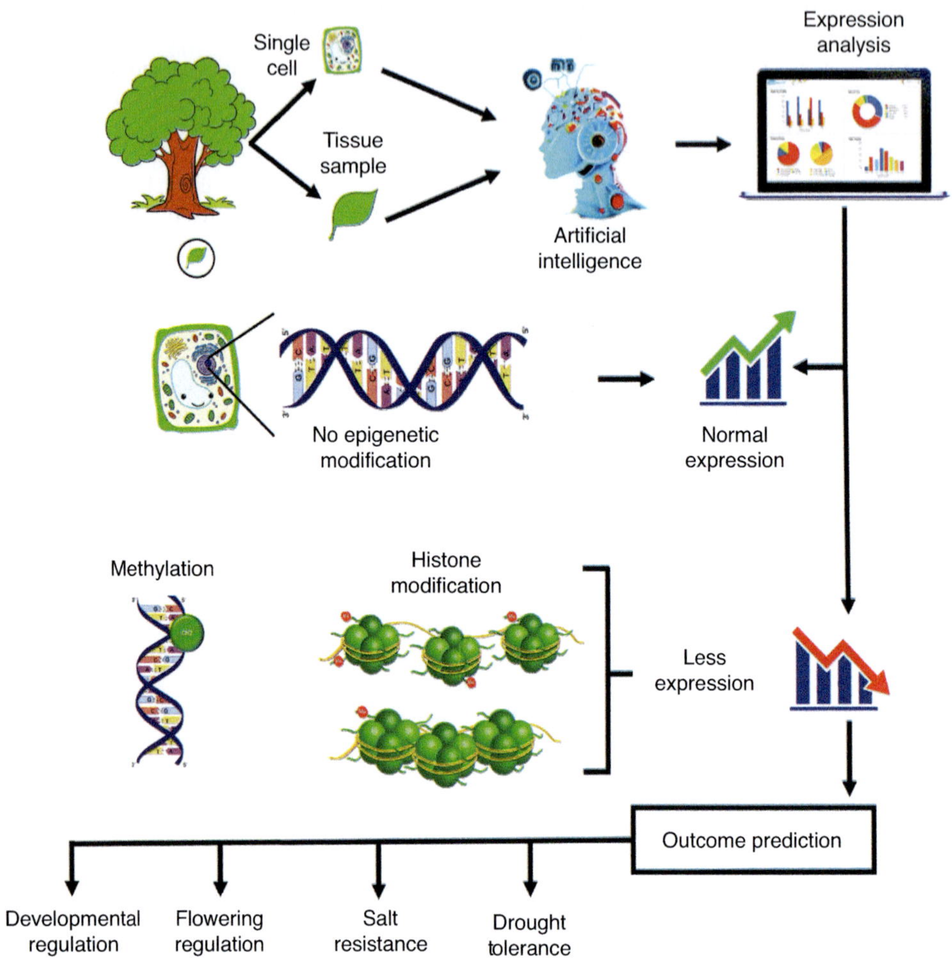

Fig. 5.2. AI for single-cell and spatial epigenomics. (Author's own image.)

function, providing insights into plant morphogenesis and development (Neumann *et al.*, 2022). Overall, these innovations will drive the exploration of novel applications and methodologies in plant research, addressing both fundamental biological questions and practical agricultural challenges (Seyfferth *et al.*, 2021; Cuperus, 2022).

AI-enabled single-cell and spatial omics in plants face several challenges that hinder their full potential. One significant issue is the complexity of plant cell structures, particularly the polysaccharide walls, which complicate single-cell analysis compared to animal cells (Cole *et al.*, 2021). While advancements in scRNA-seq have improved our knowledge of GRNs in individual plant cells, predicting these networks remains difficult due to the need for high-throughput methodologies and robust analytical frameworks (Tripathi and Wilkins, 2021). The integration of ML techniques into single-cell data analysis is still in its infancy, with existing models often lacking the ability to effectively characterize unseen conditions or integrate diverse data sources (Ji *et al.*, 2021). Finally, the setting up of centralized databases for plant single-cell data is crucial for fostering collaboration and data sharing, yet such resources are currently limited (Cole *et al.*, 2021). Addressing these challenges is important for advancing plant research and enhancing resilience to climate change.

5.9 Conclusion

By enabling a thorough examination of the molecular and functional heterogeneity of particular cells in intricate multicellular tissues, single-cell omics methods and spatial multi-omics have completely transformed the study of plant cells. The investigation of individual plant cells is made possible by single-cell omics, which offer a more accurate understanding of cell heterogeneity and the underlying molecular processes governing plant growth, development, and stress adaptation. In contrast, using AI and ML facilitated the direct observation of hundreds of cells' transcriptomes and proteomics within their natural environment, hence aiding in the comprehension of cell-to-cell connections, communication, and non-autonomous processes by providing organized data interpretation. Moreover, these technologies offer a thorough comprehension of the statistical distribution of expression of the gene, facilitating a more profound comprehension of the molecular underpinnings of functions particular to individual tissues of plants. Hence, AI in single-cell and spatial multi-omics leads to extraordinary insights into plant biology.

References

Adema, K., Schon, M.A., Nodine, M.D. and Kohlen, W. (2024) Lost in space: What single-cell RNA sequencing cannot tell you. *Trends in Plant Science* 29(9), 1018–1028. DOI: 10.1016/j.tplants.2024.03.010.

Akhter, Z., Bi, Z., Ali, K., Sun, C., Fiaz, S. *et al.* (2021) In response to abiotic stress, DNA methylation confers epigenetic changes in plants 10(6), 1096. DOI: 10.3390/plants10061096.

Aptoula, E. and Yanikoglu, B. (2013) Morphological features for leaf based plant recognition. In: *2013 IEEE International Conference on Image Processing*, pp. 1496–1499. DOI: 10.1109/ICIP.2013.6738307.

Bartels, H.C., Hameed, S., Young, C., Nabhan, M., Downey, P. *et al.* (2024) Spatial proteomics and transcriptomics of placenta accreta spectrum. *Bioinformatics*. DOI: 10.1101/2024.03.21.585167.

Bastas, K.K. and Kumar, A. (2023) Omics technologies in the plant–microbe interactions. In: *Microbial Biocontrol: Molecular Perspective in Plant Disease Management*. Springer Nature, Singapore, pp. 257–282. DOI: 10.1007/978-981-99-3947-3_13.

Bawa, G., Liu, Z., Yu, X., Tran, L.S.P. and Sun, X. (2024) Introducing single cell stereo-sequencing technology to transform the plant transcriptome landscape. *Trends in Plant Science* 29(2), 249–265. DOI: 10.1016/j.tplants.2023.10.002.

Bravo González-Blas, C., Quan, X.-J., Duran-Romaña, R., Taskiran, I.I., Koldere, D. *et al.* (2020) Identification of genomic enhancers through spatial integration of single-cell transcriptomics and epigenomics. *Molecular Systems Biology* 16(5), e9438. DOI: 10.15252/msb.20209438.

Bressan, D., Battistoni, G. and Hannon, G.J. (2023) The dawn of spatial omics. *Science* 381(6657), eabq4964. DOI: 10.1126/science.abq4964.

Bucksch, A., Atta-Boateng, A., Azihou, A.F., Battogtokh, D. and Baumgartner, A. (2017) Morphological plant modeling: Unleashing geometric and topological potential within the plant sciences. *Frontiers in Plant Science* 8, 900. DOI: 10.3389/978-2-88945-297-2.

Caudai, C., Galizia, A., Geraci, F., Le Pera, L., Morea, V. *et al.* (2021) AI applications in functional genomics. *Computational and Structural Biotechnology Journal* 19, 5762–5790. DOI: 10.1016/j.csbj.2021.10.009.

Cembrowska-Lech, D., Krzemińska, A., Miller, T., Nowakowska, A., Adamski, C. *et al.* (2023) An integrated multi-omics and artificial intelligence framework for advance plant phenotyping in horticulture. *Biology* 12(10), 1298. DOI: 10.3390/biology12101298.

Chaki, J. and Parekh, R. (2011) Plant leaf recognition using shape based features and neural network classifiers. *International Journal of Advanced Computer Science and Applications* 2(10), 10. DOI: 10.14569/IJACSA.2011.021007.

Charles-Edwards, D.A., Doley, D. and Rimmington, G.M. (1986) Modelling plant growth and development. In: *Mathematics, Agricultural and Food Sciences, Biology*, Vol. 118019333. Academic Press.

Chen, C., Ge, Y. and Lu, L. (2023b) Opportunities and challenges in the application of single-cell and spatial transcriptomics in plants. *Frontiers in Plant Science* 14, 1185377. DOI: 10.3389/fpls.2023.1185377.

Chen, C., Zhang, X. and Yue, M. (2024) Spatial multi-omics in medicinal plants: From biosynthesis pathways to industrial applications. *Trends in Plant Science* 29(5), 510–513. DOI: 10.1016/j.tplants.2024.02.008.

Chen, M., Lv, S. and Meng, Y. (2010) Epigenetic performers in plants. *Development, Growth & Differentiation* 52(6), 555–566. DOI: 10.1111/j.1440-169X.2010.01192.x.

Chen, T.-Y., You, L., Hardillo, J.A.U. and Chien, M.-P. (2023a) Spatial transcriptomic technologies. *Cells* 12(16), 2042. DOI: 10.3390/cells12162042.

Choe, K., Pak, U., Pang, Y., Hao, W. and Yang, X. (2023) Advances and challenges in spatial transcriptomics for developmental biology. *Biomolecules* 13(1), 156. DOI: 10.3390/biom13010156.

Chuine, I., Cortazar-Atauri, I.G., Kramer, K. and Hänninen, H. (2013) Plant development models. In: *Phenology: An Integrative Environmental Science*. Springer, Netherlands, pp. 275–293. DOI: 10.1007/978-94-007-6925-0_15.

Cole, B., Bergmann, D., Blaby-Haas, C.E., Blaby, I.K., Bouchard, K.E. *et al.* (2021) Plant single-cell solutions for energy and the environment. *Communications Biology* 4(1), 962. DOI: 10.1038/s42003-021-02477-4.

Cortijo, S., Wardenaar, R., Colomé-Tatché, M., Johannes, F. and Colot, V. (2014) Genome-wide analysis of DNA methylation in *Arabidopsis* using MeDIP-chip. In: *Plant Epigenetics and Epigenomics: Methods and Protocols*. Humana Press, pp. 125–149. DOI: 10.1007/978-1-62703-773-0_9.

Cuperus, J.T. (2022) Single-cell genomics in plants: Current state, future directions, and hurdles to overcome. *Plant Physiology* 188(2), 749–755. DOI: 10.1093/plphys/kiab478.

Dalakouras, A. and Vlachostergios, D. (2021) Epigenetic approaches to crop breeding: Current status and perspectives. *Journal of Experimental Botany* 72(15), 5356–5371. DOI: 10.1093/jxb/erab227.

Dezem, F.S., Marção, M., Ben-Cheikh, B., Nikulina, N., Omotoso, A. *et al.* (2023) A machine learning one-class logistic regression model to predict stemness for single cell transcriptomics and spatial omics. *BMC Genomics* 24(1), 717. DOI: 10.1186/s12864-023-09722-6.

Dezem, F.S., Arjumand, W., DuBose, H., Morosini, N.S. and Plummer, J. (2024) Spatially resolved single-cell omics: Methods, challenges, and future perspectives. *Annual Review of Biomedical Data Science* 7(1), 131–153. DOI: 10.1146/annurev-biodatasci-102523-103640.

Dries, R., Chen, J., del Rossi, N., Khan, M.M., Sistig, A. *et al.* (2021) Advances in spatial transcriptomic data analysis. *Genome Research* 31(10), 1706–1718. DOI: 10.1101/gr.275224.121.

Duarte, J.B. and Vencovsky, R. (2005) Spatial statistical analysis and selection of genotypes in plant breeding. *Pesquisa Agropecuária Brasileira* 40(2), 107–114. DOI: 10.1590/S0100-204X2005000200002.

Epperson, B.K. (2000) Spatial genetic structure and non-equilibrium demographics within plant populations. *Plant Species Biology* 15(3), 269–279. DOI: 10.1111/j.1442-1984.2000.00046.x.

Farooq, M.A., Gao, S., Hassan, M.A., Huang, Z., Rasheed, A. *et al.* (2024) Artificial intelligence in plant breeding. *Trends in Genetics* 40(10), 891–908. DOI: 10.1016/j.tig.2024.07.001.

Ferrús, A. and Garcia-Bellido, A. (1976) Morphogenetic mutants detected in mitotic recombination clones. *Nature* 260(5550), 425–426. DOI: 10.1038/260425a0.

Fulcher, J.M., Dawar, P., Balasubramanian, V.K., Williams, S.M. and Nadeau, K.D. (2024) Single-cell proteomics of Arabidopsis leaf mesophyll identifies drought stress-related proteins. *Plant Biology*. DOI: 10.1101/2024.07.01.601628.

Gaudinier, A. and Brady, S.M. (2016) Mapping transcriptional networks in plants: Data-driven discovery of novel biological mechanisms. *Annual Review of Plant Biology* 67(1), 575–594. DOI: 10.1146/annurev-arplant-043015-112205.

Giacomello, S., Salmén, F., Terebieniec, B.K., Vickovic, S., Navarro, J.F. *et al.* (2017) Spatially resolved transcriptome profiling in model plant species. *Nature Plants* 3(6), 1–11. DOI: 10.1038/nplants.2017.61.

Goldberg, R.B. (1986) Regulation of plant gene expression. *Philosophical Transactions of the Royal Society of London. B, Biological Sciences* 314(1166), 343–353. DOI: 10.1098/rstb.1986.0057.

Graw, S., Chappell, K., Washam, C.L., Gies, A., Bird, J. *et al.* (2021) Multi-omics data integration considerations and study design for biological systems and disease. *Molecular Omics* 17(2), 170–185. DOI: 10.1039/D0MO00041H.

Gutierrez-Marcos, J.F. and Dickinson, H.G. (eds) (2012) Epigenetic reprogramming in plant reproductive lineages. *Plant and Cell Physiology* 53(5), 817–823. DOI: 10.1093/pcp/pcs052.

Harfouche, A.L., Nakhle, F., Harfouche, A.H., Sardella, O.G., Dart, E. *et al.* (2023) A primer on artificial intelligence in plant digital phenomics: Embarking on the data to insights journey. *Trends in Plant Science* 28(2), 154–184. DOI: 10.1016/j.tplants.2022.08.021.

Hase, T., Ghosh, S., Yamanaka, R. and Kitano, H. (2013) Harnessing diversity towards the reconstructing of large scale gene regulatory networks. *PLoS Computational Biology* 9(11), e1003361. DOI: 10.1371/journal.pcbi.1003361.

Hejnowicz, Z. (1984) Trajectories of principal directions of growth, natural coordinate system in growing plant organ. *Acta Societatis Botanicorum Poloniae* 53(1), 29–42. DOI: 10.5586/asbp.1984.004.

Herben, T. and Hara, T. (2003) Spatial pattern formation in plant communities. In: *Morphogenesis and Pattern Formation in Biological Systems: Experiments and Models*. Springer Japan, Tokyo, pp. 223–235. DOI: 10.1007/978-4-431-65958-7_19.

Heyndrickx, K.S. and Vandepoele, K. (2012) Systematic identification of functional plant modules through the integration of complementary data sources. *Plant Physiology* 159(3), 884–901. DOI: 10.1104/pp.112.196725.

Heywood, J.S. (1991) Spatial analysis of genetic variation in plant populations. *Annual Review of Ecology and Systematics* 22(1), 335–355. DOI: 10.1146/annurev.es.22.110191.002003.

Hong-bo, S. (2006) Plant gene regulatory network system under abiotic stress. *Acta Biologica Szegediensis* 50(1–2), 1–9.

Horan, K., Lauricha, J., Bailey-Serres, J., Raikhel, N. and Girke, T. (2005) Genome cluster database. A sequence family analysis platform for *Arabidopsis* and rice. *Plant Physiology* 138(1), 47–54. DOI: 10.1104/pp.104.059048.

Hou, N., Lin, X., Lin, L., Zeng, X., Zhong, Z. *et al.* (2024) Artificial intelligence in cell annotation for high-resolution RNA sequencing data. *TrAC Trends in Analytical Chemistry* 178, 117818. DOI: 10.1016/j.trac.2024.117818.

Isewon, I., Apata, O., Oluwamuyiwa, F., Aromolaran, O. and Oyelade, J. (2022) Machine learning algorithms: Their applications in plant omics and agronomic traits' improvement. *F1000Research* 11, 1256. DOI: 10.12688/f1000research.125425.1.

Ji, Y., Lotfollahi, M., Wolf, F.A. and Theis, F.J. (2021) Machine learning for perturbational single-cell omics. *Cell Systems* 12(6), 522–537. DOI: 10.1016/j.cels.2021.05.016.

Jin, M.Z. and Jin, W.L. (2023) Spatial epigenome-transcriptome comapping technology. *Trends in Cell Biology* 33(6), 449–450. DOI: 10.1016/j.tcb.2023.03.011.

Kabir, M., Noman, N. and Iba, H. (2010) Reverse engineering gene regulatory network from microarray data using linear time-variant model. *BMC Bioinformatics* 11(Suppl 1), 1–15. DOI: 10.1186/1471-2105-11-S1-S56.

Kakoulidou, I., Avramidou, E.V., Baránek, M., Brunel-Muguet, S., Farrona, S. *et al.* (2021) Epigenetics for crop improvement in times of global change. *Biology* 10(8), 766. DOI: 10.3390/biology10080766.

Karin, J., Mintz, R., Raveh, B. and Nitzan, M. (2024) Interpreting single-cell and spatial omics data using deep networks training dynamics. *BioRxiv*. DOI: 10.1101/2024.04.06.588373.

Kim, J.H. (2021) Multifaceted chromatin structure and transcription changes in plant stress response. *International Journal of Molecular Sciences* 22(4), 2013. DOI: 10.3390/ijms22042013.

KIlng, G.J. (1991) Morphology and growth of maize. *Agricultural and Food Sciences* 195436456.

Kumar, S. and Mohapatra, T. (2021) Dynamics of DNA methylation and its functions in plant growth and development. *Frontiers in Plant Science* 12, 596236. DOI: 10.3389/fpls.2021.596236.

Larsson, L., Frisén, J. and Lundeberg, J. (2021) Spatially resolved transcriptomics adds a new dimension to genomics. *Nature Methods* 18(1), 15–18. DOI: 10.1038/s41592-020-01038-7.

Li, Y., Pearl, S.A. and Jackson, S.A. (2015) Gene networks in plant biology: Approaches in reconstruction and analysis. *Trends in Plant Science* 20(10), 664–675. DOI: 10.1016/j.tplants.2015.06.013.

Lin, M. and Wu, R. (2006) A joint model for nonparametric functional mapping of longitudinal trajectory and time-to-event. *BMC Bioinformatics* 7(1), 1–12. DOI: 10.1186/1471-2105-7-138.

Liu, C., Che, D., Liu, X. and Song, Y. (2013) Applications of machine learning in genomics and systems biology. *Computational and Mathematical Methods in Medicine* 2013, 1–1. DOI: 10.1155/2013/587492.

Liu, M., Yadav, R.K., Roy-Chowdhury, A. and Reddy, G.V. (2010) Automated tracking of stem cell lineages of *Arabidopsis* shoot apex using local graph matching. *The Plant Journal* 62(1), 135–147. DOI: 10.1111/j.1365-313X.2009.04117.x.

Liu, Q., Chen, S., Jiang, R. and Wong, W.H. (2021) Simultaneous deep generative modelling and clustering of single-cell genomic data. *Nature Machine Intelligence* 3(6), 536–544. DOI: 10.1038/s42256-021-00333-y.

Lloyd, J.P.B. and Lister, R. (2022) Epigenome plasticity in plants. *Nature Reviews. Genetics* 23(1), 55–68. DOI: 10.1038/s41576-021-00407-y.

Lu, C., Wei, Y., Abbas, M., Agula, H., Wang, E. *et al.* (2024) Application of single-cell assay for transposase-accessible chromatin with high throughput sequencing in plant science: Advances, technical challenges, and prospects. *International Journal of Molecular Sciences* 25(3), 1479. DOI: 10.3390/ijms25031479.

Lv, T., Zhang, Y., Liu, J., Kang, Q. and Liu, L. (2024) Single-cell multi-omics and spatial multi-omics data integration via dual-path graph attention auto-encoder. *BioRxiv*. DOI: 10.1101/2024.06.03.597266.

McDonald, T. and Chen, Y.R. (1990) Application of morphological image processing in agriculture. *Transactions of the ASAE* 33(4), 1346–1352. DOI: 10.13031/2013.31479.

Montes, C., Zhang, J., Nolan, T.M. and Walley, J.W. (2024) Single-cell proteomics differentiates Arabidopsis root cell types. *BioRxiv*. DOI: 10.1101/2024.04.09.588771.

Munné-Bosch, S. (2022) Spatiotemporal limitations in plant biology research. *Trends in Plant Science* 27(4), 346–354. DOI: 10.1016/j.tplants.2021.10.005.

Murmu, S., Sinha, D., Chaurasia, H., Sharma, S., Das, R. *et al.* (2024) A review of artificial intelligence-assisted omics techniques in plant defense: Current trends and future directions. *Frontiers in Plant Science* 15, 1292054. DOI: 10.3389/fpls.2024.1292054.

Nabwire, S., Suh, H.K., Kim, M.S., Baek, I. and Cho, B.K. (2021) Application of artificial intelligence in phenomics. *Sensors* 21(13), 4363. DOI: 10.3390/s21134363.

Neumann, M., Xu, X., Smaczniak, C., Schumacher, J., Yan, W. *et al.* (2022) A 3D gene expression atlas of the floral meristem based on spatial reconstruction of single nucleus RNA sequencing data. *Nature Communications* 13(1), 2838. DOI: 10.1038/s41467-022-30177-y.

Ni, P., Huang, N., Nie, F., Zhang, J., Zhang, Z. *et al.* (2021) Genome-wide detection of cytosine methylations in plant from nanopore data using deep learning. *Nature Communications* 12(1), 5976. DOI: 10.1038/s41467-021-26278-9.

Noe, J.P. and Campbell, C.L. (1985) Spatial pattern analysis of plant-parasitic nematodes. *Journal of Nematology* PMID(2), 86–93.

Nolan, T.M. and Shahan, R. (2023) Resolving plant development in space and time with single-cell genomics. *Current Opinion in Plant Biology* 76, 102444. DOI: 10.1016/j.pbi.2023.102444.

Ohtsu, K., Takahashi, H., Schnable, P.S. and Nakazono, M. (2007) Cell type-specific gene expression profiling in plants by using a combination of laser microdissection and high-throughput technologies. *Plant & Cell Physiology* 48(1), 3–7. DOI: 10.1093/pcp/pcl049.

Ohyanagi, H., Takano, T., Terashima, S., Kobayashi, M., Kanno, M. *et al.* (2015) Plant omics data center: An integrated web repository for interspecies gene expression networks with NLP-based curation. *Plant and Cell Physiology* 56(1), e9–e9. DOI: 10.1093/pcp/pcu188.

Paczkowska, M., Barenboim, J., Sintupisut, N., Fox, N.S., Zhu, H. *et al.* (2020) Integrative pathway enrichment analysis of multivariate omics data. *Nature Communications* 11(1), 735. DOI: 10.1038/s41467-019-13983-9.

Perry, G.L.W., Miller, B.P. and Enright, N.J. (2006) A comparison of methods for the statistical analysis of spatial point patterns in plant ecology. *Plant Ecology* 187(1), 59–82. DOI: 10.1007/s11258-006-9133-4.

Picard, M., Scott-Boyer, M.P., Bodein, A., Périn, O. and Droit, A. (2021) Integration strategies of multi-omics data for machine learning analysis. *Computational and Structural Biotechnology Journal* 19, 3735–3746. DOI: 10.1016/j.csbj.2021.06.030.

Ramakrishnan, M., Satish, L., Kalendar, R., Narayanan, M., Kandasamy, S. *et al.* (2021) The dynamism of transposon methylation for plant development and stress adaptation. *International Journal of Molecular Sciences* 22(21), 11387. DOI: 10.3390/ijms222111387.

Rao, A., Barkley, D., França, G.S. and Yanai, I. (2021) Exploring tissue architecture using spatial transcriptomics. *Nature* 596(7871), 211–220. DOI: 10.1038/s41586-021-03634-9.

Rautenstrauch, P. and Ohler, U. (2024) Liam tackles complex multimodal single-cell data integration challenges. *Nucleic Acids Research* 52(12), e52. DOI: 10.1093/nar/gkae409.

Reddy, G.V., Gordon, S.P. and Meyerowitz, E.M. (2007) Unravelling developmental dynamics: Transient intervention and live imaging in plants. *Nature Reviews. Molecular Cell Biology* 8(6), 491–501. DOI: 10.1038/nrm2188.

Redmond, E.J., Ronald, J., Davis, S.J. and Ezer, D. (2024) Single-plant-omics reveals the cascade of transcriptional changes during the vegetative-to-reproductive transition. *The Plant Cell* 36(10), 4594–4606. DOI: 10.1093/plcell/koae226.

Rhaman, M.S., Ali, M., Ye, W. and Li, B. (2024) Opportunities and challenges in advancing plant research with single-cell omics. *Genomics, Proteomics & Bioinformatics* 22(2), qzae026. DOI: 10.1093/gpbjnl/qzae026.

Robertson, S.M. and Wilkins, O. (2023) Spatially resolved gene regulatory networks in Asian rice (*Oryza sativa* cv Nipponbare) leaves. *The Plant Journal* 116(1), 269–281. DOI: 10.1111/tpj.16375.

Rogers, E.D., Jackson, T., Moussaieff, A., Aharoni, A. and Benfey, P.N. (2012) Cell type-specific transcriptional profiling: Implications for metabolite profiling. *The Plant Journal* 70(1), 5–17. DOI: 10.1111/j.1365-313X.2012.04888.x.

Rudge, T. and Haseloff, J. (2005) A computational model of cellular morphogenesis in plants. In: *European Conference on Artificial Life*, Springer, pp. 78–87. DOI: 10.1007/11553090_9.

Schaar, A.C., Tejada-Lapuerta, A., Palla, G., Gutgesell, R. and Halle, L. (2024) Nicheformer: A foundation model for single-cell and spatial omics. *BioRxiv*. DOI: 10.2139/ssrn.4803291.

Seyfferth, C., Renema, J., Wendrich, J.R., Eekhout, T., Seurinck, R. *et al.* (2021) Advances and opportunities in single-cell transcriptomics for plant research. *Annual Review of Plant Biology* 72(1), 847–866. DOI: 10.1146/annurev-arplant-081720-010120.

Sodhi, P., Vijayarangan, S. and Wettergreen, D. (2017) In-field segmentation and identification of plant structures using 3D imaging. In: *2017 IEEE/RSJ International Conference on Intelligent Robots and Systems (IROS)*, IEEE, Vancouver, BC, pp. 5180–5187. DOI: 10.1109/IROS.2017.8206407.

Solano, C.J., Barba-Espín, G., Suardíaz, J. and Hernández, J.A. (2022) Basic integration of artificial intelligence of a plant experimentation chamber with leds and sensors through connection to the IoT with node-RED and securing access to data. In: *International Online Conference on Agriculture*, p. 5 (Vol. 10). DOI: 10.3390/IOCAG2022-12228.

Song, Y.-C., Das, D., Zhang, Y., Chen, M.-X., Fernie, A.R. *et al.* (2023) Proteogenomics-based functional genome research: Approaches, applications, and perspectives in plants. *Trends in Biotechnology* 41(12), 1532–1548. DOI: 10.1016/j.tibtech.2023.05.010.

Sorrells, M.E. and Wilson, W.A. (1997) Direct classification and selection of superior alleles for crop improvement. *Crop Science* 37(3), 691–697. DOI: 10.2135/cropsci1997.0011183X003700030001x.

Srivatsan, S.R., Regier, M.C., Barkan, E., Franks, J.M., Packer, J.S. *et al.* (2021) Embryo-scale, single-cell spatial transcriptomics. *Science* 373(6550), 111–117. DOI: 10.1126/science.abb9536.

Stassen, S.V., Yip, G.G.K., Wong, K.K.Y., Ho, J.W.K. and Tsia, K.K. (2021) Generalized and scalable trajectory inference in single-cell omics data with VIA. *Nature Communications* 12(1), 5528. DOI: 10.1038/s41467-021-25773-3.

Sugiyama, M. (2000) Genetic analysis of plant morphogenesis in vitro. *International Review of Cytology* 196, 67–84. DOI: 10.1016/s0074-7696(00)96002-9.

Sun, C., Ali, K., Yan, K., Fiaz, S., Dormatey, R. *et al.* (2021) Exploration of epigenetics for improvement of drought and other stress resistance in crops: A review 10(6), 1226. DOI: 10.3390/plants10061226.

Tenorio Berrío, R. and Dubois, M. (2024) Single-cell transcriptomics reveals heterogeneity in plant responses to the environment: A focus on biotic and abiotic interactions. *Journal of Experimental Botany* 75(17), 5188–5203. DOI: 10.1093/jxb/erae107.

Tenorio Berrío, R., Verstaen, K., Vandamme, N., Pevernagie, J., Achon, I. *et al.* (2022) Single-cell transcriptomics sheds light on the identity and metabolism of developing leaf cells. *Plant Physiology* 188(2), 898–918. DOI: 10.1093/plphys/kiab489.

Terpstra, I.R., Snoek, L.B., Keurentjes, J.J., Peeters, A.J. and Van den Ackerveken, G. (2010) Regulatory network identification by genetical genomics: Signaling downstream of the arabidopsis receptor-like kinase ERECTA. *Plant Physiology* 154(3), 1067–1078. DOI: 10.1104/pp.110.159996.

Thornton, C.A., Mulqueen, R.M., Torkenczy, K.A., Nishida, A., Lowenstein, E.G. *et al.* (2021) Spatially mapped single-cell chromatin accessibility. *Nature Communications* 12(1), 1274. DOI: 10.1038/s41467-021-21515-7.

Tripathi, R.K. and Wilkins, O. (2021) Single cell gene regulatory networks in plants: Opportunities for enhancing climate change stress resilience. *Plant, Cell & Environment* 44(7), 2006–2017. DOI: 10.1111/pce.14012.

Vachier, C. (2001) Morphological scale-space analysis and feature extraction. In: *Proceedings 2001 International Conference on Image Processing*, IEEE, pp. 676–679 (Vol. 3). DOI: 10.1109/ICIP.2001.958209.

Vignes, M., Vandel, J., Allouche, D., Ramadan-Alban, N., Cierco-Ayrolles, C. *et al.* (2011) Gene regulatory network reconstruction using Bayesian networks, the dantzig selector, the lasso and their meta-analysis. *PloS One* 6(12), e29165. DOI: 10.1371/journal.pone.0029165.

Vu, A.H., Kang, M., Wurlitzer, J., Heinicke, S., Li, C. *et al.* (2024) Quantitative single-cell mass spectrometry provides a highly resolved analysis of natural product biosynthesis partitioning in plants. *Journal of the American Chemical Society* 146(34), 23891–23900. DOI: 10.1021/jacs.4c06336.

Wan, X., Xiao, J., Tam, S.S.T., Cai, M., Sugimura, R. *et al.* (2023) Integrating spatial and single-cell transcriptomics data using deep generative models with spatialscope. *Nature Communications* 14(1), 7848. DOI: 10.1038/s41467-023-43629-w.

Wang, H., Xu, Y., Zhang, Z., Zhang, G., Tan, C. *et al.* (2024) Development and application of transcriptomics technologies in plant science. *Crop Design* 3(2), 100057. DOI: 10.1016/j.cropd.2024.100057.

Wang, Y., Zhang, P., Guo, W., Liu, H., Li, X. *et al.* (2021) A deep learning approach to automate whole-genome prediction of diverse epigenomic modifications in plants. *New Phytologist* 232(2), 880–897. DOI: 10.1111/nph.17630.

Weckwerth, W. (2008) Integration of metabolomics and proteomics in molecular plant physiology--coping with the complexity by data-dimensionality reduction. *Physiologia Plantarum* 132(2), 176–189. DOI: 10.1111/j.1399-3054.2007.01011.x.

Wenzel, G. (2006) Molecular plant breeding: Achievements in green biotechnology and future perspectives. *Applied Microbiology and Biotechnology* 70(6), 642–650. DOI: 10.1007/s00253-006-0375-9.

Xu, L., Pierroz, G., Wipf, H.M.-L., Gao, C., Taylor, J.W. *et al.* (2021) Holo-omics for deciphering plant-microbiome interactions. *Microbiome* 9(1), 69. DOI: 10.1186/s40168-021-01014-z.

Yang, Y., Saand, M.A., Huang, L., Abdelaal, W.B., Zhang, J. *et al.* (2021) Applications of multi-omics technologies for crop improvement. *Frontiers in Plant Science* 12, 563953. DOI: 10.3389/fpls.2021.563953.

Yin, R., Xia, K. and Xu, X. (2023) Spatial transcriptomics drives a new era in plant research. *The Plant Journal* 116(6), 1571–1581. DOI: 10.1111/tpj.16437.

Zamanighomi, M., Lin, Z., Daley, T., Chen, X., Duren, Z. *et al.* (2018) Unsupervised clustering and epigenetic classification of single cells. *Nature Communications* 9(1), 2410. DOI: 10.1038/s41467-018-04629-3.

Zeng, Y., Chen, X., Gao, H. and Wang, X. (2018) An artificial intelligent single cell is part of the cell dream world. *Cell Biology and Toxicology* 34(4), 247–249. DOI: 10.1007/s10565-018-9433-1.

Zhang, J., Ahmad, M. and Gao, H. (2023) Application of single-cell multi-omics approaches in horticulture research. *Molecular Horticulture* 3(1), 18. DOI: 10.1186/s43897-023-00067-y.

Zhang, J.M., Fan, J., Fan, H.C., Rosenfeld, D. and Tse, D.N. (2018) An interpretable framework for clustering single-cell RNA-seq datasets. *BMC Bioinformatics* 19(1), 93. DOI: 10.1186/s12859-018-2092-7.

Zhang, X., Luo, Z., Marand, A.P., Yan, H., Jang, H. *et al.* (2024) A spatially resolved multiomic single-cell atlas of soybean development. *Plant Biology*. DOI: 10.1101/2024.07.03.601616.

Zhao, H., Tu, Z., Liu, Y., Zong, Z., Li, J. *et al.* (2021) PlantDeepSEA, a deep learning-based web service to predict the regulatory effects of genomic variants in plants. *Nucleic Acids Research* 49(W1), W523–W529. DOI: 10.1093/nar/gkab383.

Zhu, B., Gao, S., Chen, S., Yeung, J., Bai, Y. *et al.* (2024) Cross-domain information fusion for enhanced cell population delineation in single-cell spatial-omics data. *BioRxiv* 2024.05.12.593710. DOI: 10.1101/2024.05.12.593710.

Zuo, C., Dai, H. and Chen, L. (2021) Deep cross-omics cycle attention model for joint analysis of single-cell multi-omics data. *Bioinformatics* 37(22), 4091–4099. DOI: 10.1093/bioinformatics/btab403.

6 Artificial Intelligence for Reimagining Crop Design

Siddhartha Das[1]* and Priyanka Paul[2]

[1]Department of Plant Pathology, M.S Swaminathan School of Agriculture, Centurion University of Technology and Management, Paralakhemundi, India; [2]Department of Agronomy and Agroforestry, M.S Swaminathan School of Agriculture, Centurion University of Technology and Management, Paralakhemundi, India

Abstract

Artificial intelligence (AI) is transforming agriculture through its integration into crop design, fostering precision, sustainability, and resilience. This chapter focuses on the revolutionary possibilities of AI within crop improvement, focusing on its ability to predict growth, health, and yield based on minimal input data. Advanced machine learning algorithms, developed from huge quantities of crop pictures and phenotypic traits, will simulate future growth trajectories and optimize resource allocation. For example, AI tools can evaluate one image of a young plant to predict its development under varying environmental conditions, offering actionable insights for fertilization, irrigation, and pest control strategies. Through applications such as genomic selection and phenotypic prediction, AI accelerates breeding cycles by identifying superior genotypes and linking genetic markers to desirable traits. Moreover, AI-driven systems enhance decision making by processing current information from Internet of Things sensors, allowing exact assessment of soil conditions and nutrient levels, as well as climatic factors. These innovations not only increase crop output but also decrease ecological impact by minimizing resource waste. The chapter also highlights challenges in adopting AI for crop design, including the difficulties of integrating diverse datasets with the necessity for user-friendly interfaces for landowners. Collaborative approaches, such as participatory AI frameworks, are proposed to ensure inclusivity and adaptability across different agricultural contexts. By bridging the separation across computational power and applications in reality, AI holds immense promise for reimagining crop design to address the needs of a world population that is growing sustainably.

Keywords: crop design, AI, soil health, Internet of Things (IoT), phenotypic prediction

6.1 Introduction

Diseases of plants have a substantial impact on both the yield and supply of crops, posing a major threat to food security. Early detection is crucial to minimize crop loss and boost productivity. Agriculture accounts for about 17% of India's GDP (Agarwal *et al.*, 2019), and the country is a leading producer of crops such as eggplant, potatoes, tomatoes, and pepper. Environmental components and infection contribute to the increase of plant diseases (Kodama and Hata, 2018), while pests are responsible for a 30–33% reduction in yield

*Corresponding author: siddhartha.das10@gmail.com

DOI: 10.1079/9781800628649.0006

annually (Kumar *et al.*, 2019). Infections caused by fungi, bacteria, and viruses require improved management practices to safeguard crop output and quality (Ayoub Shaikh *et al.*, 2022). Despite technological advances, many producers continue to depend on conventional techniques. Slight condition symptoms and the need for specialized tools make detection difficult. Machine learning (ML) and artificial intelligence (AI) offer promising solutions, including early disease identification, yield forecasting, and labor reduction. These technologies, central to Agriculture 5.0, use sensors, satellite imaging, and smart devices for precision farming. A bibliometric study of more than 120 ML models identified commonly used AI and deep learning (DL) algorithms for detecting infections, pests, and weeds. These include support vector machines (SVMs), decision trees, and convolutional neural networks (CNNs), which support actual analysis of data and allow educated choices. Alongside plant protection, plant development has evolved significantly—from traditional seed selection in Breeding 1.0 to genomic, AI, biotechnology, precision, and data-driven methods. Historically, breeding advanced through the discovery of inbreeding depression and scientific crossing methods of marker-assisted selection, genetic improvement techniques to the merging of genetic information in genetic modification, and improved traits. Today, AI accelerates breeding processes through highly characterized functional genomic sequencing, leading to the development of high-yield, drought-tolerant, heat-resistant, adaptive, sustainable plants. With a rising population, limited arable land, and climate change, AI is critical for developing crops that withstand pests, diseases, pathogens, drought, temperature, and flooding, while maintaining productivity and health benefits (Zohry and Ouda, 2022). AI simulates human intelligence through algorithms capable of learning, predicting, and optimizing (Amit, 2018). It organizes vast agricultural datasets into actionable insights, simplifying breeding decisions and enhancing productivity.

6.2 Connecting the Development of Agricultural Technology to Crop Protection

Crop protection plays an important part in providing agricultural profitability as well as environmental long-term viability. Over time, agriculture has progressed through various stages. Farming 1.0, until the 1950s, relied on manual labor to manage pests, resulting in modest profits but adequate nutrition. Farming 2.0 began in the late 1950s with the introduction of synthetic pesticides and mechanized tools, shifting agriculture toward industrialized food production. In the final decades of the 20th century, Farming 3.0 emerged, incorporating information-based technologies for smarter decision making in crop management. With Agriculture 4.0, precision farming became a reality through the use of telematics, global navigation satellite systems (GNSS), sensor technologies, Internet of Things (IoT), mobile devices, and cloud computing. These innovations enabled automated equipment, smart nozzles, and precision pesticide application, reducing chemical use and improving food quality. Agriculture 5.0 (Ag5.0) pushes the boundaries further by implementing, ML, AI, and autonomous systems to facilitate data-driven, low-intervention farming (Saiz-Rubio and Rovira-Más, 2020). Future farming must tackle major global issues like environmental change and availability limitations, as well as the emergence of more aggressive or new pest species due to rising temperatures and unpredictable weather (Mulla *et al.*, 2020). Maintaining a growing community while limiting adverse ecological effects harm demands innovation for crop protection. Ag5.0 supports this goal through technologies that reduce agrochemical use and enhance precision in pest and disease management, underlining the shift toward sustainable, resilient, and intelligent agriculture (van Dijk *et al.*, 2020).

6.3 Perception, Analysis, and Action are the Phases of Precision Crop Protection

Modern crop protection involves three key stages: perception; analysis and decision making; and actuation (Behmann *et al.*, 2015). The perception stage involves detecting and collecting plant data, often using imaging or sensor technology on ground-based as well as remote-sensing devices. This data may include images for identifying crops, weeds, or disease symptoms. In the decision-making phase, ML models examine huge volumes of data related to farming to identify trends evaluate conditions, as well as generate treatment maps for targeted interventions (Wang *et al.*, 2019). These models are instrumental in identifying crop-specific pests, weeds, and diseases by processing visual and location-based data under varying environmental and socioeconomic factors (Pätzold *et al.*, 2020). The actuation stage applies site-specific treatments using smart machinery, often equipped with GNSS for accurate location tracking. Technologies like unmanned aerial vehicles (UAVs), autonomous machines, and decision support systems are increasingly used for large-scale precision crop protection (Dainelli *et al.*, 2021). UAVs capture real-time data using RGB, multispectral, hyperspectral, thermal, and active sensors such as LiDAR and radar, significantly improving pest and weed monitoring. ML algorithms, supported by high-performance central processing units (CPUs), graphics processing units (GPUs), and tensor processing units (TPUs), play a critical role in rapid, accurate detection and diagnosis (Shafi *et al.*, 2019). These innovations allow for immediate or near-real-time treatment through advanced equipment like smart nozzles and autonomous sprayers, enhancing efficiency and sustainability in site-specific crop protection (Lowenberg-DeBoer *et al.*, 2021).

6.4 ML Taxonomy and Applications

ML methods are categorized based on factors like model type, argumentation, data accessibility, and oversight. During accuracy agriculture, ML is commonly used for regression analysis, grouping, classification, discovering anomalies, reduced dimensionality and rules of association building. In the accurate protection of crops, ML may be usefully applied to tasks such as regression, classification, clustering, abnormality detection, decrease in dimensionality, and rule-based association learning. Conventional ML algorithms may address these tasks, while advanced models like DL methods and artificial neural networks are particularly useful for tasks like regression and identification.

6.4.1 Conventional machine learning methods

Conventional ML techniques analyze input data through optimized architectures suitable for standard computing hardware, producing generally adequate outcomes; however, they often lack the precision and flexibility offered by complex artificial neural networks. Among the most commonly used ML tasks is classification, which applies supervised learning methods to categorize structured or unstructured data into binary or multi-class forms. In agriculture, such methods are essential for identifying plant pathogens, weeds, and insects from visual data. Before classification, phases like image/video preprocessing, segmentation, and feature extraction may involve other ML techniques. These can be constructed using metric methods, such as linear regression, or more flexible, non-metric approaches. Regularization techniques like LASSO, Ridge, and Elastic Net are critical for preventing overfitting, and improving model generalization during unseen data evaluation. Kernel-based regression uses similar input patterns to estimate outputs based on relationships with known data points. Heterogeneous techniques, that combine those predictions of numerous base algorithms, have shown improved performance over single models in both classification and regression tasks. These methods reduce bias, and variance, and improve model stability (Sagi and Rokach, 2018). Technologies like Bagging, Gradient Boosting Machines, Boosting, CatBoost and Extremely Randomized Trees are notable ensemble approaches. Ensemble regression models further enhance precision (Ren *et al.*, 2016). Techniques like partitional grouping optimize group assignments based on objective functions (Nanda and Panda, 2014), especially in agricultural datasets involving multiple sensors and image features. This can be achieved using feature selection (which retains original variables) or feature extraction (which creates new variables). Both methods aim to reduce computational complexity while maintaining essential information (Chhikara *et al.*, 2020). These are critical in crop protection for detecting unusual plant symptoms or pest activity (Guansong *et al.*, 2022). Finally, association rule learning uncovers patterns or regularities in datasets and applies them to predict future trends. This is valuable for understanding co-occurring agricultural phenomena and guiding decision making. Techniques in this category include descriptive and predictive rule mining (Fürnkranz and Kliegr, 2015).

6.4.2 Models for deep learning and artificial neural networks

Artificial neural networks (ANNs) are adaptable algorithms motivated by actual neural networks. They utilize nodes that are interlinked (synapses) to generate and learn complex patterns as well as solve non-linear problems through general function approximation. ANNs adaptively learn from large datasets using sequential calculations and can predict outcomes when presented with new data. These models are widely used in classification and regression tasks, including proximity functions (mapping inputs to outputs), pattern recognition, associative memory (identifying patterns from partial data), and trend generation for pattern reconstruction (LeCun *et al.*, 2015). DL, built on multi-layered ANNs, has significantly advanced numerous fields, particularly precision agriculture. DL models automatically extract hierarchical features, reducing manual development efforts and enhancing computational efficiency, even with small datasets (Kamilaris and Prenafeta-Boldú, 2018). Specialized DL architectures address different tasks—CNNs for image recognition, recurrent neural networks (RNNs) for temporal predictions, autoencoders for dimensionality reduction, and generative adversarial networks (GANs) for image synthesis (Sarker, 2021). CNNs are most widely used in precision crop protection for image classification. They learn visual features without manual intervention, unlike traditional ML (Hong *et al.*, 2020). CNN performance depends on input volume, depth, and computational resources (Table 6.1). Common CNN variants used in crop analysis include YOLO, FPN, Faster R-CNN, RetinaNet, R-CNN, SSD, and Fast R-CNN. These models aid in detecting and classifying crop diseases, pests, and stress conditions, contributing to sustainable agriculture.

6.5 Impact of ML on Precise Crop Protection

The implementation of AI and ANNs in precision crop protection has expanded rapidly between 2010 and 2022, with a comprehensive review identifying 107 ML and 18 ANN models

Table 6.1. Features among the most basic deep learning models utilized in agriculture. (Author's own table.)

CNN design	Depth	Variables (millions)	Top five efficiency percentage	References
Xception	126	22.8	94.5	Chollet, 2017
MobileNet	55	4.3	89.5	Howard *et al.*, 2017
GoogLeNet	22	4.0	92.2	Szegedy *et al.*, 2015
DenseNet	402	20.2	93.6	Huang *et al.*, 2016
VGG-Net	16	138.4	90.1	Simonyan and Zisserman, 2014
ResNet	152	60.4	93.1	He *et al.*, 2016

utilized across multiple agricultural domains. Of these, 105 ML algorithms and 17 ANN models have been successfully implemented, particularly in detecting and managing crop diseases, weeds, and pests. SVMs lead classical ML methods with more than 1700 documented applications. Other widely adopted algorithms include Random Forests (RFs), Decision Trees and Principal Component Analysis (PCA), each exceeding 1100 uses (Zhu *et al.*, 2017). Techniques like k-nearest neighbors (k-NN), logistic regression, k-means clustering, and bagging are also prominent, especially in weed identification and spectral data analysis. CNNs dominate ANN applications, particularly for image-based tasks such as disease and weed detection, with more than 1200 studies using architectures like ResNet, GoogLeNet, VGGNet, MobileNet, and Xception (Selvaraj *et al.*, 2020a). Regression and classification remain the primary tasks, constituting more than 80% of all agricultural applications. Dimensionality compression algorithms like PCA and partial least squares discrimination analysis (PLS-DA) are increasingly more frequently utilized for agriculture than in other sectors. Since 2018, ANNs have increasingly outperformed classical ML methods, especially in high-complexity cases, signaling a shift toward DL models in precision crop protection (Sarker, 2021).

6.5.1 Crop diseases

Several methods involving ML are being applied to the early detection and forecasting of agricultural diseases utilizing hyperspectral imaging and sensor data (Table 6.2). For example:

- Stepwise discriminant analysis (SDA), Fisher linear discriminant analysis (FLDA), and k-NN were used to categorize *Colletotrichum*-induced anthracnose in strawberries (Lu *et al.*, 2017).

Table 6.2. Suitable research on machine learning crop-disease-related technologies. (Author's own table.)

Detection technological advances	Crop/ pathogen type	Primary objective	Problem to be solved	ML algorithm	References
On-ground hyperspectral camera	X Sugar beet/ fungal	Initial identification of Rhizoctonia root and crown rot (*Rhizoctonia solani*) in the foliage.	Ensemble regression and classification	Linear SVM, radial SVM, and k-NN	Barreto *et al.*, 2020
Field spectroradiometer	Wheat/ fungal	Detecting as well as tracking powdery mildew (*Erysiphe graminis*)	Ensemble and regression	RF, SVM, and PLSR	Feng *et al.*, 2022
UAV-based hyperspectral images	Wheat/ fungal	Yellow rust (*Puccinia striiformis* f.Sp. Tritici) can be identified throughout the growing season.	Regression and classification	RF and ResNet	Zhang *et al.*, 2019a
UAV-based multispectral images	Apple trees/ bacterial	Identification of apple fire blight (*Erwinia amylovora*)	Classification, anomaly detection, and dimensionality reduction	RF, SVM, DT, and Isolation Forest	Xiao *et al.*, 2022

- PLS-DA and RF detected pre-symptomatic *Phytophthora* infestation infections in potatoes (Gold *et al.*, 2020).
- A combination of PLS regression, support vector regression (SVR), and RF regression models were used to monitor *Erysiphe graminis* in wheat (Feng *et al.*, 2022).
- Use of SVM, RF, back propagation neural network (BPNN), and least squares (LS)-SVM led to early detection of tobacco mosaic virus (Zhu *et al.*, 2017).

6.5.2 Weeds

Weed science has profited significantly from spectrum evaluation and data mining. ML and ANN models are able to identify weeds (Table 6.3). Scientists have addressed:

- SVM, RF, and SPA for discovering wavelength relationships among types of weeds, like Barnyard grass and weedy rice (Zhang *et al.*, 2019b).
- The identification of herbicide-resistant strains, including *Amaranthus palmeri*, using FLDA and RBF kernels (Huang *et al.*, 2022).

6.5.3 Insects

ML and ANN models are able to identify and simulate infestations of insects (Table 6.4). For example:

- Combined ANNs use NIR and e-nose data to predict aphid numbers (Fuentes *et al.*, 2021).
- MLP, SVM, XGBoost, and Naïve Bayes to protect cotton from armyworm infestations (Ramos *et al.*, 2022).

Remote sensing, particularly with UAVs, has made the identification of pests easier to scale up. For example:

- CNNs allowed for the early detection of Huanglongbing in citrus fruits (Garcia-Ruiz *et al.*, 2013).
- AlexNet, MobileNet, and ShuffleNet were used to detect fungal grapevine diseases (Tang *et al.*, 2020).

The great spatial resolution of UAV imagery makes it easier to discover weed seedlings early on and manage them effectively. ML approaches tackled field challenges including:

- K-means, SVM, RF, and ensemble models were used to detect weeds inside and

Table 6.3. Research on ML algorithms of agriculture weed types. (Author's own table.)

Detection technological advances	Crop/weed species	Primary objective	Problem to be solved	ML algorithm	References
UAV-based multi spectral and/or RGB images	Wheat/ blackgrass weed	Spectrum evaluation and determination of blackgrass weeds	Dimensionality reduction and classification	RF with Bayesian optimization	Su *et al.*, 2022
Satellite multi spectral images	Winter wheat/ cruciferous weeds	Monitoring cruciferous weed areas in many areas on bigger dimensions	Classification	MLC	de Castro *et al.*, 2013
On-ground RGB imagery	Tomato/several weeds	Identification and designation of different types of weeds	Classification	Retina Net, YOLOv7 Lopez and Faster RCNN	López López-Correa *et al.* (2022)
Field spectroradiometer	No crop/*Sorghum halepense*	Separating between glyphosate-resistant and susceptible Johnson grass	Regression, ensemble, and classification	SVM with FLDA and RF	Huang *et al.*, 2022

Table 6.4. Suitable research on ML algorithms in the field of crop insects. (Author's own table.)

Detection technological advances	Crop/insect type	Primary objective	Problem to be solved	ML algorithm	References
On-ground RGB imagery	Tomato and pepper/pest	Vision-based programmed recognition and quantification of Bemisiatabaci and *Trialeurodes vaporariorum*	Classification	MLP, Faster-RCNN and k-NN	Gutierrez *et al.*, 2019
UAV-based multispectral imagery	Cotton/spidermite	Identifying two-spotted spider mites in crop fields	Classification	Alex Net and SVM	Huang *et al.*, 2018
Repository of insect images	No crop/multi-class plagues	Identifying as well as categorizing multi-class effects on biodiversity in insect photos	Classification	SSD, Fast RCNN and VGG19	Xia *et al.*, 2018
Portable NIR spectroscopy and e-nose sensors	Wheat/aphid	Identifying and projecting insect numbers	Regression and classification	ANN-based regression models	Fuentes *et al.*, 2021

outside crop oilseed and cereals (de Castro *et al.*, 2013).

- MLP-based classification of grass and broadleaf (Torres-Sánchez *et al.*, 2021).
- Bayesian-optimized RF and DT to map invasive species in patches (Su *et al.*, 2022).

UAV-based disease classification has also proven to be effective. For example:

- Banana disorders were classified using RetinaNet, ResNet50, and VGG16 (Selvaraj *et al.*, 2020b).
- AlexNet and RNNs were used to detect fungal infections in wheat using ground-based hyperspectral imagery (Jin *et al.*, 2018).

Recent improvements in CNN have resulted in vastly enhanced real-time field vision and remote diagnostic applications. Examples are:

- GoogleLeNet, EfficientNet, YOLOv7, and RetinaNet were used to identify weed species in potato and tomato fields (López-Correa *et al.*, 2022).
- ZF, VGG16, and ResNet50/101 were used to classify pest photos into many classes (Liu *et al.*, 2019).

6.6 New Precision Crop Safety Technologies Integrated Using AG5.0

Recent advancements in the protection of crops have been significantly influenced by and ML and AI aligning with the concepts of digital agricultural production under Ag4.0 and progressing toward Ag5.0. AI, defined by its ability to learn, reason, and make decisions, is becoming increasingly integrated into agricultural practices, particularly in the form of intelligent systems for crop protection. ML algorithms have been widely applied across agricultural tasks and are evolving to meet the more complex goals of Ag5.0 (Secretariat, 2021). Ag5.0 emphasizes two primary objectives: (i) early identification of pests, diseases, and weeds; and (ii) real-time autonomous task execution. Early detection enables proactive intervention before substantial crop damage occurs. This is increasingly important as climate change accelerates the emergence and spread of new pests (Shankar *et al.*, 2020). Ag5.0 technologies enhance detection by integrating data from climate records, satellite images, soil and crop sensors, and farm management systems to accurately assess pest outbreaks (Ramcharan *et al.*, 2019). These systems use localized data to apply precise management strategies tailored to specific scenarios. Artificial General Intelligence, version 5.0 integrates advanced ML algorithms (Coulibaly *et al.*, 2022), smart devices, communication infrastructure and agricultural machinery (Albiero *et al.*, 2022), forming a cohesive ecosystem to tackle future agricultural challenges efficiently.

6.6.1 Hardware options for accurate crop protection

Agriculture is experiencing an important change with the introduction of Ag5.0, moving away from traditional, labor-intensive methods to a data-driven, intelligent, and automated system. This evolution is being powered by technological advancements in hardware, including smart sensors, IoT ecosystems, high-performance computing (HPC), and emerging neuromorphic computing. These innovations aim to increase crop yields, reduce input use, and enhance precision in agricultural practices. A key development is the integration of AI directly into sensors, enabling real-time processing at the edge. Unlike conventional sensors, AI-powered smart sensors collect and analyse data on-site, reducing latency and dependence on cloud infrastructure. Sony's IMX500 and IMX501 exemplify this innovation by processing lightweight AI models like MobileNet V1 directly on-chip, extracting semantic metadata from images within milliseconds (Sony, 2020). This facilitates bandwidth-efficient real-time weed and pest detection, boosting the effectiveness of decision support systems (DSS) while minimizing pesticide usage and environmental harm (Muhammad *et al.*, 2019). However, deploying AI-enabled sensors in agriculture presents challenges such as energy efficiency, hardware durability in harsh conditions, and maintaining model accuracy across diverse crops. To address the growing data demands, HPC technologies including GPUs, TPUs, and NVMe (non-volatile memory express) storage are being used to accelerate AI training and the development of artificial twins—virtual duplicates of the reality agricultural systems. These twins simulate crop growth and disease dynamics, offering predictive insights for management (Glaessgen and Stargel, 2012). FPGAs (field programmable gate arrays) are also gaining traction in Ag5.0. Their custom logic and low power consumption make them ideal for edge computing tasks like CNN inference, enabling mobile and IoT devices to function independently of constant internet access. Cybersecurity is an emerging concern in Ag–IoT ecosystems. Inadequate security in IoT devices can compromise sensitive agricultural data and disrupt automated processes. The adoption of application-specific integrated circuits (ASICs) and ML-optimized gateways helps mitigate these threats by enhancing data protection, optimizing routing, and maintaining privacy via local storage (Oñate and Sanz, 2023). Looking forward, neuromorphic computing represents a revolutionary leap in agricultural hardware and software. Inspired by the human brain, this approach uses memristors—two-terminal devices capable of simultaneous data maintenance and computing. These are essential for spiking neural networks (SNNs), which mimic biological synapses and transmit data as discrete spikes, unlike traditional ANNs. SNNs offer advantages such as low power consumption, reduced latency, and improved real-time performance (Davies *et al.*, 2018). Though still in experimental stages, SNNs are promising for energy-efficient agricultural robots, sensor networks, and mobile crop monitoring systems.

6.6.2 Precision crop safety with telecommunications

Customized protection of crops is rapidly evolving into a grouping of organizations framework that integrates diverse technologies—including ground-level, proximal, and remote sensing—to create a comprehensive approach to crop health monitoring and management. Central to this integration is telecommunication technology, which enables seamless connectivity among sensors, processors, and actuators, facilitating real-time data collection, analysis, and autonomous crop treatment. Wireless sensor networks (WSNs) provide low-power, short-range communication, with Zigbee offering up to 100 m of coverage and Bluetooth enabling higher data rates (1–24 Mbps) for shorter distances (Zeadally *et al.*, 2019). For broader coverage, technologies like Wi-Fi, WiMAX, and increasingly LoRa/LoRaWAN are employed. LoRaWAN is especially beneficial in rural settings because of its low use of electricity and long coverage, and secure data transmission in unlicensed frequency bands (Lavric, 2019). Real-time agricultural interventions are executed through actuator platforms such as UAVs, unmanned ground vehicles (UGVs), and automated sprayers. These systems require rapid data processing and actuation, which is enabled by cloud computing, edge computing, and Edge AI. Cloud services support complex ML models but are hindered by latency issues (Mahmoud and Xia, 2019). With the emergence of 5G/6G networks, latency for applications like drone-based crop spraying is expected to decrease significantly (Ullah *et al.*, 2020). However, privacy and connectivity limitations encourage the adoption of fog computing and edge processing (Sun, 2020). Edge AI further enhances this by deploying ML models locally on embedded devices, reducing dependency on continuous internet access and enabling fast, autonomous decisions (García-Valls *et al.*, 2018). This makes Edge AI a promising solution for the next generation of precision agricultural robots.

6.6.3 Precision crop protection with robotics

Autonomous mobile robots (AMRs) are revolutionizing precision agriculture within Ag5.0 frameworks by enhancing productivity, reducing labor dependence, and enabling actual time, decisions based on data. These machines have been fitted with modern features locomotion, perception, and intellectual skills that are ideal for complex agricultural tasks such as weed detection and targeted pest control—key elements of sustainable crop management (Shamshiri *et al.*, 2018). Several AMRs like RIPPA and AgBot-II utilize AI-based vision and DL models such as Inception-v3 to identify and manage weeds with high precision (McCool *et al.*, 2018). Others, including Robotti, AVO, and BONIROB, apply computer vision and ML for real-time weed management using chemical, mechanical, or hybrid approaches (Grimstad and From, 2017). Robots such as Kilter AX1, DINO, and Odd.bot demonstrate centimeter-level accuracy in micro-dosing herbicides or performing mechanical weeding via RTK GPS and smart nozzle systems, reducing pesticide use by up to 95%. Platforms like Titan FT35 and FARMING GT combine ML and internal actuators for high-accuracy weed removal across crop types. Another innovation is collaborative robots designed for organic farming. These co-bots work alongside humans to perform labor-intensive tasks like harvesting and mechanical weeding, enhancing efficiency in non-chemical operations (Zhang *et al.*, 2021). Future precision farming will benefit from multi-robot fleet systems (MFSs), integrating UAVs and UGVs under decentralized control for superior adaptability and output (de Santos *et al.*, 2017). UAVs equipped with RGB, multispectral, and hyperspectral sensors, along with onboard AI, enable accurate pest and disease detection (Floreano and Wood, 2015). ML techniques also support GPS-independent UAV tracking using artificial neural networks and Kalman filters (Belhajem *et al.*, 2016). Collectively, the fusion of AMRs, ML, and robotic fleet management in Ag5.0 enhances sustainability, efficiency, and precision in modern agriculture (Emmi *et al.*, 2014).

6.7 Software Conceptual Foundations

6.7.1 Knowledge required and expected outcomes

Agricultural design for management can take two ways, each with its own set of information. The first involves assigning a specialist to address this issue. Collaborating with multiple experts who specialize in specific aspects of crop cultivation is ideal. Experts use their own knowledge to justify their decisions, while also considering the findings of others. The “blackboard” technique involves specialists writing and reviewing one other's views for enhancement of their own advice. Our second method involves using frameworks to convey an understanding of wheat. Models represent a hypothetical relationship involving two factors, which might be numerical or conceptual. The level of fertilizer used affects the percentage of nitrogen absorbed by the plant, the total amount of cereals per square centimeter, yield, as well as other factors. In general, there are three types of variables: (i) the development factors and growth variables; (ii) factors affecting production; and (iii) decision-making variables.

6.7.2 Model for formulating and solving problems

Mathematics-based approaches to this issue involve various minimization strategies, and solve a system of (in)equations containing n unknown variables. Yet, this approach has some disadvantages., like a lack of user guidance and a preference for linear computations over the ones we used. To overcome these obstacles, replicate genuine human thinking. The method of problem solving is modeled as follows. Fix certain variables and transmit their values under known conditions to assess unknown arguments. This process continues until each of the network parameters has consistent values. Value transmission involves restricting or filtering the distribution of variables. If the limitation is insufficient to assign an individual number for every factor, an amount is assumed and transmitted over the network of variables. If this assumption violates a constraint, return to the previous assumption and propose a different value for the exact same parameter. The diffusion method may result in many sets of values that satisfy all restrictions. If no set has been identified, it indicates that the very first set of restrictions was incongruous. In this case, a non-imperative obligation can be temporarily suspended and the entire procedure can be restarted to find alternatives for the slightly different challenge.

6.7.3 Four basic mechanisms for constraint satisfaction

Assertion. At first, every parameter is assigned an assigned area di. The yield ranges from 2000 to 12,000 kg/ha. An assertion assigns a collection of statistics (S) to a property from the domain di. Initial claims include agricultural plan parameters (e.g. yield level) and restrictions (e.g. earliest sowing date). They establish an initial environment where transmission can begin. R may compute a parameter “Wis” if each of its variables (111, V2,... Vn) has an exclusive value. To compute W for a tuple relation R, follow these steps: (i) select the list of pairs containing the present values of V1, V2,... Vn; (ii) append the numbers of W to S; and (iii) specify that the actual value of W is S. Assume R is rational. In this connection, we use a mathematical equation with W on the leftmost side along with V^, V2,... Vn on the side closest to us. We then evaluate S as this side's quantity. (iv) We claim that the actual amount of W is S. The technique estimates the lowest or greatest amount of W based on V~, V2,... Vn, as opposed to computing W for many values. When S represents a single value, W is believed to be known. Once triggered, R cannot be used to compute an additional response.

6.8 Software Description

6.8.1 Knowledge representation

Separating universal design knowledge derived from domain-dependent information needs declarative and adaptable information modelling. Object-oriented development methods of programming address both of these requirements. Class items contain fundamental information. They are a key component in software packages that can be reused for comparable challenges. To use the generic bundle to a specific domain, for example, the quantity of programs must be built. Items include information about the grain crop administration. Examples learn general information from their classes. There are two types of things that require attention: parameters and interaction, which clearly structure the knowledge that is accessible. Objects (classes or instances) typically include both descriptive and procedural information, which is organized through characteristics. Descriptive attributes describe a thing's status or organization, such as the present state of a parameter or the parameters of a relationship. Procedural characteristics specify how to conduct algorithmic processes on objects, such as equations or constraints.

6.8.1.1 Variables

Computable factors, unlike non-computable factors, include an INFO.SOURCE property containing an index of relationships for calculation and a COMPUTATION attribute containing the processing code. Diverse factors, and different ongoing variables, have a RE-STRICT.TUPLES attribute with the law to enter the limited mechanism. Additionally, the computer's code differs depending on if the data point is constant or discontinuous. Constant characteristics can be classed based on what they do. (i) Net architecture representation with INFO.SOURCE as well as PARAME-TER.OF characteristics. The last one includes connections that prohibit constant. (ii) The currently active spreading stage is indicated by the CURRENT.VALUES characteristic, and it includes a selected portion of the first domain from originally processed development systems. The VALUE characteristic stores the information of CURRENT.VALUES after it shrinks to one number. (iii) The procedure script includes the INITIALIZATION property, that offers for initial claims, as well as COMPUTATION and RESTRICT. (iv) This section includes TUPLES characteristics for individual parameters used in assumption systems and clarification establishments with the TRACE link for recording actions as well as outcomes.

6.8.1.2 Relations

There are two types of relation things like tuple relations and functional relationships. Their characteristics perform two types of processes: (i) The ARGUMENTS property provides a list of variables related to the entire network structure, including computed and non-computed parameters. (ii) Managing dispersion dynamics: The Importance of the Unknown. ARG. NB begins with the total amount of parameters specified of AR-GUMENTS assigned and reduces by 1 when a parameter of the connection is given an unusual value (which is, in its value characteristic).

6.8.1.2.1 DESCRIPTION OF THE RELATIONSHIP FOR FUNCTIONALS. FORMULAS property provides a list of formulations. Each equation estimates one of the data relation's inputs as an extension of the others. This requires an association to be recoverable. Formulas must be written to calculate NG as an indicator of P 1G and PIG. POT, and P 1G is a variable of NG and P I G.POT. It is important to highlight that P 1G. POT is not a comprehensible logic under RF.CR.234.

6.8.1.2.2 AN EXPLANATION OF THE TUPLE RELATIONS CONNECTION. The TUPLES variable includes several tuples. A tuple represents a collection of equally valid values that constitute the relation's parameters. Working items are procedures that update variables and relations to determine actions during propagation. When a parameter's CURRENT.VALUES trait changes, and the present value associated with it is instantly enabled. If the changed value is an empty value, the process assigns the value to the variable's VALUE attribute and, if discrete, adds its name to the RESTRICTIVE. ARG attribute of each relation where it appears. Restricting the current scope of a single variable constrains the domains of additional variables in the same tuple relation.

6.8.2 Algorithm for addressing problems

The system controls three data frameworks: (i) a not-yet-triggered interaction (Q), containing all the network's interactions at the start of the procedure; (ii) the triggerable interactions (T), which have only a single unidentified contention; and (iii) the connections that can be restricted (U). T and U are initially vacant. Initial statements involve setting the initial value of a collection of properties.

6.8.3 Principles of algorithm implementation

The procedure is carried out using a method based on rules. The foundations and findings of rules affect the properties of two basic object groups: variables and relations. The processes primarily rely on there are two sets of regulations: action and reaction. TYPE. RULES AND ACTION. RULES—and by an identifiable object—the controllers are in operation right now.

6.8.4 Supposition drivers and strategies

ACTION.RULES, the filter's condition leads to this decision: Activate (supposition? V?x), where?x has been removed by an extra assumption and represents a test response for? V. The select.var.supposition procedure selects the attribute of the variable based on the following assumption. Multiple requirements for choosing to consider the variable's structure (attainable or non-computable, discontinuous or continuous), network composition (the total amount of connections affecting the unit), and search behavior (number of currently unknown disputes).

6.9 AI Developments That Help with Crop Breeding

The use of AI is a transformative technology in the field of computers that simulates human reasoning and resolving issues capabilities through machines (Dwivedi *et al.*, 2021). AI encompasses various subfields, including natural language processing (NLP), conversational systems, computer vision and neural networks, (Li *et al.*, 2021). These technologies function through classical computing techniques and CMOS hardware, enabling computers may do things autonomously and mimic the intellect of humans. AI also contributes to the process of "big data," which involves the managing and analysis of vast, complex datasets that are beyond the capacity to perform in traditional analytical ways. In agriculture, AI presents limitless possibilities, especially in areas like computerized breeding and precision farming. It supports automation and real-time monitoring. AI technologies such as quantities sequencing and phenomics have significantly accelerated the breeding of freshly discovered kinds of plants (Esposito *et al.*, 2020), while ML models are increasingly employed in genomic prediction with marker-based choosing. Major agricultural companies like Monsanto and John Deere are leveraging AI to optimize farming operations by integrating soil, seed, and weather data to enhance yields and reduce costs (Faulkner *et al.*, 2014). These systems often utilize weather forecasts, satellite data, and real-time equipment monitoring via cloud-based solutions (Stergiou and Psannis 2017). Precision farming initiatives, such as those conducted by NEC and Dacom in Romania, have demonstrated the effectiveness of environmental sensors and data analytics in increasing productivity (El Bilali and Allahyari, 2018). The growing complexity of agricultural datasets poses challenges that necessitate the development of advanced algorithms, systems, and analytical methods to extract meaningful insights. AI techniques, combining deep neural networks and mathematical modeling., are being developed to improve capabilities in planning, learning, and decision making across the agricultural sector (Shaw *et al.*, 2019). Researchers are also creating technologies to understand plant responses to diverse climatic conditions, enhancing resilience and productivity (Jeong *et al.*, 2016). With advancements in computer power usage, including making use of GPUs, AI implementation has become more efficient and accessible (Huntingford *et al.*, 2019). Supercomputers like Summit, capable of supporting 27,000 GPUs, represent the future of AI's capacity to address global agricultural challenges and the safety of food. As a result, Intelligence is prepared for use to revolutionize agriculture by optimizing resources, increasing yields, and contributing to sustainable food production.

6.10 Addressing the Phenomics Bottlenecks with AI Technologies

Recent advancements in plant phenomics offer greatly enhanced accuracy of breeding programs through high-volume characterization technologies that assess complex plant traits. These innovations have been accelerated by the integration of AI, which includes ML, DL, and computer vision, into non-invasive imaging techniques. AI-based image analysis enables efficient data collection and management, supporting robust phenotyping under both controlled and open-field conditions. AI plays an important part in field phenotyping by transforming raw sensor data into meaningful phenotypic information. It aids in modeling genotype–phenotype interactions, maintaining large-scale phenomics databases, and promoting open-source tools that facilitate data sharing and community-driven research (Nabwire *et al.*, 2021). Despite progress, field-based phenotyping still lags behind indoor systems, highlighting the need for more advanced infrastructure and real-time analytics solutions. ML techniques like k-NN, RF, and supporting vectors. Machines have shown promising results in agricultural prediction tasks (Selvaraj *et al.*, 2020a). DL has also proven effective in analyzing composite materials, indicating its potential in phenotyping image analysis (Bang *et al.*, 2020). Developments highlight AI's revolutionary significance in modern phenomics, particularly in understanding plant traits in diverse environmental conditions and enhancing breeding efficiency.

6.11 Investigating Potential of AI for Gene Function Analysis

Recent advancements in high-throughput biological technologies have generated vast quantities of data, particularly in the field of omics, which involves large-scale analysis of genetic, proteomic, and metabolomic data (Jung *et al.*, 2021). The complexity and volume of such datasets make traditional analysis methods inadequate. As a result, ML and AI have become essential resources for interpreting omics data and managing biological information efficiently (Grapov *et al.*, 2018). AI is increasingly applied in plant genomics to explore gene functions, predict traits, and analyze large datasets. ML algorithms are utilized over a broad-spectrum bioinformatics activity, especially comprising protein-coding gene recognition, gene expression profiling, identifying cis-regulatory elements (CREs), intracellular localization, phenotype prediction as well as metabolic pathway analysis (Mahood *et al.*, 2020). Tools like DeepSEA and DeepBind exemplify the application of DL in the prediction of DNA and RNA binding sites and other genomic features (Esposito *et al.*, 2020). AI also supports comparative genomic studies and facilitates knowledge transfer from model organisms to target crops (Caudai *et al.*, 2021). The rapid growth of sequencing data presents significant challenges in storage and analysis, but AI-based bioinformatics approaches offer scalable solutions. These technologies enable simultaneous gene expression analysis under varying conditions, helping researchers validate hypotheses and uncover biological insights (Dasari *et al.*, 2021).

6.12 Using AI to Connect Crop Genomes to Phenomes

Modern agricultural engineering is increasingly focused on bridging the gap between genotype and phenotype to accelerate precision breeding and crop improvement. A major challenge in this field is effectively linking vast genomic information with high-throughput phenotyping capabilities, which is critical for advancing breeding strategies and utilizing omics technologies (Harfouche *et al.*, 2019). AI offers powerful tools for overcoming these barriers by facilitating germplasm acquisition and single nucleotide polymorphism (SNP) detection, which contribute to improving crop performance (Mochida *et al.*, 2019). AI-based models can analyze large, complex datasets to predict gene functions, crop yield, and plant traits. These models are particularly valuable in high-throughput phenotyping and genomic characterization, where the goal is to quickly identify genes linked to desired agricultural traits. This integration of AI with genomics and phenomics significantly accelerates breeding programs by streamlining data interpretation and enhancing model accuracy. AI was utilized for genomic assembly, gene expression cascade analysis, identification of key SNPs in polyploid plants, and analysis of multiplex biological challenges in metabolomics, proteomics, transcriptomics, and systems biology (Esposito *et al.*, 2020). For instance, AI-driven methods such as DeepBind and DeepSEA have proven successful in predicting genomic traits, including transcription factor binding sites. These tools help decode complex regulatory mechanisms and the effects of genetic variations on phenotypes. Crop stress characterization is another domain where AI shines. It enables large-scale data processing for various stress conditions, including heat, drought, and disease pressure. By triggering genomic algorithms, AI supports high-throughput crop profiling for both natural and computerized situations (Razzaq *et al.*, 2021). Moreover, AI can evaluate molecular patterns like transcription factor binding sites (Long *et al.*, 2020), long non-coding RNAs, microRNAs, coding genes, epigenetic modifications, and CREs (Ernst and Kellis, 2017). These functions make AI an essential tool for genome interpretation and crop trait improvement. Crop databases enriched with genomic and phenotypic information can be mined using AI to uncover novel traits and genes, offering a new standard for managing and analyzing big data in plant science (Tong and Nikoloski, 2021). Next-generation sequencing (NGS) has enhanced gene mapping precision and genome-wide association studies (GWAS) in agriculture. For example, in soybean research, genetic loci have been identified for key traits like seed weight (Jing *et al.*, 2018), seed oil and protein content, pod dehiscence, nitrogen fixation, plant height, branching (Sonah *et al.*, 2015), disease resistance and tocopherol content. AI has supported the analysis of these genomic traits by streamlining the search for functional genes and linking them with agronomic phenotypes, including use of bulk segregant analysis (BSA), especially when integrated with NGS. AI-enhanced approaches such as DeepBSA have also emerged, using DL to map quantitative trait loci (QTLs) and clone functional genes in crops efficiently (Li *et al.*, 2019). Additionally, AI technologies are increasingly applied to maintain the surrounding area variables like heat, UV radiation, wind, and hail that directly affect crop productivity and stress response (Crane-Droesch, 2018). AI enables breeders and researchers to assess and improve crop resistance to various environmental stressors.

6.13 AI Providing Access to Envirotyping Data for Crop Breeding

Environmental degradation continues to pose a major threat to both natural ecosystems and crop productivity. To address this, a third "typing" method called envirotyping has been developed, complementing genotyping and phenotyping. Envirotyping captures environmental variables that influence crop development, such as phenotypic responses. This technique is vital for crop modeling and climate-resilient breeding strategies. A growing research area, genome–environment associations (GEAs), integrates ecological and climate data with evolutionary genomic sequences to identify genomic estimated adaptation values (GEAVs). These values enhance genomic predictions and can be modeled using advanced ML to assess polygenic adaptive traits (Cortés *et al.*, 2022). The integration of multi-omics data, AI, and big data within genomic–enviromics breeding pipelines enables a more precise selection of traits linked to environmental adaptation. Climate-smart breeding strategies aim to develop resilient crops that optimize limited resources while adapting to their environments (Resende *et al.*, 2021). Using AI tools like ML and genomic prediction (GP), breeders can identify adaptive traits in wild relatives, such as drought tolerance, and incorporate them via genome editing or advanced selection (Cortés and López-Hernández, 2021). These systems assist breeders in selecting climate-adapted traits using real-time, data-driven tools. Automated phenotyping technologies capture indoor and field data and transmit it through HTTP servers, facilitating genotype-by-environment interaction (GEI) analysis across various conditions. AI-driven models, including DL, enable the selection of cultivars optimized for specific or multiple environments (Shafiekhani *et al.*, 2017). This integrative breeding approach enhances efficiency, ensuring cultivars are suitable for future climates and contributing to sustainable agriculture (Fig. 6.1).

6.14 Limitations and Potential Applications of AI in Disease Detection

AI has shown considerable potential in plant disease identification, but its drawbacks include data bias, limited generalizability, and reliance on high-quality, labeled datasets. Most AI models struggle with real-world variables such as illumination, background noise, and rare disease signs. Furthermore, diseases with comparable visual characteristics can confound AI systems. Despite these obstacles, AI has the potential to improve early disease diagnosis,

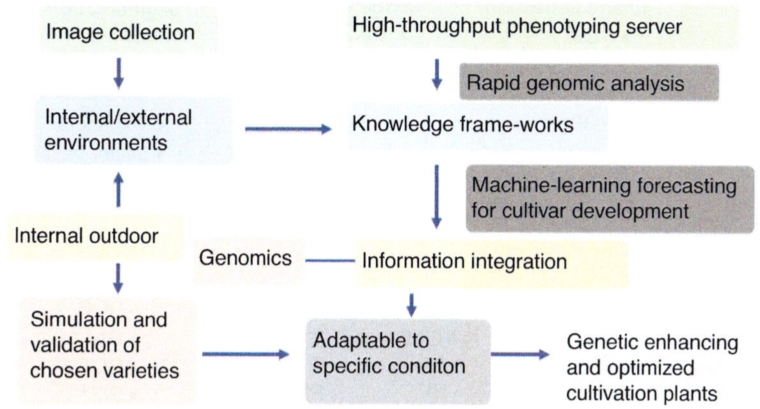

Fig. 6.1. AI-enabled internet functions and knowledge processing for trait improvement and model development. (Author's own image.)

large-scale monitoring with drones and satellite photography, and combining environmental data for greater accuracy.

6.14.1 Limitations

In the past, we explored how AI may be used to help agriculture, specifically diagnosing diseases in crops like vegetables. We examined numerous automation models and methods developed by scholars around the world and presented them in the literature. AI has great potential in agriculture, especially for detecting plant diseases. To enhance disease identification, it is critical to address the constraints of these models. This section describes the major barriers to the correct implementation of automatic plant disease recognition and categorization.

6.14.2 Analysis of background and noise

Scientific images of plant disorders can include annoying sounds, colored backdrops, and extra components like roots, vegetation, and mud. It is crucial to identify and separate these components from collected photos. To effectively identify unhealthy plants in real-time, the proposed automatic method should remove all extraneous components from the image and isolate only the relevant segment.

6.14.3 Factors influencing image acquisition

Current archives typically feature images taken in controlled spaces such as labs. Getting an equivalent image in the field may be difficult due to elements such as sunshine intensity, humidity levels, or weather conditions. To meet exploration goals, images of one identical plant object from diverse angles, times, and circumstances in nature are required. The selection of imaging tools has a considerable impact on system performance. The reliability of forecasts is affected by a number of variables, including the sampling instrument, light magnitude, date and time of day, and humidity level. To successfully solve these difficulties, we must pair training with the implementation of an autonomous disease prediction algorithm immediately (Fig. 6.2).

6.14.4 Identify and isolate condition signs

Modern processing of pictures is critical for research on agriculture, as it allows for identifying

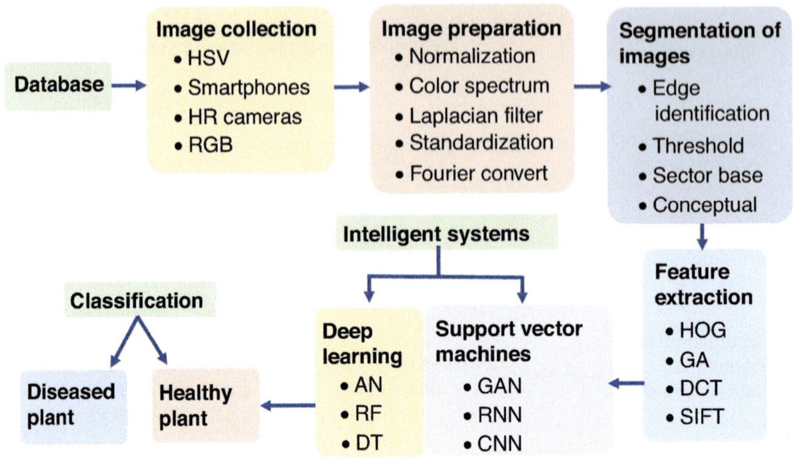

Fig. 6.2. Systematic data flow from data collection, image analysis, disease detection, and model establishment. (Author's own image.)

and isolating diseases with comparable symptoms. Recognizing and classifying symptoms of disorders that share similar characteristics increases efficiency. Yet, this effort gets challenging when multiple conditions present the same symptoms and influences from the environment. Different classification strategies are required to distinguish vegetable illnesses with diverse appearances.

6.14.5 The effect of the amount of data on the accuracy of models

Educating methods for particular diseases of agriculture necessitates more data than the standard few hundred images utilized by writers. The DL-based data enhancement technique enhances the overall training image quality. A covariate shift occurs when the training data do not match the data used to apply the model. Using huge datasets can improve model performance, but it may additionally improve computational effort.

6.14.6 Data imbalance for various diseases

Disproportionate structures in training data pose a challenge for automatic detection computation. Plant infections have limited data and are not uniform between classes, as previously mentioned. To enable accurate analysis and estimation, diseases must be represented by equal-sized sample sets that include both sick and healthy plants.

6.14.7 Multiple concurrent diseases

The idea that each picture represents just a single disease is not always correct, as many conditions, nutritional deficiencies, and pests can exist in a single image. This makes recognizing and tracking a particular condition more difficult, and the expression of signs can differ depending on the geographic area. As a result, it is critical to consider these elements while interpreting photographs.

6.14.8 Disease with similar symptoms

Agriculture-related diseases are difficult to detect due to their similar symptoms and patterns. Scientists routinely investigate across the optical spectrum. Using wavelengths of light to differentiate diseases can be complicated, expensive, and difficult. Novel methods have the ability to reduce the requirement for massive datasets and the risk of errors in the farming process.

6.15 Conclusion

Correct detection and classification of diseased plants are critical for agricultural improvement, productivity, and quality. Traditional disease diagnosis demands significant resources, knowledge, and operational considerations, especially due to overlapping symptoms and various disease groupings. Automating disease detection and classification in plants, primarily through the analysis of plant leaves, is critical for agricultural improvement,. DL and ML models have shown promise in this area, improving accuracy and efficiency in real-time agricultural applications. Supportive ML algorithms like SVMs, as well as RF and CNNs, are recommended in DL for image-based disease categorization. However, challenges such as varied lighting conditions, limited datasets, and inconsistent sample quality can affect model performance. To enhance reliability, strategies like using diverse and large-scale datasets, real-time testing environments, and attention to environmental variables are necessary. Between 2010 and 2022, around 125 ML algorithms were applied across multiple agricultural domains—122 specifically targeting plant diseases, weeds, and pests. This progress marks the transition from Ag4.0 to Ag5.0, emphasizing autonomous, AI-driven crop management with minimal human intervention. Ag5.0 integrates robotics, AI, and ML to automate decision making, pest detection, and multi-task execution in real time. Additionally, exploring pharmaceutical and pesticide interventions based on AI-diagnosed diseases opens new research pathways. Ethical and inclusive implementation is crucial, requiring companies to develop human-centered, socially responsible technologies. The integration of AI into agriculture has led to increased profitability and global economic growth, encouraging industries to innovate through sustainable and adaptive production methods.

References

Agarwal, A., Sarkar, A. and Dubey, A.K. (2019) Computer vision-based fruit disease detection and classification. In: Trivedi, S.T.M.C., Mishra, K.K., Misra, A.K. and K.K, K. (eds) *Smart Innovations in Communication and Computational Sciences*. Springer, pp. 105–115. DOI: 10.1007/978-981-13-2414-7_11.

Albiero, D., Pontin Garcia, A., Kiyoshi Umezu, C. and Paulo, R. (2022) Swarm robots in mechanized agricultural operations: A review about challenges for research. *Computers and Electronics in Agriculture* 193, 106608. DOI: 10.1016/j.compag.2021.106608.

Amit, K. (2018) *Artificial Intelligence and Soft Computing: Behavioral and Cognitive Modeling of the Human Brain*. CRC Press, Boca Raton, Florida, USA.

Ayoub Shaikh, T., Rasool, T. and Rasheed Lone, F. (2022) Towards leveraging the role of machine learning and artificial intelligence in precision agriculture and smart farming. *Computers and Electronics in Agriculture* 198, 107119. DOI: 10.1016/j.compag.2022.107119.

Bang, H.T., Park, S. and Jeon, H. (2020) Defect identification in composite materials via thermography and deep learning techniques. *Composite Structures* 246, 112405. DOI: 10.1016/j.compstruct.2020.112405.

Barreto, A., Paulus, S., Varrelmann, M. and Mahlein, A.K. (2020) Hyperspectral imaging of symptoms induced by *Rhizoctonia solani* in sugar beet: Comparison of input data and different machine learning algorithms. *Journal of Plant Diseases and Protection* 127(4), 441–451. DOI: 10.1007/s41348-020-00344-8.

Behmann, J., Mahlein, A.-K., Rumpf, T., Römer, C. and Plümer, L. (2015) A review of advanced machine learning methods for the detection of biotic stress in precision crop protection. *Precision Agriculture* 16, 239–260. DOI: 10.1007/s11119-014-9372-7.

Belhajem, I., Ben Maissa, Y. and Tamtaoui, A. (2016) A robust low cost approach for real time car positioning in a smart city using Extended Kalman Filter and evolutionary machine learning. In: *2016 4th*

IEEE International Colloquium on Information Science and Technology (CIST), pp. 806–811. DOI: 10.1109/CIST.2016.7804998.

Caudai, C., Galizia, A., Geraci, F., Le Pera, L., Morea, V. *et al.* (2021) AI applications in functional genomics. *Computational and Structural Biotechnology Journal* 19, 5762–5790. DOI: 10.1016/j.csbj.2021.10.009.

Chhikara, P., Jain, N., Tekchandani, R. and Kumar, N. (2020) Data dimensionality reduction techniques for industry 4.0: Research results, challenges, and future research directions. *Software: Practice and Experience* 52(3), 658–688. DOI: 10.1002/spe.2876.

Chollet, F. (2017) Xception: Deep learning with depthwise separable convolutions. In: *IEEE Conference on Computer Vision and Pattern Recognition (CVPR)*, pp. 1251–1258. DOI: 10.1109/CVPR.2017.195.

Cortés, A.J. and López-Hernández, F. (2021) Harnessing crop wild diversity for climate change adaptation. *Genes* 12(5), 783. DOI: 10.3390/genes12050783.

Cortés, A.J., López-Hernández, F. and Blair, M.W. (2022) Genome-environment associations, an innovative tool for studying heritable evolutionary adaptation in orphan crops and wild relatives. *Frontiers in Genetics* 13, 910386. DOI: 10.3389/fgene.2022.910386.

Coulibaly, S., Kamsu-Foguem, B., Kamissoko, D. and Traore, D. (2022) Deep learning for precision agriculture: A bibliometric analysis. *Intelligent Systems with Applications* 16, 200102. DOI: 10.1016/j.iswa.2022.200102.

Crane-Droesch, A. (2018) Machine learning methods for crop yield prediction and climate change impact assessment in agriculture. *Environmental Research Letters* 13(11), 114003. DOI: 10.1088/1748-9326/aae159.

Dainelli, R., Toscano, P., Gennaro, S.F. and Matese, A. (2021) Recent advances in unmanned aerial vehicles forest remote sensing—A systematic review. part II: Research applications. *Forests* 12(4), 397. DOI: 10.3390/f12040397.

Dasari, C.M., Amilpur, S. and Bhukya, R. (2021) Exploring variable-length features (motifs) for predicting binding sites through interpretable deep neural networks. *Engineering Applications of Artificial Intelligence* 106, 104485. DOI: 10.1016/j.engappai.2021.104485.

Davies, M., Srinivasa, N., Lin, T.-H., Chinya, G., Cao, Y. *et al.* (2018) Loihi: A neuromorphic manycore processor with on-chip learning. *IEEE Micro* 38(1), 82–99. DOI: 10.1109/MM.2018.112130359.

de Castro, A.I., López-Granados, F. and Jurado-Expósito, M. (2013) Road-scale cruciferous weed patch classification in winter wheat using QuickBird imagery for in-season site-specific control. *Precision Agriculture* 14(4), 392–413. DOI: 10.1007/s11119-013-9304-y.

de Santos, P.G., Ribeiro, A., Fernandez-Quintanilla, C., Lopez-Granados, F., Brandstoetter, M, *et al.* (2017) Fleets of robots for environmentally-safe pest control in agriculture. *Precision Agriculture* 18(4), 574–614. Available at: https://doi.org/10.1007/s11119-016-9476-3

Dwivedi, Y.K., Hughes, L., Ismagilova, E., Aarts, G., Coombs, C. *et al.* (2021) Artificial intelligence (AI): Multidisciplinary perspectives on emerging challenges, opportunities, and agenda for research, practice and policy. *International Journal of Information Management* 57, 101994. DOI: 10.1016/j.ijinfomgt.2019.08.002.

El Bilali, H. and Allahyari, M.S. (2018) Transition towards sustainability in agriculture and food systems: Role of information and communication technologies. *Information Processing in Agriculture* 5(4), 456–464. DOI: 10.1016/j.inpa.2018.06.006.

Emmi, L., Gonzalez-de-Soto, M., Pajares, G. and Gonzalez-de-Santos, P. (2014) New trends in robotics for agriculture: Integration and assessment of a real fleet of robots. *The Scientific World Journal* 1–21. DOI: 10.1155/2014/404059.

Ernst, J. and Kellis, M. (2017) Chromatin-state discovery and genome annotation with ChromHMM. *Nature Protocols* 12(12), 2478–2492. DOI: 10.1038/nprot.2017.124.

Esposito, S., Carputo, D., Cardi, T. and Tripodi, P. (2020) Applications and trends of machine learning in genomics and phenomics for next-generation breeding. *Plants* 9(1), 34. DOI: 10.3390/plants9010034.

Faulkner, A., Cebul, K. and McHenry, G. (2014) *Agriculture Gets Smart: The Rise of Data and Robotics*. Cleantech Group.

Feng, Z.H., Wang, L.Y., Yang, Z.Q., Zhang, Y.Y., Li, X. *et al.* (2022) Hyperspectral monitoring of powdery mildew disease severity in wheat based on machine learning. *Frontiers in Plant Science* 13, 828454. DOI: 10.3389/fpls.2022.828454.

Floreano, D. and Wood, R.J. (2015) Science, technology and the future of small autonomous drones. *Nature* 521(7553), 460–466. DOI: 10.1038/nature14542.

Fuentes, S., Tongson, E., Unnithan, R.R. and Gonzalez Viejo, C. (2021) Early detection of aphid infestation and insect-plant interaction assessment in wheat using a low-cost electronic nose (e-nose), near-infrared spectroscopy and machine learning modeling. *Sensors* 21(17), 5948. DOI: 10.3390/s21175948.

Fürnkranz, J. and Kliegr, T. (2015) A brief overview of rule learning. In: Appice, A. (ed.) *Lecture Notes in Computer Science: Discovery Science*, Vol. 9202. Springer, pp. 54–69. DOI: 10.1007/978-3-319-21542-6_5.

Garcia-Ruiz, F., Sankaran, S., Maja, J.M., Lee, W.S., Rasmussen, J. *et al.* (2013) Comparison of two aerial imaging platforms for identification of huanglongbing-infected citrus trees. *Computers and Electronics in Agriculture* 91, 106–115. DOI: 10.1016/j.compag.2012.12.002.

García-Valls, M., Dubey, A. and Botti, V. (2018) Introducing the new paradigm of social dispersed computing: Applications, technologies and challenges. *Journal of Systems Architecture* 91, 83–102. DOI: 10.1016/j.sysarc.2018.05.007.

Glaessgen, E. and Stargel, D. (2012) The digital twin paradigm for future NASA and U.S. air force vehicles. In: *53rd AIAA/ASME/ASCE/AHS/ASC Structures, Structural Dynamics and Materials Conference, American Institute of Aeronautics and Astronautics*, American Institute of Aeronautics and Astronautics, Honolulu, Hawaii. DOI: 10.2514/6.2012-1818.

Gold, K.M., Townsend, P.A., Herrmann, I. and Gevens, A.J. (2020) Investigating potato late blight physiological differences across potato cultivars with spectroscopy and machine learning. *Plant Science* 295, 110316. DOI: 10.1016/j.plantsci.2019.110316.

Grapov, D., Fahrmann, J., Wanichthanarak, K. and Khoomrung, S. (2018) Rise of deep learning for genomic, proteomic, and metabolomic data integration in precision medicine. *OMICS: A Journal of Integrative Biology* 22(10), 630–636. DOI: 10.1089/omi.2018.0097.

Grimstad, L. and From, P.J. (2017) The thorvald II agricultural robotic system. *Robotics* 6(4), 24. DOI: 10.3390/robotics6040024.

Guansong, P., Chunhua, S., Longbing, C. and van den Hengel, A. (2022) Deep learning for anomaly detection: a review. *ACM Computing Surveys* 54(2), 1–38. Available at: https://doi.org/10.1145/3439950

Gutierrez, A., Ansuategi, A., Susperregi, L., Tubío, C., Rankić, I. *et al.* (2019) A benchmarking of learning strategies for pest detection and identification on tomato plants for autonomous scouting robots using internal databases. *Journal of Sensors* 1–15. DOI: 10.1155/2019/5219471.

Harfouche, A.L., Jacobson, D.A., Kainer, D., Romero, J.C., Harfouche, A.H. *et al.* (2019) Accelerating climate resilient plant breeding by applying next-generation artificial intelligence. *Trends in Biotechnology* 37(11), 1217–1235. DOI: 10.1016/j.tibtech.2019.05.007.

He, K., Zhang, X., Ren, S. and Sun, J. (2016) Deep residual learning for image recognition. In: *2016 IEEE Conference on Computer Vision and Pattern Recognition (CVPR)*, IEEE, pp. 770–778. DOI: 10.1109/CVPR.2016.90.

Hong, S.J., Kim, S.Y., Kim, E., Lee, C.H., Lee, J.S. *et al.* (2020) Moth detection from pheromone trap images using deep learning object detectors. *Agriculture* 10(5), 170. DOI: 10.3390/agriculture10050170.

Howard, A.G., Zhu, M., Chen, B., Kalenichenko, D., Wang, W. *et al.* (2017) MobileNets: Efficient convolutional neural networks for mobile vision applications. *arXiv Preprint arXiv* 1704, 04861. DOI: 10.48550/arXiv.1704.04861.

Huang, G., Liu, Z., Maaten, L. and Weinberger, K.Q. (2016) Densely connected convolutional networks. In: *Proceedings of the IEEE Conference on Computer Vision and Pattern Recognition*, pp. 4700–4708. DOI: 10.48550/arXiv.1608.06993.

Huang, H., Deng, J., Lan, Y., Yang, A., Deng, X. *et al.* (2018) A two-stage classification approach for the detection of spider mite- Infested cotton using UAV multispectral imagery. *Remote Sensing Letters* 9(10), 933–941. DOI: 10.1080/2150704X.2018.1498600.

Huang, Y., Zhao, X., Pan, Z., Reddy, K.N. and Zhang, J. (2022) Hyperspectral plant sensing for differentiating glyphosate-resistant and glyphosate-susceptible johnsongrass through machine learning algorithms. *Pest Management Science* 78(6), 2370–2377. DOI: 10.1002/ps.6864.

Huntingford, C., Jeffers, E.S., Bonsall, M.B., Christensen, H.M., Lees, T. *et al.* (2019) Machine learning and artificial intelligence to aid climate change research and preparedness. *Environmental Research Letters* 14(12), 124007. DOI: 10.1088/1748-9326/ab4e55.

Jeong, J.H., Resop, J.P., Mueller, N.D., Fleisher, D.H., Yun, K. *et al.* (2016) Random forests for global and regional crop yield predictions. *PLOS ONE* 11(6), e0156571. DOI: 10.1371/journal.pone.0156571.

Jin, X., Jie, L., Wang, S., Qi, H.J. and Li, S.W. (2018) Classifying wheat hyperspectral pixels of healthy heads and fusarium head blight disease using a deep neural network in the wild field. *Remote Sensing* 10(3), 395. DOI: 10.3390/rs10030395.

Jing, Y., Zhao, X., Wang, J., Teng, W., Qiu, L. *et al.* (2018) Identification of the genomic region underlying seed weight per plant in soybean (Glycine max L. Merr.) via high-throughput single-nucleotide polymorphisms and a genome-wide association study. *Frontiers in Plant Science* 9, 1392. DOI: 10.3389/fpls.2018.01392.

Jung, J., Maeda, M., Chang, A., Bhandari, M., Ashapure, A. *et al.* (2021) The potential of remote sensing and artificial intelligence as tools to improve the resilience of agriculture production systems. *Current Opinion in Biotechnology* 70, 15–22. DOI: 10.1016/j.copbio.2020.09.003.

Kamilaris, A. and Prenafeta-Boldú, F.X. (2018) Deep learning in agriculture: A survey. *Computers and Electronics in Agriculture* 147, 70–90. DOI: 10.1016/j.compag.2018.02.016.

Kodama, T. and Hata, Y. (2018) Development of classification system of rice disease using artificial intelligence. In: 2018 IEEE International Conference on Systems, Man, and Cybernetics (SMC). Miyazaki, Japan, pp. 3699–3702. DOI: 10.1109/SMC.2018.00626.

Kumar, A., Sarkar, S. and Pradhan, C. (2019) Recommendation system for crop identification and pest control technique in agriculture. In: *2019 International Conference on Communication and Signal Processing (ICCSP), Chennai, India*, pp. 185–189. DOI: 10.1109/ICCSP.2019.8698099.

Lavric, A. (2019) LoRa (long-range) high-density sensors for internet of things. *Journal of Sensors* 1–9. DOI: 10.1155/2019/3502987.

LeCun, Y., Bengio, Y. and Hinton, G. (2015) Deep learning. *Nature* 521, 436–444. DOI: 10.1038/nature14539.

Li, D., Zhao, X., Han, Y., Li, W. and Xie, F. (2019) Genome-wide association mapping for seed protein and oil contents using a large panel of soybean accessions. *Genomics* 111(1), 90–95. DOI: 10.1016/j.ygeno.2018.01.004.

Li, Y., Zhou, X. and Li, G. (2021) Bridging natural language and graphical user interfaces. In: *Artificial Intelligence for Human Computer Interaction: A Modern Approach*. Springer, pp. 463–493. DOI: 10.1007/978-3-030-82681-9_18.

Liu, L., Wang, R., Xie, C., Yang, P., Wang, F. *et al.* (2019) PestNet: An end-to-end deep learning approach for large-scale multi-class pest detection and classification. *IEEE Access* 7, 45301–45312. DOI: 10.1109/ACCESS.2019.2909522.

Long, P., Zhang, L., Huang, B., Chen, Q. and Liu, H. (2020) Integrating genome sequence and structural data for statistical learning to predict transcription factor binding sites. *Nucleic Acids Research* 48(22), 12604–12617. DOI: 10.1093/nar/gkaa1134.

López-Correa, J.M., Moreno, H., Ribeiro, A. and Andújar, D. (2022) Intelligent weed management based on object detection neural networks in tomato crops. *Agronomy* 12(12), 2953. DOI: 10.3390/agronomy12122953.

Lowenberg-DeBoer, J., Franklin, K., Behrendt, K. and Godwin, R. (2021) Economics of autonomous equipment for arable farms. *Precision Agriculture* 22(6), 1992–2006. DOI: 10.1007/s11119-021-09822-x.

Lu, J., Ehsani, R., Shi, Y., Abdulridha, J., Castro, A.I. *et al.* (2017) Field detection of anthracnose crown rot in strawberry using spectroscopy technology. *Computers and Electronics in Agriculture* 135, 289–299. DOI: 10.1016/j.compag.2017.01.017.

Mahmoud, M.S. and Xia, Y. (2019) *Networked Control Systems: Cloud Control and Secure Control.* Butterworth-Heinemann.

Mahood, E.H., Kruse, L.H. and Moghe, G.D. (2020) Machine learning: A powerful tool for gene function prediction in plants. *Applications in Plant Sciences* 8(7), e11376. DOI: 10.1002/aps3.11376.

McCool, C., Beattie, J., Firn, J., Lehnert, C., Kulk, J. *et al.* (2018) Efficacy of mechanical weeding tools: A study into alternative weed management strategies enabled by robotics. *IEEE Robotics and Automation Letters* 3, 1184–1190. DOI: 10.1109/LRA.2018.2794619.

Mochida, K., Koda, S., Inoue, K., Hirayama, T., Tanaka, S. *et al.* (2019) Computer vision-based phenotyping for improvement of plant productivity: A machine learning perspective. *GigaScience* 8(1), giy153. DOI: 10.1093/gigascience/giy153.

Muhammad, A., Mohammad, A.-U., Zubair, S., Ali, M. and Aggoune, E.-H.M. (2019) Internet-of-things (iot)-based smart agriculture: toward making the fields talk. *IEEE Access* 7, 129551–129583. Available at: https://doi.org/10.1109/ACCESS.2019.2932609

Mulla, S., Singh, S.K., Singh, K.K. and Praveen, B. (2020) Climate change and agriculture: A review of crop models. In: *Global Climate Change and Environmental Policy: Agriculture Perspectives*. Springer, Singapore, pp. 423–435.

Nabwire, S., Suh, H.K., Kim, M.S., Baek, I. and Cho, B.K. (2021) Review: Application of artificial intelligence in phenomics. *Sensors* 21(13), 4363. DOI: 10.3390/s21134363.

Nanda, S.J. and Panda, G. (2014) A survey on nature inspired metaheuristic algorithms for partitional clustering. *Swarm and Evolutionary Computation* 16, 1–18. DOI: 10.1016/j.swevo.2013.11.003.

Oñate, W. and Sanz, R. (2023) Analysis of architectures implemented for Iiot. *Heliyon* 9(1), e12868. DOI: 10.1016/j.heliyon.2023.e12868.

Pätzold, S., Hbirkou, C., Dicke, D., Gerhards, R. and Welp, G. (2020) Linking weed patterns with soil properties: A long-term case study. *Precision Agriculture* 21(3), 569–588. DOI: 10.1007/s11119-019-09682-6.

Ramcharan, A., McCloskey, P., Baranowski, K., Mbilinyi, N., Mrisho, L. *et al.* (2019) A mobile-based deep learning model for cassava disease diagnosis. *Frontiers in Plant Science* 10, 272. DOI: 10.3389/fpls.2019.00272.

Ramos, A.P.M., Gomes, F.D.G., Pinheiro, M.M.F., Furuya, D.E.G., Gonçalvez, W.N. *et al.* (2022) Detecting the attack of the fall armyworm (*Spodoptera frugiperda*) in cotton plants with machine learning and spectral measurements. *Precision Agriculture* 23(2), 470–491. DOI: 10.1007/s11119-021-09845-4.

Razzaq, A., Kaur, P., Akhter, N., Wani, S.H. and Saleem, F. (2021) Next-generation breeding strategies for climate-ready crops. *Frontiers in Plant Science* 12, 620420. DOI: 10.3389/fpls.2021.620420.

Ren, Y., Zhang, L. and Suganthan, P.N. (2016) Ensemble classification and regression-recent developments, applications and future directions. *IEEE Computational Intelligence Magazine* 11(1), 41–53. DOI: 10.1109/MCI.2015.2471235.

Resende, R.T., Piepho, H.-P., Rosa, G.J.M., Silva-Junior, O.B., Silva, F.F. *et al.* (2021) Enviromics in breeding: Applications and perspectives on envirotypic-assisted selection. *Theoretical and Applied Genetics* 134(1), 95–112. DOI: 10.1007/s00122-020-03684-z.

Sagi, O. and Rokach, L. (2018) Ensemble learning: A survey. *WIREs Data Mining and Knowledge Discovery* 8(4), e1249. DOI: 10.1002/widm.1249.

Saiz-Rubio, V. and Rovira-Más, F. (2020) From smart farming towards agriculture 5.0: A review on crop data management. *Agronomy* 10(2), 207. DOI: 10.3390/agronomy10020207.

Sarker, I.H. (2021) Deep learning: A comprehensive overview on techniques, taxonomy, applications and research directions. *SN Computer Science* 2(6), 420. DOI: 10.1007/s42979-021-00815-1.

Secretariat, I. (2021) Scientific review of the impact of climate change on plant pests: A global challenge to prevent and mitigate plant-pest risks in agriculture, forestry and ecosystems. In: *FAO on Behalf of the IPPC Secretariat*.

Selvaraj, G.M., Vergara, A., Montenegro, F., Alonso Ruiz, H., Safari, N. *et al.* (2020b) Detection of banana plants and their major diseases through aerial images and machine learning methods: A case study in DR Congo and Republic of Benin. *ISPRS Journal of Photogrammetry and Remote Sensing* 169, 110–124. DOI: 10.1016/j.isprsjprs.2020.08.025.

Shafi, U., Mumtaz, R., García-Nieto, J., Hassan, S.A., Zaidi, S.A.R. *et al.* (2019) Precision agriculture techniques and practices: From considerations to applications 19(17), 3796. DOI: 10.3390/s19173796.

Selvaraj, M.G., Valderrama, M., Guzman, D., Valencia, M., Ruiz, H. *et al.* (2020a) Machine learning for high-throughput field phenotyping and image processing provides insight into the association of above and below-ground traits in cassava (*Manihot esculenta* Crantz). *Plant Methods* 16, 87. DOI: 10.1186/s13007-020-00625-1.

Shafiekhani, A., Kadam, S., Fritschi, F.B. and DeSouza, G.N. (2017) Vinobot and vinoculer: two robotic platforms for high-throughput field phenotyping. *Sensors* 17(1), 214. DOI: 10.3390/s17010214.

Shamshiri, R.R., Weltzien, C., Hameed, I.A., Yule, I.J. and Grift, E. (2018) Research and development in agricultural robotics: A perspective of digital farming. *International Journal of Agricultural and Biological Engineering* 11(4), 1–11. DOI: 10.25165/j.ijabe.20181104.4278.

Shankar, P., Johnen, A. and Liwicki, M. (2020) Data fusion and artificial neural networks for modelling crop disease severity. In: *2020 IEEE 23rd International Conference on Information FUSION (FUSION), Rustenburg, South Africa*, pp. 1–8. DOI: 10.23919/FUSION45008.2020.9190211.

Shaw, J., Rudzicz, F., Jamieson, T. and Goldfarb, A. (2019) Artificial intelligence and the implementation challenge. *Journal of Medical Internet Research* 21(7), e13659. DOI: 10.2196/13659.

Simonyan, K. and Zisserman, A. (2014) Very deep convolutional networks for large-scale image recognition. *arXiv Preprint arXiv* 1409, 1556. DOI: 10.48550/arXiv.1409.1556.

Sonah, H., O'Donoughue, L., Cober, E., Rajcan, I. and Belzile, F. (2015) Identification of loci governing eight agronomic traits using a GBS-GWAS approach and validation by QTL mapping in soyabean. *Plant Biotechnology Journal* 13(2), 211–221. DOI: 10.1111/pbi.12249.

Sony (2020) First Intelligent Vision Sensors with AI Processing Functionality. IMX500, Japan.

Stergiou, C. and Psannis, K.E. (2017) Recent advances delivered by mobile cloud computing and internet of things for big data applications: a survey. *International Journal of Network Management* 27(4), e1930. Available at: https://doi.org/10.1002/nem.1930

Su, J., Yi, D., Coombes, M., Liu, C., Zhai, X. *et al.* (2022) Spectral analysis and mapping of blackgrass weed by leveraging machine learning and UAV multispectral imagery. *Computers and Electronics in Agriculture* 192, 106621. DOI: 10.1016/j.compag.2021.106621.

Sun, P. (2020) Security and privacy protection in cloud computing: Discussions and challenges. *Journal of Network and Computer Applications* 160, 102642. DOI: 10.1016/j.jnca.2020.102642.

Szegedy, C., Liu, W., Jia, Y., Sermanet, P., Reed, S. *et al.* (2015) Going deeper with convolutions. In: *2015 IEEE Conference on Computer Vision and Pattern Recognition (CVPR),* Boston, MA, USA, pp. 1–9. DOI: 10.1109/CVPR.2015.7298594.

Tang, Z., Yang, J., Li, Z. and Qi, F. (2020) Grape disease image classification based on lightweight convolution neural networks and channelwise attention. *Computers and Electronics in Agriculture* 178, 105735. DOI: 10.1016/j.compag.2020.105735.

Tong, H. and Nikoloski, Z. (2021) Machine learning approaches for crop improvement: Leveraging phenotypic and genotypic big data. *Journal of Plant Physiology* 257, 153354. DOI: 10.1016/j.jplph.2020.153354.

Torres-Sánchez, J., Mesas-Carrascosa, F.J., Jiménez-Brenes, F.M., Castro, A.I. and López-Granados, F. (2021) Early detection of broad-leaved and grass weeds in wide row crops using artificial neural networks and UAV imagery. *Agronomy* 11(4), 749. DOI: 10.3390/agronomy11040749.

Ullah, Z., Al-Turjman, F. and Mostarda, L. (2020) Cognition in UAV-aided 5G and beyond communications: A survey. *IEEE Transactions on Cognitive Communications and Networking* 6(3), 872–891. DOI: 10.1109/TCCN.2020.2968311.

van Dijk, M., Gramberger, M., Laborde, D., Mandryk, M., Shutes, L. *et al.* (2020) Stakeholder-designed scenarios for global food security assessments. *Global Food Security* 24, 100352. DOI: 10.1016/j.gfs.2020.100352.

Wang, Y.E., Wei, G.-Y. and Brooks, D. (2019) Benchmarking TPU, GPU, and CPU platforms for deep learning. *arXiv Preprint arXiv* 1907, 10701. DOI: 10.48550/arXiv.1907.10701.

Xia, D., Chen, P., Wang, B., Zhang, J. and Xie, C. (2018) Insect detection and classification based on an improved convolutional neural network. *Sensors* 18(12), 4169. DOI: 10.3390/s18124169.

Xiao, D., Pan, Y., Feng, J., Yin, J., Liu, Y. *et al.* (2022) Remote sensing detection algorithm for apple fire blight based on UAV multispectral image. *Computers and Electronics in Agriculture* 199, 107137. DOI: 10.1016/j.compag.2022.107137.

Zeadally, S., Siddiqui, F. and Baig, Z. (2019) 25 years of bluetooth technology. *Future Internet* 11(9), 194. DOI: 10.3390/fi11090194.

Zhang, K., Lammers, K., Chu, P., Li, Z. and Lu, R. (2021) System design and control of an apple harvesting robot. *Mechatronics* 79, 102644. DOI: 10.1016/j.mechatronics.2021.102644.

Zhang, X., Han, L., Dong, Y., Shi, Y., Huang, W. *et al.* (2019a) A deep learning-based approach for automated yellow rust disease detection from high-resolution hyperspectral UAV images. *Remote Sensing* 11(13), 1554. DOI: 10.3390/rs11131554.

Zhang, Y., Gao, J., Cen, H., Lu, Y., Yu, X. *et al.* (2019b) Automated spectral feature extraction from hyperspectral images to differentiate weedy rice and barnyard grass from a rice crop. *Computers and Electronics in Agriculture* 159, 42–49. DOI: 10.1016/j.compag.2019.02.018.

Zhu, H., Chu, B., Zhang, C., Liu, F., Jiang, L. *et al.* (2017) Hyperspectral imaging for presymptomatic detection of tobacco disease with successive projections algorithm and machine-learning classifiers. *Scientific Reports* 7(1), 4125. DOI: 10.1038/s41598-017-04501-2.

Zohry, A.E.H. and Ouda, S. (2022) Climate-resilient crops. In: *Climate-Smart Agriculture: Reducing Food Insecurity*. Springer International Publishing, pp. 115–135. DOI: 10.1007/978-3-030-93111-7_6.

7 Artificial Intelligence-Assisted Omics Techniques in Plant Defense: Recent Advancements and Future Prospects

Mohd Anas[1]*, Sana Basri[2], Neha Chaurasiya[2] and Mohammad Ali[3]

[1]Department of Botany, Faculty of Life Sciences, Aligarh Muslim University, Aligarh, India; [2]Proteomics and Plant Biotechnology Lab, Department of Botany, Faculty of Life Sciences, Aligarh Muslim University, Aligarh, India; [3]Department of Computer Science, Aligarh Muslim University, Aligarh, India

Abstract

Physiological and biochemical processes in plants are used to address different biotic and abiotic threatening factors. Multi-omics approaches including proteomics, transcriptomics, genomics, and metabolomics have transformed the understanding of these immune responses by providing detailed molecular clues on how organisms respond to stressors. Nevertheless, because of highly dimensional and complicated omics information, it is often challenging to decipher them systematically. This chapter focuses on the application of AI algorithms, primarily deep learning and machine learning, as effective techniques for analyzing high-dimensional omics information in plant immune responses. It offers a brief introduction to the essential omics methods, describes the general difficulties and the present restraints for AI integration into omics, and contemplates possible advances in this rising field. Finally, integrating an application of AI in omics could provide a strong weapon in understanding the molecular basis of plant defense mechanisms and hence comes with the potential of improving crop protection methods under changing climate conditions and newly emerging pathogens.

Keywords: AI algorithms, crop protection, high-dimensional omics, machine learning, pathogens

7.1 Introduction

Throughout their life cycle, plants, having a sessile nature, encounter various harsh environmental situations. They regulate their gene activity to preserve their essential metabolic equilibrium. To increase agricultural output in these adverse circumstances, it is necessary to first comprehend the complexities of these plant responses to environmental stress. Moreover, it has been documented that these environmental challenges never happen singly; instead, the combination of these factors adds to the total stress the plant faces in ways that are not entirely predictable (Panjabi-Sabharwal *et al.*, 2010). This also means that sustaining global food security depends critically on the defense of plants against biotic stress, which results from organisms including pathogens and pests, besides abiotic stress originating from environmental conditions like salt, drought, and extreme temperatures. Crop needs are increasing because the world's population continues to rise,

*Corresponding author: mohdanas147177@gmail.com

DOI: 10.1079/9781800628649.0007

so crop protection techniques must be increasingly effective. Safeguarding agricultural output and ensuring world supplies of food requires protection for crops against the destructive effects of pests, diseases, and unfavorable environmental conditions. The multi-billion-dollar loss emphasizes the immense economic losses arising as a result of these stresses and the decline in crop production.

Pests and diseases are some of the examples of biotic stresses that constantly threaten the health of crops. Pathogens, such as bacteria, viruses, and fungi, may annihilate the whole population of plants, thereby leading to a severe shortage of food and heavy financial losses. Similarly, pests—that is, organisms ranging from nematodes to insects—may demolish crops, leading to lower output and reduced quality (Murmu *et al.*, 2024). The diseases of plants can be extremely detrimental, leading to reductions of yield of as much as 50% in some areas. The most affected are the small-scale farmers, with disease tending to pose severe economic challenges to them. Moreover, plant diseases have an adverse effect on species variety, raise the price of disease control procedures, and even harm human well-being (Ristaino *et al.*, 2021). When new diseases of plants and pest outbursts emerge, it has severe economic consequences on agriculture and jeopardizes public health, national security, and food security (Anderson *et al.*, 2004). Environmental stress factors are irreversible, affecting the growth and survival of plants. Among them are protracted droughts, high temperatures, soil salinity, and pollution by heavy metals. Abiotic stresses gradually and insidiously affect agricultural yields, nutritional value, and general health of plants. Conventional breeding, chemical-based remedies, and agronomic techniques have been utilized together in efforts to counter abiotic and biotic stresses. In addition, most of them are reactive and, hence, they cannot offer enough protection, particularly against infections that are newly emerging or environmental conditions that change rapidly (Murmu *et al.*, 2024). Advanced omics approaches have transfigured how molecular-level plant stress processes are investigated. These techniques yield a wealth of information by exposing complex networks including proteins, genes, and metabolites through plant defense against biotic stress. Nonetheless, the large amounts of information these omics techniques produce make analysis and interpretation extremely difficult, which calls for creating potent computational tools (Fig. 7.1) (Shen *et al.*, 2022). Artificial Intelligence (AI) is emerging as a formidable technique for untangling huge omics datasets and thus gaining

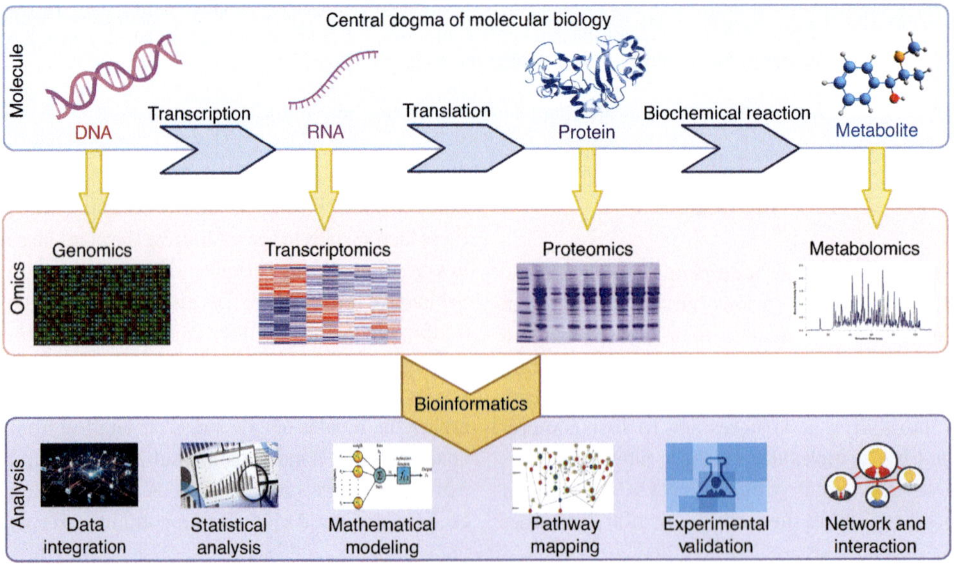

Fig. 7.1. Various omics techniques and their applications. (Author's own image.)

better comprehension of the complex mechanisms underpinning plant responses to stress. New approaches based on knowledge regarding genetic, physiological, phenotypic, and molecular components of plant defense offered by such methodologies can be developed to enhance crop resilience and minimize stress-induced defacement. Indeed, the complicated relationships of proteins and genes, most importantly, metabolites and defense responses, meant that the whole process was going to be extremely complex by traditional plant defense research standards; however, recent advances in the implementation of AI-assisted omics methods for research have managed to find unique ways to simplify this vast terrain of data. These methods cover a very wide range of AI approaches and effectively decipher and simulate the complex connections between molecular constituents and stress reactions (Murmu *et al.*, 2024).

7.2 Types of Omics in Plant Defense Research

In general, the word omics implies biological sciences. Several fields have endings with omics, such as genomics, transcriptomics, proteomics, and metabolomics. Table 7.1 comprises the information related to these branches. In molecular biology, omics is generally used for a set of molecules evaluated worldwide thoroughly or extensively. In a nutshell, these sciences aim to specify and evaluate every biomolecule participating in the construction, formation, purpose, kinetics, and activities of the basic unit of life, the organism, or tissue. Numerous omics methods are used in plant defense research, and each one provides a distinct perspective on the molecular processes that underlie plant reactions to environmental stressors and pathogens.

7.2.1 Genomics

Omics data can rapidly escalate in size and complexity, making it challenging to analyze using traditional statistical correlations or visual methods. This has led to the adoption of AI, also called machine intelligence, which can handle enormous datasets that surpass human cognitive abilities. It can also extract information that goes beyond what we currently know about the system being studied and, most importantly, can automatically improve through expertise gained on training data (McCarthy *et al.*, 2006). Since plants are stationary organisms, they have developed complex defense systems to fend off biotic and abiotic stresses. AI and genomics have combined to become potent solutions for determining the genetic underpinnings

Table 7.1. Different omics branches, study and aim. (Author's own table.)

Branch	Study	Aim
Genomics	Genome	To study the construction, purpose, development, genome mapping, gene quantification and their characterization.
Transcriptomics	Transcriptome	To evaluate the number of individual RNA molecules and their molecular identities.
Proteomics	Proteome	To study biochemical attributes and influential functions and how concentrations, modulations, and switching construction occurs during development and in reaction to a stimulus.
Metabolomics	Metabolome	Study of each small-molecule metabolite profile and some distinct cellular phenomena established during metabolite activity.

of plant defense. Genome-wide association studies (GWAS) are essential for improving crop resilience to abiotic stress. A number of investigations have shown the value of GWAS in evaluating changes in sunflower oil content brought on by abiotic stress (Mangin *et al.*, 2017). Prior research found quantitative trait loci (QTLs) linked to maize yield under water and heat stress (Millet *et al.*, 2016). In GWAS investigations, ecological factors found single nucleotide polymorphisms (SNPs) associated with sorghum water shortage and 213 genomic regions associated with drought resilience (Lasky *et al.*, 2015; Spindel *et al.*, 2018). The new science of epigenomics combines genomics and epigenetics, which deals with heritable changes other than DNA sequences. With epigenetic mechanisms reacting to stressors and environmental conditions, this integration reveals gene regulation in cellular responses to stress. To examine these occurrences throughout developmental stages or evaluate abnormalities brought on by plant diseases, genome-level studies are required (Callinan and Feinberg, 2006; Muthamilarasan *et al.*, 2019).

Crossbreeding, a traditional crop breeding technique, is a popular and successful way to alter different plant species. Crop breeding operations can effectively enhance crop types and production. Recombinant DNA breeding, mutagenesis breeding and gene editing (GE) associated breeding for plant advancement are the main breeding techniques utilized in contemporary agriculture. It takes years for genetic recombination and crossbreeding to introduce desired alleles and boost diversity (Scheben *et al.*, 2017). Transgenic breeding is simple and eminent; it significantly reduces the breeding time by enhancing crop attributes through external genetic transformation to produce economically viable superior cultivars. However, this technique uses plant transformation to introduce particular genes into the genome at arbitrary locations, producing types with foreign DNA. GE-mediated crop breeding is quicker, more accurate, and more efficient than mutagenesis breeding, hybridization, and conventional recombinant DNA breeding. By precisely altering the targeted regulatory sequence or gene or altering the RNA and/or DNA bases in superior types, GE enhances a desired feature in a relatively short amount of time. Clustered regularly interspaced short palindromic repeats (CRISPR)-Cas9, zinc-finger nucleases (ZFNs), transcription activator-like effector nucleases (TALENs), and meganuclease (MegN) are all components of current GE technology (Manghwar *et al.*, 2019, 2020; Khalil, 2020; Zhang *et al.*, 2020). CRISPR/Cas-mediated GE is a revolutionary technique that can produce targeted edits for quick crop plant improvement, elevate quality and yield in less time, and produce variations in required traits like abiotic and biotic resistance. It is also easy to apply, highly efficient, and inexpensive. Globally, plant diseases and harsh weather conditions result in significant crop losses. The development of GE techniques has created new opportunities for the creation of crop types with various resistances (Wang *et al.*, 2022). The groundbreaking CRISPR-Cas9 technology has revolutionized genome editing. Bacteria first developed as a defense mechanism against viruses. Eukaryotic genomes are extensively edited by CRISPR-Cas systems, opening the door to crop plant engineering for enhanced ability to withstand abiotic and biotic stressors. Genomic research is being transformed by AI, with particularly exciting potential implications in plant defense. AI systems can find genes and processes involved in plant defense mechanisms by analyzing enormous volumes of genomic data, which will help create more resilient crops (Biswas *et al.*, 2023).

7.2.2 Transcriptomics

The dynamic field of transcriptomics, which studies an organism's whole collection of RNA transcripts within specific cells or tissues (transcriptome), allows for the analysis of reactions of gene expression to various stimuli across predefined periods of time (Raza *et al.*, 2021). All the RNA present in cells is called the transcriptome. While much of this RNA probably will not be translated into proteins, it is still involved in determining the phenotype of cells. The technique used in this field, transcriptome profiling, permits the detection of differences in the expression of genes and provides information about the functions of some genes. The primary technology utilized for precise quantification of gene expression is transcriptome sequencing, commonly known as RNA-seq, which utilizes next-generation sequencing (both short and long reads). Notable advancements have been documented in various fields such as plants, humans, animals, and clinical research through RNA-seq, focusing on the gene expression of mRNA, small non-coding RNAs, long non-coding RNAs, alternative splicing variations, isoform variations, gene fusions, and single-nucleotide variations, which vary based on the objectives of the study, reproducibility, platforms, and methodologies employed. By integrating transcriptome sequencing with additional techniques like chromatin immunoprecipitation, methylation analysis, and GWAS, insight into genetic and epigenetic regulation can be attained. Long-read sequencing can also be employed to explore the epi-transcriptome, which encompasses RNA methylation, modifications, and substituted polyadenylation events. Transcriptome investigations have revealed gene sets with changed expression in a variety of crops, inclusive of rice and sorghum, in response to osmotic stress, heat, drought, and hormone treatments (Dugas *et al.*, 2011; Jin *et al.*, 2013; Johnson *et al.*, 2014). The analyses are vital for functional research and are essential for understanding the variations in gene expression throughout growth and stress responses. For example, as RNA-seq research in sweet potatoes and foxtail millet reveals biotic and abiotic stress-related gene expression responses in different tissues, transcriptomics is important in explaining responses related to stress and growth and developmental processes in crops (Li *et al.*, 2017). Research on rice, rapeseed oil, and maize has used RNA-seq to help discover genes that respond to drought stress (Bhardwaj *et al.*, 2015). From crosstalk pathways towards the disclosure of complex genes along with similar expression patterns, comparative transcriptome research allows research on unique expression patterns of numerous genes across some crop species facing stress (Li *et al.*, 2013; Zhu *et al.*, 2013). These research outcomes focus on stress tolerance through controlled gene regulatory networks and offer an optimistic view for augmenting crop attributes by genetic modifications (Murmu *et al.*, 2024).

7.2.3 Proteomics

Understanding how plants react to biotic and abiotic stimuli depends heavily on proteomics, a thorough method of researching proteins. The quantitative evaluation of complex protein samples is hampered by the fact that proteins are far more diverse chemically and physically than nucleic acids. Furthermore, it is expected that the number of protein species for a given organism would be several orders of magnitude greater than the number of genes based on RNA splicing and post-translational modifications. Yet, another layer of complexity appears if one considers the probable amount of protein–protein interactions in an organism that has been shaped by developmental events and physiological constraints (Van Wijk, 2001).

Based on Aizat and Hassan (2018), the four fundamental areas of proteomics—structural, functional, expression, and sequence proteomics—provide a deeper insight into the complex interactions occurring within the plant cell. Through amino acid sequence determination with more advanced techniques like high-performance liquid chromatography (HPLC), sequence proteomics provides data regarding the protein-building blocks (Twyman *et al.*, 2013). Employing several approaches, including X-ray diffraction, computer modeling, nuclear magnetic resonance (NMR), crystallization, and electron microscopy, structural proteomics attempts to understand three-dimensional protein structures and functions. In a bid to determine the specific functions the different proteins have in the living environment, functional proteomics determines the protein activities using Y2H assays and protein microarray profiling. As illustrated by the iTRAQ method, which provides critical insights into the way plants respond to environmental stresses, scientists are able to utilize quantitative proteomics for the identification of protein changes in expression levels induced by stressors (Liu *et al.*, 2015; Zhu *et al.*, 2018; Yang *et al.*, 2020).

Proteomics has shown itself to be crucial in the larger picture of plant reactions to biotic stress. Research on *Vitis* species and other crops demonstrates how proteome analysis can detect translational changes like phosphorylation and ubiquitination as well as stress-responsive proteins. This knowledge facilitates a comprehension of the complex molecular processes that underlie pathogen virulence and plant defense mechanisms (Mosa *et al.*, 2017). Similarly, proteomics identifies proteins linked to stress response in plants like wheat under abiotic stressors like drought. The combination of proteomics and phosphoproteomics investigates a range of responses to different stressors, helping to identify crop cultivars that are vulnerable and resistant to these difficulties. Furthermore, integrating proteomics with other omics fields such as metabolomics and functional genomics improves the understanding of the biology of stress and makes it easier to find molecular markers for breeding initiatives (Margaria *et al.*, 2013; Yang *et al.*, 2013; Zhang *et al.*, 2014).

Proteomic studies of plant tissues or organs were used to distinguish lines with variable genetic backgrounds as well as to monitor changes in development or the effects of environmental stressors on protein patterns. In a study of the maize leaf proteome that relied on 2-D electrophoresis for protein separation, when a pH 4–7 range was used for isoelectric focusing, more than 900 spots were resolved, and 200 spots were identified with gels covering the pH gradient of 6–11. Protein and expressed sequence tag (EST) databases served as the source of mass spectrometry (MS) protein spot identification. Surprisingly, for more than half of the spots found, no function has been given to prevalent leaf proteins, or only a potential one (Porubleva *et al.*, 2001). The assessment of barley seed development using proteomics showed that one set of proteins was present at every stage of this process, whereas other sets were linked to prior grain filling, the afterward desiccation phase, or temporary cumulation throughout this developmental phenomenon. The identification of pertinent locations revealed the build-up of enzymes of the antioxidant defense system, serine protease inhibitors, and low-Mr α-amylase/trypsin inhibitors. For the first time, the existence of several proteins within this process was shown experimentally, and cultivar-specific spot fluctuations were tracked. The outcomes show how proteome analysis can provide fresh perspectives on the cellular processes underlying seed development (Finnie *et al.*, 2002). Numerous proteomic techniques, such as liquid chromatography (LC)–MS/MS, matrix-assisted laser desorption/ionization–time-of-flight (MALDI-TOF), SDS-PAGE, and iTRAQ, are widely used in a variety of crops to examine how they react to biotic and abiotic stressors. The understanding of the molecular processes through which plants respond to environmental stressors is greatly aided by proteomic research, which eventually advances crop output and stress tolerance (Mohammadi *et al.*, 2012; Ramalingam *et al.*, 2015).

7.2.4 Metabolomics

Metabolomics, the study of metabolites in systems of biology, which provides the basis to analyze the plant metabolome and identify regulatory processes following stress, has potential to provide a better understanding of biochemical processes involved in gene activity by merging next-generation sequencing. Metabolites that have been confirmed by gas chromatography–mass spectrometry (GC–MS) as biological markers in rice types susceptible to the GMB1 pathogen have been proven as crucial for the plant's resistance to stress and infections. In this regard, the importance of metabolomics is highlighted by parallel approaches that elucidate an increased level of the metabolite related to various types of diseases of rice and barley (Murmu *et al.*, 2024). The goal of metabolite profiling is to measure all or a group of metabolites in a sample at the same time.

A variety of analytical methods can be used for metabolite profiling. These methods include FT-IR spectroscopy, capillary electrophoresis–mass spectrometry (CE–MS), LC–MS, NMR, and GC–MS (Sumner *et al.*, 2003; Shulaev, 2006). The most advanced analytical tool used in the characterization of plant metabolites is GC–MS. Some years back, it was considered one of the earliest approaches toward high-throughput plant metabolite profiling. A time-of-flight mass spectrometer or electron impact quadrupole is used commonly for GC–MS. A few hundred molecules belonging to different classes of chemicals, including sugars, sugar alcohols, organic acids, aromatic amines, fatty acids, and amino acids, can be profiled with the help of GC–MS (Fiehn *et al.*, 2000; Roessner *et al.*, 2000). The crucial advantage of GC–MS for metabolomics is the accessibility of both commercially and openly accessible EI spectrum libraries. GC–MS profiling can only determine volatile molecules or compounds that are volatilized after chemical derivatization and, hence, this is a constraint (Halket *et al.*, 2005).

LC–MS and CE–MS are good alternatives for non-volatile chemicals. The introduction of ultra-performance LC technology, which can significantly raise separation efficiency and reduce time, has continuously increased the use of LC–MS in metabolomics over the years (Giri *et al.*, 2007; Granger *et al.*, 2007). Because of its excellent resolving power, minimal sample volume requirements, and capacity to simultaneously separate cations, anions, and uncharged molecules, CE–MS offers a competitive alternative for metabolite profiling (Soga *et al.*, 2003). A study used metabolite profiling to investigate the interactions of the *Arabidopsis* responses to temperature stress. The authors recognized a set of well-known metabolites and unknown mass spectral tags specific to either heat, cold, or both stressors through GC–MS profiling of the plant *Arabidopsis* under heat and cold stress (Kaplan *et al.*, 2004). The metabolic response to cold stress was more impressive, yet it was more surprising that most of the metabolites caused by heat shock were also elicited by cold shock. Indeed, most of the metabolites elaborated by the two stresses had not been previously associated with any temperature stress condition. These findings were combined with transcript profiling data in later work to identify the mechanism behind cold adaptation in *Arabidopsis*. The authors concluded that mechanisms of regulation independent of transcript abundance could be crucial in the metabolic changes of plants during cold acclimation because they found that transcript abundance correlated with a metabolite plethora for some metabolic processes but did not correlate for others (Kaplan *et al.*, 2007). Wheat crops also respond to biotic stress by producing phenolic and phenylpropanoid metabolites.

Metabolomics is an important area because of the high volume of metabolite production in plant systems. Environmental metabolomics has gained further significance in the discovery that rice crops accumulate secondary metabolites such as polyamines under drought stress conditions, and therefore plants can react in many different ways to abiotic stressors. Rice, tomato, maize, and soybean have also been extensively studied using many metabolomics techniques, such as LC/GC–MS, GC/EI–TOF-MS, HPLC, and NMR. These techniques seem to offer important insights into how these crops react to abiotic and biotic stressors (Murmu *et al.*, 2024).

7.3 Basics of AI Techniques

7.3.1 Unsupervised and supervised machine learning

Machine learning (ML) is becoming a backbone to AI because it is regarded as a powerful tool for data analysis in various complex, high-dimensional plant-omics data (Mahesh, 2020). Such ML techniques can be divided broadly into categories of supervised and unsupervised learning applying different methodologies to applications in a field of study on plant defenses. Supervised learning actually relates to labeled datasets with input–output pairs predefined (Nasteski, 2017). The models train to map inputs to respective outputs; thus, they are highly effective for predictive tasks. Applications of supervised learning in plant-omics include:

- **Stress identification:** It can be predicted whether a plant exhibits a stress response through support vector machines (SVMs), random forest, and so on, using environmental as well as molecular data. For example, SVMs are used to classify plant genotypes resistant or susceptible to specific pathogens (Salcedo-Sanz *et al.*, 2014) while the random forest (Breiman, 2001) and gradient boosting (Friedman, 2001) machines perform well in feature importance analysis, thereby identifying key biomarkers in stress responses.
- **Feature selection and biomarker discovery:** These algorithms find the most relevant genes, proteins, or metabolites that are involved in stress responses, allowing for targeted interventions in crop improvement.

- **Trait prediction:** Using neural networks and ensemble methods, including gradient boosting machines, it is possible to predict traits related to agriculture such as yield, pest resistance, and drought tolerance from genomics. Shallow architectures are well-suited to yield traits with multi-omics data.

The objective of unsupervised learning is to discover the underlying patterns or structures in data, and it is thus very applicable in the exploratory analysis of plant-omics because many large datasets are not yet well defined for a predefined set of labels (Usama *et al.*, 2019). Unsupervised methods include:

- **Clustering:** Techniques such as k-means or hierarchical clustering group genes/metabolites that share similar expression profiles, thus depicting co-regulated networks or metabolic pathways (Rokach and Maimon, 2005). K-means clustering is the algorithm widely employed to group genes or proteins based on the expression profiles, that are encountered under different kinds of stress (Likas *et al.*, 2003). Hierarchical clustering provides dendrogram-based visualizations for understanding the relationships among stress-response components (Murtagh and Contreras, 2012).
- **Dimensionality reduction:** This uses methods like principal component analysis (PCA) and t-distributed stochastic neighbor embedding (t-SNE) to reduce dimensions of omics data without losing any important biological variance by making visualization and interpretability possible.

The integration of both supervised and unsupervised approaches allows researchers to build robust models that can illustrate plant responses under biotic and abiotic stresses, thereby linking raw omics data to a biological interpretation.

7.3.2 Major deep learning architectures

Deep learning (DL) thus forms a special domain in the ML area, since it has brought about a complete shift in the treatment of large, high-dimensional datasets produced in plant-omics research. DL models learn hierarchical representations that help them to particularly excel at performing various complex problems.

7.3.2.1 Convolutional neural networks

Convolutional neural networks (CNNs) are very popular in analyzing image-based phenomics data (Gu *et al.*, 2018). In plant defense research, they can be used to detect early symptoms of stress, like lesions caused by pathogens or drought-induced leaf wilting, from hyperspectral or thermal images. CNNs have strong spatial feature extraction and pattern extraction capability, enabling stress response to be detected accurately and monitored.

7.3.2.2 RNNs and LSTMs

Recurrent neural networks (RNNs), with their derivative known as long short-term memory (LSTM), permit the analysis of a time series (Sherstinsky, 2020). LSTMs focus precisely on capturing the long-range dependency in predicting possible variations of plant gene expression against stresses over time and offer views over the actual time courses governing these plant response patterns.

7.3.2.3 Autoencoders

Autoencoder is an unsupervised network with the purpose of reducing the dimensionality of omics data (Baldi, 2012). This acts as a noise filter, removing unwanted information while keeping essential features of interest. A typical example would be metabolomics and proteomics, where complex data complexity has blurred the unclear hidden patterns. Autoencoders do reduce the process of feature extraction but improve further analyses of cluster or classification.

7.3.2.4 Graph neural networks

As new architecture for the analysis of biological interaction networks, for example, protein–protein interactions (PPIs) or gene regulatory networks, graph neural networks

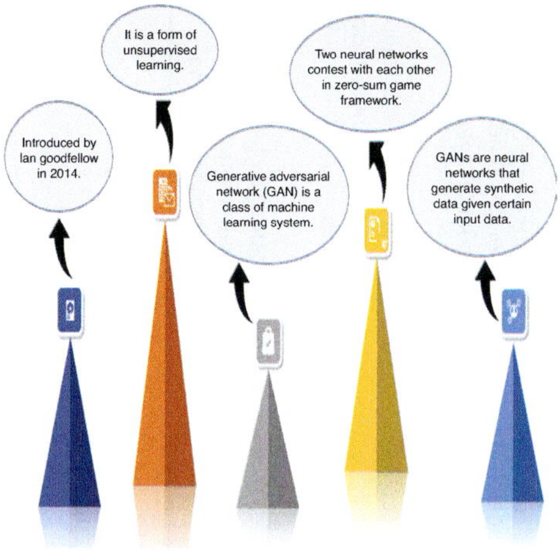

Fig. 7.2. The basics of GANs. (Author's own image.)

(GNNs) represent such interactions graphically and then model interdependencies between molecules involved with insight into how molecular mechanisms actually drive plant defense responses (Scarselli *et al.*, 2009).

7.3.2.5 Transformers

Initially designed for the tasks in natural language processing, BERT and other ViT like vision transformers find applicability in supporting analyses of plant-omics data. Their use to focus only on the important features of the data allows them to be used very easily for the task of multi-omics integration, with context-awareness in analyses.

7.3.2.6 Generative adversarial networks

Generative adversarial networks (GANs) (Goodfellow *et al.*, 2020) recently gained popularity in plant-omics, as did their use for the generation of synthetic data and modeling of complex distributions (Fig. 7.2). They consist of a generator and a discriminator (Nguyen *et al.*, 2017; Liu and Hsieh, 2019; Mao *et al.*, 2019); GANs operate by adversarial training, which allows:

- **Synthetic data generation:** GANs create realistic omics datasets that overcome small sample sizes typically seen in most plant studies (Eilertsen *et al.*, 2021; (Jordon *et al.*, 2022).
- **Data augmentation:** GANs improve training datasets by producing diverse synthetic samples, thereby enhancing the robustness and generalizability of models.
- **Image-to-image translation:** GANs allow for different imaging modalities, such as converting RGB images to thermal or hyperspectral representations, which in turn improves the stress detection capability.
- **Cross-modality integration:** GANs allow integrating several omics layers, for example, transcriptomics and metabolomics, to obtain a comprehensive picture of plant defense systems.

7.3.2.7 Comparison of deep learning architectures

The use of DL has opened up new avenues in plant defense research—from the early detection of stress symptoms to modeling complex molecular networks. DL provides a strong foundation for predictive and exploratory analyses, driving innovation in plant sciences. Table 7.2 compares key DL architectures, with a view to their original purpose, strengths, weaknesses, and major applications in different domains.

The comparison brings out the differences in the respective roles as well as the capabilities of each architecture, giving an insight into what can be applied to which task and, by default, the areas for which the DL applications may be enhanced.

7.3.3 Validation strategies

Validation of these models is needed for their use in plant-omics; they must be consistent, precise, and applicable. Without robust validation, insights resulting from such models could well be biologically irrelevant or simply not reproducible. There are two major cross-validation methods: K-fold cross-validation and leave-one-out cross-validation (LOOCV).

Table 7.2. Major deep learning architectures and their applications. (Author's own table.)

Architecture	Primary use	Strengths	Weaknesses	Key applications
Convolutional neural networks (CNNs)	Image and spatial data processing	Excellent for extracting spatial features, robust for image recognition and classification	Limited to spatial data, struggles with sequential or temporal data	Image recognition, medical imaging, video analysis, object detection
RNNs and LSTMs	Sequential and temporal data processing	Handles temporal dependencies, captures long-term relationships (LSTMs)	Prone to vanishing/exploding gradients (RNNs), computationally intensive (LSTMs)	Time-series forecasting, natural language processing (NLP), and speech recognition
Autoencoders	Dimensionality reduction, unsupervised learning	Effective for feature extraction, denoising, and reconstruction	Sensitive to overfitting, lacks interpretability compared to supervised models	Anomaly detection, data compression, feature extraction, image denoising
Graph neural networks (GNNs)	Modeling relationships in graph-structured data	Captures dependencies in structured data, effective for network and graph-based analysis	Computationally expensive for large graphs, scalability challenges	Social network analysis, protein interaction networks, recommendation systems
Transformers	Sequential data processing with attention mechanisms	Handles long-range dependencies, parallelizable, excels in capturing global context	Requires large datasets for training, high computational cost	NLP (translation, summarization), vision transformers, multi-modal data analysis
Generative adversarial networks (GANs)	Data generation and augmentation	Generates high-quality synthetic data, excellent for creative tasks	Difficult to train, sensitive to mode collapse	Image synthesis, style transfer, anomaly detection, synthetic data generation for training

K-fold cross-validation is a very often-used technique wherein the dataset is divided into k subsets (Anguita *et al.*, 2012). The model then trains on k-1 subsets and tests it on the rest of the subsets. By iterating over all subsets, it reduces bias and variance and gives an accurate estimate of model performance. LOOCV, however, is particularly useful for small datasets, where it iteratively trains the model on all but one sample, so every data point is tested (Vehtari *et al.*, 2017). Testing the developed models on new, independent datasets that were gathered either in a varying experimental condition or with other varieties of the very same crop also validates that generalization. An example is one trained with the performance of a rice dataset in maize, wheat, and others. To avoid overfitting, DL models usually use regularization techniques such as dropout regularization, which can randomly deactivate neurons during training. The network has to generalize rather than memorize and conduct early stopping in which the training is stopped by stopping criteria when the model has stabilized on validation data. Regarding evaluation metrics, the quantitative metrics are a must for determining the performance of the model. Supervised models include metrics like accuracy, precision, recall, F1-score, and AU-ROC, which measure the quality of the classification and predictions. Unsupervised models including metrics used for clustering output include silhouette scores and Davies–Bouldin Index to determine the meaningful biological groups. AI-based predictions must match up with biological evidence. For example, genes reported to be pivotal for stress responses must be tested through experimental means such as quantitative PCR (qPCR) or functional genomic tools such as CRISPR/Cas9. State-of-the-art AI architectures now more routinely use explainability tools, which include SHAP (Shapley additive explanations) and LIME (local interpretable model-agnostic explanations). Such tools allow users to understand how the model reached its predictions, hence making it easier for plant biologists to use. Through such validation strategies, AI models in plant-omics can be made to strike a balance between having high computational accuracy and biological relevance, hence being useful to yield actionable insights in plant defense research.

7.4 Why Analyze Plant-Omics Data Using Machine Learning Rather Than Traditional Methods?

ML, in its ability to handle high-dimensional, large-scale datasets, is believed to be one of the most ingenious tools to deal with every plant-omic challenge. According to Singh *et al.* (2016), the emergence of next-generation sequencing has precipitated a flood of data, using which researchers have begun to mine for complex relationships, unwrap the mystery of the plant stress response, and explain the complexity of the genomic response. Data generated using plant-omics have other issues, including high dimensionality, uncertainty, and interdependence between variables. Traditional statistical models are incapable of properly handling these complexities (Altman and Krzywinski, 2018; Niazian and Niedbała, 2020). ML, especially DL, has been shown to be one of the most efficient solutions to those problems, allowing the analysis of plant traits affected by genotype–environment interactions (Arsenovic *et al.*, 2019). Unsupervised and semi-supervised ML algorithms applied in plant systems biology would permit large dataset analyses without extensive labeled training sets (Yan and Wang, 2022). The use of ML is growing for improving plant agronomic traits with the large volume of omics data available (Isewon *et al.*, 2022).

Studies indicate the advantages of different ML models; thus, decision tree-based ensemble models show an efficient way to develop genomics predictions and integrated analysis of plant-omics data (Gokalp and Tasci, 2019; Silva *et al.*, 2019; Farooq *et al.*, 2022; Isewon *et al.*, 2022). ML has played an increasing role in elucidating complex functions of plant molecular biology in research involving pathogen effector genes and plant immunity (Silva *et al.*, 2019). ML methods have shown promise in transcriptomics for enhancing the detection of differential gene expression with increased sensitivity (Wang *et al.*, 2018). However, nonlinear ML approaches in differential expression analysis could have some intrinsic limitations, indicating that alternative explainable AI models must be developed for better interpretation of results in identifying relevant gene sets (Sabbatini and Calegari, 2023). These may synergize ML with aspects of classical biological knowledge to obtain a potent tool for uncovering biological dynamics in massive datasets, complementing traditional modeling methods (Xu and Jackson, 2019; Gilpin *et al.*, 2020). Different ML techniques—using tree-based methods, Bayesian models, network-based fusion methods, kernel methods, matrix factorization models, and deep neural networks—have enabled major progress in analyzing multi-view biological data (Li *et al.*, 2016). Through ML, multivariate analysis affords multiple variables to be considered together in the search for novel biomarkers and predictive models (Reel *et al.*, 2021). This elevates the platform for plant science research in leveraging omics towards improving agronomic traits.

Nevertheless, there are still some challenges that have to be overcome to fully realize the integrated potential of multi-omics data, particularly concerning scalability issues that represent a considerable barrier (Noor *et al.*, 2019). With the high complexity of multidimensional multi-omics data, sophisticated approaches are required to perform feature selection and information extraction. The core strength of ML lies in its ability to mine large amounts of data, and identify and rank the most pertinent features. For example, Du *et al.* (2019) employed the ML-based feature selection methods of PCA and LASSO on RNA-sequencing datasets towards understanding the response to salt stress in rice. The power of ML to conduct prediction and classification makes it ideal for scientists to create responsive situations in phenotypic outcomes and identify important potential biomarkers. The techniques have a dramatic advantage over traditional techniques fraught with computer limitations in classifying large-scale omics data. The wide use of ML in plant-omics research is largely thanks to its unrivaled capability in handling the intricacies of omics data. ML has brought about a major change in predictive modeling, allowing researchers to adopt exploratory methods for data analysis.

Another notable strength of ML is in finding patterns and relationships from complex datasets—an asset that is particularly useful in the field of omics when subtle biological mechanisms become clear to the researcher. In addition, ML adapts to the inherent heterogeneity of the omics dataset—heterogeneity that can arise from biological variability and from technical differences. Its flexible general approach does not make many assumptions about the data distribution, qualifying it for a wider range of applications of different data types and experimental designs. This adaptability enables ML to unveil hidden patterns and make predictions that can lead to new areas of investigation. The construction of gene regulatory networks (GRNs) is one of the important goals of multi-omics analysis. Although relatively few ChIP-seq experiments have addressed transcription factor (TF) binding sites in plants, most inference of transcriptional regulatory networks (GRNs) has, by and large, been based on expression data (Bubb and Deal, 2020). Traditional algorithms such as correlation and mutual information have often failed in establishing the directionality of regulatory processes or in accounting for time lags in gene expression (Banf and Rhee, 2017; Redekar *et al.*, 2017; Haque *et al.*, 2019). Certain probabilistic graphical models, such as GENIST and JRmGRN, were developed to overcome these drawbacks, but their requirements are set on data resolution in the spatial and temporal dimension in respect to expression (de Luis Balaguer *et al.*, 2017; Deng *et al.*, 2018). In contrast, ML provides a powerful tool for the GRN inference through the application of multi-omics, effectively elevating the prediction accuracy (Walley *et al.*, 2016; Ko and Brandizzi, 2020). For example, iDREM interacts with a hidden Markov model to reconstruct temporal GRNs in relation to biotic and abiotic stresses from transcriptomic, proteomic, and epigenomic datasets (Ding *et al.*, 2018). Further, single-cell RNA sequencing (scRNA-seq) has yielded GRNBoost2 on top of GENIE3, thus allowing for cell-specific GRN inference (Moerman *et al.*, 2019). Other tools such as SCENIC facilitate large dataset analysis and can even run 50,000 cells and 10,000 genes in under 2 hours (Van de Sande *et al.*, 2020). Thus, it is evident that the ongoing embrace of ML by plant-omics research stems from the fact that ML could manage difficult subtleties typical of what Avatar would characterize as omics data. ML has proved to be crucial to the advancement of plant-omics by bringing covariate information, supporting a broad range of datasets, offering predictive modeling and categorization, and permitting exploratory analysis.

7.5 The Application of AI in Plant-Omics Against Stress

The study of plant defense combines high-level molecular technologies with AI to elucidate and expand knowledge on how stress affects plant survival. AI-supported omics provide a route towards understanding the much more complicated gene–protein–metabolite networks involved in plant defense than has been previously achieved by time-consuming traditional methods. AI-driven techniques provide *in-silico* identification of important components within the lists of defense pathways, biomarker discovery, and previously hidden pathways that add details to our ever-increasing understanding of plant defense. Multi-omics information integration—genomics, transcriptomics, proteomics, and metabolomics—offers a holistic appreciation of complex biological processes engaged in stress responses (Fig. 7.3). AI methods also improve this by the detection of patterns and relationships inherent in these multifaceted datasets, and ultimately the detection of potentially significant biomarkers and research development targets. This combined method promises to develop stress-resistant crops and optimization of agriculture practices, hence readying sound food production systems (Arabnia and Tran, 2011; Yan and Wang, 2023). This application eventually may revolutionize the integration and analysis of multi-omics factors that address plant defense and open new avenues of developing productive crops resistant to multiple environmental stresses.

7.5.1 AI-assisted genomics

ML algorithms play a central role in facilitating gene identification associated with stress responses in plants with a view to helping breeders and researchers to optimize crop yields. For instance, Liang *et al.* (2011) performed a study using a variation of SVM to reveal major drought-resistance-related genes in *Arabidopsis thaliana*. Similarly, Shikha and colleagues (2017) established the basis for Bayes algorithms to identify significant SNPs

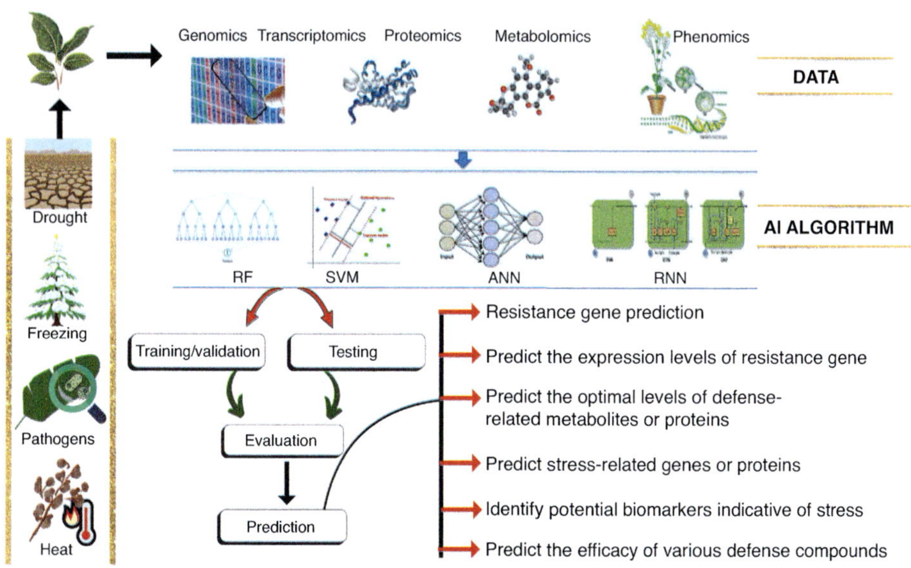

Fig. 7.3. Omics types in plant defense research. (Author's own image.)

associated with drought resistance in maize. SVMs for stress factors other than drought resistance have also been established. Ravari *et al.* (2015) evaluated salt tolerance in Iran wheat genotypes and identified potent indices to predict salt-resistant varieties by artificial neural network analysis; Wang *et al.* (2013) predicted salt-resistant genes in rice using an SVM-based model. The cis-regulatory code controlling the reaction of Arabidopsis roots to iron deprivation was investigated using ML methods (Schwarz *et al.*, 2020). According to a series of high-accuracy experiments conducted by Pal *et al.* (2016) in the field of plant disease-resistance research, SVM and its variants have also proven to be quite successful in predicting disease-resistance proteins. With the development of EFFECTORP, the first ML classifier for fungal effectors, by Sperschneider *et al.* (2016), ML has also made major strides in the prediction of pathogen effector proteins. Although disease resistance genes have received a lot of attention, plant disease susceptibility genes might be studied using ML methods, which would considerably aid agriculture (Yang and Guo, 2017). Furthermore, ML-based plant single-cell genomics has enormous potential for identifying new cell types, deciphering regulatory networks, and discovering cellular heterogeneity—all of which will have an influence on almost every facet of human health and agriculture. Studies by Silva *et al.* (2019) and Raimundo *et al.* (2021) have shown that ML can create low-dimensional representations, identify cell types, predict trajectories, create gene regulatory networks, and integrate multimodal data. Other ML methods, such neural network models (Lin *et al.*, 2017) and SIMLR algorithms (Wang *et al.*, 2017), have drawbacks, such as amplification artifacts during single-cell RNA sequencing and limited sequencing coverage. Each of these might provide information to help properly understand the complexities of single-cell genomics. To unlock ML potential in plant single-cell genomics, more investigation is needed. The fact that ML has only been used to identify genes in a small number of crops thus far highlights the need to expand its use to additional commercially significant plants in order to improve our understanding of stress resistance mechanisms and expedite breeding applications.

7.5.2 AI-assisted transcriptomics

Asefpour Vakilian (2020) conducted an extensive review on the role of microRNAs (miRNAs) in plant responses to abiotic and biotic stress factors. In particular, the author described the roles of specific miRNAs in response to drought, salinity, cold, and heat in *Arabidopsis thaliana* using feature selection algorithms. Key miRNAs including miRNA-169, miRNA-159, miRNA-396, and miRNA-393 were shown to be significant transcriptional regulators of plant stress responses by feature selection based on information theory. The analysis of correlation was performed through regression models such as decision trees, SVMs, and Naive Bayes, while the SVM model particularly with a Gaussian kernel provided fantastic predictive power, indicated by a very high R^2 value of 0.96 for predicting plant stress responses based on miRNA concentrations. Shaik and Ramakrishna (2014) showed significant connections between biotic and abiotic stress types on rice through intricate interactions, which is in contrast to the traditional belief that different stress types have distinct signaling pathways. These researchers discovered co-stress-responsive genes by meta-analyzing microarray data, and they discovered that over 70% of the common differentially expressed genes have conserved expression patterns under both biotic and abiotic stressors. Advanced ML, such as recursive-support vector machines and random forests, was used to classify abiotic vs biotic stress responses based on gene expression profiling. The recursive-support vector machine discovered 196 genes crucial for this categorization and classified these stressors with 100% accuracy. Meng *et al.* (2021) used supervised classification algorithms in a different investigation to forecast how genes will react to cold stress. The study found that models built on genomic knowledge performed as well as did models trained on wider datasets. Cold-stress response models developed with only one plant species showed impressive capabilities in predicting the gene response of other types of plants; they also successfully demarcated between cold-sensitive and cold-resistant species. Surprisingly, multi-species models provide superior predictions across species, suggesting that ML has potential for accelerating the comprehension of how gene expression reacts to environmental stressors in many plant species.

Zhou *et al.* (2022) undertook research into transcriptomic changes in maize genotypes for heat and cold stress. Significant alterations in gene expression were found. In particular, they noted enrichment of motifs near transcription start sites of thermally responsive genes and that models that utilized these motifs performed better still with respect to predicting gene expression responses. However, applying these models generated difficulties when transferring to other maize genotypes due to diminished performance. Pradhan *et al.* (2023a) have had the daunting task of predicting, considering abiotic stress-challenged elements, long non-coding RNA, which holds great importance in enhancing the ability of crops yield. The authors manufactured computational models for predicting the abiotic responsiveness of stress associated with long non-coding RNA with the help of ML. An impressive cross-validating accuracy of 68.84% was achieved, and an independent test set improved the accuracy to 76.23%. As the model is made easily available, the researchers have developed an online resource known as ASLncR. In a related study, Pradhan and colleagues (2023b) also investigated predicting miRNAs involved in responses to abiotic stresses like cold, drought, heat, and salt. They developed a support vector machine-based model that gave accurate results from 87.71% for cold to 90.15% for other stress conditions. To make this tool also readily available to the general botanical community, the authors developed an online server termed AsmiR. Meher *et al.* (2022a) created an ML-based technique to predict which miRNAs are responsive to abiotic stress. They used three datasets: miRNA, pre-miRNA, and a combined miRNA + pre-miRNA dataset. With area under the precision-recall curve (auPRC) and receiver operating characteristic curve (auROC) values ranging from 65.64 to 77.94% and prediction accuracies between 62.33 and 69.21%, the use of SVM in the model performed well. To aid in the discovery of pre-miRNAs and miRNAs that respond to abiotic stress, the authors also launched the ASRmiRNA online prediction server. These studies show great promise for the virtue ML has to offer in plant stress-responsive gene discovery.

7.5.3 AI-assisted proteomics

Meher *et al.* (2022a), using ML computational approaches, helped identify the abiotic stress-responsive genes (SRGs) under different hostile conditions. A precision of 60–78% was achieved using the SVM model. They also developed ASRpro—an integrated web-based prediction tool for such purposes. In contrast, Yang *et al.* (2019) used random forest algorithms to forecast PPI for known plant pathogens. Such predictions were found to be more accurate in a PPI prediction study that merged sequence data and network properties for a plant–pathogen interaction. as The introduction of InterSPPI web server continues ongoing research works in this area. In a study by Karan *et al.* (2023), the focus was the prediction of plant–pathogen interactions, taking special consideration of the rice–fungus blast pairing and modeling state-of-the-art ML algorithms to predict interactions leading to as much as 95% sensitivity on experimental data. By comparing results from other pathogen-hosted datasets to those of the model, the specificity for *Magnaporthe grisea and Oryza sativa* was confirmed. For the purpose of classifying unknown abiotic stress protein sequences, Ahmed *et al.* (2023) created a DL model and a novel activation function called the Gaussian Error Linear Unit with Sigmoid (SIELU). They attained an accuracy range between 80.78 and 95.11%, which was superior to any of its counterparts.

7.5.4 AI-assisted metabolomics

Liu *et al.* (2017) performed an analysis of 216 plant species based on an incomplete metabolite profile. They grouped metabolites with identical structures collectively using a network method called clustering and treated the plants as binary vectors in a hierarchical clustering classification. Quite remarkably, these results matched known phylogenetic relationships; despite the lack of complete datasets, metabolite composition is shown to have great potential as a taxonomic identifier. The research also suggested that metabolite content could also act as a reflection of the dietary and therapeutic qualities of the plant–secondary metabolite connections that were otherwise previously unexplored. In a different study, Fürtauer *et al.* (2018) monitored the metabolic regulation due to abiotic stress in plants. The tests were conducted on mutant lines that lacked sufficient starch or sucrose metabolism as well as Arabidopsis wild-type plants. For these experiments, high-light and cold stresses were applied to plants. Changes in the primary metabolomes and proteomes were quantified, and an ML model was trained to distinguish mutant lines in stress and control states. The method recognized a core set of 23 proteins that could effectively predict these stress conditions associated with high light and temperature. Significantly, 18 of those proteins were associated with PPIs, providing insights into the biochemical regulating mechanisms that govern dynamic responses from environmental changes.

7.5.5 AI-assisted phenomics

DL methods have been a success story in many crop management applications, with specific emphasis on plants like wheat, tomato, rice, and potato. Researchers in this domain have built a DL-based video detection system to recognize plant diseases and insects for automated and fast detection through continuous video analysis employing cutting-edge models like Faster R-CNN and YOLOv3 for real-time recognition. Video sequences were transformed into frames and further evaluated using the Faster R-CNN framework. In assessing identification quality, video-based assessment methods were incorporated in which a custom-designed backbone outperformed existing systems on an untrained rice video dataset. Mi *et al.* (2020) built a specialized deep-nested architecture of C-DenseNet for wheat stripe rust identification and grading. C-DenseNet was shown to be better than the traditional DenseNet and ResNet frameworks by scoring 97.99% accuracy during the test using models equipped with the convolution block attention module (CBAM) along the DenseNet architecture.

In a similar context, Wang and Liu (2021) reported an improved process of detecting tomato leaf spots early by grafting the MobileNetv2-YOLOv3 model. A very significant improvement in recognition accuracy was made possible using the general intersection over union-based bounding box regression loss. Some results showed that under some conditions, this lightweight model could reach an impressive F1 score of up to 94.13%. For the detection of viruses in seed potatoes, Polder *et al.* (2019) pioneered a hyperspectral imaging-driven in-field disease detection technique. Their system employed a hyperspectral line-scan camera and was trained on field data using a CNN. The technique proved to have very good accuracy and recall, reinforcing its application for present world potato illness recognition. Moreover, a DL pipeline was created by Chen and Yuan (2019) to identify and quantify pests that affect agriculture in images. Their approach filtered out conflicting discoveries by combining CNN and region proposal network (RPN) with non-maximum suppression (NMS). The advanced framework performed with an accuracy score of 0.93 and a modest miss rate of 0.10, emphasizing its efficacy for pest detection. Focusing on plant responses to salinity stress, Feng *et al.* (2020) fused DL and image processing techniques for non-destructive monitoring of plant characteristics. Using high-throughput phenotyping technologies, they applied hyperspectral imaging to study 13 okra genotypes under salt stress. For the most part, the augmented segmentation of plants and leaves joined forces with fine-tuned DL models to devise a good model for the extraction of plant and leaf architectural features. The physiological and biochemical responses of salinity-stressed okra matched their changes in the spectral data. The results led to the development of predictive models for the key traits, which had correlation coefficients from 0.588 to 0.835, demonstrating very promising results in the assessment of salinity stress.

7.6 Challenges

Considering more significant biological scientific developments, fundamental and applied plant studies have moved towards data-driven research methodologies during the past 20 years (Leonelli *et al.*, 2017; Leonelli, 2019). Significant advancements in high-efficiency genomics, the setting up of open-access genomic repositories, the advancement of high-efficiency phenomic analysis tools for implementation in labs, fields, and greenhouses, and the broad implementation of remote sensing technologies in farming have improved our capacity to gauge complexity in biology at the cellular, organismal, and ecological levels (Tardieu *et al.*, 2017). The agricultural AI sector is expected to reach a value of US$1.55 billion by 2025 because of the swift expansion spurred by these advancements. Large-scale agrarian enterprises have profited most from this AI rebellion, especially in the agricultural industry, where investments in research and development have been directed mainly towards precious crops for agriculture like tropical fruits and enormous-scale orchard and grapevine management, as well as staple crops like rice and maize (Carbonell, 2016). In addition, new avenues for the exploration of new international issues become available for plant researchers as well as stakeholders of agriculture with the increasing volume of data combined with the emerging AI methods (Tsiligiridia and Ainali, 2018). A notable instance from genomics that dates back to the 1990s is the implementation of computer algorithms to discover genes and estimate their activities using learning information from many species (Hayes and Borodovsky, 1998; Birney *et al.*, 2004; Zou *et al.*, 2019). This method is still very applicable to studies on orphan and non-model plants, where it may be too expensive or impossible to utilize testing methods such as CRISPR knockouts, which are used to discover and verify gene functions for particular species. Under such circumstances, research on model species can be used to predict gene function (Zou *et al.*, 2019).

Another problem in genomic studies is that computational methods can include determining pathogen virulence effector genes from genomic sequences (Sperschneider, 2020) and inferring gene controlling processes (Mochida *et al.*, 2018). Up until now, the most common AI technique used in plant and crop research has been ML. Using artificial neural networks and profile hidden Markov models like SMART and Pfam, ML has been majorly applied in sequence data evaluation, including the identification of functional domains and signal peptides in amino acid sequences (Larrañaga *et al.*, 2006; El-Gebali *et al.*, 2019). Cracking complicated biological processes has become significantly more possible thanks to the quick advancements in ML and biological data creation. Nevertheless, there are unique difficulties in using ML in plant molecular research. ML techniques are very contextually dependent, much like conventional molecular approaches, highlighting the importance of proper research design. It is important to realize that, despite the fact that ML aims at generating predictive models, each approach has strengths and weaknesses that may influence the ability of an algorithm to predict outcomes in specific situations. As a result of innate technical and biological differences, an ML model optimized for one dataset might not be applicable to the next. Thanks to the plethora of omics databases, an enormous amount of data is available. Still, a significant fraction of these databases bears the marks of sparseness and disturbance. This provides a considerable problem regarding reliably detecting biological traits, particularly when incorporating diverse omics datasets (Joyce and Palsson, 2006). Maimon and Rokach (2005) reported that a skewed dataset, in which sample quantities differ between categories, is common in ML. Investigators have used resampling techniques, including under- and over-sampling.

In many omics databases, ambiguity and complexity are introduced due to the degree of noise and sparseness, which can make it challenging for scientists to detect relevant biological trends or characteristics. The responsibility lies with scientists to distinguish accurate biological signals from noise in surroundings, thus underlining the importance of using strong analytical techniques to ensure accurate results. Additionally, overfitting is a serious issue, particularly in DL. When a predictive model gets rigidly adapted to the details of a single dataset, it is said to be overfitting and has a reduced capacity to generalize to new, untested data. This could lessen the capacity of the model to anticipate outcomes and make it less applicable in actual situations. Dropout is one strategy used to reduce this problem (Scholz *et al.*, 2004). Adaptive techniques are needed to manage the sheer number, diversity, validity, worth, and pace of massive information in plant systems biology. For handling multi-omics data efficiently, ML techniques must be improved to consider the distinctive knowledge each omics layer provides. Managing complex data problems such as overfitting, multicollinearity, and sparsity presents significant hurdles. This complexity necessitates using specialized methodologies and cooperative efforts to integrate data for relevant and reliable analysis. The 'black box' aspect of complicated models makes them difficult to comprehend, particularly in sophisticated ML techniques like DL. Investigators frequently give more weight to a prediction model's biological significance and understanding than its absolute precision. To do this and guarantee that the conclusions from the model are both practical and supported by science, meticulous data handling and matching model results with known biological concepts are needed (Murmu *et al.*, 2024). Notwithstanding these difficulties, the research described here demonstrates the useful applications of AI in plant omics. Creating reliable, scalable algorithms capable of revealing significant biological knowledge is essential to realizing all the possibilities of AI. It is just as crucial to confirm the precision and dependability of these computational results through testing before using them in practical settings. Furthermore, optimizing the advantages of AI in plant defense and broader agricultural procedures requires cultivating partnerships among omics investigators, data analysts, and agronomic specialists (Murmu *et al.*, 2024).

Monitoring plant stress with machine vision has several drawbacks and difficulties. While AI methods have shown encouraging outcomes in various programs, as shown in Fig. 7.3, several challenges must be handled for practical use in plant stress sensing. Plant phenotypes are changing, making it challenging to use photos to identify stress indicators in the early stages. This is one of the key challenges associated with crop development (Islam *et al.*, 2024). Moreover, comparable physiological alterations may transpire in reaction to distinct abiotic stressors, hence complicating the capacity of machine vision systems to discriminate among the various origins of plant aberrations. The other major challenge is developing a general model that works well across various plant types and growing conditions. Resolution of images and the differences in imaging methods also affect the performance of machine vision algorithms in image-based plant stress identification. Besides, it is difficult to find labeled datasets that precisely capture various stress situations, making it hard to train models that can adapt well to different stress situations (Islam *et al.*, 2024). Even though AI has many possibilities for use in agriculture, farmers worldwide still need to become more familiar with sophisticated ML technologies. Because agriculture is so dependent on outside variables like weather, soil quality, and insect activity, it is challenging to adhere to a set crop management approach for the duration of the planting season. Given that these outside factors are constantly changing, a plan created at the beginning of the season might no longer be functional by the end of the season. Even though AI is quite robust, it needs considerable information to train models to make correct forecasts and judgements properly. Massive farming operations can gather geographical data or knowledge of the features of the land. However, collecting temporal data—which records variations over time—presents considerably more difficulty. This restricts the capacity to act quickly and decisively in various ecological circumstances based on facts. Obtaining crop-specific knowledge is generally limited to a single growing season gathering, which causes an extensive holdup in developing a robust AI ML model. AI is used more often for agronomic items like seeds, fertilizers, and pesticides than on-field accurate solutions, partly due to the extended data maturation process. In conclusion, using cognitive technology will become increasingly crucial to agriculture in future generations. Even though there is much active investigation and several AI tools are accessible, the agricultural industry is still neglected (Khandelwal and Chavhan, 2019).

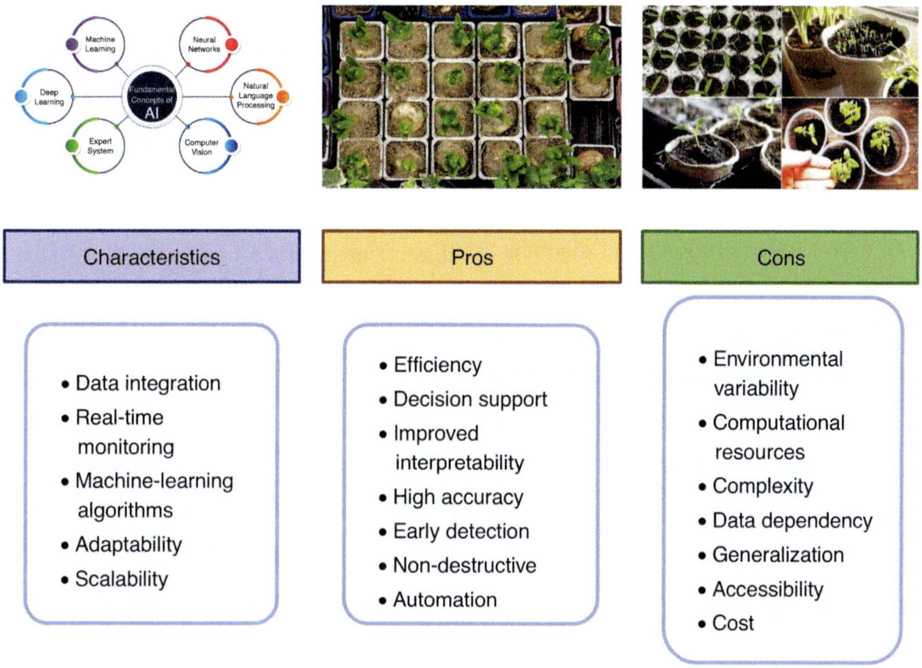

Fig. 7.4. A schematic representation of the characteristics and pros and cons of AI technologies for plant stress detection. (Author's own image.)

AI is still in the early stages of solving the problems and expectations that farmers encounter in reality. Applications must become more resistant to frequent changes in external conditions for AI to ultimately realize its abilities in farming. This will allow for real-time decision-making and the efficient use of data context. Again, another serious constraint is the high cost of the cognitive agricultural equipment available on the market. An overall lower price should be provided for AI solutions to facilitate greater availability. Commercial availability of AI technologies on open-source platforms may make such solutions more affordable and quicker to deploy, enabling farmers to access valuable insights. Eventually, this would accelerate AI acceptance into agriculture and enhance sustainability and productivity levels in farming communities (Khandelwal and Chavhan, 2019). The characteristics, pros, and cons of using AI in investigating plant stress responses are summarized in Fig. 7.4.

7.7 Future Scope

7.7.1 AI and omics data integration for early disease detection

This would finally target the integration of AI with data related to proteomics, metabolomics, transcriptomics, and genomics, so as to be able to identify the onset of crop ailments. Next-generation AI models can analyze extremely intricate multi-layered omics information as inputs to enable the recognition of minor molecular modifications that usually indicate disease onset, often weeks or even months before any visual defect starts to form.

7.7.2 Advanced phenotyping Integrated with AI-driven multi-omics

The integration of large-scale phenotyping approaches with AI-powered omics tools offers an interesting frontier. This fusion enables real-time plant wellness surveillance by correlating apparent phenotypic features with molecular reactions. By integrating phenomics data into analytical pipelines powered by AI, scientists can learn more about plant defense systems and how different stresses affect plants.

7.7.3 Predictive modelling of disease dynamics

Predictive models may be developed for predicting the dynamics of diseases in plant communities by using omics and AI data. DL and ML algorithms can be utilized to incorporate genetic, biological, and ecological factors to predict the outbreak of diseases and assess the effectiveness of prevention efforts.

7.7.4 Customized crop breeding

In the near future, one might expect scientists to focus on changing crop breeding strategies toward increased resistance to disease by employing the use of AI-assisted omics approaches. Genetic markers and processes that are connected to disease resistance can be identified, so breeders can create crops with enhanced defense systems that are better suited to certain situations.

7.7.5 AI-guided sustainable disease management

AI can lead to extremely high advances in sustainable disease management approaches. There is the potential for high accuracy in saving expenses by reduced environmental impacts and proper handling of plant diseases with the help of AI-driven insights and omics data for precise pesticide application.

7.7.6 Managing multiple stressors at once

With the increase in the risk of multiple stressors accompanying global warming and diversified farming models, it is important to know how plants react to biotic and abiotic stress. Future studies should focus more on how plants deal with such a variety of challenges altogether, based on practical farming conditions. AI-driven omics approaches will ease the deciphering of the many intricate relationships among stress variables and their combined effects on plant defense systems. Gaining such knowledge is vital in developing flexible, holistic solutions for plant protection under complex stress situations and ensuring food security for generations to come.

7.7.7 AI-enhanced omics and remote sensing synergy

AI-driven omics approaches in combination with sensor technology allow for effective monitoring and reduction of adverse conditions of plant. Aerial imaging data provide critical location-based and time-dependent information on the well-being of plants, adverse conditions, and external variables. The coupling of such remote sensing data with omics datasets allows investigators to understand comprehensively how genetic responses interplay with environmental stressors. This all-inclusive strategy promotes plant defense functions and decreases crop injury by allowing more precise and effective responses (Murmu *et al.*, 2024). Moreover, there will be a tremendous increase in new applications of machine vision in plant stress monitoring shortly, as multidisciplinary cooperation, research results, and innovation are improving. Advanced imaging devices, hyperspectral and multispectral imaging, for example, will allow a more detailed characterization of plant traits when further advanced. Such devices capture more wavelengths, and thus the ability to distinguish changes brought about by stress is improved. In addition, machine vision equipment is expected to be integrated into more extensive sensing systems that aggregate information from sources such as drones, ground-based sensors, and environmental sensors. This integration will enhance the accuracy of stress diagnosis through a more holistic understanding of the factors affecting plants and the environment.

More advancements in DL methods and neural network topologies will also significantly boost image identification and feature extraction capacities, enhancing the ability of machine vision systems to identify subtle and complex symptoms of stress precisely (Islam *et al.*, 2024). The immediate actions that minimize the effects of stress and maximize crop productivity will be reacted upon in precision agriculture. Real-time decision-support technologies are expected to facilitate quick reactions toward identified stress situations. Robotics in agriculture will gain more importance in the near future, encompassing machine vision-equipped robotic devices and self-governing drones. Such systems will effectively monitor large agricultural regions, recognize stressful situations, and carry out necessary activities with much human assistance. This partnership between human specialists and robotic systems for vision will continue being crucial. Their precision and understanding of findings will be improved using collaborative tools that allow input and direction from farmers and plant researchers to the machine learning model. This will facilitate a more successful amalgamation of technological innovations and farming expertise (Islam *et al.*, 2024).

7.8 Conclusion

In conclusion, this chapter includes insights into the transformation by ML in the translation of omics research to enforce plant defense against biotic and abiotic stresses. The incorporation of AI into plant research has resulted in ML and DL techniques altering the effective treatment of datasets from metabolomics, genomics, proteomics, and transcriptomics. AI-guided techniques are far superior to those traditionally conceived in their capacity to use high-dimensional data to gain more insight into how crops react to stressors. In this chapter, we underscore the vast opportunities that ML affords now to the development of studies concerned with plant defenses, from the identification of stress-resistance genes to understanding the processes regulated by gene expression to permit plant survival in difficult ecological conditions. With continuous exponential growth in plant sequencing data, ML methods are becoming more indispensable for analyzing and interpreting this information, enabling the characterization of new biomarkers and biological processes important for enhanced crop resistance. The positive results of AI applications across diverse omics applications, such as genomics, metabolomics, and proteomics, are translated into faster, more accurate insights into plant responses to environmental stressors.

This chapter also discusses the challenges of using ML in plant omics studies, especially in connection with the complexity of the omics data and the need for more sophisticated algorithms and computational power to exploit AI. However, the chapter also champions the positive aspects of AI, which could reshape plant defense research with better precision and specific strategies for improved crop protection. In addition, AI can also optimize agricultural practices toward crop improvement, enhancing the resistance of crops against diseases, pests, and environmental stress, thereby enhancing both quality and yield. In addition, the collaboration between data science, computer vision, and plant biology is paramount in tackling the multifaceted challenges that contemporary agriculture faces. The synergy between these streams of research could develop integrated solutions for crop protection, sustainable farming, and food security. With ongoing technological advancement, image-processing AI and sensor-oriented technologies will, in the coming years, provide another dimension of revolution to agricultural practices through real-time crop precision monitoring and management, contributing to sustainable farming methods amid climate change while improving resilience in crops as well as yield prediction, thereby enhancing global food production. As AI continues to develop in its role with omics data, it could lead to greater, yet to be determined possibilities for agriculture. AI-armed research could allow novel perspectives for plant defense mechanisms with more effective and sustainable strategies brought by various big-data measures for crop protection. Thus, these would address food problems, reducing agricultural losses and promoting green-agricultural practices to accommodate the demands of an ever-changing global population.

References

Ahmed, B., Haque, M.A., Iquebal, M.A., Jaiswal, S., Angadi, U.B. *et al.* (2023) DeepAProt: Deep learning based abiotic stress protein sequence classification and identification tool in cereals. *Frontiers in Plant Science* 13, 1008756. DOI: 10.3389/fpls.2022.1008756.

Aizat, W.M. and Hassan, M. (2018) Proteomics in systems biology. *Omics Applications for Systems Biology* 31–49. DOI: 10.1007/978-3-319-98758-3_3.

Altman, N. and Krzywinski, M. (2018) The curse(s) of dimensionality. *Nature Methods* 15(6), 399–400. DOI: 10.1038/s41592-018-0019-x.

Anderson, P.K., Cunningham, A.A., Patel, N.G., Morales, F.J., Epstein, P. *et al.* (2004) Characteristics of the pathogens and drivers of emerging infectious diseases of plants. *Trends in Ecology & Evolution* 10(19), 535–544. DOI: 10.1016/j.tree.2004.07.021.

Anguita, D., Ghelardoni, L., Ghio, A., Oneto, L. and Ridella, S. (2012, April) The "k" in k-fold cross validation. In: *Proceedings of the 20th European Symposium on Artificial Neural Networks, Computational Intelligence and Machine Learning (ESANN 2012)*, Bruges, Belgium, pp. 441–446.

Arabnia, H. and Tran, Q.N. (2011) *Software tools and algorithms for biological systems*. Springer, New York, NY.

Arsenovic, M., Karanovic, M., Sladojevic, S., Anderla, A. and Stefanovic, D. (2019) Solving current limitations of deep learning based approaches for plant disease detection. *Symmetry* 11(7), 939. DOI: 10.3390/sym11070939.

Asefpour Vakilian, K. (2020) Machine learning improves our knowledge about miRNA functions towards plant abiotic stresses. *Scientific Reports* 10(1), 3041. DOI: 10.1038/s41598-020-59981-6.

Baldi, P. (2012) Autoencoders, unsupervised learning, and deep architectures. In: *Proceedings of ICML Workshop on Unsupervised and Transfer Learning*, JMLR Workshop and Conference Proceedings, pp. 37–49.

Banf, M. and Rhee, S.Y. (2017) Computational inference of gene regulatory networks: Approaches, limitations and opportunities. *Biochimica et Biophysica Acta (BBA) - Gene Regulatory Mechanisms* 1860(1), 41–52. DOI: 10.1016/j.bbagrm.2016.09.003.

Bhardwaj, A.R., Joshi, G., Kukreja, B., Malik, V., Arora, P. *et al.* (2015) Global insights into high temperature and drought stress regulated genes by RNA-seq in economically important oilseed crop *Brassica juncea*. *BMC Plant Biology* 15, 1–15. DOI: 10.1186/s12870-014-0405-1.

Birney, E., Clamp, M. and Durbin, R. (2004) GeneWise and genomewise. *Genome Research* 14(5), 988–995. DOI: 10.1101/gr.1865504.

Biswas, A., Kumari, A., Gaikwad, D.S. and Pandey, D.K. (2023) Revolutionizing biological science: The synergy of genomics in health, bioinformatics, agriculture, and artificial intelligence. *OMICS: A Journal of Integrative Biology* 27(12), 550–569. DOI: 10.1089/omi.2023.0197.

Breiman, L. (2001) Random forests. *Machine Learning* 45(1), 5–32. DOI: 10.1023/A:1010933404324.

Bubb, K.L. and Deal, R.B. (2020) Considerations in the analysis of plant chromatin accessibility data. *Current Opinion in Plant Biology* 54, 69–78. DOI: 10.1016/j.pbi.2020.01.003.

Callinan, P.A. and Feinberg, A.P. (2006) The emerging science of epigenomics. *Human Molecular Genetics* 15(suppl_1), R95–R101. DOI: 10.1093/hmg/ddl095.

Carbonell, I.M. (2016) The ethics of big data in big agriculture. *Internet Policy Review* 5(1). DOI: 10.14763/2016.1.405.

Chen, L. and Yuan, Y. (2019) Agricultural disease image dataset for disease identification based on machine learning. In: *Big Scientific Data Management: First International Conference, BigSDM 2018, Beijing, China, November 30–December 1, 2018*, Springer International Publishing, pp. 263–274. DOI: 10.1007/978-3-030-28061-1_26.

de Luis Balaguer, M.A., Fisher, A.P., Clark, N.M., Fernandez-Espinosa, M.G., Möller, B.K. *et al.* (2017) Predicting gene regulatory networks by combining spatial and temporal gene expression data in arabidopsis root stem cells. *Proceedings of the National Academy of Sciences of the United States of America* 114(36), E7632–E7640. DOI: 10.1073/pnas.1707566114.

Deng, W., Zhang, K., Liu, S., Zhao, P.X., Xu, S. *et al.* (2018) JRmGRN: Joint reconstruction of multiple gene regulatory networks with common hub genes using data from multiple tissues or conditions. *Bioinformatics* 34(20), 3470–3478. DOI: 10.1093/bioinformatics/bty354.

Dugas, D.V., Monaco, M.K., Olsen, A., Klein, R.R., Kumari, S. *et al.* (2011) Functional annotation of the transcriptome of sorghum bicolor in response to osmotic stress and abscisic acid. *BMC Genomics* 12, 1–21. DOI: 10.1186/1471-2164-12-514.

Ding, J., Hagood, J.S., Ambalavanan, N., Kaminski, N. and Bar-Joseph, Z. (2018) iDREM: Interactive visualization of dynamic regulatory networks. *PLoS Computational Biology* 14(3), e1006019. DOI: 10.1371/journal.pcbi.1006019.

Du, Q., Campbell, M., Yu, H., Liu, K., Walia, H. *et al.* (2019) Network-based feature selection reveals substructures of gene modules responding to salt stress in rice. *Plant Direct* 3(8), e00154. DOI: 10.1002/pld3.154.

Eilertsen, G., Tsirikoglou, A., Lundström, C. and Unger, J. (2021) Ensembles of gans for synthetic training data generation. *arXiv Preprint arXiv*. DOI: 10.48550/arXiv.2104.11797.

El-Gebali, S., Mistry, J., Bateman, A., Eddy, S.R., Luciani, A. *et al.* (2019) The Pfam protein families database in 2019. *Nucleic Acids Research* 47(D1), D427–D432. DOI: 10.1093/nar/gky995.

Farooq, A., Mir, R.A., Sharma, V., Pakhtoon, M.M., Bhat, K.A. *et al.* (2022) Crop proteomics: Towards systemic analysis of abiotic stress responses. In: *Advancements in Developing Abiotic Stress-Resilient Plants*. CRC Press, pp. 265–285.

Feng, X., Zhan, Y., Wang, Q., Yang, X., Yu, C. *et al.* (2020) Hyperspectral imaging combined with machine learning as a tool to obtain high-throughput plant salt-stress phenotyping. *The Plant Journal* 101(6), 1448–1461. DOI: 10.1111/tpj.14597.

Fiehn, O., Kopka, J., Dörmann, P., Altmann, T., Trethewey, R.N. *et al.* (2000) Metabolite profiling for plant functional genomics. *Nature Biotechnology* 18(11), 1157–1161. DOI: 10.1038/81137.

Finnie, C., Melchior, S., Roepstorff, P. and Svensson, B. (2002) Proteome analysis of grain filling and seed maturation in barley. *Plant Physiology* 129(3), 1308–1319. DOI: 10.1104/pp.003681.

Friedman, J.H. (2001) Greedy function approximation: A gradient boosting machine. *The Annals of Statistics* 29(5), 1189–1232. DOI: 10.1214/aos/1013203451.

Fürtauer, L., Pschenitschnigg, A., Scharkosi, H., Weckwerth, W. and Nägele, T. (2018) Combined multivariate analysis and machine learning reveals a predictive module of metabolic stress response in *Arabidopsis thaliana*. *Molecular Omics* 14(6), 437–449. DOI: 10.1039/c8mo00095f.

Gilpin, W., Huang, Y. and Forger, D.B. (2020) Learning dynamics from large biological data sets: Machine learning meets systems biology. *Current Opinion in Systems Biology* 22, 1–7. DOI: 10.1016/j.coisb.2020.07.009.

Giri, S., Krausz, K.W., Idle, J.R. and Gonzalez, F.J. (2007) The metabolomics of (±)-arecoline 1-oxide in the mouse and its formation by human flavin-containing monooxygenases. *Biochemical Pharmacology* 73(4), 561–573. DOI: 10.1016/j.bcp.2006.10.017.

Gokalp, O. and Tasci, E. (2019) Weighted voting based ensemble classification with hyper-parameter optimization. In: *2019 Innovations in Intelligent Systems and Applications Conference (ASYU)*, Izmir, Turkey, IEEE, pp. 1–4. DOI: 10.1109/ASYU48272.2019.8946373.

Goodfellow, I., Pouget-Abadie, J., Mirza, M., Xu, B., Warde-Farley, D. *et al.* (2020) Generative adversarial networks. *Communications of the ACM* 63(11), 139–144. DOI: 10.1145/3422622.

Granger, J.H., Williams, R., Lenz, E.M., Plumb, R.S., Stumpf, C.L. *et al.* (2007) A metabonomic study of strain- and age-related differences in the *Zucker rat*. *Rapid Communications in Mass Spectrometry* 21(13), 2039–2045. DOI: 10.1002/rcm.3059.

Gu, J., Wang, Z., Kuen, J., Ma, L., Shahroudy, A. *et al.* (2018) Recent advances in convolutional neural networks. *Pattern Recognition* 77, 354–377. DOI: 10.1016/j.patcog.2017.10.013.

Halket, J.M., Waterman, D., Przyborowska, A.M., Patel, R.K.P., Fraser, P.D. *et al.* (2005) Chemical derivatization and mass spectral libraries in metabolic profiling by GC/MS and LC/MS/MS. *Journal of Experimental Botany* 56(410), 219–243. DOI: 10.1093/jxb/eri069.

Haque, S., Ahmad, J.S., Clark, N.M., Williams, C.M. and Sozzani, R. (2019) Computational prediction of gene regulatory networks in plant growth and development. *Current Opinion in Plant Biology* 47, 96–105. DOI: 10.1016/j.pbi.2018.10.005.

Hayes, W.S. and Borodovsky, M. (1998) How to interpret an anonymous bacterial genome: Machine learning approach to gene identification. *Genome Research* 8(11), 1154–1171. DOI: 10.1101/gr.8.11.1154.

Isewon, I., Apata, O., Oluwamuyiwa, F., Aromolaran, O. and Oyelade, J. (2022) Machine learning algorithms: Their applications in plant omics and agronomic traits' improvement. *F1000Research* 1256. DOI: 10.12688/f1000research.125425.1.

Islam, S., Elmekki, H., Elsebai, A., Bentahar, J., Drawel, N. *et al.* (2024) A comprehensive survey on applications of transformers for deep learning tasks. *Expert Systems with Applications* 241, 122666. DOI: 10.1016/j.eswa.2023.122666.

Jin, H., Huang, F., Cheng, H., Song, H. and Yu, D. (2013) Overexpression of the GmNAC2 gene, an NAC transcription factor, reduces abiotic stress tolerance in tobacco. *Plant Molecular Biology Reporter* 31(2), 435–442. DOI: 10.1007/s11105-012-0514-7.

Johnson, J.M., Alex, T. and Oelmüller, R. (2014) *Piriformospora indica*: The versatile and multifunctional root endophytic fungus for enhanced yield and tolerance to biotic and abiotic stress in crop plants. *Journal of Tropical Agriculture* 52(2), 103–122.

Jordon, J., Szpruch, L., Houssiau, F., Bottarelli, M., Cherubin, G. *et al.* (2022) Synthetic data--what, why and how. *arXiv Preprint arXiv*. DOI: 10.48550/arXiv.2205.03257.

Joyce, A.R. and Palsson, B.Ø. (2006) The model organism as a system: Integrating 'omics' data sets. *Nature Reviews Molecular Cell Biology* 7(3), 198–210. DOI: 10.1038/nrm1857.

Kaplan, F., Kopka, J., Haskell, D.W., Zhao, W., Schiller, K.C. *et al.* (2004) Exploring the temperature-stress metabolome of *Arabidopsis*. *Plant Physiology* 136(4), 4159–4168. DOI: 10.1104/pp.104.052142.

Kaplan, F., Kopka, J., Sung, D.Y., Zhao, W., Popp, M. *et al.* (2007) Transcript and metabolite profiling during cold acclimation of *Arabidopsis* reveals an intricate relationship of cold-regulated gene expression with modifications in metabolite content. *The Plant Journal* 50(6), 967–981. DOI: 10.1111/j.1365-313X.2007.03100.x.

Karan, B., Mahapatra, S., Sahu, S.S., Pandey, D.M. and Chakravarty, S. (2023) Computational models for prediction of protein-protein interaction in rice and *Magnaporthe grisea*. *Frontiers in Plant Science* 13, 1046209. DOI: 10.3389/fpls.2022.1046209.

Khalil, A.M. (2020) The genome editing revolution: Review. *Journal, Genetic Engineering & Biotechnology* 18(1), 68. DOI: 10.1186/s43141-020-00078-y.

Khandelwal, P.M. and Chavhan, H. (2019) Artificial intelligence in agriculture: An emerging era of research. *Research Gate Publication* 1(1), 56–60.

Ko, D.K. and Brandizzi, F. (2020) Network-based approaches for understanding gene regulation and function in plants. *The Plant Journal* 104(2), 302–317. DOI: 10.1111/tpj.14940.

Larrañaga, P., Calvo, B., Santana, R., Bielza, C., Galdiano, J. *et al.* (2006) Machine learning in bioinformatics. *Briefings in Bioinformatics* 7(1), 86–112. DOI: 10.1093/bib/bbk007.

Lasky, J.R., Upadhyaya, H.D., Ramu, P., Deshpande, S., Hash, C.T. *et al.* (2015) Genome-environment associations in sorghum landraces predict adaptive traits. *Science Advances* 1(6), e1400218. DOI: 10.1126/sciadv.1400218.

Leonelli, S. (2019) The challenges of big data biology. *Elife* 8, e47381. DOI: 10.7554/eLife.47381.

Leonelli, S., Davey, R.P., Arnaud, E., Parry, G. and Bastow, R. (2017) Data management and best practice for plant science. *Nature Plants* 3(6), 17086. DOI: 10.1038/nplants.2017.86.

Li, L., Snyder, J.C., Pelaschier, I.M., Huang, J., Niranjan, U. *et al.* (2016) Understanding machine-learned density functionals. *International Journal of Quantum Chemistry* 116(11), 819–833. DOI: 10.1002/qua.25040.

Li, P., Cao, W., Fang, H., Xu, S., Yin, S. *et al.* (2017) Transcriptomic profiling of the maize (*Zea mays* L.) leaf response to abiotic stresses at the seedling stage. *Frontiers in Plant Science* 8, 290. DOI: 10.3389/fpls.2017.00290.

Li, Y.F., Wu, Y., Hernandez-Espinosa, N. and Peña, R.J. (2013) Heat and drought stress on durum wheat: Responses of genotypes, yield, and quality parameters. *Journal of Cereal Science* 57, 398–404. DOI: 10.1016/j.jcs.2013.01.005.

Liang, Y., Zhang, F., Wang, J., Joshi, T., Wang, Y. *et al.* (2011) Prediction of drought-resistant genes in *Arabidopsis* thaliana using SVM-RFE. *PLoS ONE* 6(7), e21750. DOI: 10.1371/journal.pone.0021750.

Likas, A., Vlassis, N. and J. Verbeek, J. (2003) The global k-means clustering algorithm. *Pattern Recognition* 36(2), 451–461. DOI: 10.1016/S0031-3203(02)00060-2.

Lin, C., Jain, S., Kim, H. and Bar-Joseph, Z. (2017) Using neural networks for reducing the dimensions of single-cell RNA-seq data. *Nucleic Acids Research* 45(17), e156–e156. DOI: 10.1093/nar/gkx681.

Liu, X. and Hsieh, C.J. (2019) Rob-gan: Generator, discriminator, and adversarial attacker. In: *2019 IEEE/CVF Conference on Computer Vision and Pattern Recognition (CVPR)*, Long Beach, CA, USA, pp. 11234–11243. DOI: 10.1109/CVPR.2019.01149.

Liu, K., Abdullah, A.A., Huang, M., Nishioka, T., Altaf-Ul-Amin, M. *et al.* (2017) Novel approach to classify plants based on metabolite-content similarity. *BioMed Research International* 1–12. DOI: 10.1155/2017/5296729.

Liu, N., Dai, Q., Zheng, G., He, C., Parisien, M. *et al.* (2015) N6-methyladenosine-dependent RNA structural switches regulate RNA–protein interactions. *Nature* 518(7540), 560–564. DOI: 10.1038/nature14234.

Mahesh, B. (2020) Machine learning algorithms - a review. *International Journal of Science and Research* (1), 381–386. DOI: 10.21275/ART20203995.

Maimon, O. and Rokach, L. (eds) (2005) *Data Mining and Knowledge Discovery Handbook*, Vol. 2, No. 2005. Springer, New York, p. 2005.

Manghwar, H., Lindsey, K., Zhang, X. and Jin, S. (2019) CRISPR/cas system: Recent advances and future prospects for genome editing. *Trends in Plant Science* 24(12), 1102–1125. DOI: 10.1016/j.tplants.2019.09.006.

Manghwar, H., Li, B., Ding, X., Hussain, A., Lindsey, K. *et al.* (2020) CRISPR/cas systems in genome editing: Methodologies and tools for sgRNA design, off-target evaluation, and strategies to mitigate off-target effects. *Advanced Science* 7(6), 1902312. DOI: 10.1002/advs.201902312.

Mangin, B., Casadebaig, P., Cadic, E., Blanchet, N., Boniface, M. *et al.* (2017) Genetic control of plasticity of oil yield for combined abiotic stresses using a joint approach of crop modelling and genome-wide association. *Plant, Cell & Environment* 40(10), 2276–2291. DOI: 10.1111/pce.12961.

Mao, X., Su, Z., Tan, P.S., Chow, J.K. and Wang, Y.H. (2019) Is discriminator a good feature extractor? *arXiv Preprint arXiv*. DOI: 10.48550/arXiv.1912.00789.

Margaria, P., Abbà, S. and Palmano, S. (2013) Novel aspects of grapevine response to phytoplasma infection investigated by a proteomic and phospho-proteomic approach with data integration into functional networks. *BMC Genomics* 14, 1–15. DOI: 10.1186/1471-2164-14-38.

McCarthy, J., Minsky, M.L., Rochester, N. and Shannon, C.E. (2006) A proposal for the dartmouth summer research project on artificial intelligence, August 31. *AI Magazine* 27(4), 12–12. DOI: 10.1609/aimag.v27i4.1904.

Meher, P.K., Begam, S., Sahu, T.K., Gupta, A., Kumar, A. *et al.* (2022a) ASRmiRNA: Abiotic stress-responsive miRNA prediction in plants by using machine learning algorithms with pseudo K-tuple nucleotide compositional features. *International Journal of Molecular Sciences* 23(3), 1612. DOI: 10.3390/ijms23031612.

Meng, X., Liang, Z., Dai, X., Zhang, Y., Mahboub, S. *et al.* (2021) Predicting transcriptional responses to cold stress across plant species. *Proceedings of the National Academy of Sciences* 118(10), e2026330118. DOI: 10.1073/pnas.2026330118.

Mi, Z., Zhang, X., Su, J., Han, D. and Su, B. (2020) Wheat stripe rust grading by deep learning with attention mechanism and images from mobile devices. *Frontiers in Plant Science* 11, 558126. DOI: 10.3389/fpls.2020.558126.

Millet, E., Welcker, C., Kruijer, W., Negro, S., Nicolas, S. *et al.* (2016) Genome-wide analysis of yield in Europe: Allelic effects as functions of drought and heat scenarios. *Plant Physiology* 172(2), 00621. DOI: 10.1104/pp.16.00621.

Mochida, K., Koda, S., Inoue, K. and Nishii, R. (2018) Statistical and machine learning approaches to predict gene regulatory networks from transcriptome datasets. *Frontiers in Plant Science* 9, 1770. DOI: 10.3389/fpls.2018.01770.

Moerman, T., Aibar Santos, S., Bravo González-Blas, C., Simm, J., Moreau, Y. *et al.* (2019) GRNBoost2 and *Arboreto*: Efficient and scalable inference of gene regulatory networks. *Bioinformatics* 35(12), 2159–2161. DOI: 10.1093/bioinformatics.

Mohammadi, P.P., Moieni, A. and Komatsu, S. (2012) Comparative proteome analysis of drought-sensitive and drought-tolerant rapeseed roots and their hybrid F1 line under drought stress. *Amino Acids* 43(5), 2137–2152. DOI: 10.1007/s00726-012-1299-6.

Mosa, K.A., Ismail, A., Helmy, M., Mosa, K.A., Ismail, A. *et al.* (2017) Omics and system biology approaches in plant stress research. In: *Plant Stress Tolerance: An Integrated Omics Approach*. pp. 21–34.

Murmu, S., Sinha, D., Chaurasia, H., Sharma, S., Das, R. *et al.* (2024) A review of artificial intelligence-assisted omics techniques in plant defense: Current trends and future directions. *Frontiers in Plant Science* 15, 1292054. DOI: 10.3389/fpls.2024.1292054.

Murtagh, F. and Contreras, P. (2012) Algorithms for hierarchical clustering: An overview. *WIREs Data Mining and Knowledge Discovery* 2(1), 86–97. DOI: 10.1002/widm.53.

Muthamilarasan, M., Singh, N.K. and Prasad, M. (2019) Multi-omics approaches for strategic improvement of stress tolerance in underutilized crop species: A climate change perspective. *Advances in Genetics* 103, 1–38. DOI: 10.1016/bs.adgen.2019.01.001.

Nasteski, V. (2017) An overview of the supervised machine learning methods. *HORIZONS.B* 4(51–62), 51–62. DOI: 10.20544/HORIZONS.B.04.1.17.P05.

Nguyen, T., Le, T., Vu, H. and Phung, D. (2017) Dual discriminator generative adversarial nets. In: *Advances in Neural Information Processing Systems 30*. DOI: 10.48550/arXiv.1709.03831.

Niazian, M. and Niedbała, G. (2020) Machine learning for plant breeding and biotechnology. *Agriculture* 10(10), 436. DOI: 10.3390/agriculture10100436.

Noor, E., Cherkaoui, S. and Sauer, U. (2019) Biological insights through omics data integration. *Current Opinion in Systems Biology* 15, 39–47. DOI: 10.1016/j.coisb.2019.03.007.

Pal, T., Jaiswal, V. and Chauhan, R.S. (2016) DRPPP: A machine learning based tool for prediction of disease resistance proteins in plants. *Computers in Biology and Medicine* 78, 42–48. DOI: 10.1016/j.compbiomed.2016.09.008.

Panjabi-Sabharwal, V., Karan, R., Khan, T. and Pareek, A. (2010) Abiotic stress responses: Complexities in gene expression. In: *Abiotic Stress Adaptation in Plants: Physiological, Molecular and Genomic Foundation*. pp. 177–198. DOI: 10.1007/978-90-481-3112-9_9.

Polder, G., Blok, P.M., De Villiers, H.A.C., van der Wolf, J.M. and Kamp, J. (2019) Potato virus Y detection in seed potatoes using deep learning on hyperspectral images. *Frontiers in Plant Science* 10, 209. DOI: 10.3389/fpls.2019.00209.

Porubleva, L., Vander Velden, K., Kothari, S., Oliver, D.J. and Chitnis, P.R. (2001) The proteome of maize leaves: Use of gene sequences and expressed sequence tag data for identification of proteins with peptide mass fingerprints. *Electrophoresis* 22(9), 1724–1738. DOI: 10.1002/1522-2683(200105)22:93.0.CO;2-2.

Pradhan, U.K., Meher, P.K., Naha, S., Rao, A.R. and Gupta, A. (2023a) ASLncR: A novel computational tool for prediction of abiotic stress-responsive long non-coding rnas in plants. *Functional & Integrative Genomics* 23(2), 113. DOI: 10.1007/s10142-023-01040-0.

Pradhan, U.K., Meher, P.K., Naha, S., Rao, A.R., Kumar, U. *et al.* (2023b) ASmiR: A machine learning framework for prediction of abiotic stress-specific miRNAs in plants. *Functional & Integrative Genomics* 23(2), 92. DOI: 10.1007/s10142-023-01014-2.

Raimundo, F., Meng-Papaxanthos, L., Vallot, C. and Vert, J.P. (2021) Machine learning for single-cell genomics data analysis. *Current Opinion in Systems Biology* 26, 64–71. DOI: 10.1016/j.coisb.2021.04.006.

Ramalingam, A., Kudapa, H., Pazhamala, L.T., Weckwerth, W. and Varshney, R.K. (2015) Proteomics and metabolomics: Two emerging areas for legume improvement. *Frontiers in Plant Science* 6, 1116. DOI: 10.3389/fpls.2015.01116.

Ravari, S.Z., Dehghani, H. and Naghavi, H. (2015) Evaluation of relationship between salinity stress tolerance indices and some physiological traits in bread wheat. *Iranian Journal of Field Crop Science* 46(3), 423–432.

Raza, A., Su, W., Hussain, M.A., Mehmood, S.S., Zhang, X. *et al.* (2021) Integrated analysis of metabolome and transcriptome reveals insights for cold tolerance in rapeseed (*Brassica napus* L.). *Frontiers in Plant Science* 12, 721681. DOI: 10.3389/fpls.2021.721681.

Redekar, N., Pilot, G., Raboy, V., Li, S. and Saghai Maroof, M.A. (2017) Inference of transcription regulatory network in low phytic acid soybean seeds. *Frontiers in Plant Science* 8. DOI: 10.3389/fpls.2017.02029.

Reel, P.S., Reel, S., Pearson, E., Trucco, E. and Jefferson, E. (2021) Using machine learning approaches for multi-omics data analysis: A review. *Biotechnology Advances* 49, 107739. DOI: 10.1016/j.biotechadv.2021.107739.

Ristaino, J.B., Anderson, P.K., Bebber, D.P., Brauman, K.A., Cunniffe, N.J. *et al.* (2021) The persistent threat of emerging plant disease pandemics to global food security. *Proceedings of the National Academy of Sciences* 118(23), e2022239118. DOI: 10.1073/pnas.2022239118.

Roessner, U., Wagner, C., Kopka, J., Trethewey, R.N. and Willmitzer, L. (2000) Simultaneous analysis of metabolites in potato tuber by gas chromatography–mass spectrometry. *The Plant Journal* 23(1), 131–142. DOI: 10.1046/j.1365-313x.2000.00774.x.

Rokach, L. and Maimon, O. (2005) Clustering methods. In: *Data Mining and Knowledge Discovery Handbook*. pp. 321–352.

Sabbatini, F. and Calegari, R. (2023) Explainable clustering with CREAM. In: *20th International Conference on Principles of Knowledge Representation and Reasoning {KR-2023}*, Rhodes, Greece, pp. 593–603 (Vol. 19). DOI: 10.24963/kr.2023/58.

Salcedo-Sanz, S., Rojo-Álvarez, J.L., Martínez-Ramón, M. and Camps-Valls, G. (2014) Support vector machines in engineering: An overview. *Wiley Interdisciplinary Reviews: Data Mining and Knowledge Discovery* 4(3), 234–267. DOI: 10.1002/widm.1125.

Scarselli, F., Gori, M., Tsoi, A.C., Hagenbuchner, M. and Monfardini, G. (2009) The graph neural network model. *IEEE Transactions on Neural Networks* 20(1), 61–80. DOI: 10.1109/TNN.2008.2005605.

Scheben, A., Wolter, F., Batley, J., Puchta, H. and Edwards, D. (2017) Towards crispr /cas crops – bringing together genomics and genome editing. *New Phytologist* 216(3), 682–698. DOI: 10.1111/nph.14702.

Scholz, M., Gatzek, S., Sterling, A., Fiehn, O. and Selbig, J. (2004) Metabolite fingerprinting: Detecting biological features by independent component analysis. *Bioinformatics* 20(15), 2447–2454. DOI: 10.1093/bioinformatics/bth270.

Schwarz, B., Azodi, C.B., Shiu, S.H. and Bauer, P. (2020) Putative cis-regulatory elements predict iron deficiency responses in *Arabidopsis roots*. *Plant Physiology* 182(3), 1420–1439. DOI: 10.1104/pp.19.00760.

Shaik, R. and Ramakrishna, W. (2014) Machine learning approaches distinguish multiple stress conditions using stress-responsive genes and identify candidate genes for broad resistance in rice. *Plant Physiology* 164(1), 481–495. DOI: 10.1104/pp.113.225862.

Shen, D., Wang, Y., Chen, X., Srivastava, V. and Toffolatti, S.L. (2022) Editorial: Advances in multi-omics study of filamentous plant pathogens. *Frontiers in Microbiology* 13, 998501. DOI: 10.3389/fmicb.2022.998501.

Sherstinsky, A. (2020) Fundamentals of recurrent neural network (RNN) and long short-term memory (LSTM) network. *Physica D: Nonlinear Phenomena* 404, 132306. DOI: 10.1016/j.physd.2019.132306.

Shikha, M., Kanika, A., Rao, A.R., Mallikarjuna, M.G., Gupta, H.S. *et al*. (2017) Genomic selection for drought tolerance using genome-wide SNPs in maize. *Frontiers in Plant Science* 8, 550. DOI: 10.3389/fpls.2017.00550.

Shulaev, V. (2006) Metabolomics technology and bioinformatics. *Briefings in Bioinformatics* 7(2), 128–139. DOI: 10.1093/bib/bbl012.

Silva, J.C.F., Teixeira, R.M., Silva, F.F., Brommonschenkel, S.H. and Fontes, E.P.B. (2019) Machine learning approaches and their current application in plant molecular biology: A systematic review. *Plant Science* 284, 37–47. DOI: 10.1016/j.plantsci.2019.03.020.

Singh, A., Ganapathysubramanian, B., Singh, A.K. and Sarkar, S. (2016) Machine learning for high-throughput stress phenotyping in plants. *Trends in Plant Science* 21(2), 110–124. DOI: 10.1016/j.tplants.2015.10.015.

Soga, T., Ohashi, Y., Ueno, Y., Naraoka, H., Tomita, M. *et al*. (2003) Quantitative metabolome analysis using capillary electrophoresis mass spectrometry. *Journal of Proteome Research* 2(5), 488–494. DOI: 10.1021/pr034020m.

Sperschneider, J., Gardiner, D.M., Dodds, P.N., Tini, F., Covarelli, L. *et al*. (2016) EffectorP: Predicting fungal effector proteins from secretomes using machine learning. *The New Phytologist* 210(2), 743–761. DOI: 10.1111/nph.13794.

Sperschneider, J. (2020) Machine learning in plant-pathogen interactions: Empowering biological predictions from field scale to genome scale. *The New Phytologist* 228(1), 35–41. DOI: 10.1111/nph.15771.

Spindel, J.E., Dahlberg, J., Colgan, M., Hollingsworth, J., Sievert, J. *et al*. (2018) Association mapping by aerial drone reveals 213 genetic associations for Sorghum bicolor biomass traits under drought. *BMC Genomics* 19(1), 679. DOI: 10.1186/s12864-018-5055-5.

Sumner, L.W., Mendes, P. and Dixon, R.A. (2003) Plant metabolomics: Large-scale phytochemistry in the functional genomics era. *Phytochemistry* 62(6), 817–836. DOI: 10.1016/S0031-9422(02)00708-2.

Tardieu, F., Cabrera-Bosquet, L., Pridmore, T. and Bennett, M. (2017) Plant phenomics, from sensors to knowledge. *Current Biology* 27(15), R770–R783. DOI: 10.1016/j.cub.2017.05.055.

Tsiligiridia, T. and Ainali, K. (2018) Remote sensing Big Agridata for food availability. In: *2018 International Conference on Image and Video Processing, and Artificial Intelligence*, SPIE, pp. 302–307 (Vol. 27).

Twyman, R.M., Schillberg, S. and Fischer, R. (2013) Optimizing the yield of recombinant pharmaceutical proteins in plants. *Current Pharmaceutical Design* 19(31), 5486–5494. DOI: 10.2174/1381612811319310004.

Usama, M., Qadir, J., Raza, A., Arif, H., Yau, K.A. *et al*. (2019) Unsupervised machine learning for networking: Techniques, applications and research challenges. *IEEE Access* 7, 65579–65615. DOI: 10.1109/ACCESS.2019.2916648.

Van de Sande, B., Flerin, C., Davie, K., De Waegeneer, M., Hulselmans, G. *et al.* (2020) A scalable SCENIC workflow for single-cell gene regulatory network analysis. *Nature Protocols* 15(7), 2247–2276. DOI: 10.1038/s41596-020-0336-2.

Van Wijk, K.J. (2001) Challenges and prospects of plant proteomics. *Plant Physiology* 126(2), 501–508. DOI: 10.1104/pp.126.2.501.

Vehtari, A., Gelman, A. and Gabry, J. (2017) Practical Bayesian model evaluation using leave-one-out cross-validation and WAIC. *Statistics and Computing* 27(5), 1413–1432. DOI: 10.1007/s11222-016-9696-4.

Walley, J.W., Sartor, R.C., Shen, Z., Schmitz, R.J., Wu, K.J. *et al.* (2016) Integration of omic networks in a developmental atlas of maize. *Science* 353(6301), 814–818. DOI: 10.1126/science.aag1125.

Wang, X. and Liu, J. (2021) Multiscale parallel algorithm for early detection of tomato gray mold in a complex natural environment. *Frontiers in Plant Science* 12, 620273. DOI: 10.3389/fpls.2021.620273.

Wang, Y., Zafar, N., Ali, Q., Manghwar, H., Wang, G. *et al.* (2022) CRISPR/cas genome editing technologies for plant improvement against biotic and abiotic stresses: Advances, limitations, and future perspectives. *Cells* 11(23), 3928. DOI: 10.3390/cells11233928.

Wang, B., Zhu, J., Pierson, E., Ramazzotti, D. and Batzoglou, S. (2017) Visualization and analysis of single-cell RNA-seq data by kernel-based similarity learning. *Nature Methods* 14(4), 414–416. DOI: 10.1038/nmeth.4207.

Wang, J., Chen, L., Wang, Y., Zhang, J., Liang, Y. *et al.* (2013) A computational systems biology study for understanding salt tolerance mechanism in rice. *PLoS ONE* 8(6), e64929. DOI: 10.1371/journal.pone.0064929.

Wang, M., White, N., Grimm, V., Hofman, H., Doley, D. *et al.* (2018) Pattern-oriented modelling as a novel way to verify and validate functional-structural plant models: A demonstration with the annual growth module of avocado. *Annals of Botany* 121(5), 941–959. DOI: 10.1093/aob/mcx187.

Xu, C. and Jackson, S.A. (2019) Machine learning and complex biological data. *Genome Biology* 20(1), 76. DOI: 10.1186/s13059-019-1689-0.

Yan, J. and Wang, X. (2022) Unsupervised and semi-supervised learning: The next frontier in machine learning for plant systems biology. *The Plant Journal* 111(6), 1527–1538. DOI: 10.1111/tpj.15905.

Yan, J. and Wang, X. (2023) Machine learning bridges omics sciences and plant breeding. *Trends in Plant Science* 28(2), 199–210. DOI: 10.1016/j.tplants.2022.08.018.

Zhang, D., Hussain, A., Manghwar, H., Xie, K., Xie, S. *et al.* (2020) Genome editing with the CRISPR-cas system: An art, ethics and global regulatory perspective. *Plant Biotechnology Journal* 18(8), 1651–1669. DOI: 10.1111/pbi.13383.

Yang, X. and Guo, T. (2017) Machine learning in plant disease research. *European Journal of BioMedical Research* 3(1), 6. DOI: 10.18088/ejbmr.3.1.2017.pp6-9.

Yang, D.L., Yang, Y. and He, Z. (2013) Roles of plant hormones and their interplay in rice immunity. *Molecular Plant* 6(3), 675–685. DOI: 10.1093/mp/sst056.

Yang, S., Li, H., He, H., Zhou, Y. and Zhang, Z. (2019) Critical assessment and performance improvement of plant–pathogen protein–protein interaction prediction methods. *Briefings in Bioinformatics* 20(1), 274–287. DOI: 10.1093/bib/bbx123.

Yang, Z., Li, J.L., Liu, L.N., Xie, Q. and Sui, N. (2020) Photosynthetic regulation under salt stress and salt-tolerance mechanism of sweet sorghum. *Frontiers in Plant Science* 10, 1722. DOI: 10.3389/fpls.2019.01722.

Zhang, Q., Chen, Q., Wang, S., Hong, Y. and Wang, Z. (2014) Rice and cold stress: Methods for its evaluation and summary of cold tolerance-related quantitative trait loci. *Rice* 7(1), 1–12. DOI: 10.1186/s12284-014-0024-3.

Zhou, P., Enders, T.A., Myers, Z.A., Magnusson, E., Crisp, P.A. *et al.* (2022) Prediction of conserved and variable heat and cold stress response in maize using cis-regulatory information. *The Plant Cell* 34(1), 514–534. DOI: 10.1093/plcell/koab267.

Zhu, G., Xiao, H., Guo, Q., Zhang, Z., Zhao, J. *et al.* (2018) Effects of cadmium stress on growth and amino acid metabolism in two Compositae plants. *Ecotoxicology and Environmental Safety* 158, 300–308. DOI: 10.1016/j.ecoenv.2018.04.045.

Zhu, Y.N., Shi, D.Q., Ruan, M.B., Zhang, L.L., Meng, Z.H. *et al.* (2013) Transcriptome analysis reveals crosstalk of responsive genes to multiple abiotic stresses in cotton (*Gossypium hirsutum* L.). *PLoS ONE* 8(11), e80218. DOI: 10.1371/journal.pone.0080218.

Zou, Q., Sangaiah, A.K. and Mrozek, D. (2019) Editorial: Machine learning techniques on gene function prediction. *Frontiers in Genetics* 10, 938. DOI: 10.3389/fgene.2019.00938.

8 AI-Enhanced Omics for Advancing Medicinal Plant Research and Drug Discovery

Esha Kainat[1], Muhammad Majeed[1*], Shumaila Anam[2], Aftab Umar[2], Mehrab Khalil[3], Maryam Afzaal[4], Jameel M. Al-Khayri[5*], Abid Mehmood Alvi[6], Muhammad Kamran[6] and Ajaz Ahmed[6]

[1]Department of Botany, University of Gujrat, Punjab, Pakistan; [2]Department of Botany, University of Agriculture, Faisalabad, Pakistan; [3]Zoology, Wildlife and Fisheries, University of Agriculture, Faisalabad, Pakistan; [4]Department of Biochemistry, Government College University, Faisalabad, Pakistan; [5]Department of Agricultural Biotechnology, King Faisal University, Al-Ahsa, Saudi Arabia; [6]Department of Plant Protection, Ghazi University, Dera Ghazi Khan, Pakistan

Abstract

The integration of artificial intelligence (AI) technology with omics platforms improves herbal medicine research by facilitating superior discovery, characterization, and validation of bioactive components. The exploration of natural sources, particularly ethno-medicinal plants, has historically been a cornerstone in medication discovery. The enhancement in healthcare and quality of life is mostly attributed to the emergence of modern medicine, which relies on chemically manufactured pharmaceuticals. Centuries of traditional knowledge have produced a plethora of bioactive chemicals, serving as a crucial foundation for pharmaceutical advancement. Healthcare transformation meets its highest potential through AI technology while the combination of Ayurveda and AI creates new traditional medicine solutions. Researchers in traditional medicine have created a prescription assistance system with AI methodologies that adhere to established decision-making protocols. Investigation of herbal cures and their constituents relies on AI techniques for analysis. The integration of advanced technology, particularly AI, has revolutionized the process of discovering new pharmaceuticals from ethno-medical plants. As processing power increases and neural network techniques advance, AI is progressing swiftly. Nearly every scientific subject is employing AI to enhance research and derive advantages, including the healthcare sector. The implementation of AI within drug discovery strategies produces significant opportunities to develop fresh solutions against diseases with no present treatments. New drug discovery operations will become more productive through decreased research and development expenses and enhanced discovery success rates.

Keywords: Artificial intelligence, bioinformatics, herbs, machine learning, multiple omics, traditional medicine

*Corresponding authors: m.majeed@uog.edu.pk, majeedmohal@gmail.com and jkhayri@kfu.edu.sa

DOI: 10.1079/9781800628649.0008

8.1 Introduction

Artificial intelligence (AI), a subset of computer science, encompasses the creation of algorithms and computer systems capable of executing activities that usually necessitate human intelligence, including solving problems, making decisions, and pattern recognition. In healthcare, AI is emerging as a formidable instrument capable of revolutionizing numerous facets of the sector, encompassing diagnostics, treatment planning, patient involvement, and administrative functions. AI technologies, encompassing machine learning (ML) and natural language processing, have already demonstrated their efficacy in healthcare via applications such as medical image analysis, predictive modeling, and chatbots for patient engagement. These breakthroughs have the capacity to diminish diagnostic inaccuracies, enhance therapeutic results, and elevate the overall patient experience. As AI progresses, it is crucial to examine its contribution to the concepts of traditional, complementary, and integrative medicine (TCIM), which prioritizes personalized care and a comprehensive perspective on health.

The notion of AI originated in the 1950s, but its advancement was significantly enhanced in the early 21st century due to government financing and increased public awareness. Researchers have pinpointed various specific domains requiring enhancements to replicate human intelligence, such as search and optimization capabilities, computer vision, natural language processing, knowledge representation, ML, probabilistic reasoning, neural networks, and planning and decision making. The conventional drug development process commences with the identification of disease pathophysiology, wherein extensive biomolecule screening may be employed in meticulously constructed population-based case-control research utilizing omics, succeeded by a systems biology approach. Subsequently, additional functional and mechanistic assessments utilizing suitable disease models and molecular biology techniques must be conducted to discover therapeutic targets.

The identification of a new pharmaceutical compound is consistently arduous. Upon identifying a therapeutic target, the primary challenge is to create a highly selective lead molecule that minimizes the likelihood of off-target effects. Many chemically produced lead compounds exhibit various pharmacological toxicities, rendering them inappropriate for final US Food and Drug Administration (FDA) approval. This constraint not only hinders the discovery process but also results in significant attrition costs. This obstacle can be mitigated with the significant utilization of plant-based traditional resources. The incorporation of AI in drug discovery possesses significant potential to produce positive outcomes for developing novel treatments for untreatable diseases. It will also diminish research and development (R&D) expenses and enhance the success rate of medicine discovery. Nonetheless, the necessity for ongoing enhancement of AI platforms is unequivocal. Recent findings indicated that AI constructed using convolutional neural networks (CNN) demonstrates superior accuracy when used to structured internal medical data for pneumonia detection. Multi-omics encompasses a collection of research methodologies reliant on high-throughput analytical and detection techniques within contemporary biological research, including genomics, transcriptomics, proteomics, and metabolomics, among others. These tools enable a thorough comprehension of molecular intricacies at several levels, assisting in the investigation of the deep connections between human health and disorders (Zhang and Qiu, 2023).

8.1.1 Background

In Western nations, medicinal herbs are gaining prominence and significance within both public and scientific spheres. Herbal medications are classified as dietary supplements in the USA, unlike their regulated position in China and other nations. Consequently, global research must persist in identifying and enhancing the efficacy of herbal active principles, both individually and in combination, encompassing active components, active fractions, and active herbal formulations. Western medicine predominantly utilizes pure, singular compounds, whether derived from natural or synthetic sources, whereas herbal medicine has historically employed various combinations of compounds through processed natural products, chiefly medicinal herbs, to address and alleviate the symptoms of several human ailments. Herbal medicine may exhibit fewer and less severe adverse effects compared to individual pure pharmaceuticals, rendering it particularly appealing to consumers. The emphasis of herbal medicine on combination therapy effectively integrates both ancient and contemporary beliefs. None the less, contemporary analytical and chemical methodologies are essential to ascertain efficacy and safety, deliver qualitative and quantitative assessments for dietary supplements, and formulate innovative, effective, and safe pharmaceuticals of international standards. Pharmaceutical design is a cyclical process. Bioactivity-directed fractionation and isolation discern active natural components from individual plants or mixtures (Zhang *et al.*, 2023). These lead compounds can be chemically altered and enhanced by an understanding of structure–activity relationships, mechanisms of action, drug metabolism, molecular modeling, and combinatorial chemistry research. The assessment of efficacy and toxicity, along with clinical trials, can facilitate the development of new pharmaceuticals derived from herbal medicine. To perpetuate the tradition of herbal medicine and the global applications of many medicinal plants, ongoing research into active formulations, bioactive fractions, and isolated chemicals is essential for medication development in the 21st century.

8.1.2 Scope

Multi-omics is a holistic approach, exhibiting traits including dynamism, spatiotemporal fluctuations, and complexity, which correspond effectively with this holistic idea. The integration of AI technology with omics platforms improves herbal medicine research by facilitating superior discovery, characterization, and validation of bioactive components. The combined efficacy of these advancements facilitates a thorough exploration of herbal medication interactions, leading to enhanced field research in the domain of herbal medicine (Yang *et al.*, 2024). The amalgamation yields enhanced therapeutic applications from a better understanding of the interconnections within herbal medicine.

8.1.3 Limitations

Despite the swift progress in sequencing and spectrometry technologies, along with the accessibility of computational tools, a comprehensive review from a multi-omics perspective summarizing the current advancements in identifying genes, proteins, and key/secondary metabolites involved in biosynthesis pathways in medicinal plants remains absent.

The majority of medicinal plants encounter inconsistencies or deficiencies in data quality and availability, hindering the development of AI models. The development of standardized databases for metabolomics, gene expression, and phytochemicals is ongoing. AI encounters challenges in predicting the therapeutic effects of herbal remedies due to the intricate interactions of their biological constituents with various elements (Vera and Palaoag, 2023). AI systems have significant challenges in evaluating the interplay of opposing and complementary effects among chemicals. The training of AI algorithms on synthetic drugs results in suboptimal performance when evaluating herbal systems with various constituents. The absence of standardized computing systems hinders complete reproducibility across researchers.

The comprehension of AI technology is complicated due to the qualitative and context-dependent nature of traditional herbal medicine knowledge. The complete integration of historical medical data into AI systems encounters conditional obstacles to owing to linguistic and cultural diversity challenges. Traditional herbal remedies derived from natural sources are increasingly favored due to their numerous health benefits and minimal harmful effects. The absence of standardized information for the classification and evaluation of herbal treatments constitutes a significant limitation. The primary observed restrictions are the following. The identification process and evaluation assessment are complicated by compatibility issues. The comprehension of action processes encounters delays owing to protracted procedures. Medicinal plants and their derivatives are produced commercially on a limited scale. Scientific investigation of the phytochemical components of therapeutic herbs and their associated health benefits is also restricted. End-users must address the data management issues and ethical dilemmas arising from AI implementation (Amabie *et al.*, 2023). It is imperative to close the technology divide between various generations. Securing high-quality raw materials is challenging. The accurate validation of medicinal herbs presents a challenging endeavor. The adoption of Ayurveda necessitates a collaborative effort among politicians, practitioners, and experts.

8.2 Drug Discovery and Development and Artificial Intelligence

Herbal medications are highly sought after in primary healthcare in both developed and developing nations. The exploration of natural sources, particularly ethnomedicinal plants, has historically been a cornerstone in medication discovery. Centuries of traditional knowledge have produced a plethora of bioactive chemicals, serving as a crucial foundation for pharmaceutical advancement. The integration of advanced technology, particularly AI, has revolutionized the process of discovering new pharmaceuticals from ethnomedicinal plants. Ethno-pharmacological research has revealed numerous plants with therapeutic properties, and Indigenous communities have safeguarded invaluable knowledge regarding the traditional uses of these plants. In the field of medical quality control, AI plays a crucial role in enhancing the accuracy and efficiency of various quality-assurance processes. The application of AI methodologies, including deep learning models, ML algorithms, and data analytics, has significantly expedited the identification, isolation, and characterization of bioactive compounds from these plants. Advancements in AI facilitate the swift evaluation of large chemical libraries, prediction of molecular interactions, and enhancement of drug–target interactions. The integration of AI with molecular modeling techniques and high-throughput screening expedites the identification of prospective drug candidates, hence streamlining the protracted drug discovery process (Sharma *et al.*, 2025). Moreover, AI algorithms enhance the analysis of complex chemical structures, predict pharmacokinetic properties, and optimize formulations, hence improving the effectiveness and success rate of drug discovery.

8.2.1 Integration of AI and traditional medicine in drug discovery

AI is a computer-based emulation of human cognition that utilizes gathered data to address intricate problems through the use and development of algorithms. Human development of new algorithms is a laborious and occasionally intimidating endeavor, which can be expedited through computer execution. The neural network, an ML system that loosely emulates the human brain, can formulate task-specific algorithms. The neural network epitomizes AI; yet, its functionalities remain inadequately delineated, and the training procedure necessitates human intervention. As processing power increases and neural network techniques advance, AI is progressing swiftly. Nearly every scientific subject is employing AI to enhance research and derive advantages, including the healthcare sector. Consequently, numerous major pharmaceutical companies have begun to implement this system incrementally in their drug development initiatives. Pfizer uses the IBM Watson platform for their immuno-oncology drug discovery. Sanofi has implemented Exscientia's AI platform for the discovery of drugs targeting metabolic illnesses, while Genentech is utilizing the GNS Healthcare AI platform for cancer drug discovery. Sumitomo Dainippon Pharma (Japan) recently developed a therapeutic molecule for obsessive–compulsive disorder within one year utilizing Exscientia's AI-assisted platform. This molecule is the inaugural AI-generated compound for human trials. Remarkably, the appropriate therapeutic candidate was selected for clinical trials within a mere year, which is exceptionally brief relative to the standard drug discovery timeline. With suitable adjustments, AI presents a viable method in several forms of drug discovery. The integration of AI with a chemical library derived from plant-based traditional medicine could advance the drug discovery process (McGee, 2024). This review examines the concept of AI integration throughout many stages of contemporary drug research. Furthermore, we offer a framework for an AI-enhanced drug discovery platform focused on plant-based traditional medicine.

8.2.2 The effective and safer lead compound discovery with AI

Ongoing endeavors are being undertaken to identify alternate strategies for the efficient and safer discovery of lead compounds. One method for doing this is through plant-based traditional medicine, a field that could provide a vast reservoir of knowledge for investigation; however, the active compounds in plant-derived traditional remedies remain largely unidentified in contrast to contemporary medicine. In instances where the active ingredient is partially identified, the conventional dosage form is a slurry of the entire plant or leaf extract, which remains undefined due to potential variations in concentration based on the size and age of the plant or leaf. Consequently, the dose is mostly arbitrary, and the risk of toxicity is elevated due to the presence of various unidentified non-medicinal chemicals. Despite the aforementioned limitations of plant-based traditional medicine, it served as the primary method for treating all ailments before the advent of modern medicine (Ranade, 2024). The historical prevalence of plant-based traditional medicine indicates that medicinal plant extracts may serve as valuable sources for innovative drug development when combined with contemporary technologies and concepts. Considering that each plant extract comprises a multitude of chemical constituents and several old medicinal texts detailing their use, a sophisticated self-learning automated system, such as an AI, will be necessary to interpret all accessible data. The incidence of numerous major diseases, including cardiovascular disorders, different malignancies, and infectious diseases, has markedly declined since 1970. Consequently, the global average lifetime has markedly risen. The enhancement in healthcare and quality of life is mostly attributed to the emergence of modern medicine, which relies on chemically manufactured pharmaceuticals. The process of chemically synthesizing a medicine commences with the identification of a lead compound for a specific pharmacological target, followed by optimization, preclinical research involving a minimum of two animal models to assess efficacy and toxicity, and subsequent clinical trials in humans (Majeed *et al.*, 2023c). The procedure for lead compound identification, refinement, and execution of preclinical investigations spans 5–6 years, with an average approval rate of 1 in 5000 lead compounds for human trials. Achieving final approval from the FDA for clinical use necessitates an average of 12 years and incurs a cost of US$2.6 billion for each molecule. In addition to the protracted discovery process involving substantial financial depletion, numerous contemporary pharmaceuticals exhibit considerable undesirable effects, necessitating substitution with alternative, safer pharmacological compounds. This contemporary drug development strategy remains suboptimal for identifying cures for numerous chronic diseases, including diabetes and cancer.

8.3 Recognition and Quality Mapping of Traditional Herbal Drugs

AI has garnered significant industrial attention owing to its versatile applications across various business sectors. Researchers in traditional medicine have created a prescription assistance system with AI methodologies that adhere to established decision-making protocols. Investigation of herbal cures and their constituents relies on AI techniques for analysis. Individuals lacking botanical knowledge and planning skills can utilize image recognition to learn about herbal plants. The area has developed as an amalgamation of botanical taxonomy and computer technology, and academics now regard it as an interdisciplinary discipline (Mall *et al.*, 2025). Researchers are presently utilizing a computer vision-assisted method to swiftly discern and classify herbal flora in response to the increasing demand for such competencies. Numerous research studies have been conducted in the past decade to develop methods for the accurate identification of medicinal plants.

8.3.1 Classification techniques using machine learning models

Dimensionality reduction techniques can enhance the visuals of herbal medications by eliminating unattractive features. Thereafter, image augmentation methods may be employed to enhance the resultant image dataset. Diverse ML classification methods have been employed on herbal photos to extract essential morphological features and achieve successful results. ML classification methods employ diverse algorithms to sort and discern patterns, as well as the coloration of herbal medications, into predetermined categories based on specific criteria, frequently utilizing training datasets. Predictive computations employ classification methods to assign data into predefined groupings or subpopulations. Classification is the procedure of allocating incoming data to a specific category. These algorithms will classify disorganized data into certain categories depending on the training. Classification algorithms are affected by various parameters, such as Euclidean distance, hyperplane design, non-linear relationships, and activation functions. These factors influence the manner in which algorithms process input and generate output (Majeed *et al.*, 2023b). This section delineates descriptions of frequently utilized identification and classification models for herbal medications based on ML approaches.

8.3.1.1 Recurrent neural networks

Recurrent neural networks (RNNs) are a particular type of artificial neural network (ANN) designed to process data, including text, audio, or time series data. These technologies are widely employed for applications such as natural language processing, machine translation, and other procedures involving sequential data. In contrast to feed-forward neural networks, which handle input data in a singular traversal, RNNs have loops that enable the retention and temporal processing of information. The output from each buried layer in an RNN is reintroduced as input to the subsequent layer, in conjunction with the present input. Herbal medications may possess intricate properties and qualities that may not be accurately detected using a single-pass neural network. In these instances, RNNs can be effectively utilized to recognize and analyze intricate characters in a layer-by-layer approach. This can help alleviate the issues related to color gradients on the surfaces of herbal medications in many instances (Majeed *et al.*, 2023a). RNNs possess multiple varieties, such as long short-term memory (LSTM) and gated recurrent unit (GRU) networks. These versions of RNNs are designed to overcome the issue of vanishing gradients present in standard RNN models.

8.3.1.2 Multi-layer perceptron

The multi-layer perceptron (MLP) is an augmentation of the feed-forward neural network. The three types of layers in a neural network are the input layer, output layer, and hidden layer. The input layer acquires the signal intended for processing as outlined. The output level is responsible for functions such as prediction and classification. The essential computational element of the MLP comprises an unlimited number of hidden layers situated between the input and output layers (Majeed *et al.*, 2021c). This neural network can be utilized for herbal medications, where most attributes are shared, and the specific nuances of surface morphology must be examined to effectively differentiate between herbal drugs and their adulterants.

8.3.1.3 Validation of the developed model

It is feasible to develop an ML model capable of identifying photos. This model might subsequently be employed to pinpoint the requisite botanical pictures. To ensure a model's reliability and efficacy, it is essential to validate it by evaluating its suitability for the intended purpose. The validation of a created ML model entails evaluating metrics including accuracy rate, precision rate, false positive rate, true positive rate, false negative rate, specificity, sensitivity, recall, CPU elapsed time, and F1 score (Majeed *et al.*, 2022b). Accuracy is defined as the ratio of correct predictions to the total experimental data. The accuracy can be readily calculated by dividing the number of correct forecasts by the total number of predictions. This can be calculated using Equation (1):

$$\text{Accuracy}(\%) = (\text{Correct Image Identification} / \text{Total Number of Image Identifications}) \times 100 \quad (1)$$

Typically, any ML model with an accuracy of 70% is regarded as a proficient model. Precision is the ratio of true positive cases to the total expected instances within a specific class. The calculation can be expressed using Equation (2).

$$\text{Precision} = \frac{\text{True Positive}}{\text{True Positive+False Position}} \quad (2)$$

True positives occur when an observation is accurately classified as belonging to a specific category. True negatives occur when you correctly identify that an observable is not part of a specific class. False positives occur when an observation is incorrectly assumed to belong to a class when it does not. False negatives arise when an observation is incorrectly classified as not belonging to a specific class, although actually belonging to it. Recall denotes the ratio of instances in which a model accurately identifies membership in a specific class, compared to the total number of cases that genuinely belong to that class. The F1 score denotes the harmonic mean of precision and recall.

8.4 Artificial Intelligence in Ayurveda

Healthcare transformation can meet its highest potential through AI technology while the combination of Ayurveda and AI creates new traditional medicine solutions. Ayurveda serves as a traditional medical system based in India which places equal importance on complete body healing through individual health evaluation and the study of their unique physiological state. Ayurvedic diagnosis gets better when AI is applied because the system produces more precise assessments, which enable doctors to create individualized patient treatment strategies. The article investigates modern AI techniques used for Ayurvedic diagnosis by analyzing both their prospects and implementation barriers in this novel approach (Chen *et al.*, 2023). The main objective of this research was to determine how AI technology might elevate the precision level of Ayurvedic diagnosis systems. The secondary outcome assessment focused on two aspects: first, it evaluated the diagnostic process streamlining potential of AI; and second, it examined how it can benefit Prakriti (~constitution) analysis. Sophisticated algorithms within Ayurvedic diagnosis enable the analysis of massive patient data collection from clinical histories combined with symptoms and diagnostic reports using AI technology. AI algorithms detect hidden patterns from varied datasets which do not reveal themselves to medical professionals. The accurate treatment of individual patients becomes possible when Ayurvedic physicians receive precise diagnoses through this system. The implementation of AI brings advantages to Ayurvedic diagnosis through its ability to improve Prakriti analysis techniques. AI systems can process information from extensive data sources including genomic information and life habits together with environmental effects thus building comprehensive knowledge of individual health standing. Ayurvedic practitioners utilize multiple assessment methods to determine Prakriti more precisely to create individualized care plans that address patients' unique health requirements and imbalances (Majeed *et al.*, 2022a). Through its diagnostic capabilities, AI offers substantial value to Ayurvedic medical diagnosis. Traditional diagnostic practices, including pulse examination and tongue diagnosis, need lengthy practitioner experience before mastery can be achieved. The implementation of AI-based image recognition systems and pattern recognition technology allows these diagnostic practices to become digitized, enabling standardized Ayurveda medical services to become available to practitioners all over the world. The standardization process aids both the maintenance and popular spread of Ayurvedic diagnostic procedures.

8.5 Integration of High-Throughput Omics Technologies

Multi-omics research methods have established extensive databases across a single medical plant species or various species for their genomic, transcriptomic, proteomic, metabolomic, and phytochemical information sets (Fig. 8.1). The Ginseng Genome Database stands as the most thorough herbal database because its authors obtained well-defined genome information. The Global Pharmacopoeia Genome Database (GPGD) stores genetic information of about

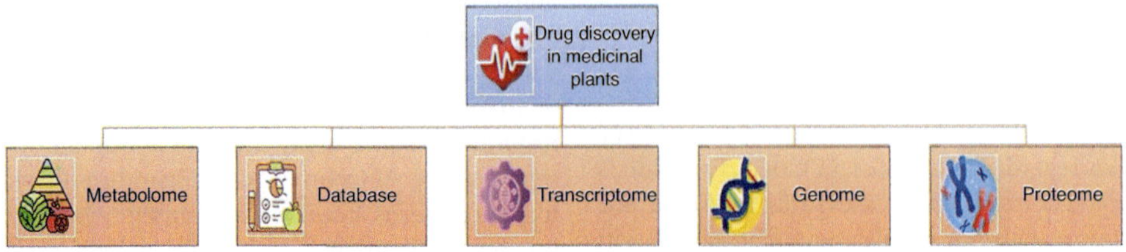

Fig. 8.1. Medicinal plant multi-omics research to facilitate drug discovery.

674 species including 9682 transcriptome datasets as well as 55 whole genomes among 49 species and 2203 organelle genomes. The MepmiRDB database stands as the initial repository dedicated to storing medicinal plant microRNA information of 29 species including ginseng, wolfberry and red sage. The publicly accessible databases IMPPAT, NPACT, NuBBEDB, and NANPDB hold information about medicinal plant phytochemicals. Phytochemical databases display two types of information, which include traditional data such as common name, taxonomy, location, medicinal parts, and application together with specific data about medicinal properties (Muhammad *et al.*, 2024a). Through cooperation, these databases enable researchers to execute in-depth data mining of pharmacological property gene annotation and expression profiles as well as microRNA (miRNA) regulation of secondary metabolite biosynthesis and accumulation processes while providing DNA barcodes for medicinal material identification. A study of saponin production genes through heterologous biosynthesis took place after Chinese ginseng (*Panax notoginseng*) genome research became available. Scientists discovered information about benzylisoquinoline alkaloid (BIA) biosynthesis regulation and transport through landraces of Chinese opium poppy following the completion of its genome project.

8.6 Technology for Demystifying Herbal Medicine

The omics detection tools have become predominant for medicinal herb analyses in recent times. The multi-omics integration technique serves as a main investigative tool that enables us to understand natural products obtained from herbal medicine along with discovering regulatory biological processes and targeted findings. Research conducted under a multi-omics framework examined the anti-tumor activity of *Marsdenia tenacissima* (Roxb.) Moon. The research examined active components to discover anti-tumor targets in *Marsdenia tenacissima* (Roxb.) Moon. Hepatocellular carcinoma patients receive beneficial multi-target and multi-component therapy from the laboratory-derived substances Tenacissosides I, H, and G that have demonstrated specific benefits for angiogenesis inhibition and apoptosis promotion and immune function enhancement (Li *et al.*, 2022). Multi-omics research showed that pills of the patented drug). Multi-omics research showed that pills of the patented drug *Tripterygium wilfordii* Hook F actively regulate most genes during immunological non-responder treatment, while its bioactive compound triptolide stops both IFN-γ production and immune cell activation and downstream gene expression. Another study has applied multi-omics analysis to investigate metabolome as well as proteome characteristics for understanding how gross saponins of *Tribulus terrestris* L. fruit protect against ischemic stroke along with discovering their therapeutic mechanisms. Researchers identified 359 differential proteins together with 110 metabolites that are primarily linked to sphingolipid metabolism and complement and coagulation cascades and glycerophospholipid metabolism.

Total flavonoids extracted from Astragali Radix proved its leucopenia-blocking impact against cyclophosphamide through a comprehensive systems study. Total flavonoids modified both PI3K-Akt signaling and Jak-STAT signaling pathways at the metabolomics level. Total flavonoids functioned as protective agents by supporting cell proliferation together with their regulatory role at the immunologic system level (Qu *et al.*, 2021). The low polarity fraction from Bupleuri Radix revealed its antidepressant-like mechanisms of action by multi-omics investigation. The screening of 16 biomarkers associated with antidepressant effects demonstrated participation from taurine and hypotaurine metabolism and glyoxylate and dicarboxylate metabolism as well as primary bile acid biosynthesis pathways and showed correlation with gut microbiota. Researchers applied metabolomic alongside proteomic and transcriptomic analyses to establish systematic knowledge about the therapeutic effects and mechanisms of Bufei Yishen Formula (BYF) for chronic obstructive pulmonary disease (COPD) patients through their study on lung tissue samples from COPD rats treated with BYF (Majeed The low polarity fraction from Bupleuri Radix revealed its antidepressant-like mechanisms of action by multi-omics investigation. The screening of 16 biomarkers associated with antidepressant effects demonstrated participation from taurine and hypotaurine metabolism and glyoxylate and dicarboxylate metabolism as well as primary bile acid biosynthesis pathways and showed correlation with gut microbiota. Researchers applied metabolomic alongside proteomic and transcriptomic analyses to establish systematic knowledge about the therapeutic effects and mechanisms of Bufei Yishen Formula (BYF) for chronic obstructive pulmonary disease (COPD) patients through their study on lung tissue samples from COPD rats treated with BYF (Majeed *et al.*, 2021c; Sinha *et al.*, 2023). Through combining transcriptomics with proteomics and metabolomics, researchers evaluated the effect and mechanism of Bufei Jianpi formula (BJF). Focal adhesion as well as antioxidant activity and lipid metabolism served as long-term anti-COPD effect targets that researchers identified in BJF studies (Zhao *et al.*, 2017). Research methods combined miRNAome with metabolome and proteome components to determine what fundamental processes geniposide uses for hepatoprotection. A total of 28 miRNAs were identified together with 7 metabolites and 20 proteins among the statistically significant differentially expressed entities. Geniposide showed the capability to regulate citrate cycle pathways via dehydrogenase activation, and it improved functional recovery (Qiu *et al.*, 2017). The study utilized multi-omics techniques to determine how Zhen-Wu-Bu-Qi Decoction demonstrated its anti-inflammatory effects while protecting colon tissue by affecting the PI3KMAPK/NF-κB signaling pathway.

A multi-omics analysis is used to explore anti-osteoporosis mechanism of the root of *Achyranthes bidentata* Blume. A total of 22 biomarkers were identified and indicated that glycine and threonine metabolism, glyoxylate and dicarboxylate metabolism were major intervention pathways. The abundance of *Oscillospira* was negatively correlated with an anti-osteoporosis effect, whereas *Turicibacter*, *Anaerofilum* and *Rothia* bacteria were positively correlated with the effect. System pharmacology analysis could reveal the effective mechanism of effective ingredients from herbs against rheumatoid arthritis (Majeed *et al.*, 2021b; Tassadduq *et al.*, 2022; Tariq *et al.*, 2023). It provided a systematic method to screen active components and optimize their proportion, revealing the overall effects of medicinal herbs via multi-omics integration.

8.7 Metabolomics: A Recent Advanced Omics Technology in Herbal Medicine Research

A medicinal plant's metabolome represents valuable resources that allow the development of new phototherapeutic treatments for different diseases. Since the last decade, metabolomics has advanced into a quickly expanding dynamic technology that enables high-throughput identification and profiling of targeted and untargeted metabolites in medicinal plants. Through this metabolomics platform, researchers can conduct herbal drug discovery along with ensuring medicinal plant quality control and they can use it to analyze complex biological systems through endogenous metabolic studies. Various new analytical platforms and bioinformatics technologies have started to emerge and advance significantly the study of plant metabolomics. Plant metabolomics research relies mainly on gas chromatography–mass spectrometry (GC–MS) with liquid chromatography (LC)– MS and LC–solid phase extraction–nuclear magnetic resonance (LC–SPE–NMR). The analysis of plant metabolomics mainly occurs through GC–MS and LC–MS equipment because these platforms provide sensitive detection capabilities alongside extensive parallel metabolite analysis (Majeed *et al.*, 2021a; Ullah *et al.*, 2022). By combining comparative metabolomics technologies with plant biotechnology approaches researchers will receive major advancements for medical research. This chapter includes complete insights into untargeted medicinal plant metabolite analysis through NMR and MS operating systems which present concepts about fundamental approaches and workflow and identification techniques and data analysis procedures and sample preparation processes. The information explains how metabolomics applications manage medical plant quality and presents obstacles during research along with possible solutions and outlined advantages of metabolomics for herbal medicine investigations. Approaches and Challenges Plant Metabolites as New Leads to Herbal Drug Discovery.

Pharmacotherapy has been consistently using natural products alongside structural counterparts for a long period. The design of new drug discovery leads through natural chemical compounds remains a complex scientific problem. The approach examines phytochemical analysis combined with natural resource-derived bioactive substance characterization along with pharmacological studies of these compounds. The wealth of information about medicinal plants with various properties in indigenous wisdom stems from easily accessible plant-derived metabolites as well as bioactive chemicals which cost little and cause minimal immune system reactions. They represent a workable choice instead of man-made drugs because they serve well as therapeutic medications. Drug discovery research calls for the development of secure lead molecules through two key steps which include the transformation of screening hits to drug candidates via GC–MS and NMR and IR laboratory analyses along with HPLC and HPTLC structure identification (Khan *et al.*, 2024). Research for discovering new medications has benefited greatly from the implementation of advanced technology that enables natural compound screening. We will perform new molecule screening research through software and database systems to establish natural products as leading pharmaceutical development resources. The process finally produces lead structures as its end result. Increased research will enable scientists to properly evaluate problems for the purpose of manufacturing therapeutic plant metabolites at scale for industrial applications. We analyze modern technological developments that enable natural product-based drug discovery and introduce their specific applications with an emphasis on future opportunities.

8.8 An Omics Database for Herbal Medicine Plants

Plant natural products (PNPs) provide crucial raw materials to human nutrition while supplying industrial usage for medicinal agents and producing approximately half of all anticancer drugs including paclitaxel and vinblastine in addition to ginsenoside. The biosynthesis method produces PNPs effectively, yet the growing medicinal phyto-omics information enables researchers to understand PNP synthetic pathways. Genetic resources function as a foundation for molecular breeding programs of medicinal plants (Khan *et al.*, 2022; Khoja *et al.*, 2022). The MPOD database accumulates genomic and transcriptomic information released about medicinal plants starting from January 2018. We conducted six genome analyses together with 28 transcriptomic and five metabolomic sequence experiments as part of this research. The MPOD database provides complete access to its genomic and transcriptomic sequences for orthologous gene searching as well as homology analysis through blasts across different species gene families. The MPOD system performed vital correlation investigations between metabolite patterns and gene expression across tissue metabolite content analysis and Pearson correlation evaluation of metabolic pathway genes and transcriptional profiles. MPOD extends the functionality of HMOD by providing overview explanations regarding metabolic pathways for the groups of metabolites including flavonoids, alkaloids and terpenoids. MPOD includes 'the biosynthetic tools' technical module to enhance synthetic biology through the utilization of well-known bioinformatics programs such as SynVisio, heatmap and enrichment

(Hassan *et al.*, 2022b; Jamil *et al.*, 2022; Nasar *et al.*, 2024a, b). MPOD implements a database structure built on MySQL and ThinkPHP and FastAdmin and split into four distinct functional areas known as genomics transcriptomics pathways and biosynthetic tools. This genomics section comprises the four components of genomes and genome size with re-sequencing and genes. MPOD includes 154 published genomes and six unpublished genome assemblies of *Synsepalum dulcificum*, *Antirrhinum majus*, *Platycodon grandiflorus*, *Codonopsis pilosula*, *Panax vietnamensis*, and *Gynostemma pentaphyllum*, which are part of this project. The web interface of species contains four sections for species introduction, sequencing data, assembly results, and data source links and reference information. Users can access MPOD to link the published GCA data on NCBI as well as download FASTA formatted files for assembly, CDS, and protein sequences from this database for unpublished genomic data.

Further functionality of MPOD includes the integration of regularly used bioinformatics software that includes BLAST, Search, Heatmap, and JBrowse. The complete set of MPOD genomes and gene models exists as part of JBrowse user interface. SynVisio illustrates the chromosome-level reference genomes with gene synteny relationship displays. The Co-expression analysis method produces networks of genetic sets that show strong correlations in expression patterns.

8.9 Bioinformatics for Metabolic Pathways and Medicinal Ingredients

The initiation of genetic improvement and metabolic engineering for medicinal plants starts with a complete comprehension of their metabolic regulatory sequences. The following section will describe the databases while demonstrating the applicable software tools that researchers need to construct metabolic pathways. The Internet platform MPGR delivers genomic details of 14 medicinal plant types for user access. The database delivers extensive details about bioactive plant components by presenting their chemical data and enzyme profiles as well as genetic information for biosynthesis, in addition to listing reference material. HMOD functions as an all-inclusive database platform for various omics information derived from different medicinal herbal species. HMOD has two operational modules for metabolomics research, which include: the 'Pathways' segment that connects to KEGG PATHWAY Database while showing information on biosynthetic routes for medicinal substances like glucosinolate and betalain together with their catalyst enzymes; and the 'Metabolomics' feature, which compiles published reports on medicinal plant metabolomics work (Hassan *et al.*, 2022a; Hussain *et al.*, 2024; Shafqat *et al.*, 2024). In the 'Metabolomics' module researchers compiled published metabolomics studies regarding medicinal plants by incorporating laboratory findings along with experimental setups and research results.

The plant *Catharanthus roseus* produces pharmaceutical important terpenoid indole alkaloids (TIAs). Scientists have created the CathaCyc database, which displays detailed information regarding 390 enzymatic pathways for TIAs and their precursors synthesized in *C. roseus* cells through 1347 enzymes. The expression patterns of genes that produce enzymes and key regulators for *C. roseus* metabolic processes originated from RNA-seq data analysis and were presented as bar diagram information. The web server PathPred enables users to identify the multiple-step synthetic route of specific query compounds. According to the assumption of common pathways for similar compounds, PathPred performs similarity searches against KEGG compound to enhance predicted chemical reactions of provided compounds (Hussain *et al.*, 2023; Rani *et al.*, 2024; Rasool *et al.*, 2024; Majeed *et al.*, 2024a). The method based on similarity analysis fails to clarify the specific pathways for particular derivative compounds.). The method based on similarity analysis fails to clarify the specific pathways for particular derivative compounds.

Scientists aim to complete metabolomics research by establishing relationships between natural plant substances and their therapeutic properties. Standard preparations of traditional prescriptions contain crude plant mixtures which results in complex scenarios of pharmacological targets being matched by distinct metabolites with varying chemical properties. The public databases that document bioactive compounds found in medicinal plants need to be established first to improve the precision of traditional medicine industry standards. To date, several repositories accommodating the chemical structures and/or the therapeutic values of the bioactive ingredients identified from the medicinal plants have been publicly available, such as SACPD (Saudi anti-human cancer plants database), MAPS (medicinal plant activities, phytochemicals & structural database), DIACAN (the antidiabetic and anti-cancer medicinal plants database), AromaDb, NeMedPlant, IMPPAT, Phytochemica, SerpentinaDB, InDiaMed, KNApSAcK, and HIT (herb ingredients' targets). The TCM Database@ Taiwan offers complete 2D and 3D structure data about more than 20,000 pure medicinal plant extracts used in traditional Chinese medicine research (Haq *et al.*, 2022; Majeed *et al.*, 2024b). Text mining enabled TCM-Gene-DIT to acquire disease data and information about TCM-target genes together with their associated signaling pathways and PPIs for gene expression regulatory effect deduction by researchers.

8.10 Challenges and Limitations

The implementation of ML models in AI-based identification methods helps achieve faster results with higher accuracy levels for herbal medications. Numerous restrictions arise from using ML for identifying and authenticating herbal drugs and they include the following. The quantity of high-quality labeled datasets pertaining to herbal medications remains scarce at present. A model struggles to detect data patterns when data are limited because of under-fitting. The wide variability in data emerges because herbal drugs originate from multiple locations. Numerous ML models including the dimensionality reduction class receive the label of 'black box' due to their handling of complex information, which resists interpretation. The lack of transparency during interpretation became a major restriction in this particular area. The complex nature of herbal medicine structures makes it difficult for ML models to establish distinct patterns needed for identification of these substances. The data noises found in herbal medicine records present barriers that reduce the practicality of implementing ML models. ML models need high-performance computing platforms, which create substantial challenges for their practical implementation (Bhattacharya *et al.*, 2023; Haq *et al.*, 2022). The absence of digital resources needed for ML models leads to scalability issues when dealing with larger scales of operation.

8.11 Ethical Considerations During the Application of AI

Traditional herbal medicines stand as important foundations in medical healthcare systems. The integration of AI for traditional herbal medicine identification leads to ethical problems regarding data privacy together with concerns about relying on AI decision-making capabilities.

Algorithm audits must occur regularly as well as audits of its system integration and formation of programming groups that represent various populations (Bashir *et al.*, 2023; Barwant *et al.*, 2024). AI technology assists medical professionals in making improved clinical decisions, whereas complete clinical judgment remains under human control.

AI systems face the risk of data breaches as well as illegal access to their collected sensitive data. The establishment of strong protective data frameworks together with policy structures stands essential for protecting processed information. During training any bias will result in unfair and biased output from an AI system. Policies regarding AI model training need development in order to prevent discriminatory outcomes of AI-based decision systems. AI models function as black boxes because they make it hard for users to comprehend the decision-making methods implemented by these systems. Users and healthcare providers should experience better practice clarity from AI models to enhance their trust in them (Aziz *et al.*, 2023; Rehman *et al.*, 2023; Naveed *et al.*, 2024a, b). The occurrence of mistakes or inaccuracies within AI model operation leads to mistaken user identification or authentication that puts users at substantial risk. A strong testing system, validation phase, as well as continuous monitoring coupled with clear AI system policies, are needed.

8.12 Future Recommendations

Traditional herbal drugs sourced from natural origins gain relevance in holistic healthcare because they offer both advantageous results and minimal adverse effects for wellness and better health practices. The biggest challenge in identifying natural herbal drugs arises from a lack of general understanding about their methods of identification and quality evaluation. AI-based machines introduce an innovative solution to tackle current challenges by enabling quality-based commercialization of raw herbal drugs (Arshad *et al.*, 2022; Naveed *et al.*, 2024c). The article explores real-world artificial learning implementations in this field while explaining distinct methods of feature extraction and imaging processing tools that aid ML model processing.

This chapter presents different global herbal image databases as well as detailed information about identification models used to detect both herbal plants and drugs, exploring the possible implementation of ML models such as ANN along with CNN and suitable validation parameters for developing AI-based smart authentication methods for herbal drugs. The identification and quality mapping domain of traditional herbal drugs will experience significant advancement through AI integration, which leads to improved accurate drug identification that reduces human mistakes and maintains proper quality control standards. The future will advance sophisticated AI models as well as real-time on-site identification tools, while establishing complete traditional herbal databases for robust AI training operations. A significant amount of progress has occurred regarding ML-based medicinal plant quality assessment, yet specific aspects require additional advancement. The proper approach requires targeted information gathering through compatible data channels that merge different types of information. More research is needed because existing data lacks adequate investigation. The selection process for data preprocessing methods depends on lengthy and exhausting trial and error evaluations, even though this approach risks missing crucial additional data. Complete regulations and precise recommendations must exist to define the selection process of preprocessing methods for building ML models. The successful prevention of over-fitting during data fusion requires the implementation of feature extraction methods for addressing dimensionality problems. The main objective of data analysis requires determining proper variable selection methods or feature extraction approaches. The utilization of a single ML method will not work effectively on all datasets (Arif *et al.*, 2021; Muhammad *et al.*, 2024b; Rasool *et al.*, 2025). The selected ML techniques should depend on both specific data problems alongside the characteristics of the available dataset. The strong data assessment capabilities of ML techniques for medicinal plant data analysis need to improve continually because analytical technology continues to advance. Further studies must be conducted across the four specific areas (multi-source data, preprocessing and feature extraction, data fusion, and ML) to address existing research problems. This review suggests important strategies to expand the implementation of herbal drug identification and authentication methods as well as ways to increase their technical quality.

8.13 Conclusion

The implementation of AI within drug discovery strategies produces significant opportunities to develop fresh solutions against diseases with no present treatments. New drug discovery operations will become more productive through decreased R&D expenses and enhanced discovery success rates. AI platforms demand an ongoing process of improvement without any dispute. The CNN-built AI system has recently been demonstrated to be accurate. The necessary union between AI specialists and TCIM practitioners and policy makers represents a vital response for resolving these obstacles while maximizing new market potentials. The forthcoming studies on this intersection will likely examine developments in AI-assisted medicinal therapy medicine (including AI reinforcement of herbal formulations) along with Patient-Medicare, AI relationship research and AI discrimination reduction techniques alongside critical patient trust assessments of AI-centric TCIM healthcare delivery systems. HMOD delivers a complete platform that combines omics data assortment and KEGG pathway information about herbal medicine plants. HMOD will receive regular updates as new datasets are added to the platform and its functionality will evolve for providing enhanced value in biological genomic analysis, transcriptomic investigation, and data analysis.

Acknowledgments

The authors extend their appreciation for the support of the Deanship of Scientific Research, Vice Presidency for Graduate Studies and Scientific Research, King Faisal University, Saudi Arabia [Grant No. KFU254024].

References

Amabie, T., Izah, S.C., Ogwu, M.C. and Hait, M. (2023) Harmonizing tradition and technology. In: *The Synergy of Artificial Intelligence in Traditional Medicine*. Springer International Publishing, pp. 1–23. DOI: 10.1007/978-3-031-21973-3_70-1.

Arif, U., Bhatti, K.H., Ajaib, M., Wagay, N.A., Majeed, M. *et al.* (2021) Ethnobotanical indigenous knowledge of Tehsil Charhoi, District Kotli, Azad Jammu and Kashmir, Pakistan. *Ethnobotany Research and Applications* 22. DOI: 10.32859/ERA.22.50.1-24.

Arshad, F., Waheed, M., Harun, N., Fatima, K., Khan, B.A. *et al.* (2022) Indigenous farmer's perception about fodder and foraging species of Semi-arid lowlands of Pakistan: A case study of district kasur, pakistan. *Taiwania* 67(4), 510–523. DOI: 10.6165/tai.2022.67.510.

Aziz, A., Anwar, M.M., Majeed, M., Fatima, S., Mehdi, S.S. *et al.* (2023) Quantifying landscape and social amenities as ecosystem services in rapidly changing Peri-Urban landscape. *Land* 12(2). DOI: 10.3390/land12020477.

Barwant, M.M., Rasool, A., Mannam, V.K., Hussain, A., Abdullah, M. *et al.* (2024) Nanotechnology applications in livestock farming and animal health. In: *Harnessing NanoOmics and Nanozymes for Sustainable Agriculture*. IGI Global, pp. 413–442. DOI: 10.4018/979-8-3693-1890-4.ch020.

Bashir, S.M., Altaf, M., Hussain, T., Umair, M., Majeed, M. *et al.* (2023) Vernacular taxonomy, cultural and ethnopharmacological applications of avian and mammalian species in the vicinity of Ayubia National Park, Himalayan Region. *Biology* 12(4), 609. DOI: 10.3390/biology12040609.

Bhattacharya, R., Bhattacharya, R., Majeed, M., Bhandari, S., Aziz, R. *et al.* (2023) Environmental rehabilitation of industrial waste dumping site. In: *Biohydrometallurgical Processes*. CRC Press, pp. 44–89. DOI: 10.1201/9781003451457-4.

Chen, Z., Peng, P., Wang, M., Deng, X. and Chen, R. (2023) Bioinformatics-based and multiscale convolutional neural network screening of herbal medicines for improving the prognosis of liver cancer: A novel approach. *Frontiers in Medicine* 10, 1218496. DOI: 10.3389/fmed.2023.1218496.

Haq, S.M., Tariq, A., Li, Q., Yaqoob, U., Majeed, M. *et al.* (2022) Influence of edaphic properties in determining forest community patterns of the Zabarwan mountain range in the Kashmir Himalayas. *Forests* 13(8), 1214. DOI: 10.3390/f13081214.

Hassan, M., Haq, S.M., Ahmad, R., Majeed, M., Sahito, H.A. *et al.* (2022a) Traditional use of wild and domestic fauna among different ethnic groups in the western Himalayas—a cross cultural analysis. *Animals* 12(17), 2276. DOI: 10.3390/ani12172276.

Hassan, M., Haq, S.M., Majeed, M., Umair, M., Sahito, H.A. *et al.* (2022b) Traditional food and medicine: Ethno-traditional usage of fish Fauna across the valley of Kashmir: A western Himalayan Region. *Diversity* 14(6), 455. DOI: 10.3390/d14060455.

Hussain, F., Salam, I.U., Memon, Z.N., Abdullah, M., Abbas, G. *et al.* (2024) Occurrence of fungal microbial contamination in drinking water of megacity of Karachi (Pakistan) and their physico-chemical control. *Heliyon* 10(7), e28926. DOI: 10.1016/j.heliyon.2024.e28926.

Hussain, S., Raza, A., Abdo, H.G., Mubeen, M., Tariq, A. *et al.* (2023) Relation of land surface temperature with different vegetation indices using multi-temporal remote sensing data in Sahiwal region, Pakistan. *Geoscience Letters* 10(1), 33. DOI: 10.1186/s40562-023-00287-6.

Jamil, M.D., Waheed, M., Akhtar, S., Bangash, N., Chaudhari, S.K. *et al.* (2022) Invasive plants diversity. In: *Ecological Status, and Distribution Pattern in Relation to Edaphic Factors in Different Habitat Types of District Mandi Bahauddin*, Vol. 14(20). Sustainability (Switzerland), Punjab, Pakistan. DOI: 10.3390/su142013312.

Khan, A.M., Li, Q., Saqib, Z., Khan, N., Habib, T. *et al.* (2022) MaxEnt modelling and impact of climate change on habitat suitability variations of economically important Chilgoza Pine (Pinus gerardiana Wall.) in South Asia. *Forests* 13(5), 715. DOI: 10.3390/f13050715.

Khan, A.M., Altaf, M., Hussain, T., Hamed, M.H., Safdar, U. *et al.* (2024) Ethnopharmacological uses of fauna among the people of central Punjab, Pakistan. *Frontiers in Veterinary Science* 11, 1351693. DOI: 10.3389/fvets.2024.1351693.

Khoja, A.A., Haq, S.M., Majeed, M., Hassan, M., Waheed, M. *et al.* (2022) Diversity, ecological and traditional knowledge of pteridophytes in the Western Himalayas. *Diversity* 14(8). DOI: 10.3390/d14080628.

Li, S., Pei, W., Yuan, W., Yu, D., Song, H. *et al.* (2022) Multi-omics joint analysis reveals the mechanism of action of the traditional Chinese medicine Marsdenia tenacissima (Roxb.) Moon in the treatment of hepatocellular carcinoma. *Journal of Ethnopharmacology* 293, 115285. DOI: 10.1016/j.jep.2022.115285.

Majeed, M., Bhatti, K.H. and Amjad, M.S. (2021a) IMPACT of climatic variations on the flowering phenology of plant species in jhelum district, punjab, pakistan. *Applied Ecology and Environmental Research* 19(5), 3343–3376. DOI: 10.15666/aeer/1905_33433376.

Majeed, M., Bhatti, K.H., Pieroni, A., Sõukand, R., Bussmann, R.W. *et al.* (2021b) Gathered wild food plants among diverse religious groups in Jhelum District, Punjab, Pakistan. *Foods* 10(3), 594. DOI: 10.3390/foods10030594.

Majeed, M., Tariq, A., Anwar, M.M., Khan, A.M., Arshad, F. *et al.* (2021c) Monitoring of land use–land cover change and potential causal factors of climate change in Jhelum district, Punjab, Pakistan, through GIS and multi-temporal satellite data. *Land* 10(10), 1026. DOI: 10.3390/land10101026.

Majeed, M., Khan, A.M., Habib, T., Anwar, M.M., Sahito, H.A. *et al.* (2022a) Vegetation analysis and environmental indicators of an arid tropical forest ecosystem of Pakistan. *Ecological Indicators* 142, 109291. DOI: 10.1016/j.ecolind.2022.109291.

Majeed, M., Lu, L., Haq, S.M., Waheed, M., Sahito, H.A. *et al.* (2022b) Spatiotemporal distribution patterns of climbers along an abiotic gradient in Jhelum district, Punjab, Pakistan. *Forests* 13(8), 1244. DOI: 10.3390/f13081244.

Majeed, M., Lu, L., Anwar, M.M., Tariq, A., Qin, S. *et al.* (2023a) Prediction of flash flood susceptibility using integrating Analytic Hierarchy Process (AHP) and Frequency Ratio (FR) algorithms. *Frontiers in Environmental Science* 10. DOI: 10.3389/fenvs.2022.1037547.

Majeed, M., Muhammad, M., Hussain, T., Ali, M., Naz, T. *et al.* (2023b) Nanopriming-based management of biotic stresses for sustainable agriculture. In: *Nanopriming Approach to Sustainable Agriculture*. IGI Global, pp. 263–289. DOI: 10.4018/978-1-6684-7232-3.ch012.

Majeed, M., Muhammad, M., Nawaz, S., Naz, T., Iqbal, M.M. *et al.* (2023c) Nanopriming for crop management for sustainable agriculture: An overview. *Nanopriming Approach to Sustainable Agriculture* 110–141. DOI: 10.4018/978-1-6684-7232-3.ch005.

Majeed, M., Gulshan, A.B., Muhammad, M., Hussain, M., Aziz, R. *et al.* (2024a) An investigation into the reactions of biochar in heavy metal polluted soils. In: *Biochar-Assisted Remediation of Contaminated Soils Under Changing Climate*. Elsevier, pp. 247–276. DOI: 10.1016/B978-0-443-21562-9.00010-4.

Majeed, M., Noor, W., Rasool, A., Saleem, M., Bashir, I. *et al.* (2024b) Nanotechnology for soil remediation and land restoration. In: *Harnessing NanoOmics and Nanozymes for Sustainable Agriculture*. IGI Global, pp. 60–77. DOI: 10.4018/979-8-3693-1890-4.ch004.

Mall, P., Chamoli, V. and Raina, D. (2025) Artificial intelligence -powered precision: Predicting standard curves for herbal medicines. *Social Science Research Network Electronic Journal*. DOI: 10.2139/ssrn.5083338.

McGee, R.W. (2024) Using Chinese herbal medicine to treat cancer patients: A study incorporating artificial intelligence. *Biomedical Journal of Scientific & Technical Research* 56(5). DOI: 10.26717/BJSTR.2024.56.008924.

Muhammad, M., Basit, A., Arooj, A., Dixit, G., Majeed, M. *et al.* (2024a) Roles of gibberellins in plant defense against biotic and abiotic stress. In: *Plant Growth Regulators to Manage Biotic and Abiotic Stress in Agroecosystems*. CRC Press, pp. 183–210.

Muhammad, M., Basit, A., Majeed, M., Shah, A.A., Ullah, I. *et al.* (2024b) Bacterial pigments and their applications. In: *Bacterial Secondary Metabolites*. Elsevier, pp. 277–298. DOI: 10.1016/B978-0-323-95251-4.00019-3.

Nasar, S., Muhammad, M., Siddiqi, E.H., Majeed, M., Ameen, N. *et al.* (2024a) OMICS-based knowledge for achieving food and nutritional security. *OMICS-Based Techniques for Global Food Security* 67–90. DOI: 10.1002/9781394209156.ch4.

Nasar, S., Rasool, A., Ikram, W., Shujat, S., Majeed, M. *et al.* (2024b) Nano-osmics: A new approach to plant breeding. *Sustainable Agriculture: Nanotechnology and Biotechnology for Crop Production and Protection* 129. DOI: 10.1515/9783111234694-008.

Naveed, M., Majeed, M., Jabeen, K., Hanif, N., Naveed, R. *et al.* (2024a) Plant disease diagnosis with Artificial Intelligence (AI). In: *Microbial Data Intelligence and Computational Techniques for Sustainable Computing*. Springer Nature Singapore, Singapore, pp. 217–234. DOI: 10.1007/978-981-99-9621-6_15.

Naveed, M., Memon, Z.N., Abdullah, M., Makhdoom, S.I., Azeem, A. *et al.* (2024b) Use cases and future aspects of intelligent techniques in microbial data analysis. In: *Microbial Data Intelligence and Computational Techniques for Sustainable Computing*. Springer Nature Singapore, Singapore, pp. 259–280. DOI: 10.1007/978-981-99-9621-6_17.

Naveed, M., Zahid, N., Fatima, I., Saleem, A., Majeed, M. *et al.* (2024c) Sustainable AI-driven applications for plant care and treatment. In: *Microbial Data Intelligence and Computational Techniques for Sustainable Computing*. Springer Nature Singapore, Singapore, pp. 235–258. DOI: 10.1007/978-981-99-9621-6_16.

Qiu, S., Zhang, A., Zhang, T., Sun, H., Guan, Y, *et al.* (2017) Dissect new mechanistic insights for geniposide efficacy on the hepatoprotection using multiomics approach. *Oncotarget* 8(65), 108760. DOI: 10.18632/oncotarget.21897.

Qu, T., Gao, Y., Li, A., Li, Z. and Qin, X. (2021) Systems biology analysis of the effect and mechanism of total flavonoids of astragali radix against cyclophosphamide-induced leucopenia in mice. *J. Pharm Biomed Anal* 205, 114357. DOI: 10.1016/j.jpba.2021.114357.

Ranade, M. (2024) Artificial intelligence in Ayurveda: Current concepts and prospects. *Journal of Indian System of Medicine* 12(1), 53–59. DOI: 10.4103/jism.jism_60_23.

Rani, J., Rasool, A., Sher, H., Fatima, R., Hussain, J. *et al.* (2024) Precision farming with nanoscale sensors: Monitoring soil, water, and crop health. In: *Harnessing NanoOmics and Nanozymes for Sustainable Agriculture*. IGI Global, pp. 78–99. DOI: 10.4018/979-8-3693-1890-4.ch005.

Rasool, A., Tariq, M., Asif, M., Shah, G.M., Fatima, R. *et al.* (2024) Nanotechnology for improved crop resilience in challenging environments. In: *Harnessing NanoOmics and Nanozymes for Sustainable Agriculture*. IGI Global, pp. 149–175. DOI: 10.4018/979-8-3693-1890-4.ch008.

Rasool, A., Feroz, E., Qayyum Khan, Z., Qayyum Khan, M., Majeed, M. *et al.* (2025) Chapter 19 sustainable agriculture: A nanoosmic-based management. In: Rajput, V., Singh, A., Ghazaryan, K., Minkina, T. and Al-Tawaha, A. (eds) *Sustainable Agriculture: Nanotechnology, Biotechnology, Management and Food Security*. De Gruyter, Boston, pp. 411–430. DOI: 10.1515/9783111369754-019.

Rehman, A., Aziz, A., Anwar, M.M., Majeed, M., Albanai, J.A. *et al.* (2023) Quantifying the impacts of urbanization on urban green, evidences from maga city, lahore pakistan. *Discover Sustainability* 4(1), 1–11. DOI: 10.1007/s43621-023-00169-z.

Sinha, D., Maurya, A.K., Abdi, G., Majeed, M., Agarwal, R. *et al.* (2023) Integrated genomic selection for accelerating breeding programs of climate-smart cereals. *Genes* 14(7). DOI: 10.3390/genes14071484.

Shafqat, A., Rasool, A., Fatima, R., Majeed, M., Rahman, K.U. *et al.* (2024) Nanomaterials for sustainable pest and disease management in agriculture. *Harnessing NanoOmics and Nanozymes for Sustainable Agriculture* 100–125. DOI: 10.4018/979-8-3693-1890-4.ch006.

Sharma, S., Naman, S. and Baldi, A. (2025) Recognition and quality mapping of traditional herbal drugs: Way forward towards artificial intelligence. *Traditional Medicine Research* 10(1), 2. DOI: 10.53388/TMR20240416001.

Tariq, A., Mumtaz, F., Majeed, M. and Zeng, X. (2023) Spatio-temporal assessment of land use land cover based on trajectories and cellular automata Markov modelling and its impact on land surface temperature of Lahore district Pakistan. *Environmental Monitoring and Assessment* 195(1). DOI: 10.1007/s10661-022-10738-w.

Tassadduq, S.S., Akhtar, S., Waheed, M., Bangash, N., Nayab, D. *et al*. (2022) Ecological distribution patterns of wild grasses and abiotic factors. *Sustainability* 14(18), 11117. DOI: 10.3390/su141811117.

Ullah, I., Aslam, B., Shah, S.H.I.A., Tariq, A., Qin, S. *et al*. (2022) An integrated approach of machine learning, remote sensing, and GIS data for the landslide susceptibility mapping. *Land* 11(8), 1265. DOI: 10.3390/land11081265.

Vera, M.C.S. and Palaoag, T.D. (2023) Implementation of a smarter herbal medication delivery system employing an AI-powered chatbot. *International Journal of Advanced Computer Science and Applications* 14(3). DOI: 10.14569/IJACSA.2023.0140358.

Yang, J.S., Tsai, S.C., Hsu, Y.M., Bau, D.T., Tsai, C.W. *et al*. (2024) Integrating natural product research laboratory with artificial intelligence: Advancements and breakthroughs in traditional medicine. *BioMedicine* 14(4), 1–14. DOI: 10.37796/2211-8039.1475.

Zhang, A. and Qiu, S. (2023) Multi-omics technology for demystifying herbal medicine. *Advanced Materials Letters* 14(1), 0–0. DOI: 10.5185/amlett.2023.011712.

Zhang, W., Zeng, Y.F., Jiao, M., Ye, C., Li, Y. *et al*. (2023) Integration of high-throughput omics technologies in medicinal plant research: The new era of natural drug discovery. *Frontiers in Plant Science* 14. DOI: 10.3389/fpls.2023.1073848.

Zhao, P., Li, J., Li, Y., Tian, Y., Yang, L, *et al*. (2017) Integrating transcriptomics, proteomics, and metabolomics profiling with system pharmacology for the delineation of long-term therapeutic mechanisms of bufei jianpi formula in treating COPD. *Biomed Res Int* 7091087. DOI: 10.1155/2017/7091087.

9 Plant Whole Genome Duplication Mechanisms and Their Breeding Applications

Macharla Srinidhi[1], Mani Manoj[1], Mahankali Sravanalakshmi[1], Priti Kumari[1], Sasikumar Janani[1], Paneerselvam Swarnamughi[2], Esakkimuthu Balaji[3] and Arumugam Vijaya Anand[1]*

[1]Department of Human Genetics and Molecular Biology, Bharathiar University, Coimbatore, India; [2]Department of Genetics, University of Madras, Chennai, India; [3]Prashant Fertility and Research Centre, Chennai, India

Abstract

Whole genome duplication (WGD) and pangenomic diversity are critical evolutionary mechanisms that drive genetic variation, adaptation, and functional innovation in plant lineages. WGD, involving autopolyploidy and allopolyploidy, reshapes genomic architecture and chromatin organization, contributing to species diversity and plant evolution. These events promote gene redundancy, facilitating sub-functionalization and neofunctionalization, which lead to functional divergence, genomic rearrangements, and enhanced adaptability. Alongside tandem duplications, transposable elements, and horizontal gene transfer, WGDs significantly amplify plant diversity, adaptability, and genomic complexity. Model plants such as *Arabidopsis thaliana* and *Brassicaceae* demonstrate how WGD and diploidization increase genetic diversity, phenotypic plasticity, and evolutionary success across varied environments. Comparative genomics, transcriptomics, and proteomics provide insights into gene retention patterns and expression divergence that underlie these adaptive features. In agriculture, WGD-derived traits enhance crop resilience by improving stress tolerance, resource efficiency, and environmental adaptability. Combining WGD and pangenome studies with artificial intelligence (AI) and machine learning (ML) offers unprecedented opportunities for precision breeding. AI-driven tools can analyze complex genomic datasets, identify retention patterns, predict functional roles, and model evolutionary scenarios. These capabilities accelerate the development of climate-resilient crops by identifying key genomic regions linked to desirable traits. Despite challenges such as data integration, computational complexity, and regulatory concerns, AI and ML offer solutions through automated analysis, predictive modeling, and efficient data integration frameworks. Leveraging AI alongside genome editing and synthetic biology can maximize the benefits of WGD and pangenomics, fostering sustainable agriculture and enhancing global food security.

Keywords: Whole genome duplication (WGD), machine learning, autopolyploidy, allopolyploidy, horizontal gene transfer, sub-functionalization, neo-functionalization, artificial intelligence

*Corresponding author: avahgmb@buc.edu.in

DOI: 10.1079/9781800628649.0009

9.1 Introduction

9.1.1 Overview of Whole Genome Duplication

Whole genome duplication (WGD) refers to the process where an organism's entire genome is duplicated, resulting in a doubling of its genetic material. This phenomenon is significant in evolutionary biology as it provides raw genetic material that can lead to increased genome complexity and the emergence of novel phenotypic traits. WGD can facilitate functional innovation by allowing duplicated genes to diverge in regulatory control while maintaining similar biochemical functions, thus expanding the functional scope of gene networks (Wapinski *et al.*, 2007). The evolutionary implications of WGD are profound, as it can drive diversification and adaptation to new environments, contributing to the emergence of evolutionary novelties (Moriyama and Koshiba-Takeuchi, 2018). Additionally, WGD has been observed in various species, including plants like apple, where it influences the evolution of ohnologous chromosomes (Lallemand *et al.*, 2020).

WGDs have played a significant role in the evolutionary history of plants, occurring repeatedly across various lineages. In the *Brassicales* order, numerous WGDs have been identified, particularly within the *Brassicaceae* family, with additional events noted in the *Cleomaceae* and *Resedaceae* families, highlighting the complexity of WGD placement and inference methods in phylogenetic studies (Mabry *et al.*, 2020). Ferns, another major plant group, exhibit a history of 19 WGDs, including three ancient events that are shared among a large proportion of extant species, suggesting a strong correlation between WGDs and diversification rates, despite a decline in fern biodiversity during the Cretaceous (Huang *et al.*, 2020a). Furthermore, WGDs are linked to significant climatic events, such as the Paleocene–Eocene Thermal Maximum, which may have facilitated adaptive responses in species like *Ceratopteris thalictroides* (Zhang *et al.*, 2019).

9.1.2 Introduction to pangenomes

A pangenome is defined as the complete set of genetic information encompassing both common and unique genomes within a species, derived from a collection of genetic sequences from diverse populations (Barh *et al.*, 2020). This concept initially emerged in bacterial studies but has since been extended to eukaryotic organisms, leading to two primary definitions: the gene-based approach, which considers the union of all genes, and the sequence-based approach, which focuses on all nonredundant genomic sequences (Hu *et al.*, 2024a). Pangenomics facilitates the exploration of within-species genetic diversity, overcoming reference bias associated with single genome studies (Hutchens and DePerno, 2009). It also provides insights into evolutionary histories and genomic variations, enhancing our understanding of biological adaptability and phenotypic diversity (Walia *et al.*, 2024b).

Pangenomes encompass the complete gene repertoire of a species, distinguishing between core, dispensable, and unique genes. Core genes are universally present across all individuals of a species, while dispensable genes are absent in at least one individual, indicating variability within the gene pool. Recent advancements in machine learning (ML) have facilitated the classification of genes as core or dispensable using a single annotated reference genome, potentially reducing the need for extensive pangenome construction (Yocca and Edger, 2022). Moreover, the concept of dispensability has evolved, with new methods like QUOD quantifying gene dispensability rather than relying solely on binary classifications. This approach assigns a dispensability score to genes, highlighting their role in lineage-specific adaptations (Sielemann *et al.*, 2020). In microbial studies, such as those on the human gut microbiota, core genes define species, while unique genes contribute to strain-specific traits and adaptability, illustrating the dynamic nature of pangenomes (Shoer *et al.*, 2024).

9.1.2.1 AI role in pangenome construction and annotation

Artificial intelligence (AI)-driven approaches for plant pangenome construction and annotation significantly enhance crop breeding and improvement programs by leveraging advanced data analysis and predictive modeling. The integration of AI tools, such as the AgroNT model, facilitates the characterization of genetic diversity and the identification of functional variants across multiple plant genomes, thereby improving the precision of genomic predictions and trait enhancements (Mendoza-Revilla *et al.*, 2024). These AI methodologies streamline the breeding process by optimizing parent selection and predicting the combined effects of traits, which is crucial for developing resilient crop varieties tailored to future environmental challenges (Farooq *et al.*, 2024). Also, the role of AI in accelerating genetic improvements through techniques like CRISPR-Cas9 exemplifies its potential to address global issues such as food security and climate change, while also raising important ethical considerations regarding equitable access and genetic diversity preservation (Feng *et al.*, 2024).

Methodological considerations significantly influence pangenome construction, with variations in assembly techniques affecting gene presence and diversity outcomes (Glick and Mayrose, 2023). Furthermore, pangenomes compiled from diverse cultivars and wild relatives provide a rich resource for identifying novel genes associated with desirable traits, which can be reintroduced into modern crops through various breeding techniques (Naithani *et al.*, 2023a). This integrative approach not only enhances the understanding of plant genetics but also offers new avenues for targeted breeding strategies to address challenges posed by climate change.

9.1.3 Predictive evolutionary modeling using AI

By combining cutting-edge data analysis methods with farming methods, AI-powered predictive evolutionary modeling has the potential to greatly increase crop resistance to environmental stresses. The production of stress-tolerant agricultural varieties is made easier by the discovery of stress-responsive genes and regulatory networks and also the integration of AI and bioinformatics (Zhang *et al.*, 2024). Furthermore, it has been demonstrated that AI methods like ML increase the precision of production forecasts under various climatic conditions, empowering farmers to take well-informed decisions that promote sustainability and food security. For instance, AI-driven genomic prediction models, such as SoyDNGP, allow for accurate forecasting of trait combinations and environmental interactions, thus refining breeding outcomes (Feng *et al.*, 2024; Na and Na, 2024). Furthermore, by offering real-time monitoring and customized treatments, AI technologies in precision agriculture maximize resource use and reduce environmental effects and also using predictive modeling in disease management, AI may proactively treat plant health concerns, enhancing agricultural resilience overall (González-Rodríguez *et al.*, 2024; Gul and Banday, 2024).

9.1.4 Relevance of whole genome duplication in plant evolution

WGD is a significant driver of plant diversification, influencing both microevolutionary and macroevolutionary processes. Research indicates that WGD events have repeatedly occurred throughout plant evolution, providing genetic redundancy that facilitates the emergence of new functions and phenotypes, thereby enhancing adaptive potential and morphological innovation (Qiao *et al.*, 2022). Specifically, the expansion and functional divergence of gene families, such as the HD-Zip I transcription factors, illustrate how WGD contributes to the regulatory complexity necessary for diverse developmental processes and stress responses (Żyła and Babula-Skowrońska, 2023). Additionally, in gymnosperms, WGD correlates with major phenotypic innovations and diversification patterns, suggesting that genomic changes are foundational to the evolution of major clades (Stull *et al.*, 2021). Furthermore, WGD has been linked to the diversification of biosynthetic pathways, as seen in the apple tribe, where it underpins the development of distinct medicinal properties (Su *et al.*, 2021).

WGD events are prevalent across various plant lineages, with notable examples found in the *Brassicales* and ferns. In the *Brassicales*, multiple WGDs have been identified, particularly within the *Brassicaceae* family, as well as in its sister family, *Cleomaceae*, which has a specific WGD event termed Th-α. Additionally, evidence suggests a shared WGD event among members of the *Resedaceae* family, named Rs-α, highlighting the complexity of WGD occurrences in this order (Mabry *et al.*, 2020). Ferns also exhibit significant WGD events, with a study identifying 19 WGDs across their phylogeny, including three ancient events that are shared by a majority of extant ferns. These duplications are thought to have contributed to the high chromosome numbers observed in many fern species (Huang *et al.*, 2020b. The integration of these findings underscores the evolutionary significance of WGDs in shaping plant diversity and adaptation across different lineages.

9.2 Mechanisms of Whole Plant Genome Duplication

9.2.1 Polyploidization

In general, polyploidy means the involvement of two or more than two complete sets of chromosomes in an organism. The species undergoing polyploidy show differential changes in which structural and functional alterations are involved. It is seen in all of the major eukaryotic families and is also mostly found in plants as well (Leitch and Bennett, 1997; Otto and Whitton, 2000). There is a great impact on polyploidy in which genomic and genetic tools are widely used to determine the significance on the basis of criteria (Bourke *et al.*, 2018). In polyploidization, some of the species-rich plants are seen in recent and ancient times, and it is noted that there are many consequences arising in polyploidy and playing a vital role in evolution. These structural and functional changes can be seen through two different types of polypoidy in plants: autopolyploidy and allopolyploidy (Doyle *et al.*, 2008; Leitch and Leitch, 2008).

Autopolyploidy and allopolyploidy both play a major role but are different in certain aspects. Autopolyploidy mean that, as the name suggests, it arises within the same species in WGD events. In contrast, allopolyploidy arises from genome duplication showing interspecific hybridization and doubling of nonhomologous genomes (Akama *et al.*, 2014). They both play an evolutionary role in nature; recent data show that although allopolyploidy has been a focus, autopolyploidy is more common than previously assumed and may be more frequent than allopolyploidy (Parisod *et al.*, 2010). These two processes have a special type of inheritance pattern in general. Autopolyploidy results in a genome set of homologous chromosomes, which can be called a polysomic inheritance, but for allopolyploidy, a disomic inheritance is explained with nonhomologous combination states that do not pair. In autopolyploidy, homologous chromosomes are randomly paired and segregate at random to form only bivalents, whereas segmental allopolyploidy forms are multivalent in which natural hybrids are seen and also differentiated genomes are involved (Soltis *et al.*, 1993; Jackson and Jackson, 1996; Landergott *et al.*, 2006; Stift *et al.*, 2008).

9.2.2 AI-based prediction of polyploidization events

AI-based prediction of polyploidization events holds significant potential across agriculture, medicine, and conservation biology. In agriculture, polyploidy enhances crop resilience and yield by improving tolerance to biotic and abiotic stresses, as evidenced by its application in developing superior cultivars through techniques like flow cytometry and chromosome counting (Miri, 2020; Islam *et al.*, 2022). In the realm of medicinal plants, artificial polyploidization can increase the vigor and secondary metabolite content, addressing the growing global demand for these resources (Salma *et al.*, 2017). Furthermore, understanding the evolutionary implications of polyploidization can inform conservation strategies by elucidating how polyploidy contributes to plant adaptation and diversity, potentially aiding in the preservation of endangered species (Paterson, 2005; Burgess, 2015).

Key genetic and epigenetic factors influencing AI-based predictions of polyploidization events encompass both DNA sequence characteristics and epigenetic modifications. Genetic factors include the DNA sequence of promoter regions, which can predict gene expression and sub-genome dominance in ancient polyploids, as demonstrated by deep learning (DL) models applied to various plant systems (Guo *et al.*,

2024). Epigenetically, changes in DNA methylation play a crucial role, with studies showing that global cytosine methylation levels tend to increase with ploidy, impacting gene regulation and survival of newly formed polyploids (Li *et al.*, 2011; Alonso *et al.*, 2016). Additionally, phenomena such as gene silencing, nucleolar dominance, and transposable element activation contribute to the epigenetic landscape, facilitating rapid adaptation to genomic changes post-polyploidization (Yang and Peng, 2005).

ML algorithms, particularly DL and modified genomic best linear unbiased prediction (GBLUP), significantly enhance the accuracy of AI-based predictions of polyploidization events across various species. DL techniques excel in capturing complex genetic interactions, such as dominance and epistasis, which are crucial in polyploid organisms, as demonstrated in studies involving strawberries and blueberries (Zingaretti *et al.*, 2020). In sugarcane, ML methods like random forest and DL showed comparable or improved predictive accuracies over traditional models, particularly when accounting for non-additive genetic variations (Chen *et al.*, 2023). Furthermore, incorporating allele dosage information into genomic selection models has been shown to increase predictive abilities by up to 140% in simulated datasets, highlighting the importance of considering genetic complexity in polyploid species (Batista *et al.*, 2022). The mechanisms of WGD in plants are illustrated in Fig. 9.1.

9.2.3 Genome doubling

Genome doubling is a crucial process in plant evolution and development. In a nucleus of a plant cell when the number of chromosomes gets doubled it is said to be genome doubling. There are developmental and physiological changes associated with genome doubling arising in polyploidy due to an increase in cell size. The production of macro-mutants by chromosome doubling may provide a population with new ways to adapt to environmental demands. Within a microevolutionary time scale, the byproducts of chromosomal doubling serve as the foundation for punctuated evolution (Levin, 1983).

The molecular pathways that could lead to the long-term stabilization of polyploids or their adaption to a condition where their genomes are doubled are mostly unknown. The homologous pairing suppressor (Ph1) from allohexaploid wheat is the only gene in plants that has been molecularly described and is crucial for polyploid stabilization. Wheat and other allopolyploids are hybrid plants with somewhat diverse chromosomal groups. Ph1 promotes bivalent pairing and stable meiosis by strengthening chromosomal pairing preferences for more similar (homologous) chromosomes over less similar (homologous) ones (Griffiths *et al.*, 2006). An essential molecular insight into the meiotic stabilization process in allopolyploids is provided by this work.

In plants, polyploidy, also known as genome doubling, is a frequent and important evolutionary phenomenon. It affects a variety of features in both the short and long term. WGD first causes physical features and cells to enlarge, but it also reduces cell numbers and fitness (Clo and Kolář, 2021). The benefits of doubling may outweigh the drawbacks of hybridity, and vice versa; nevertheless, further research is needed to confirm the generality of this pattern. Numerous non-Mendelian genetic and genomic effects caused by doubling and hybridization either independently or cooperatively have also been discovered by some investigations (Qiu *et al.*, 2020). WGD can have profound consequences on population life history, ecology, and genetics. The impact of polyploidization per se on heritable variance can be counteracted by alterations to other population characteristics. An overview of the theoretical and empirical research on the various effects of genome doubling on quantitative trait heritable variance and the evolutionary potential of polyploid populations relative to their diploid progenitors has been provided by Clo (2022).

9.2.4 Chromosomal rearrangements

The terms "genetic architecture" and "genomic locations" describe how many genes there are, how much of an effect they have, and how much of an effect additive, dominant, and epistatic gene effects contribute (Holland, 2007). The

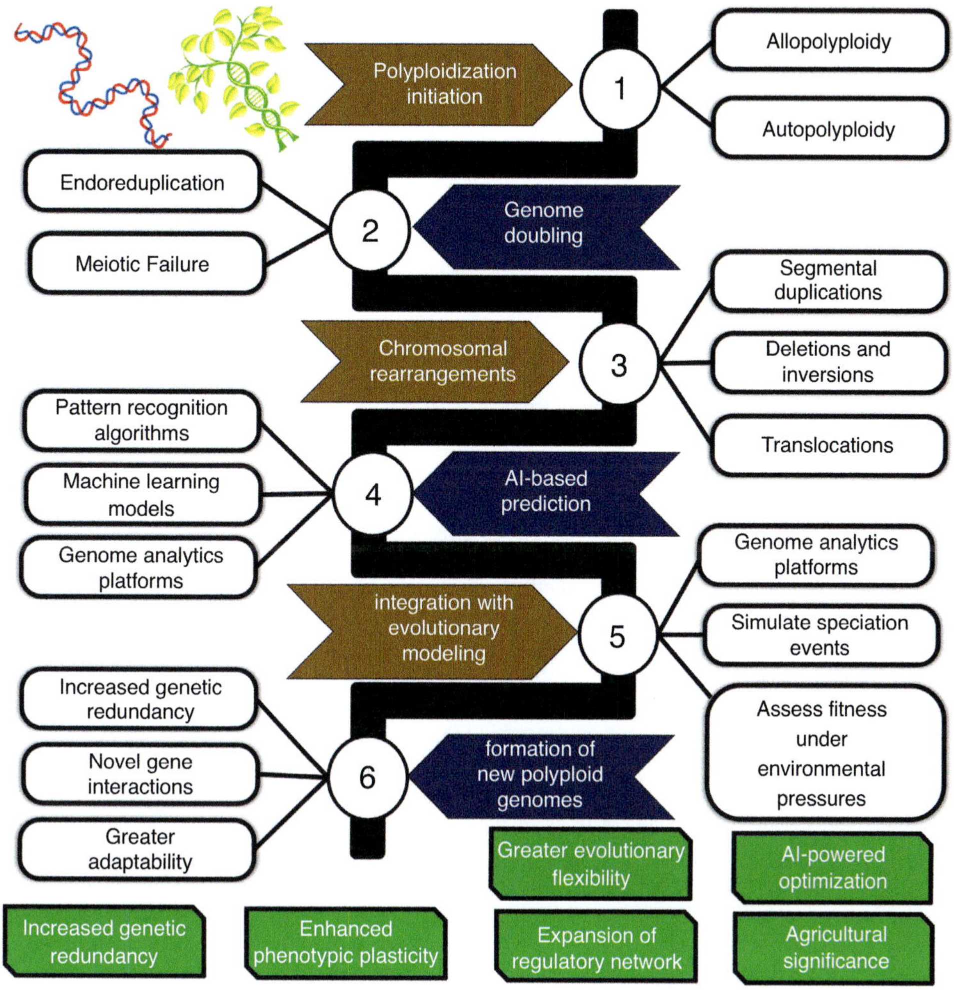

Fig. 9.1. Mechanisms of whole genome duplication in plants. (Author's own image.)

intricate three-dimensional organization of plant genomes and how it affects different cellular processes have been made clear by recent studies. Technologies for chromosomal conformation capture have shed light on the hierarchical structure of chromatin, revealing everything from topologically associated domains to gene loops (Ouyang *et al.*, 2020).

Using cutting-edge methods like single-cell multi-omics and proximity ligation-independent approaches, future research directions will examine RNA–chromatin interactions, liquid–liquid phase separation, and dynamic chromatin alterations. The increased resolution of these methods has made it possible to depict chromatin compartmentalization and interactions at the gene scale. These investigations have uncovered astounding knowledge gaps in this field as well as the extremely intricate mechanisms controlling plant nuclear architecture. The current review covers the state-of-the-art in plant chromosome architecture, including what we know about other nuclear components and the hierarchical organization of the genome in three dimensions (Huang *et al.*, 2020c. Research on genetic architecture and enhanced comprehension of plant population structure will influence our comprehension of plant evolution and the development of crop enhancement tactics.

9.3 Evolutionary Consequences of WGD in Plants

9.3.1 Gene redundancy and retention

Genes that are duplicated during WGD in plants have different outcomes. A variety of factors, including evolutionary rate, structural complexity, and GC content, cause some duplicates to be preferentially retained while many are rapidly lost through fractionation (Jiang *et al.*, 2013). The retained duplicates frequently experience functional divergence, with neofunctionalization (NF) and subfunctionalization (SF) being frequent results (Li *et al.*, 2015). Functional divergence happens more slowly than expression pattern divergence in duplicates (Shi *et al.*, 2020). The current theories (i.e. the Increased Gene Dosage Hypothesis, Gain of Function Hypothesis, the Gene Balance Hypothesis, Functional Buffering Model, and the SF Hypothesis) address the fate of nuclear genes following duplication events in light of these recent observations. Various dosage-compensation mechanisms have been developed to mitigate detrimental dosage imbalances and a recently presented extension of the Gene Balance Hypothesis aims to explain why a particular functional class of genes is shared by all flowering plants in terms of single-copy status (Edger and Pires, 2009). The evolution of duplicated genes may follow a sequence of function relaxation, diversifying selection, and purifying selection (Guo *et al.*, 2013). The percentage of dosage-balanced gene pairs declines with the age of the WGD event (Guo *et al.*, 2013).

SF and NF are the two primary processes that account for the persistence of duplicated genes. Whereas NF causes one copy to acquire new functions, SF involves dividing ancestral functions among duplicates. According to the NF hypothesis, one daughter gene after duplication keeps its ancestral function while the other acquires new ones. The SF hypothesis, on the other hand, contends that degenerate mutations in duplicate genes cause their joint levels and patterns of activity to be reduced to those of a single ancestral gene (He and Zhang, 2005). The divergence of duplicate expression data is observed under both a neutral evolution model and a regulatory NF model across several species. If duplicate divergence is conserved across multiple species sharing the WGD, the test's power will increase and it will yield the strongest evidence of NF. Lastly, we ignore the fact that selection acts differently on various individual genes when we solely consider evolution patterns of global expression (Felsenstein, 1985). To assess whether the data supports SF or NF on a gene-by-gene basis, it is crucial to modify current phylogenetic methods and create new ones (Martins and Hansen, 1997).

9.3.2 Genome size and complexity

As our understanding of the implications and relevance of this amazing variation in genome sizes among land plants improves, it is becoming clear that this feature is crucial in defining plant genome evolution as well as the assemblages of plant communities at the ecosystem level. Critical insights into the genomic and epigenetic mechanisms underlying variations in genome size can be obtained through the use of analytical methods (Pellicer *et al.*, 2018). Plant genomes and evolution are significantly impacted by WGDs. WGD has the ability to quickly expand genome size, yet size thresholds and genome shrinking cause the mean genome size across ploidies to remain constant (Zenil-Ferguson *et al.*, 2016). Increased cell growth and stress resistance are among the initial benefits of WGD (Bomblies, 2020). In comparison to diploids, neo-tetraploids have greater morphological characteristics and cells, but lower fitness and fewer cells overall, according to a meta-analysis. In certain plant lineages, ancient WGD episodes are linked to higher rates of species diversification, which may act as accelerators for important discoveries and evolutionary novelty (Soltis and Soltis, 2016). These results demonstrate the intricate interactions that shape plant variety and adaptation between WGD, genome size, and evolutionary processes.

Polyploidy events, which have repeatedly occurred throughout plant evolution, are the driving force behind genomic reorganization and gene loss in plants (Adams and Wendel, 2005). Following these WGDs, there are significant genomic reorganizations and deliberate gene loss, resulting in intricate networks of symbiosis among different species (Ku *et al.*, 2000). Some plant species have shown dramatic genome rearrangement following WGD, which has led many to hypothesize that genome restructuring may be a common outcome of WGD. It is possible to see notable variations in the way the genome of a single plant species changes following WGD, suggesting that there are no hard-and-fast guidelines for predicting how a genome will respond to WGD (Hufton and Panopoulou, 2009). It is unclear what the underlying mechanisms are that cause these changes. We briefly review the forms and origins of polyploidy and summarize the insights gained from genome-wide gene expression analyses in newly synthesized autopolyploids and allopolyploids in order to advance knowledge of genomic and gene expression changes in polyploidy. We demonstrate transcriptome divergence in newly formed allopolyploids as well as in their progenitors has been demonstrated (Chen and Ni, 2006). The authors also suggest models for how chromatin modification, transcriptional regulation, and RNA-mediated pathways establish the locus-specific expression of homologous and orthologous genes during the formation and evolution of allopolyploids (Chen and Ni, 2006).

9.3.3 Speciation and adaptation

Structural differences in the nuclear genome contribute to plant speciation and diversification. Large-scale changes like WGD (polyploidy) and smaller-scale variations like aneuploidy and dysploidy are also possible. A study highlights the importance of chromosomal instability in genome evolution and adaptation by concentrating on somatic and meiotic abnormalities that lead to structural changes or polyploidy (De Storme and Mason, 2014). The study also examines the effects of stress on ploidy and chromosome integrity, arguing that these changes are critical for adaptive genome evolution, especially in severe conditions (De Storme and Mason, 2014). Soltis and Soltis (2016) explore the connection between WGD, the emergence of new traits, and noteworthy developments of plant evolution. The significant influence that WGD, or polyploidy, has had on plant evolution—particularly in angiosperms was examined. WGD events are common across plant lineages, especially in families like *Asteraceae*, and they are linked to increased species diversity. WGD in the ancient Equisetum lineage has been studied, and it was discovered that two WGD occurrences took place in the Carboniferous and Triassic periods (Clark *et al.*, 2019). The analysis shows that these WGDs have no macroevolutionary influence, challenging assumptions about the role of WGD in innovation and survival in both angiosperms and non-angiosperms. The findings reevaluate WGD's evolutionary significance (Clark *et al.*, 2019).

In order to clearly demonstrate how WGD supports evolutionary success and adaptation in times of environmental instability, it should be possible to "replay the duplication tape of life" in model organisms by evaluating how WGD affects fitness under various selection regimes. Research on yeast demonstrates that polyploidy can hasten evolutionary adaption to difficult conditions through complicated effects on the fitness or frequency of advantageous mutations (Selmecki *et al.*, 2015). Regarding the importance of niche differentiation in polyploid evolution, there is still disagreement, despite the fact that WGD is a significant speciation force. Additionally, it is still unclear how the development of polyploid niches is impacted by genome doubling in and of itself vs subsequent divergence. This may be due in part to the fact that interploid gene flow and the intraspecific genetic structure of polyploid complexes are frequently overlooked in ecological research (Padilla-García *et al.*, 2022).

9.4 WGD and Plant Pangenomes

9.4.1 Influence of WGD on core and dispensable genomes

WGD was important for the expansion, evolution, and diversification of the genome in polyploids. After polyploidization, many gene families become larger, and the duplicated genes undergo functional diversification through NF or SF. These systems probably helped plants survive during evolution and may aid in adaptation (Hong *et al.*, 2021). The *Portulacaceae*, *Cactaceae*, and *Anacampserotaceae* underwent diversification shifts; the *Cactaceae*, however, underwent extensive gene duplications, while the other two families did not. With an emphasis on genes linked to environmental stress responses, we investigated gene family expansion and molecular evolutionary patterns. We discovered evidence of notable gene family expansion in clades present in extreme environments and genes with stress adaptation (Wang *et al.*, 2019). Tandem duplication has also helped plants expand their gene families, especially those related to responses to environmental stress. Due to their propensity for asymmetric expansion and lineage-specific selection, these tandem duplicates may be crucial for adaptive evolution in response to shifting environmental conditions (Hanada *et al.*, 2008). In general, gene family expansion and functional diversification in plants have been greatly impacted by both WGD and tandem duplication processes.

In order to predict the fate of endangered plants and to enable the proposal of conservation guidelines for their rescue, preliminary investigations into population genetics, demographic history, inbreeding depression, and the genetic load associated with it should also be included (Yang *et al.*, 2018). Furthermore, we anticipate that more research will use genomic data from threatened plants to examine the genetics of conservation and to guide conservation policy as whole-genome sequencing becomes more widely accessible. Cultivars, landraces, wild species closely related to cultivated varieties, elite lines of breeders, and mutants are examples of plant genetic resources (PGR). Much work has focused on gathering PGR because of the loss of genetic diversity brought about by agricultural practices and the accessibility of genetic data. Before molecular profiling, accessions were analyzed morphologically in germplasm collections (Van *et al.*, 2011).

The most effective way to close the current taxonomic gaps may be to concentrate on the genomes of plant species that face significant extinction risks or are essential to the restoration of ecosystems. A review summarizes recent developments in plant genomics, emphasizing their significance for plant diversity conservation, and providing an overview of how to incorporate genomics into plant conservation (Schneider, 2023). The complexity of the factors that cause plant extinction include decreased genetic diversity, increased genome erosion, and hybridization-induced loss of species integrity (Schneider, 2023). Every facet of genomics, from identifying phenotypic traits to utilizing next-generation sequencing (NGS) on entire genomes, will significantly influence not only PGR conservation but also plant breeding initiatives.

9.4.2 Contribution to genetic diversity

There is ongoing discussion regarding the significance of WGD in plant evolution. WGDs have been linked to benefits like better colonization, a wider range of adaptations, and larger effective populations. However, there is uncertainty about the possible relationship between WGDs and higher rates of diversification because there is not a thorough mapping of WGDs within a significant plant clade (Smith *et al.*, 2018). Additionally, polyploidy is predicted to have both short- and long-term benefits by population genetic theory, and recent empirical evidence indicates that established polyploids may serve as “sponges” to accumulate adaptive allelic diversity. The pool of genetic variants available to polyploids is further increased by introgression with other tetraploid lineages, diploid progenitors, or even other species, in addition to their increased genetic variability (Baduel *et al.*, 2018). WGD can produce significant homoeologous expression rewiring at the population level, which results in phenotypic diversity. Allelic expression differences tend to be lessened by hybridization, whereas parental differences are amplified and variable homolog content is produced by WGD (Sun *et al.*, 2017). WGD may have evolutionary benefits, but long-term success hinges on overcoming early obstacles and ecological drivers.

Although WGD disrupts conserved processes, particularly meiosis, it can also enhance adaptation. (Bohutínská *et al.*, 2021). Multiple sets of genetic material may have an evolutionary advantage for adaptive evolution, as suggested by the widespread occurrence of polyploidy. On the other hand, higher gene and genome dosages in allopolyploids (combinations of two or more divergent genomes) and autopolyploids (duplications of a single genome) frequently result in chromosome imbalances, genome instability, and incompatibilities with regulation. Plant WGD can result in genetic and epigenetic modifications that enhance phenotypic diversity and adaptability (Chen, 2007). Comparisons between recently generated polyploidy, or neopolyploids, and their parents were specifically on the focus of a study, in order to reduce the potential confounding effects of adaptation and drift that might exist in ancient polyploids (Porturas *et al.*, 2019). The findings showed that WGD had no effect on flowering phenology, reproductive output dropped, and floral morphology traits grew in size across a wide range of angiosperms. Furthermore, the authors discovered that, for the phenotypic traits under investigation, increased trait variation following WGD was unusual (Porturas *et al.*, 2019).

9.4.3 Modeling genome plasticity using machine learning

ML is increasingly utilized to model genome plasticity in plants, enhancing the understanding of structural variations and their implications for breeding. Recent advancements, such as the Smart Model for Epigenetics in Plants (SMEP), demonstrate the capability of DL to predict various epigenomic modifications with high accuracy, thereby facilitating the analysis of genome plasticity across different species (Wang *et al.*, 2021). The exploration of structural genome variations highlights the role of spontaneous changes during breeding, which can be effectively analyzed using ML techniques to identify patterns in extensive genomic and phenotypic datasets (de Maagd *et al.*, 2020; Crossa *et al.*, 2024). Supervised learning methods are adept at modeling complex relationships between genetic markers and phenotypic traits, improving prediction accuracy and optimizing breeding strategies (Xavier, 2021; Danilevicz *et al.*, 2022). This integration of ML in genomic prediction streamlines data analysis and enhances understanding of genotype-by-environment interactions, which are crucial for effective plant breeding (Crossa *et al.*, 2025).

ML algorithms have shown significant promise in accurately predicting the effects of genetic mutations on plant genome plasticity. For example, the genomic pre-trained network (GPN) learns genome-wide variant effects through unsupervised pretraining, outperforming traditional conservation score-based predictors in *Arabidopsis thaliana* (Benegas *et al.*, 2023). Similarly, AgroNT, a foundational large language model, demonstrates state-of-the-art predictions for regulatory annotations and functional variants across 48 plant species, including cassava (Mendoza-Revilla *et al.*, 2024). Deep autoregressive models have also been developed to capture inter-residue correlations in biological sequences, achieving competitive performance in mutation effect predictions (Kim and Kim, 2020).

ML-based genome plasticity models hold significant potential for enhancing plant breeding through various applications. These models optimize genome editing techniques by improving editing efficiency and specificity, facilitating guide RNA design, and enabling accurate key site detection, thereby accelerating crop development processes (Chen *et al.*, 2024). Furthermore, genomic prediction models like SoyDNGP leverage AI to improve parent selection and predict the combined effects of multiple traits, addressing challenges in crop breeding and contributing to sustainable agricultural practices (Ferguson *et al.*, 2022; Feng *et al.*, 2024). Overall, the integration of ML in plant breeding represents a transformative approach that enhances the precision and efficiency of crop improvement efforts (Eftekhari *et al.*, 2024).

9.4.4 Impact on genome plasticity

The hallmark of interactions between plant microbes is rapid (co-)evolution at multiple timescales. The characteristics of the genomes of the interacting partners provide a large part of the mechanistic basis for the rapid evolution. Here, we summarize recent discoveries about the traits and processes in the genome that allow plants and phytopathogens to evolve quickly (Frantzeskakis *et al.*, 2020). Plant polyploids exhibit swift and dynamic alterations in their genomic structure and gene expression, a reflection of the functional and genomic plasticity of their duplicate genes and genomes (Jackson and

Chen, 2010). Major advancements in morphology, stress response, and other areas have been fueled by WGD, or polyploidy, and this trend is still ongoing. When taken as a whole, these viewpoints will start to show how genomic change can result in organismal novelty and diversity (Soltis and Soltis, 2021).

Plant adaptation may benefit from contributions from transposable elements to genetic and epigenetic variation (Lisch, 2013). Our knowledge of the timing of polyploidy events and their effects on genome structure has improved with recent genomic studies (Adams and Wendel, 2005). The astounding diversity of plant forms and functions that are currently known is probably a result of this genomic flexibility.

The sharing of genetic material between organisms that are not parent–offspring relationships is known as horizontal gene transfer, or HGT (Soucy *et al.*, 2015). The vertical transfer of genetic information through reproduction assures the preservation of species integrity in plants, as it does in all other eukaryotes. Nonetheless, a number of studies published in recent years have unequivocally demonstrated that HGTs (the interspecific transfer of genetic information across reproductive barriers) affect all living things, including plants (Aubin *et al.*, 2021). According to recent research, the nuclear and organellar genomes of multicellular eukaryotes are not as active in HGT as the mitochondrial genomes of plants. Several studies have identified parasitic plants as both donors and recipients of mitochondrial genes, despite the lack of knowledge regarding the mechanisms underlying plant HGT. The majority of cases that have been found so far have involved one transferred gene per species (Richardson and Palmer, 2007). Thus, the predominant mode of HGT in plants reported so far appears to be mitochondrion to mitochondrion. Of the 40 plant-to-plant HGT cases reported so far, all but one have a mitochondrial gene transferred (encoding housekeeping respiratory or ribosomal protein). The development of NGS technologies has created new avenues for the comparative genomic analysis of HGTs. HGT is a genetic transmission mechanism that plays a major role in the long-term evolution and diversification of green plants. According to recent studies, land plants and charophyte green algae have evolved and adapted primarily through HGT events (Wang *et al.*, 2022).

9.5 Comparative Genomics of WGD Events

9.5.1 Case studies in model plants

A model organism used to identify particular defense mechanisms of plant–pathogen resistance is *Arabidopsis thaliana*. For comparing these theories, the metabolic network of the model plant *A. thaliana* is an effective tool. The evolution of flowering plants has been rife with gene and genome duplications. WGDs, in contrast to small-scale gene duplications, replicate entire pathways or networks. This makes them unique in that their duplicated pathways or networks may evolve novel functionality through the coordinated SF or NF of their constituent genes. A unique instance of coordinated divergence in gene expression in *A. thaliana* after WGDs has been demonstrated (De Smet *et al.*, 2017). With one homolog exhibiting predominant expression in aerial tissues and the other homolog displaying biased expression in tip-growth tissues, the authors identified a set of 92 homoeologous gene pairs that all exhibit a similar pattern of tissue-specific gene expression divergence following WGD (De Smet *et al.*, 2017).

Investigating consecutive duplications of the *Arabidopsis* genome, this work demonstrates that different types of duplicated enzymes survive between events, indicating roles for both absolute and relative dosage constraints. The frequency of ancient polyploids found in current genomes raises concerns about how evolution proceeded following WGD. The dosage balance hypothesis states that restrictions on the relative dosages of central network genes are a significant factor in the maintenance of duplicates. On the other hand, selection based on the absolute abundance of particular gene products may also lead to preservation (Bekaert *et al.*, 2011). The origins of lineage-specific genes (LSGs) in *A. thaliana* are thoroughly analyzed in this work. The lineage-specific status of 958 genes, which are exclusive to *A. thaliana*, can be explained by either *de novo* origination or differential gene loss and retention. A subset of these genes has alignments to intergenic regions in *Arabidopsis lyrata*. A smaller subset of LSGs is further identified as accession-specific genes, most likely of recent origin in *A. thaliana*, with an incomplete open reading frame across multiple accessions (Donoghue *et al.*, 2011). The complex nature of post-WGD evolution in *Arabidopsis* and other flowering plants is highlighted by the fact that the effects of WGD on gene expression and phenotypic plasticity can vary depending on the genotype of the progenitor and environmental context.

The plant family *Brassicaceae* includes the genus *Brassica*. The *Brassicaceae* (*Cruciferae*) family is a model for polyploidy and evolutionary research because of its remarkable species, genetic, and physiological diversity as well as its substantial economic potential (Kagale *et al.*, 2014). Events involving WGDs are thought to be significant diversification drivers. There were independent mesopolyploid WGDs followed by diploidization processes in at least 11 of the 52 *Brassicaceae* tribes. Using the internal transcribed spacers of nuclear ribosomal DNA, BEAST analyses of 48 tribes, each containing 2101 taxa, were performed. All major splits among the *Brassicaceae* lineages were used in these analyses. Using Bayesian analysis of macroevolutionary mixtures (BAMM), diversification patterns were examined on these tribe-wide chronograms and compared with species richness and genome size variation data from the entire family. The *Brassicaceae* split 29.9 million years ago during the Oligocene, and most tribes began diversifying in the Miocene, with an average age of 12.5 million years ago for their crown groups. Decoupled from previous mesopolyploid WGDs, significant rate shifts were found in 12 out of 52 tribes during the Miocene and Pliocene (Huang *et al.*, 2020b). There are numerous distinct species of Brassica plants, and each species has a wide variety of morphotypes with extreme characteristics. Comparing *Brassica* species to *A. thaliana*, the model plant, reveals an additional whole genome triplication (WGT) event. Whole genome sequencing of *Brassica rapa*, *Brassica oleracea* and other species revealed that WGT is crucial to the speciation and morphotype diversification of Brassica plants (Cheng *et al.*, 2014).

A study examined the effects of diploidization on post-polyploid gene retention and whether multiple base chromosome numbers resulted from genome diploidization following independent mesopolyploid WGDs (Mandáková *et al.*, 2017). Using comparative chromosome painting, whole-transcriptome analysis of gene age distributions, and phylogenetic analyses of gene duplications, 16 species representing ten *Brassicaceae* tribes were examined. Across the *Brassicaceae*, researchers discovered evidence for at least 13 distinct mesopolyploidies that were followed by varying degrees of diploidization. Studies of the genomic data for the *Brassicaceae* family show that clade-specific genome duplications followed by diploidization can be the cause of extant chromosome number variation in many plant groups, particularly in monophyletic taxa with multiple base chromosome numbers (Mandáková *et al.*, 2017). The synchronized diversification of *Brassicaceae* species points to the possibility that polyploid events contributed to increased tolerance and adaptability to the rapidly changing global environment, thereby promoting the radiation of new species (Kagale *et al.*, 2014).

9.5.2 AI-powered comparative pangenomics

AI-powered comparative pangenomics in plants significantly enhances the discovery of novel genetic traits by enabling detailed genomic comparisons across diverse cultivars, landraces, and wild relatives. This approach reveals a comprehensive array of genes and structural variations that may have been lost during domestication and breeding processes, thus facilitating the identification of valuable alleles associated with traits such as disease resistance and abiotic stress tolerance (Naithani *et al.*, 2023b). The pangenome concept, which encompasses both core and variable genes, allows researchers to construct high-resolution genomic frameworks that can be integrated with phenotypic data for trait dissection, as demonstrated in studies on crops like melon (Oren *et al.*, 2022). Furthermore, advanced assembly methods and tools for pangenome construction enhance the quality and accuracy of genetic analyses, ultimately guiding targeted breeding strategies to improve crop resilience and productivity in the face of climate change (Rehman *et al.*, 2022; Hou and Pang, 2024).

9.5.3 Pangenome studies in polyploid plants

For interspecific comparative studies to be accurate, expression data from different species must be comparable. When considering species with varying levels of ploidy, the relative expression per cell should be obtained (Pérez *et al.*, 2014). There are potential benefits associated with the polyploid state over the diploid condition. The novel phenotypes from polyploidy might support adaptation to new ecological niches distinct from those of their diploid ancestors or be involved in heterosis (Aversano *et al.*, 2012). Researchers have summarized what is currently known about the genetic, epigenetic, and transcriptional alterations linked to plant polyploidy and evaluated the potential role that these alterations may have had in the evolutionary success of polyploidy plants (Hegarty and Hiscock, 2008). Deeper examination of polyploid genomes has been made possible by developments in sequencing and genomics, which have shed light on the genesis, maintenance, and divergence of these genomes. These instruments enable the investigation of genome-wide alterations and variations in gene expression linked to polyploidy (Aversano *et al.*, 2012). Nevertheless, field-based research is required to ascertain whether the genetic modifications of polyploids are actually adaptive (Hegarty and Hiscock, 2008).

Development, gene function, and physiological stress responses have become more biologically understood thanks to comparative transcriptomics and proteomics. RNA-seq and high-throughput proteomics are still not widely used in plant adaptation research, though. These approaches have produced discovery tools that could greatly contribute to our growing understanding of adaptive diversification (Voelckel *et al.*, 2017). Plant–microbe interactions can be better understood by examining the proteome and metabolome, which not only reflect additional regulatory levels but also frequently contain the products of regulatory processes (Feussner and Polle, 2015).

The interaction between plants and nanoparticles (NPs) has garnered a lot of attention in recent years. More research is required on the intricate systems that plants have evolved to regulate the uptake, accumulation, and mobilization of NPs, particularly in light of proteomics and genetic level considerations. To this end, by establishing a link between gene expression and cell response, transcriptomic technologies have essentially advanced our understanding of the molecular mechanisms in plants that withstand the stress of NPs (Munir *et al.*, 2023). By identifying and measuring RNA and proteins in plant tissues, these high-throughput technologies provide a better understanding of gene function and physiological stress responses (Voelckel *et al.*, 2017). The components of cellular signaling, molecular complexes, and pathway regulation in plant development have all been clarified with the help of mass spectrometry-based proteomics (Kaufmann *et al.*, 2011). In order to address the frequently poor correlation between mRNA and protein expressions, the integration of transcriptomic and proteomic data can reveal complex regulatory behaviors that may not be evident from individual analyses (Haider and Pal, 2013).

9.6 Implications for Plant Breeding and Agriculture

9.6.1 AI for optimizing WGD-derived crop traits

AI is increasingly recognized for its potential to optimize traits derived from WGD in crop breeding, particularly in enhancing yield, disease resistance, and adaptability to climate change. By integrating AI with genomic prediction models, such as DL techniques, researchers can analyze complex datasets to identify beneficial traits and improve breeding efficiency (Farooq *et al.*, 2024; Feng *et al.*, 2024). High-throughput phenotyping (HTP) combined with AI enables rapid assessment of phenotypic traits, facilitating precise trait selection and accelerating the breeding cycle (Sheikh *et al.*, 2024). ML algorithms play a crucial role in identifying and prioritizing crop traits for optimal yield and disease resistance by analyzing complex datasets that encompass environmental factors, soil characteristics, and crop performance. These algorithms, such as random forest, support vector regression, and ensemble methods, can uncover non-linear relationships and patterns that traditional methods may overlook, thereby enhancing the accuracy of crop recommendations and disease resistance predictions (Jakarbet *et al.*, 2024). For instance, studies have demonstrated that hybrid models combining various ML techniques can achieve high accuracy rates, with some models reaching up to 99.77% in crop recommendation systems (Savaliya *et al.*, 2024). Additionally, ML facilitates the integration of diverse data sources, including weather forecasts and satellite imagery, to optimize resource allocation and improve decision making in precision agriculture (Bachu *et al.*, 2024). This comprehensive approach not only aids in selecting suitable crops but also enhances inherent disease resistance mechanisms, ultimately contributing to sustainable agricultural practices and food security (Upadhyaya *et al.*, 2024).

9.6.2 Enhancing crop resilience

WGD-derived traits significantly enhance crop resilience by improving stress tolerance and resource utilization. These traits facilitate the development of crops that can withstand various environmental challenges, thereby ensuring food security in the face of climate change. They also contribute to genetic diversity, allowing crops to better adapt to abiotic stresses such as temperature extremes and drought. For instance, specific quantitative trait loci (QTLs) associated with heat tolerance have been identified, which can be targeted in breeding programs (Zhao *et al.*, 2019). The integration of HTP technologies enables the identification of resilient traits across diverse environments, optimizing selection processes for stress-resilient varieties (Kissoudis *et al.*, 2016). Functional traits derived from WGD, such as improved root architecture and leaf morphology, enhance resource-use efficiency, allowing crops to thrive in suboptimal conditions. Trait stacking, which combines multiple beneficial traits, has been shown to improve adaptability and yield across varying environmental conditions (Chaudhary *et al.*, 2022). It increases genetic variability, which is crucial for breeding programs aimed at enhancing traits such as disease resistance and yield potential. For instance, the study on buckwheat genotypes highlights significant genetic variability that can be harnessed for disease resistance and improved agronomic traits (Koshariya *et al.*, 2024). Ancient WGDs have led to the emergence of novel traits that facilitate adaptation to diverse environments. This is evident in economically important plant families, where WGDs correlate with species diversification and the development of traits that enhance survival under changing conditions (Hofberger *et al.*, 2014). It also contributes to phenotypic plasticity, allowing crops to adjust their growth and development in response to environmental changes. This adaptability is essential for maintaining productivity in fluctuating climates (Laitinen, 2024).

WGD often leads to increased floral morphology size, which can enhance pollen transfer efficiency, potentially improving reproductive success and yield. However, WGD may also result in decreased reproductive output, indicating a complex relationship between WGD-derived traits and crop yield (Porturas *et al.*, 2019). Metabolic engineering leverages WGD to enhance photosynthetic efficiency and stress tolerance, directly contributing to improved crop yield and quality. By manipulating metabolic pathways, researchers can optimize traits that are critical for crop performance, addressing global food security challenges (Fan *et al.*, 2018). While WGD provides a foundation for resilience through genetic diversity and novel trait development, the interplay of these traits with environmental factors and breeding practices is crucial for realizing their full potential in crop resilience. They also increase genetic diversity in breeding stress-resistant crops by adding more copies of genes. This helps genetic engineers create biotic stress-resistant cultivars (Getachew *et al.*, 2021). This multifaceted approach integrates advanced genomic technologies, sustainable practices, and innovative agricultural systems, which is particularly important as drought stress is a major threat to crop yields, and genomic editing techniques like CRISPR can optimize these traits effectively (Rai *et al.*, 2023).

9.6.3 Exploiting pangenome diversity

Pangenome diversity significantly enhances genetic research outcomes by providing a comprehensive view of genomic variation across populations. This diversity allows for improved disease discovery, better understanding of complex traits, and more accurate variant interpretation (Williamson and Fatumo, 2024). WGD imparts wide-ranging and significant evolutionary repercussions. Gene duplication is one of the main outcomes, and it supplies the building blocks for evolutionary innovation. Pseudogenization (loss of function), NF (acquisition of additional functions), and SF (dividing the original function to duplicates) are the potential results of duplicated genes. These mechanisms can result in the emergence of novel characteristics and adaptations as well as functional diversification (Shen *et al.*, 2024). Genomic data from diverse populations can lead to a deeper understanding of complex diseases, as it captures variations that may be population specific (Williamson and Fatumo, 2024). Traditional reference genomes often fail to represent the full spectrum of genetic diversity. Pangenomes include both core and variable sequences, revealing population-specific genetic components that influence traits and adaptability (Gong *et al.*, 2023). It extends beyond human genetics, impacting agricultural and microbial research by elucidating evolutionary mechanisms and improving crop performance through a super-pangenome approach (Aggarwal *et al.*, 2022).

The application of pangenome diversity in precision breeding represents a transformative approach in plant genetics, enabling the identification and utilization of genetic variation across species. This method enhances breeding strategies by integrating comprehensive genomic data, which is crucial for crop improvement. Pangenomes encapsulate the complete genetic diversity of a species, allowing breeders to access a broader genetic base than a single reference genome can provide (Hu *et al.*, 2024b). The practical haplotype graph (PHG) facilitates the management of pangenomic data, enabling accurate haplotype reconstruction even from low-density sequencing, which is vital for understanding genetic diversity in crops like maize (Bradbury *et al.*, 2022). Pangenomes encompass a wide array of genes and structural variations, revealing lost alleles during domestication that can be reintroduced into high-yielding varieties through breeding techniques (Naithani *et al.*, 2023b). Studies on crops like tomatoes have shown that wild relatives harbor genes that can significantly boost yields and improve flavor, thus enhancing nutritional quality (Li *et al.*, 2016). Advances in molecular markers and NGS facilitate the identification of candidate genes linked to desirable traits, accelerating breeding programs aimed at increasing yield and nutritional content (Ramesh and Krishna, 2020).

Pangenomic approaches eliminate single-sample bias, ensuring a more representative genomic range, which is essential for breeding programs targeting specific traits (Munir *et al.*, 2020). The integration of multi-omics data into pangenome databases enhances the identification of genetic markers linked to desirable traits, thus streamlining the breeding process (Hu *et al.*, 2024a). While pangenomics offers significant advantages, challenges remain in computational management and data integration, which must be addressed to fully realize its potential in precision breeding. The on-going development of pangenomic resources promises to revolutionize breeding by providing insights into complex traits and adaptation mechanisms, potentially leading to more resilient crop varieties (Aggarwal *et al.*, 2022). The implications of WGD for agricultural plants are explained in Fig. 9.2.

9.7 Challenges and Future Directions

9.7.1 Challenges in studying WGD and pangenomes

While the potential of WGD-derived traits is substantial, challenges such as consumer acceptance and the complexity of trait interactions with environmental variables remain significant hurdles in their application (Naqvi *et al.*, 2022).

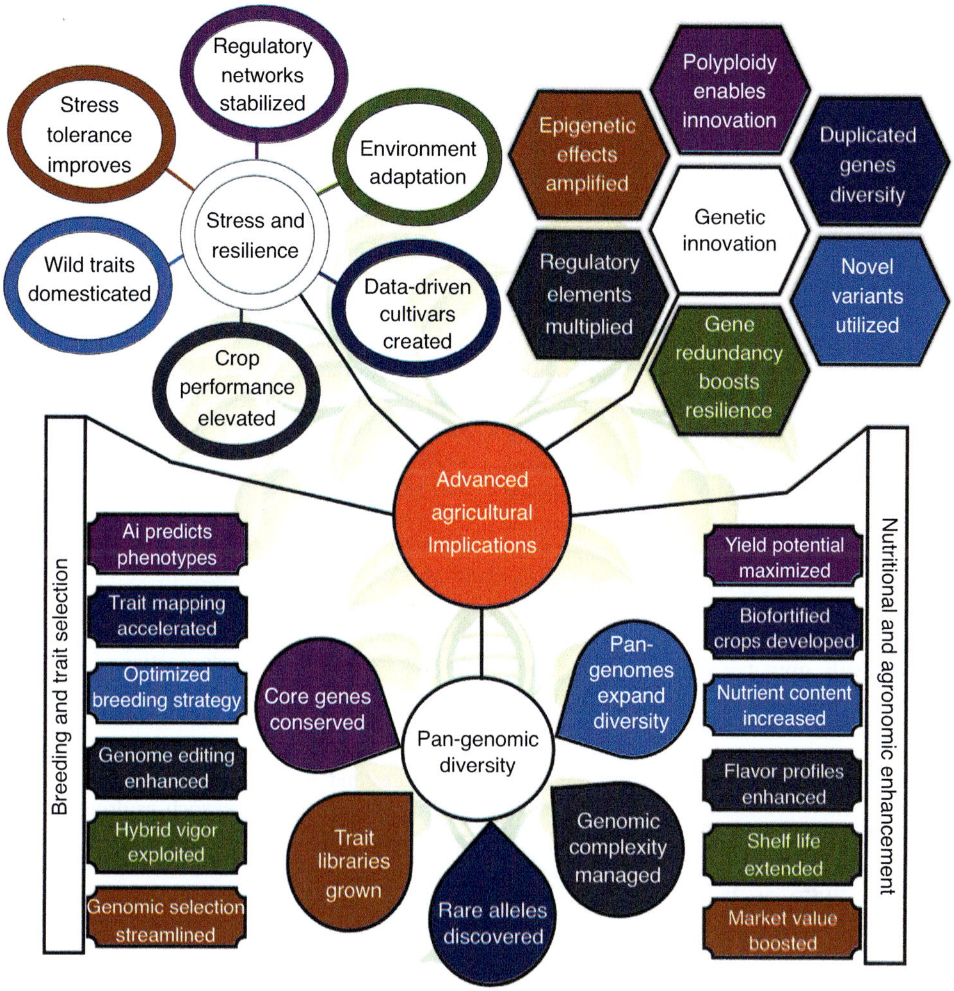

Fig. 9.2. Advanced agricultural implications. (Author's own image.)

Studying WGD and pangenomes presents significant technical and computational challenges that researchers must navigate. These challenges stem from the complexity of genomic data and the limitations of current analytical methods. Current pangenome formats struggle to efficiently represent shared evolutionary histories, leading to storage inefficiencies and scaling issues. The proposed Pan-MAN structure offers a solution by achieving substantial compression (5.2- to 680-fold) while maintaining essential genetic information, enhancing the scope of pangenomic analyses (Walia *et al.*, 2024a). Traditional genotyping methods introduce reference biases and are computationally intensive, particularly with short-read sequencing. The Pan-Genie algorithm improves efficiency, being over four times faster than mapping-based approaches, and better at identifying large insertions and variants in repetitive regions (Ebler *et al.*, 2022). Despite these advancements, the inherent complexity of pangenomic data and the evolving nature of genomic research continue to pose challenges, necessitating ongoing innovation in analytical tools and methodologies. Analyzing complex genomic regions remains challenging, particularly for repetitive and clinically relevant genes. The Pan-Genome Research Tool Kit (PGR-TK) addresses this by enabling multiscale analysis of structural variations (Chin *et al.*, 2023).

9.7.2 Emerging AI technologies for genomic research

Emerging AI technologies significantly enhance the accuracy and efficiency of plant genome assembly through advanced methodologies and innovative sequencing techniques. For instance, the deep learning-based method AutoHiC automates chromosome-level genome assembly by utilizing high-throughput chromosome conformation capture (Hi-C) data, achieving over 90% error detection accuracy and improving genome contiguity (Jiang *et al.*, 2024). Additionally, ultra-long read sequencing technologies have revolutionized the assembly process, enabling gapless, telomere-to-telomere (T2T) genome assemblies that provide high genomic accuracy and integrity, particularly beneficial for complex and polyploid plant genomes (Zhou *et al.*, 2022). Furthermore, the integration of bioinformatics tools, such as the Canu assembler and syntenic scaffolding, allows for the assembly of complex genomes with unmatched completeness and correctness, facilitating further genomic research (Ferguson *et al.*, 2022).

9.7.3 Future research directions

Integrating WGD studies with omics technologies enhances our understanding of genetic variations and their implications in various biological contexts. This integration facilitates the exploration of complex interactions within biological systems, leading to significant advancements in genomics. Comprehensive analysis using omics technologies, including genomics, transcriptomics, and proteomics, provide a multi-faceted view of the molecular landscape, crucial for understanding the effects of WGD on gene function and evolution (Quiroz, 2024). The WGD detector pipeline allows for the detection of WGD events using both genome and transcriptome annotations, making it accessible for organisms with less well-assembled genomes (Yang *et al.*, 2019). The integration of multi-omics data can be complex, often requiring robust computational tools to manage high-dimensional data (Ramos-Lopez *et al.*, 2022). While multi-omics can enhance the interpretation of findings, it does not replace the need for larger sample sizes in studies, as shown in brain-related traits (Baranger *et al.*, 2022). While integrating WGD studies with omics technologies presents exciting opportunities for discovery, challenges in data integration and sample size must be addressed to fully realize their potential in advancing genomics research.

The integration of synthetic biology and genome editing, particularly through WGD traits, presents significant prospects for enhancing crop performance. These advanced biotechnological approaches enable precise modifications that can address the challenges of food security and climate change. Genome editing technologies, such as CRISPR and TALENs, allow for targeted modifications of gene families crucial for traits like yield and disease resistance, overcoming limitations of traditional breeding methods. These techniques enhance the adaptability of crops to abiotic and biotic stresses, thereby improving overall productivity (Ruhil *et al.*, 2024). Synthetic biology expands the capabilities of genetic engineering by enabling the introduction of multiple genes in a single event, facilitating complex trait development that traditional methods cannot achieve. This approach can create novel genetic variations and improve crop resilience against environmental challenges, such as drought and pests (Sargent *et al.*, 2022).

The application of ML in genome editing optimizes editing efficiency and specificity, further accelerating crop improvement efforts. By enhancing the design of guide RNAs and key site detection, ML bridges the gap between genome editing and practical breeding applications. While these advancements hold great promise, challenges remain in the regulatory landscape and public acceptance of genetically modified organisms. Continued research and dialogue are essential to navigate these complexities and fully realize the potential of synthetic biology and genome editing in agriculture.

9.8 Conclusion

Increasing genetic diversity, adaptation, and evolutionary innovation in plants and animals depends on WGD and pangenomics. WGD encourages the development of novel traits and increases genetic diversity in crops, significantly enhancing their ability to withstand stress and adapt. This process leads to increased gene family size, functional diversity, and enhanced stress tolerance. It involves processes like autopolyploid and allopolyploid. For precision breeding applications to be effective, WGD presents additional difficulties, such as complex regulatory systems and genomic instability. WGD is enhanced by pangenomics, which highlights genetic variability within species and sheds light on how genetic variety fosters adaptation. To fully understand the complex functions of polyploidy and gene duplication in plant evolution, advances in genomic and proteomic technology, together with artificial AI and ML, are essential. AI-driven techniques provide a greater knowledge of the influence of WGD on genomic architecture and evolutionary potential by analyzing complicated datasets, predicting gene function, and modeling the evolutionary effects of WGD. Notwithstanding these technical developments, there are still obstacles to overcome before WGD and pangenomics can be effectively included into breeding plans. These obstacles include data complexity, computational administration, and regulatory issues. By facilitating effective data processing, spotting trends in genetic databases, and forecasting desired features for breeding, AI and ML algorithms provide game-changing solutions.

References

Adams, K.L. and Wendel, J.F. (2005) Polyploidy and genome evolution in plants. *Current Opinion in Plant Biology* 8(2), 135–141. DOI: 10.1016/j.pbi.2005.01.001.

Aggarwal, S.K., Singh, A., Choudhary, M., Kumar, A., Rakshit, S. *et al.* (2022) Pangenomics in microbial and crop research: Progress, applications, and perspectives. *Genes* 13(4), 598. DOI: 10.3390/genes13040598.

Akama, S., Shimizu-Inatsugi, R., Shimizu, K.K. and Sese, J. (2014) Genome-wide quantification of homeolog expression ratio revealed nonstochastic gene regulation in synthetic allopolyploid Arabidopsis. *Nucleic Acids Research* 42(6), e46–e46. DOI: 10.1093/nar/gkt1376.

Alonso, C., Balao, F., Bazaga, P. and Pérez, R. (2016) Epigenetic contribution to successful polyploidizations: Variation in global cytosine methylation along an extensive ploidy series in *Dianthus broteri* (Caryophyllaceae). *The New Phytologist* 212(3), 571–576. DOI: 10.1111/nph.14138.

Aubin, E., El Baidouri, M. and Panaud, O. (2021) Horizontal gene transfers in plants. *Funct Integr* 11(8), 857. DOI: 10.3390/life11080857.

Aversano, R., Ercolano, M.R., Caruso, I., Fasano, C., Rosellini, D. *et al.* (2012) Molecular tools for exploring polyploid genomes in plants. *International Journal of Molecular Sciences* 13(8), 10316–10335. DOI: 10.3390/ijms130810316.

Bachu, L., Kandibanda, A., Grandhi, N., Athina, D.P. and Kumar Ande, P. (2024) Machine learning for enhanced crop management and optimization of yield in precision agriculture. In: *2024 8th International Conference on I-SMAC (IoT in Social, Mobile, Analytics and Cloud) (I-SMAC)*, IEEE, Kirtipur, Nepal, pp. 1289–1293. DOI: 10.1109/I-SMAC61858.2024.10714733.

Baduel, P., Bray, S., Vallejo-Marin, M., Kolář, F. and Yant, L. (2018) The "polyploid hop": Shifting challenges and opportunities over the evolutionary lifespan of genome duplications. *Frontiers in Ecology and Evolution* 6, 117. DOI: 10.3389/fevo.2018.00117.

Baranger, D.A., Hatoum, A.S., Polimanti, R., Gelernter, J., Edenberg, H.J. *et al.* (2022) Multi-omics analyses cannot identify true-positive novel associations from underpowered genome-wide association studies of four brain-related traits. *BioRxiv* 2022–04. DOI: 10.1101/2022.04.13.487655.

Barh, D., Soares, S.C., Tiwari, S. and Azevedo, V.A.D. (eds) (2020) Pan-genomics: Applications, challenges, and future prospects.

Batista, L.G., Mello, V.H., Souza, A.P. and Margarido, G.R.A. (2022) Genomic prediction with allele dosage information in highly polyploid species. *Theoretical and Applied Genetics* 135(2), 723–739. DOI: 10.1007/s00122-021-03994-w.

Bekaert, M., Edger, P.P., Pires, J.C. and Conant, G.C. (2011) Two-phase resolution of polyploidy in the *Arabidopsis* metabolic network gives rise to relative and absolute dosage constraints. *The Plant Cell* 23(5), 1719–1728. DOI: 10.1105/tpc.110.081281.

Benegas, G., Batra, S.S. and Song, Y.S. (2023) DNA language models are powerful predictors of genome-wide variant effects. *Proceedings of the National Academy of Sciences* 120(44), e2311219120. DOI: 10.1073/pnas.2311219120.

Bohutínská, M., Alston, M., Monnahan, P., Mandáková, T., Bray, S. *et al.* (2021) Novelty and convergence in adaptation to whole genome duplication. *Molecular Biology and Evolution* 38(9), 3910–3924. DOI: 10.1093/molbev/msab096.

Bomblies, K. (2020) When everything changes at once: Finding a new normal after genome duplication. *Proceedings Biological Sciences* 287(1939), 20202154. DOI: 10.1098/rspb.2020.2154.

Bourke, P.M., Voorrips, R.E., Visser, R.G.F. and Maliepaard, C. (2018) Tools for genetic studies in experimental populations of polyploids. *Frontiers in Plant Science* 9, 513. DOI: 10.3389/fpls.2018.00513.

Bradbury, P.J., Casstevens, T., Jensen, S.E., Johnson, L.C., Miller, Z.R. *et al.* (2022) The practical haplotype graph, a platform for storing and using pangenomes for imputation 38(15), 3698–3702. DOI: 10.1093/bioinformatics/btac410.

Burgess, D.J. (2015) Polyploid gains. *Nature Reviews Genetics* 16(4), 196–196. DOI: 10.1038/nrg3925.

Chaudhary, P., Singh, S., Chaudhary, A., Sharma, A. and Kumar, G. (2022) Overview of biofertilizers in crop production and stress management for sustainable agriculture. *Frontiers in Plant Science* 13, 930340. DOI: 10.3389/fpls.2022.930340.

Chen, Z.J. (2007) Genetic and epigenetic mechanisms for gene expression and phenotypic variation in plant polyploids. *Annual Review of Plant Biology* 58(1), 377–406. DOI: 10.1146/annurev.arplant.58.032806.103835.

Chen, Z.J. and Ni, Z. (2006) Mechanisms of genomic rearrangements and gene expression changes in plant polyploids. *BioEssays* 28(3), 240–252. DOI: 10.1002/bies.20374.

Chen, C., Powell, O., Dinglasan, E., Ross, E.M., Yadav, S. *et al.* (2023) Genomic prediction with machine learning in sugarcane, a complex highly polyploid clonally propagated crop with substantial non-additive variation for key traits. *The Plant Genome* 16(4), e20390. DOI: 10.1002/tpg2.20390.

Chen, L., Liu, G. and Zhang, T. (2024) Integrating machine learning and genome editing for crop improvement. *Abiotech* 5(2), 262–277. DOI: 10.1007/s42994-023-00133-5.

Cheng, F., Wu, J. and Wang, X. (2014) Genome triplication drove the diversification of *Brassica* plants. *Horticulture Research* 1, 14024. DOI: 10.1038/hortres.2014.24.

Chin, C.S., Behera, S., Khalak, A., Sedlazeck, F.J., Sudmant, P.H. *et al.* (2023) Multiscale analysis of pangenomes enables improved representation of genomic diversity for repetitive and clinically relevant genes. *Nature Methods* 20(8), 1213–1221. DOI: 10.1038/s41592-023-01914-y.

Clark, J.W., Puttick, M.N. and Donoghue, P.C.J. (2019) Origin of horsetails and the role of whole-genome duplication in plant macroevolution. *Proceedings. Biological Sciences* 286(1914), 20191662. DOI: 10.1098/rspb.2019.1662.

Clo, J. (2022) Polyploidization: Consequences of genome doubling on the evolutionary potential of populations. *American Journal of Botany* 109(8), 1213–1220. DOI: 10.1002/ajb2.16029.

Clo, J. and Kolář, F. (2021) Short- and long-term consequences of genome doubling: A meta-analysis. *American Journal of Botany* 108(11), 2315–2322. DOI: 10.1002/ajb2.1759.

Crossa, J., Montesinos-Lopez, O.A., Costa-Neto, G., Vitale, P., Martini, J.W.R. *et al.* (2025) Machine learning algorithms translate big data into predictive breeding accuracy. *Trends in Plant Science* 30(2), 167–184. DOI: 10.1016/j.tplants.2024.09.011.

Danilevicz, M.F., Gill, M., Anderson, R., Batley, J., Bennamoun, M. *et al.* (2022) Plant genotype to phenotype prediction using machine learning. *Frontiers in Genetics* 13, 822173. DOI: 10.3389/fgene.2022.822173.

de Maagd, R.A., Wiel, C.C.M. and Schouten, H.J. (2020) *The Plasticity of Plant Genomes: Causes and Consequences: A Survey of Data on Structural Genome Variation in Plants (No.CGM 2020-04)*. Cogem.

De Smet, R., Sabaghian, E., Li, Z., Saeys, Y. and Peer, Y. (2017) Coordinated functional divergence of genes after genome duplication in *Arabidopsis thaliana* the plant cell 29(11), 2786–2800. DOI: 10.1105/tpc.17.00531.

De Storme, N. and Mason, A. (2014) Plant speciation through chromosome instability and ploidy change: Cellular mechanisms, molecular factors and evolutionary relevance. *Current Plant Biology* 1, 10–33. DOI: 10.1016/j.cpb.2014.09.002.

Donoghue, M.T., Keshavaiah, C., Swamidatta, S.H. and Spillane, C. (2011) Evolutionary origins of *Brassicaceae* specific genes in *Arabidopsis thaliana*. *BMC Evolutionary Biology* 11, 1–23. DOI: 10.1186/1471-2148-11-47.

Doyle, J.J., Flagel, L.E., Paterson, A.H., Rapp, R.A., Soltis, D.E. *et al.* (2008) Evolutionary genetics of genome merger and doubling in plants. *Annual Review of Genetics* 42(1), 443–461. DOI: 10.1146/annurev.genet.42.110807.091524.

Ebler, J., Ebert, P., Clarke, W.E., Rausch, T., Audano, P.A. *et al.* (2022) Pangenome-based genome inference allows efficient and accurate genotyping across a wide spectrum of variant classes. *Nature Genetics* 54(4), 518–525. DOI: 10.1038/s41588-022-01043-w.

Edger, P.P. and Pires, J.C. (2009) Gene and genome duplications: The impact of dosage-sensitivity on the fate of nuclear genes. *Chromosome Research* 17(5), 699–717. DOI: 10.1007/s10577-009-9055-9.

Eftekhari, M., Ma, C. and Orlov, Y.L. (2024) Editorial: Applications of artificial intelligence, machine learning, and deep learning in plant breeding. *Frontiers in Plant Science* 15, 1420938. DOI: 10.3389/fpls.2024.1420938.

Fan, X., Bi, Y., Zhang, Y., Jeffers, D., Yin, X. *et al.* (2018) Improving breeding efficiency of a hybrid maize breeding program using a three heterotic-group classification. *Agronomy Journal* 110(4), 1209–1216. DOI: 10.2134/agronj2017.05.0290.

Farooq, M.A., Gao, S., Hassan, M.A., Huang, Z., Rasheed, A. *et al.* (2024) Artificial intelligence in plant breeding. *Trends in Genetics* 40(10), 891–908. DOI: 10.1016/j.tig.2024.07.001.

Felsenstein, J. (1985) Phylogenies and the comparative method. *The American Naturalist* 125(1), 1–15. DOI: 10.1086/284325.

Feng, W., Gao, P. and Wang, X. (2024) AI breeder: Genomic predictions for crop breeding. *New Crops* 1, 100010. DOI: 10.1016/j.ncrops.2023.12.005.

Ferguson, S., Jones, A. and Borevitz, J. (2022) Plant assemble-plant de novo genome assembly: Quality assessment. DOI: 10.17504/protocols.io.e6nvw578zvmk/v2. (accessed 4 August 2025).

Feussner, I. and Polle, A. (2015) What the transcriptome does not tell - proteomics and metabolomics are closer to the plants' patho-phenotype. *Current Opinion in Plant Biology* 26, 26–31. DOI: 10.1016/j.pbi.2015.05.023.

Frantzeskakis, L., Di Pietro, A., Rep, M., Schirawski, J., Wu, C.-H. *et al.* (2020) Rapid evolution in plant-microbe interactions - a molecular genomics perspective. *The New Phytologist* 225(3), 1134–1142. DOI: 10.1111/nph.15966.

Getachew, G., Abebe, B., Chelchisa, D., Oli, S., Chebsa, T. *et al.* (2021) *Genotype by Environment Interaction by AMMI and GGE Bi-Plot Analysis for Maize (Zea Mays L.)*. Transitional High Land Agroecology of Ethiopia.

Glick, L. and Mayrose, I. (2023) The effect of methodological considerations on the construction of gene-based plant pan-genomes. *Genome Biology and Evolution* 15(7), evad121. DOI: 10.1093/gbe/evad121.

Gong, Y., Li, Y., Liu, X., Ma, Y. and Jiang, L. (2023) A review of the pangenome: How it affects our understanding of genomic variation, selection and breeding in domestic animals? *Journal of Animal Science and Biotechnology* 14(1), 73. DOI: 10.1186/s40104-023-00860-1.

González-Rodríguez, V.E., Izquierdo-Bueno, I., Cantoral, J.M., Carbú, M. and Garrido, C. (2024) Artificial intelligence: A promising tool for application in phytopathology. *Horticulturae* 10(3), 197. DOI: 10.3390/horticulturae10030197.

Griffiths, S., Sharp, R., Foote, T.N., Bertin, I., Wanous, M. *et al.* (2006) Molecular characterization of Ph1 as a major chromosome pairing locus in polyploid wheat. *Nature* 439(7077), 749–752. DOI: 10.1038/nature04434.

Gul, D. and Banday, R.U.Z. (2024) *Transforming Crop Management Through Advanced AI and Machine Learning: Insights into Innovative Strategies for Sustainable Agriculture*. AI, Computer Science and Robotics Technology. DOI: 10.5772/acrt.20240030.

Guo, H., Lee, T.H., Wang, X. and Paterson, A.H. (2013) Function relaxation followed by diversifying selection after whole-genome duplication in flowering plants. *Plant Physiology* 162(2), 769–778. DOI: 10.1104/pp.112.213447.

Guo, Z., Zhang, K., Cai, C., Li, X., Zhang, L. *et al.* (2024) Deep learning can predict subgenome dominance in ancient but not in neo/synthetic polyploidized genomes. *The Plant Journal* 120(1), 174–186. DOI: 10.1111/tpj.16979.

Haider, S. and Pal, R. (2013) Integrated analysis of transcriptomic and proteomic data. *Current Genomics* 14(2), 91–110. DOI: 10.2174/1389202911314020003.

Hanada, K., Zou, C., Lehti-Shiu, M.D., Shinozaki, K. and Shiu, S.H. (2008) Importance of lineage-specific expansion of plant tandem duplicates in the adaptive response to environmental stimuli. *Plant Physiology* 148(2), 993–1003. DOI: 10.1104/pp.108.122457.

He, X. and Zhang, J. (2005) Rapid subfunctionalization accompanied by prolonged and substantial neofunctionalization in duplicate gene evolution. *Genetics* 169(2), 1157–1164. DOI: 10.1534/genetics.104.037051.

Hegarty, M.J. and Hiscock, S.J. (2008) Genomic clues to the evolutionary success of polyploid plants. *Current Biology* 18(10), R435–R444. DOI: 10.1016/j.cub.2008.03.043.

Hofberger, J.A., Zhou, B., Tang, H., Jones, J.D.G. and Schranz, M.E. (2014) A novel approach for multi-domain and multi-gene family identification provides insights into evolutionary dynamics of disease resistance genes in core eudicot plants. *BMC Genomics* 15(1), 966. DOI: 10.1186/1471-2164-15-966.

Holland, J.B. (2007) Genetic architecture of complex traits in plants. *Current Opinion in Plant Biology* 10(2), 156–161. DOI: 10.1016/j.pbi.2007.01.003.

Hong, S., Lim, Y.P., Kwon, S.Y., Shin, A.Y. and Kim, Y.M. (2021) Genome-wide comparative analysis of flowering-time genes; insights on the gene family expansion and evolutionary perspective. *Frontiers in Plant Science* 12, 702243. DOI: 10.3389/fpls.2021.702243.

Hou, M. and Pang, S. (2024) Plant pan-genomics: Opportunities, advances, and challenges. *Journal of Data Science and Intelligent Systems*. DOI: 10.47852/bonviewJDSIS42023107.

Hu, H., Li, R., Zhao, J., Batley, J. and Edwards, D. (2024a) Technological development and advances for constructing and analyzing plant pangenomes. *Genome Biology and Evolution* 16(4), evae081. DOI: 10.1093/gbe/evae081.

Hu, H., Wang, J., Nie, S., Zhao, J., Batley, J. *et al.* (2024b) Plant pangenomics, current practice and future direction. *Agriculture Communications* 2(2), 100039. DOI: 10.1016/j.agrcom.2024.100039.

Huang, C.H., Qi, X., Chen, D., Qi, J. and Ma, H. (2020a) Recurrent genome duplication events likely contributed to both the ancient and recent rise of ferns. *Journal of Integrative Plant Biology* 62(4), 433–455. DOI: 10.1111/jipb.12877.

Huang, X.C., German, D.A. and Koch, M.A. (2020b) Temporal patterns of diversification in *Brassicaceae* demonstrate decoupling of rate shifts and mesopolyploidization events. *Annals of Botany* 125(1), 29–47. DOI: 10.1093/aob/mcz123.

Huang, Y., Rodriguez-Granados, N.Y., Latrasse, D., Raynaud, C., Benhamed, M. *et al.* (2020c) The matrix revolutions: Towards the decoding of the plant chromatin three-dimensional reality. *Journal of Experimental Botany* 71(17), 5129–5147. DOI: 10.1093/jxb/eraa322.

Hufton, A.L. and Panopoulou, G. (2009) Polyploidy and genome restructuring: A variety of outcomes. *Current Opinion in Genetics & Development* 19(6), 600–606. DOI: 10.1016/j.gde.2009.10.005.

Hutchens, S.J. and DePerno, C.S. (2009) Efficacy of sampling techniques for determining species richness estimates of reptiles and amphibians. *Wildlife Biology* 15(2), 113–122. DOI: 10.2981/08-024.

Islam, M.M., Deepo, D.M., Nasif, S.O., Siddique, A.B., Hassan, O. *et al.* (2022) Cytogenetics and consequences of polyploidization on different biotic-abiotic stress tolerance and the potential mechanisms involved 11(20), 2684. DOI: 10.3390/plants11202684.

Jackson, R.C. and Jackson, J.W. (1996) Gene segregation in autotetraploids: Prediction from meiotic configurations. *American Journal of Botany* 83(6), 673–678. DOI: 10.1002/j.1537-2197.1996.tb12756.x.

Jackson, S. and Chen, Z.J. (2010) Genomic and expression plasticity of polyploidy. *Current Opinion in Plant Biology* 13(2), 153–159. DOI: 10.1016/j.pbi.2009.11.004.

Jakarbet, U.P., Suhas, S.P., Savithramma, R.M. and Ashwini, B.P. (2024) An empirical analysis of machine learning algorithms for agricultural crop prediction. In: *2024 8th International Conference on I-SMAC (IoT in Social, Mobile, Analytics and Cloud) (I-SMAC)*, IEEE, Kirtipur, Nepal, pp. 1782–1790. DOI: 10.1109/I-SMAC61858.2024.10714677.

Jiang, W., Liu, Y., Xia, E. and Gao, L. (2013) Prevalent role of gene features in determining evolutionary fates of whole-genome duplication duplicated genes in flowering plants. *Plant Physiology* 161(4), 1844–1861. DOI: 10.1104/pp.112.200147.

Jiang, Z., Peng, Z., Wei, Z., Sun, J., Luo, Y. *et al.* (2024) A deep learning-based method enables the automatic and accurate assembly of chromosome-level genomes. *Nucleic Acids Research* 52(19), e92–e92. DOI: 10.1093/nar/gkae789.

Kagale, S., Robinson, S.J., Nixon, J., Xiao, R., Huebert, T. *et al.* (2014) Polyploid evolution of the *Brassicaceae* during the Cenozoic era. *The Plant Cell* 26(7), 2777–2791. DOI: 10.1105/tpc.114.126391.

Kaufmann, K., Smaczniak, C., de Vries, S., Angenent, G.C. and Karlova, R. (2011) Proteomics insights into plant signaling and development. *Proteomics* 11(4), 744–755. DOI: 10.1002/pmic.201000418.

Kim, H.Y. and Kim, D. (2020) Prediction of mutation effects using a deep temporal convolutional network. *Bioinformatics (Oxford, England)* 36(7), 2047–2052. DOI: 10.1093/bioinformatics/btz873.

Kissoudis, C., van de Wiel, C., Visser, R.G. and van der Linden, G. (2016) Future-proof crops: Challenges and strategies for climate resilience improvement. *Current Opinion in Plant Biology* 30, 47–56. DOI: 10.1016/j.pbi.2016.01.005.

Koshariya, A.K., Satkar, K. and Rawte, S. (2024) Novel biotechnological strategies to boost disease resistance and enhance quality parameters in horticultural plants. *Plant Cell Biotechnology and Molecular Biology* 25(7–8), 79–100. DOI: 10.56557/pcbmb/2024/v25i7-88744.

Ku, H.M., Vision, T., Liu, J. and Tanksley, S.D. (2000) Comparing sequenced segments of the tomato and Arabidopsis genomes: Large-scale duplication followed by selective gene loss creates a network of synteny. *Proceedings of the National Academy of Sciences* 97(16), 9121–9126. DOI: 10.1073/pnas.160271297.

Laitinen, R.A.E. (2024) Importance of phenotypic plasticity in crop resilience. *Journal of Experimental Botany* 75(3), 670–673. DOI: 10.1093/jxb/erad465.

Lallemand, T., Aubourg, S., Hunault, G., Celton, J.M. and Landès, C. (2020) Evolution of ohnologous chromosomes following whole genome duplication in apple. In: *JournéesOuvertes de Biologie, Informatique et Mathématique*.

Landergott, U., Naciri, Y., Schneller, J.J. and Holderegger, R. (2006) Allelic configuration and polysomic inheritance of highly variable microsatellites in tetraploid gynodioecious *Thymus praecox* agg. *Theoretical and Applied Genetics* 113(3), 453–465. DOI: 10.1007/s00122-006-0310-6.

Leitch, A.R. and Leitch, I.J. (2008) Genomic plasticity and the diversity of polyploid plants. *Science* 320(5875), 481–483. DOI: 10.1126/science.1153585.

Leitch, I.J. and Bennett, M.D. (1997) Polyploidy in angiosperms. *Trends in Plant Science* 2(12), 470–476. DOI: 10.1016/S1360-1385(97)01154-0.

Levin, D.A. (1983) Polyploidy and novelty in flowering plants. *The American Naturalist* 122(1), 1–25. DOI: 10.1086/284115.

Li, J.T., Hou, G.Y., Kong, X.F., Li, C.Y., Zeng, J.M. *et al.* (2015) The fate of recent duplicated genes following a fourth-round whole genome duplication in a tetraploid fish, common carp (*Cyprinus carpio*). *Scientific Reports* 5(1), 8199. DOI: 10.1038/srep08199.

Li, X., Zeng, R. and Liao, H. (2016) Improving crop nutrient efficiency through root architecture modifications. *Journal of Integrative Plant Biology* 58(3), 193–202. DOI: 10.1111/jipb.12434.

Li, Z., Lu, X., Gao, Y., Liu, S., Tao, M. *et al.* (2011) Polyploidization and epigenetics. *Chinese Science Bulletin* 56(3), 245–252. DOI: 10.1007/s11434-010-4290-1.

Lisch, D. (2013) How important are transposons for plant evolution? *Nature Reviews. Genetics* 14(1), 49–61. DOI: 10.1038/nrg3374.

Mabry, M.E., Brose, J.M., Blischak, P.D., Sutherland, B., Dismukes, W.T. *et al.* (2020) Phylogeny and multiple independent whole-genome duplication events in the brassicales. *American Journal of Botany* 107(8), 1148–1164. DOI: 10.1002/ajb2.1514.

Mandáková, T., Li, Z., Barker, M.S. and Lysak, M.A. (2017) Diverse genome organization following 13 independent mesopolyploid events in *Brassicaceae* contrasts with convergent patterns of gene retention. *The Plant Journal* 91(1), 3–21. DOI: 10.1111/tpj.13553.

Martins, E.P. and Hansen, T.F. (1997) Phylogenies and the comparative method: A general approach to incorporating phylogenetic information into the analysis of interspecific data. *The American Naturalist* 149(4), 646–667. DOI: 10.1086/286013.

Mendoza-Revilla, J., Trop, E., Gonzalez, L., Roller, M., Dalla-Torre, H. *et al.* (2024) A foundational large language model for edible plant genomes. *Communications Biology* 7(1), 835. DOI: 10.1038/s42003-024-06465-2.

Miri, S.M. (2020) Artificial polyploidy in the improvement of horticultural crops. *Journal of Plant Physiology and Breeding* 10(1), 1–28. DOI: 10.22034/JPPB.2020.12490.

Moriyama, Y. and Koshiba-Takeuchi, K. (2018) Significance of whole-genome duplications on the emergence of evolutionary novelties. *Briefings in Functional Genomics* 17(5), 329–338. DOI: 10.1093/bfgp/ely007.

Munir, F., Saba, N.U., Arveen, M., Siddiqa, A., Ahmad, J. *et al.* (2020) Pan-genomics of plants and its applications. In: *Pan-Genomics: Applications, Challenges, and Future Prospects*. Academic Press, pp. 285–306.

Munir, N., Gulzar, W., Abideen, Z., Hasanuzzaman, M., El-Keblawy, A. *et al.* (2023) Plant–nanoparticle interactions: Transcriptomic and proteomic insights. *Agronomy* 13(8), 2112. DOI: 10.3390/agronomy13082112.

Na, M.H. and Na, I.S. (2024) AI-powered predictive modelling of legume crop yields in a changing climate. *Legume Research - An International Journal* 47(8), 1390–1395. DOI: 10.18805/LRF-790.

Naithani, S., Deng, C.H., Sahu, S.K. and Jaiswal, P. (2023a) Exploring pan-genomes: An overview of resources and tools for unraveling structure, function, and evolution of crop genes and genomes. *Biomolecules* 13(9), 1403. DOI: 10.3390/biom13091403.

Naithani, S., Mohanty, B., Elser, J., D'Eustachio, P. and Jaiswal, P. (2023b) Biocuration of a transcription factors network involved in submergence tolerance during seed germination and coleoptile elongation in rice (*Oryza sativa*) 12(11), 2146. DOI: 10.3390/plants12112146.

Naqvi, R.Z., Siddiqui, H.A., Mahmood, M.A., Najeebullah, S., Ehsan, A. *et al.* (2022) Smart breeding approaches in post-genomics era for developing climate-resilient food crops. *Frontiers in Plant Science* 13, 972164. DOI: 10.3389/fpls.2022.972164.

Oren, E., Dafna, A., Tzuri, G., Halperin, I., Isaacson, T. *et al.* (2022) Melon pan-genome and multi-parental framework for high-resolution trait dissection. *BioRxiv*. DOI: 10.1101/2022.08.09.503186.

Otto, S.P. and Whitton, J. (2000) Polyploid incidence and evolution. *Annual Review of Genetics* 34(1), 401–437. DOI: 10.1146/annurev.genet.34.1.401.

Ouyang, W., Xiong, D., Li, G. and Li, X. (2020) Unraveling the 3D genome architecture in plants: Present and future. *Molecular Plant* 13(12), 1676–1693. DOI: 10.1016/j.molp.2020.10.002.

Padilla-García, N., Šrámková, G., Záveská, E., Šlenker, M., Clo, J. *et al.* (2022) Niche differentiation following whole-genome duplication? The importance of considering the intricate evolutionary history of polyploids when assessing climatic niche evolution. *BioRxiv*. DOI: 10.1101/2022.02.16.480351.

Parisod, C., Holderegger, R. and Brochmann, C. (2010) Evolutionary consequences of autopolyploidy. *New Phytologist* 186(1), 5–17. DOI: 10.1111/j.1469-8137.2009.03142.x.

Paterson, A.H. (2005) Polyploidy, evolutionary opportunity, and crop adaptation. In: *Genetics of Adaptation*. pp. 191–196.

Pellicer, J., Hidalgo, O., Dodsworth, S. and Leitch, I.J. (2018) Genome size diversity and its impact on the evolution of land plants. *Genes* 9(2), 88. DOI: 10.3390/genes9020088.

Pérez, R., Jouve, N. and Bustos, A. (2014) Comparative analysis of gene expression among species of different ploidy. *Molecular Biology Reports* 41(10), 6525–6535. DOI: 10.1007/s11033-014-3536-4.

Porturas, L.D., Anneberg, T.J., Curé, A.E., Wang, S., Althoff, D.M. *et al.* (2019) A meta-analysis of whole genome duplication and the effects on flowering traits in plants. *American Journal of Botany* 106(3), 469–476. DOI: 10.1002/ajb2.1258.

Qiao, X., Zhang, S. and Paterson, A.H. (2022) Pervasive genome duplications across the plant tree of life and their links to major evolutionary innovations and transitions. *Computational and Structural Biotechnology Journal* 20, 3248–3256. DOI: 10.1016/j.csbj.2022.06.026.

Qiu, T., Liu, Z. and Liu, B. (2020) The effects of hybridization and genome doubling in plant evolution via allopolyploidy. *Molecular Biology Reports* 47(7), 5549–5558. DOI: 10.1007/s11033-020-05597-y.

Quiroz, I.V. (2024) Exploring the intersection of omics technologies and biotechnology in drug interaction studies. In: *Biotechnology and Drug Development for Targeting Human Diseases*, Vol. 9. Bentham Books, Singapore, pp. 188–203. DOI: 10.2174/9789815223163124090011.

Rai, S., Omar, A.F., Rehan, M., Al-Turki, A., Sagar, A. *et al.* (2023) Crop microbiome: Their role and advances in molecular and omic techniques for the sustenance of agriculture. *Planta* 257(2), 27. DOI: 10.1007/s00425-022-04052-5.

Ramesh, S. and Krishna, T.V. (2020) Finger millet (*Eleusine coracana* (L.) Gaertn.) genetics and breeding for rapid genetic gains. *Accelerated Plant Breeding Cereal Crops* 1, 365–390. DOI: 10.1007/978-3-030-41866-3_14.

Ramos-Lopez, O., Martinez, J.A. and Milagro, F.I. (2022) Holistic integration of omics tools for precision nutrition in health and disease. *Nutrients* 14(19), 4074. DOI: 10.3390/nu14194074.

Rehman, R.S., Zafar, S.A., Ali, M., Ahmad, M., Pasha, A.N. *et al.* (2022) Plant pan-genomes: A new frontier in understanding genomic diversity in plants. *Journal of Advances in Biology & Biotechnology* 25(1), 10–22. DOI: 10.9734/jabb/2022/v25i130260.

Richardson, A.O. and Palmer, J.D. (2007) Horizontal gene transfer in plants. *Journal of Experimental Botany* 58(1), 1–9. DOI: 10.1093/jxb/erl148.

Ruhil, T., Singh, H. and Barthwal, S. (2024) Genetic engineering and gene editing for targeted trait modifications. In: *Plant Functional Traits for Improving Productivity*. Springer Nature Singapore, Singapore, pp. 217–238.

Salma, U., Kundu, S. and Mandal, N. (2017) Artificial polyploidy in medicinal plants: Advancement in the last two decades and impending prospects. *Journal of Crop Science and Biotechnology* 20(1), 9–19. DOI: 10.1007/s12892-016-0080-1.

Sargent, D., Conaty, W.C., Tissue, D.T. and Sharwood, R.E. (2022) Synthetic biology and opportunities within agricultural crops. *Journal of Sustainable Agriculture and Environment* 1(2), 89–107. DOI: 10.1002/sae2.12014.

Savaliya, L., Sapovadiya, M., Garg, D., Patel, P. and Shah, M. (2024) Agricultural analysis of machine learning algorithms for crop prediction. In: *2024 8th International Conference on I-SMAC (IoT in Social, Mobile, Analytics and Cloud) (I-SMAC)*, IEEE, Kirtipur, Nepal, pp. 1196–1203. DOI: 10.1109/I-SMAC61858.2024.10714828.

Schneider, H. (2023) Integrating genomics and conservation to safeguard plant diversity. *Integrative Conservation* 2(1), 10–18. DOI: 10.1002/inc3.15.

Selmecki, A.M., Maruvka, Y.E., Richmond, P.A., Guillet, M., Shoresh, N. *et al.* (2015) Polyploidy can drive rapid adaptation in yeast. *Nature* 519(7543), 349–352. DOI: 10.1038/nature14187.

Sheikh, M., Iqra, F., Ambreen, H., Pravin, K.A., Ikra, M. *et al.* (2024) Integrating artificial intelligence and high-throughput phenotyping for crop improvement. *Journal of Integrative Agriculture* 23(6), 1787–1802. DOI: 10.1016/j.jia.2023.10.019.

Shen, W., Huang, Y., Chen, R. and Lin, H. (2024) Impact of whole genome duplication events on the diversification of legumes. *Legume Genomics and Genetics* 15. DOI: 10.5376/lgg.2024.15.0005.

Shi, T., Rahmani, R.S., Gugger, P.F., Wang, M., Li, H. *et al.* (2020) Distinct expression and methylation patterns for genes with different fates following a single whole-genome duplication in flowering plants. *Molecular Biology and Evolution* 37(8), 2394–2413. DOI: 10.1093/molbev/msaa105.

Shoer, S., Reicher, L., Zhao, C., Pollard, K.S., Pilpel, Y. *et al.* (2024) Pangenomes of human gut microbiota uncover links between genetic diversity and stress response. *Cell Host & Microbe* 32(10), 1744–1757. DOI: 10.1016/j.chom.2024.08.017.

Sielemann, K., Weisshaar, B. and Pucker, B. (2020) Reference-based Quantification of Gene Dispensability (QUOD)-test Dataset.

Smith, S.A., Brown, J.W., Yang, Y., Bruenn, R., Drummond, C.P. *et al.* (2018) Disparity, diversity, and duplications in the caryophyllales. *New Phytologist* 217(2), 836–854. DOI: 10.1111/nph.14772.

Soltis, P.S. and Soltis, D.E. (2016) Ancient WGD events as drivers of key innovations in angiosperms. *Current Opinion in Plant Biology* 30, 159–165. DOI: 10.1016/j.pbi.2016.03.015.

Soltis, D.E., Soltis, P.S. and Rieseberg, L.H. (1993) Molecular data and the dynamic nature of polyploidy. *Critical Reviews in Plant Sciences* 12(3), 243–273. DOI: 10.1080/07352689309701903.

Soltis, P.S. and Soltis, D.E. (2021) Plant genomes: Markers of evolutionary history and drivers of evolutionary change. *Plants, People, Planet* 3(1), 74–82. DOI: 10.1002/ppp3.10159.

Soucy, S.M., Huang, J. and Gogarten, J.P. (2015) Horizontal gene transfer: Building the web of life. *Nature Reviews. Genetics* 16(8), 472–482. DOI: 10.1038/nrg3962.

Stift, M., Berenos, C., Kuperus, P. and Van Tienderen, P.H. (2008) Segregation models for disomic, tetrasomic and intermediate inheritance in tetraploids: A general procedure applied to Rorippa (yellow cress) microsatellite data. *Genetics* 179(4), 2113–2123. DOI: 10.1534/genetics.107.085027.

Stull, G.W., Qu, X.-J., Parins-Fukuchi, C., Yang, Y.-Y., Yang, J.-B. *et al.* (2021) Gene duplications and phylogenomic conflict underlie major pulses of phenotypic evolution in gymnosperms. *Nature Plants* 7(8), 1015–1025. DOI: 10.1038/s41477-021-00964-4.

Su, W., Jing, Y., Lin, S., Yue, Z., Yang, X. *et al.* (2021) Polyploidy underlies co-option and diversification of biosynthetic triterpene pathways in the apple tribe. *Proceedings of the National Academy of Sciences* 118(20), e2101767118. DOI: 10.1073/pnas.2101767118.

Sun, Y., Wu, Y., Yang, C., Sun, S., Lin, X. *et al.* (2017) Segmental allotetraploidy generates extensive homoeologous expression rewiring and phenotypic diversity at the population level in rice. *Molecular Ecology* 26(20), 5451–5466. DOI: 10.1111/mec.14297.

Upadhyaya, S.R., Danilevicz, M.F., Dolatabadian, A., Neik, T.X., Zhang, F. *et al.* (2024) Genomics-based plant disease resistance prediction using machine learning. *Plant Pathology* 73(9), 2298–2309. DOI: 10.1111/ppa.13988.

Van, K., Kim, D.H., Shin, J.H. and Lee, S.H. (2011) Genomics of plant genetic resources: Past, present and future. *Plant Genetic Resources* 9(2), 155–158. DOI: 10.1017/S1479262111000098.

Voelckel, C., Gruenheit, N. and Lockhart, P. (2017) Evolutionary transcriptomics and proteomics: Insight into plant adaptation. *Trends in Plant Science* 22(6), 462–471. DOI: 10.1016/j.tplants.2017.03.001.

Walia, S., Motwani, H., Smith, K., Corbett-Detig, R. and Turakhia, Y. (2024a) Compressive pangenomics using mutation-annotated networks. *BioRxiv*. DOI: 10.1101/2024.07.02.601807.

Walia, S.S., Dhaliwal, S.S., Gill, R.S., Kaur, T., Kaur, K. *et al.* (2024b) Improvement of soil health and nutrient transformations under balanced fertilization with integrated nutrient management in a rice-wheat system in Indo-Gangetic plains–a 34-year research outcomes. *Heliyon* 10(4), e25113. DOI: 10.1016/j.heliyon.2024.e25113.

Wang, H., Li, Y., Zhang, Z. and Zhong, B. (2022) Horizontal gene transfer: Driving the evolution and adaptation of plants. *Journal of Integrative Plant Biology*. DOI: 10.1111/jipb.13407.

Wang, N., Yang, Y., Moore, M.J., Brockington, S.F., Walker, J.F. *et al.* (2019) Evolution of portulacineae marked by gene tree conflict and gene family expansion associated with adaptation to harsh environments. *Molecular Biology and Evolution* 36(1), 112–126. DOI: 10.1093/molbev/msy200.

Wang, Y., Zhang, P., Guo, W., Liu, H., Li, X. *et al.* (2021) A deep learning approach to automate whole-genome prediction of diverse epigenomic modifications in plants. *New Phytologist* 232(2), 880–897. DOI: 10.1111/nph.17630.

Wapinski, I., Pfeffer, A., Friedman, N. and Regev, A. (2007) Natural history and evolutionary principles of gene duplication in fungi. *Nature* 449(7158), 54–61. DOI: 10.1038/nature06107.

Williamson, A. and Fatumo, S. (2024) Genomic diversity improves disease discovery for all. *Science* 385(6706), 255–256. DOI: 10.1126/science.adq7784.

Xavier, A. (2021) Technical nuances of machine learning: Implementation and validation of supervised methods for genomic prediction in plant breeding. *Crop Breeding and Applied Biotechnology* 21(spe), e381421S2. DOI: 10.1590/1984-70332021v21sa15.

Yang, J.B. and Peng, Z.S. (2005) Epigenetic phenomena of plant polyploids. *Yi Chuan = Hereditas* 27(2), 335–342.

Yang, Y., Ma, T., Wang, Z., Lu, Z., Li, Y. *et al.* (2018) Genomic effects of population collapse in a critically endangered ironwood tree *Ostrya rehderiana*. *Nature Communications* 9(1), 5449. DOI: 10.1038/s41467-018-07913-4.

Yang, Y., Li, Y., Chen, Q., Sun, Y. and Lu, Z. (2019) WGDdetector: A pipeline for detecting whole genome duplication events using the genome or transcriptome annotations. *BMC Bioinformatics* 20(1), 75. DOI: 10.1186/s12859-019-2670-3.

Yocca, A.E. and Edger, P.P. (2022) Machine learning approaches to identify core and dispensable genes in pangenomes. *The Plant Genome* 15(1), e20135. DOI: 10.1002/tpg2.20135.

Zenil-Ferguson, R., Ponciano, J.M. and Burleigh, J.G. (2016) Evaluating the role of genome downsizing and size thresholds from genome size distributions in angiosperms. *American Journal of Botany* 103(7), 1175–1186. DOI: 10.3732/ajb.1500408.

Zhang, R., Wang, F.-G., Zhang, J., Shang, H., Liu, L. *et al.* (2019) Dating whole genome duplication in *Ceratopteris thalictroides* and potential adaptive values of retained gene duplicates. *International Journal of Molecular Sciences* 20(8), 1926. DOI: 10.3390/ijms20081926.

Zhang, X., Ibrahim, Z., Khaskheli, M.B., Raza, H., Zhou, F. *et al.* (2024) Integrative approaches to abiotic stress management in crops: Combining bioinformatics educational tools and artificial intelligence applications. *Sustainability* 16(17), 7651. DOI: 10.3390/su16177651.

Zhao, C., Zhang, Y., Du, J., Guo, X., Wen, W. *et al.* (2019) Crop phenomics: Current status and perspectives. *Frontiers in Plant Science* 10, 714. DOI: 10.3389/fpls.2019.00714.

Zhou, Y., Zhang, J., Xiong, X., Cheng, Z.M. and Chen, F. (2022) De novo assembly of plant complete genomes. *Tropical Plants* 1(1), 1–8. DOI: 10.48130/TP-2022-0007.

Zingaretti, L.M., Gezan, S.A., Ferrão, L.F.V., Osorio, L.F., Monfort, A. *et al.* (2020) Exploring deep learning for complex trait genomic prediction in polyploid outcrossing species. *Frontiers in Plant Science* 11, 25. DOI: 10.3389/fpls.2020.00025.

Żyła, N. and Babula-Skowrońska, D. (2023) Evolutionary consequences of functional and regulatory divergence of HD-Zip I transcription factors as a source of diversity in protein interaction networks in plants. *Journal of Molecular Evolution* 91(5), 581–597. DOI: 10.1007/s00239-023-10121-4.

10 Pangenomics and AI-Assisted Crop Breeding for Combating Plant Diseases

Kempanna Sushmitha[1], Ganesan Megha[1], Monisha Domesa[1], Manoharan Rajesh[1]*, Asirvatham Alwin Robert[2], Mani Manoj[2], Jeyabal Philomenathan Antony Prabhu[2] and Arumugam Vijaya Anand[2]

[1]Department of Biochemistry, Bharathiar University, Coimbatore, Tamil Nadu, India; [2]Department of Human Genetics and Molecular Biology, Bharathiar University, Coimbatore, Tamil Nadu, India

Abstract

The rising threat of plant diseases, intensified by climate change and evolving pathogen resistance, necessitates innovative approaches in crop breeding to ensure global food security. Finding core and accessory genes linked to disease resistance is made possible by pangenomics, which examines the whole gene sets within a species and provides unmatched insights into genetic variation. When combined with artificial intelligence (AI) models, this method transforms the identification and use of crop disease-resistant characteristics. This chapter examines how AI and pangenomics might work together to tackle plant diseases, with a focus on how they can speed up breeding initiatives. Large genomic datasets can be analyzed more easily thanks to AI techniques like machine learning and deep learning, which allow for accurate trait prediction and selection for hardy crop types. The chapter uses case studies to show how AI-driven technologies have streamlined the production of disease-tolerant crops by improving genomic selection, optimizing breeding pipelines, and identifying resistance genes. The integration of pangenomics with AI not only increases the efficiency of crop improvement but also addresses key challenges, such as managing big data, ensuring model accuracy, and considering ethical implications. This chapter outlines a revolutionary route for sustainable agriculture by fusing cutting-edge computational techniques with genomic technology, enabling scientists and breeders to successfully reduce plant diseases. These thorough insights provided open the door to utilize the potential of cutting-edge technology to protect the world's food systems from the impending dangers of plant diseases and shifting environmental circumstances.

Keywords: Pangenomics, artificial intelligence, crop breeding, plant diseases, disease resistance, genomic diversity, machine learning

10.1 Introducing the Revolutionary Approach of Pangenomics to Crop Breeding

Pangenomics is a technique used in crop breeding to examine all the genetic diversity within a species or related group of species. This process helps in recognizing the genes that have been lost due to human activities by comparing cultivated crops with their wild relatives (Li *et al.*, 2022a). Through this method, breeders can easily identify genetic variations that are responsible for domestication traits such as single nucleotide

*Corresponding author: rajeshmanoh@gmail.com

DOI: 10.1079/9781800628649.0010

polymorphisms (SNPs), copy number variation (CNV), and presence–absence variation (PAV). This expedites the process of achieving favorable agronomic characteristics in crops. Even though there is a requirement for new data generation and analysis systems, pangenome structural variation details can enhance breeding initiatives by managing marker scoring expenses while improving genomic prediction accuracy (Della Coletta *et al.*, 2021).

The concept of a core and variable genome, represented by a pangenome, was initially described by Tettelin *et al.* (2005), setting the stage for a new understanding of genetic diversity within species. Building upon this foundation, the idea of the pangenome in plants was further developed by Italian researchers, marking a significant milestone in plant genetics research (Morgante *et al.*, 2007). This pioneering work not only introduced the concept of the pangenome in plants but also investigated the pivotal role of transposable elements in shaping it. In 2014, researchers under the direction of Li *et al.* released the first study on the pangenome of plants, building on previous discoveries and concentrating on seven wild soybean (*Glycine soja*) individuals. In addition to improving our knowledge of plant genetics and breeding, this ground-breaking work made it easier to identify genes linked to a variety of domesticated features, including organ size, biomass, seed composition, blooming and maturity time, and disease resistance (Li *et al.*, 2014).

The rapid advancement of genomic resources, including pangenomes, is evident, with numerous studies already conducted across various crop species such as pepper and walnut. These studies offer valuable insights into genetic diversity and strategies for crop improvement (Naithani *et al.*, 2023). For instance, the barley pangenome has recently been explored, revealing previously hidden genetic variations that are accessible for further genetic investigations and breeding purposes. This highlights the significant impact of pangenomics on our understanding of crop genomes and the enhancement of breeding endeavors (Jayakodi *et al.*, 2020). Pangenomics represents a transformative approach in crop improvement, expanding the arsenal of tools available for plant breeding through advancements in marker and sequencing technologies (Della Coletta *et al.*, 2021).

In recent years, comparative genomic research of several individuals within the same species has made it more and more clear that a single reference genome is insufficient to capture species-wide genetic variation (Lei *et al.*, 2021). Comparative examination of genomic areas across several inbred lines of maize (*Zea mays* L.), such as B73 and Mo17, demonstrates significant genetic divergence, with more than 25% of sequences found exclusively in one line but not in the other (Brunner *et al.*, 2005). Several genotypes of rice (*Oryza sativa* L.) have shown similar differences (Yu *et al.*, 2005). These findings highlight the drawbacks of attempting to capture a species' genomic complexity using only one genome sequence.

The idea of a pangenome has become a crucial tool for describing the genetic variation within a species in order to overcome this limitation. Pangenomes combine non-redundant genome sequences, including both core and dispensable portions, into a single annotated file. Sequences that are present in all individuals make up the core, whereas dispensable sections are only present in subsets of individuals. Advances in DNA sequencing have made it easier to put together an increasing number of genomes and pangenomes. With consequences for crop breeding, this trend presents chances to use pangenome assembly and analysis to investigate adaptation and evolution, offering insightful information about genome functionality (Danilevicz *et al.*, 2020).

Climate change and the rise of new pests and diseases are just two of the many challenges facing global agriculture. Crop yields are adversely affected by droughts, floods, and high temperatures brought on by climate change. Furthermore, because they may spread quickly and seriously harm crops, newly developing pests and diseases represent a serious threat to agricultural production and global food security (Pratap *et al.*, 2019). Sustainable methods of agricultural production are desperately needed. The creation of resilient crop types that can endure the effects of climate change and fight off pests and diseases is necessary to address these issues. Using genomics and gene-editing methods is one promising way to accomplish this goal (Yaqoob *et al.*, 2023). Integrating pangenomics into plant breeding programs is crucial for identifying genes associated with biotic stress tolerance. This comprehensive approach enables a deeper understanding of genetic diversity in crop plants, facilitating the selection of beneficial traits for improved biotic stress resistance through breeding.

10.2 AI Integration in Pangenomics-Assisted Crop Breeding

By fusing cutting-edge computational methods with genetic data to improve crop breeding and management, artificial intelligence (AI)-driven pangenomics dramatically speeds up the identification of novel crop features for sustainable agriculture. By effectively analyzing large genomic databases, AI and machine learning (ML) algorithms enable the quick discovery and selection of desired features, such as better nutritional content, drought tolerance, and disease resistance (Patil, 2023). In order to address the issues of global food security and the environment, these technologies facilitate precision breeding initiatives that accelerate the production of robust and high-yielding crop varieties (Gupta *et al.*, 2024). AI can predict catastrophic weather events by analyzing large environmental datasets, which makes it possible to develop strong, adaptive agricultural systems that can foresee and react to climate change (Ashok *et al.*, 2025). Hybrid ML models, combining techniques like linear and logistic regression, further refine the identification of beneficial traits across diverse crop species, leading to the creation of innovative cultivars that optimize productivity and sustainability (Prithiviraj *et al.*, 2024). AI-driven predictive analytics also enhance crop management by providing accurate forecasts of yield and pest outbreaks, allowing for data-driven decision making that optimizes resource use and reduces environmental impacts (Gul and Banday, 2024). Furthermore, bioinformatics technologies are essential for deciphering complicated genetic data, which makes marker-assisted breeding possible and cuts down on the time and resources needed for conventional techniques (Prajval *et al.*, 2024).

10.3 The Invisible Enemies: Understanding Biotic Stressors

Plants face a constant battle against invisible enemies that threaten their survival and productivity. These enemies, known as biotic stressors, encompass a wide range of living organisms, including nematodes, arachnids, weeds, viruses, insects, bacteria, and fungi. For instance, viral pathogens can cause stunted growth, leaf discoloration, and reduced yield, while fungal and bacterial diseases can lead to wilting, rot, and plant death (Oh *et al.*, 2005). These stressors deplete essential nutrients from plants, compromise their health, and ultimately lead to significant losses in crop yield (Kochhar and Gujral, 2020).

Food legumes, which occupy a substantial portion of arable land globally, are particularly vulnerable to biotic stresses. The yield gap between food legumes and cereals is substantial, with biotic stresses being the primary cause of the most severe losses (Basandrai *et al.*, 2011). This situation, due to reduced crop yields from biotic stresses, poses a significant threat to global food security, as approximately 800 million people worldwide experience undernourishment (Fichman and Mittler, 2020).

The challenge is further exacerbated by the emergence of new pathogens and pests, influenced by both climate and genetic factors. As the world's population continues to grow, the demand for food is expected to increase by 70% by 2050 (Hussain, 2015). However, reduced crop yields due to biotic stresses put international food security at risk, making it a pressing issue that needs to be addressed.

Although plants do not possess an adaptive immune system like animals, they have developed intricate defense strategies governed by numerous genes to withstand biotic stress (Gull *et al.*, 2019). These natural defense systems enable plants to recognize and counteract harmful biotic factors, thus promoting their survival and sustaining productivity.

Combining pangenomics with AI offers a powerful approach for enhancing crop resilience against biotic stress in an era of rapid environmental changes. By merging genomics-assisted breeding (GAB) with AI tools, researchers can expedite the identification and selection of desirable traits using extensive genomic data. ML techniques such as random forests (RF) and neural networks improve the precision of genomic predictions, helping breeders enhance traits like drought tolerance and pest resistance (Saimon *et al.*, 2023). For example, genome-wide association studies (GWAS) supported by support vector machines (SVM) and RF models have pinpointed important quantitative trait loci (QTLs) linked to pathogen resistance in soybeans, including resistance to Southern root-knot nematode (Canella Vieira *et al.*, 2022). Moreover, deep learning models like bidirectional long short-term memory (BiLSTM) networks have proven effective in analyzing gene expression in maize under biotic stress, revealing key stress-responsive genes (Nazari *et al.*, 2023).

Pangenomics enables researchers to tap into the genetic variation found in wild crop relatives, which often carry genes that help plants thrive under extreme environmental conditions. These genes can be introduced into cultivated varieties to enhance their adaptability to shifting climates (Petereit *et al.*, 2022). Technological advancements in high-throughput sequencing and phenotyping support this process by facilitating the rapid discovery of genes associated with stress tolerance (Yuan *et al.*, 2023). The evolution of plant pangenomics broadens our understanding of the genetic traits available for breeding, reinforcing the effectiveness of genomics-assisted strategies for developing resilient plant varieties (Scheben *et al.*, 2016).

A thorough comprehension of plant defense mechanisms against biotic stress is fundamental to achieve long-term sustainability in crop production. However, traditional approaches have faced challenges in effectively combating these invisible enemies.

10.4 Traditional Breeding Methods: Overcoming Roadblocks to Progress

10.4.1 Mass selection

Benefits include simplicity and affordability for improving highly heritable traits like yield (Hallauer *et al.*, 2010). It has been successfully used to improve rust disease resistance in wheat (Kang *et al.*, 2007) and downy mildew resistance in maize (Hallauer *et al.*, 2010). Limitations are lack of mating control (Abadassi, 2014), inefficiency for low heritability traits or those governed by many minor genes (Falconer, 1996).

10.4.2 Pure line selection

This method involves self-pollinating individual plants with desired traits over multiple generations to achieve genetic homogeneity and purity (Yan *et al.*, 2017; Amaral *et al.*, 2019; Sterckeman *et al.*, 2019). It allows rapid fixation of target traits (Liang and Liang, 2015) but can lead to loss of genetic diversity and inbreeding depression if not managed properly (Baktash and Hassan, 2015). Studies have shown pure line selection breeding aimed at fostering resistance to biotic stresses, including the creation of disease-resistant lines in crops like wheat, tomato, and banana (Acquaah, 2009).

10.4.3 Backcross breeding

This technique introgresses specific genes/traits from a donor parent into another variety through repeated backcrossing over generations. Benefits include targeted trait transfer and retaining the recipient's desirable genome. It has been used to incorporate drought tolerance in rice (Pardal *et al.*, 2022) and disease resistances like blast and bacterial blight in rice (Janaki Ramayya *et al.*, 2021). Limitations are the lengthy process and potential reduction in genetic diversity.

10.4.4 Recurrent selection

This cyclical method accumulates beneficial alleles from a diverse base population by selecting and intermating superior individuals over multiple generations to improve quantitative traits like yield and biotic stress tolerance (Baenziger *et al.*, 1984; Bernardo, 2002). Examples include developing downy mildew resistant sorghum (Reddy *et al.*, 1992), Fusarium head blight resistant wheat, and leaf blight resistant maize (Bentivenga *et al.*, 2016; Azad *et al.*, 2021;

Anand *et al.*, 2023). Benefits are maintaining genetic diversity, simultaneous improvement of multiple traits, and gradual accumulation of favourable alleles (Hallauer *et al.*, 2010). Limitations include time-consuming cycles, risk of genetic drift, and requirement of substantial initial diversity in the base population (Bernardo, 2002).

10.4.5 Hybridization

This process involves interbreeding genetically distinct populations to produce offspring with combined traits (Cortés-Ortiz, 2017). In crops, it can generate novel genotypes adapted to new conditions and transfer desirable genes from wild relatives. For instance, methods like induced homologous pairing and chromosomal engineering have successfully introgressed over 100 fungal resistance genes into bread wheat, enhancing its resilience against biotic stresses (Das, 2020). However, limitations include potential outbreeding depression, genomic incompatibilities between distant parents (Mitchell *et al.*, 2019), and the inadvertent transfer of undesirable traits from wild introgressions.

10.4.6 Deciphering the intricate mechanisms of plant defenses against resistance

Plants have developed sophisticated defense systems to combat biotic stress, employing both pre-existing (constitutive) and responsive (inducible) strategies. Constitutive defenses, such as the waxy cuticle layer on leaves and antimicrobial secondary metabolites, act as the first line of protection against potential threats (Serrano *et al.*, 2014). When plants detect invading organisms or signaling molecules, their immune systems trigger inducible defenses.

This immune activation involves the perception of molecular signatures like pathogen-associated molecular patterns (PAMPs), microbe-associated molecular patterns (MAMPs), and damage-associated molecular patterns (DAMPs). These triggers initiate pattern-triggered immunity (PTI), a fundamental plant defense response that limits pathogen invasion (Jones and Dangl, 2006; Newman *et al.*, 2013; Pieterse *et al.*, 2014).

PTI sets off a cascade of signaling events, including ion fluxes and the production of reactive oxygen species (ROS), which in turn activate transcription factors responsible for expressing defense-related genes. None the less, some pathogens circumvent PTI through effector-triggered susceptibility (ETS), deploying effector proteins that interfere with plant immune signaling. In defense, plants utilize resistance (R) proteins to detect these effectors, thereby initiating effector-triggered immunity (ETI). This interaction is explained by the gene-for-gene hypothesis, which posits that specific R genes in plants recognize matching avirulence (Avr) genes in the pathogen.

Beyond localized immune responses, plants can also initiate systemic defense strategies known as induced resistance (IR). These include systemic acquired resistance (SAR) and induced systemic resistance (ISR), mediated by signaling compounds such as salicylic acid and jasmonic acid (Boubakri, 2020).

Together, these layered defense mechanisms illustrate the complex responses plants deploy to fend off biotic stress, ensuring both crop health and agricultural productivity in a changing world. To further strengthen plant resilience, it is essential to dissect the genetic architecture underpinning these defence strategies. Cutting-edge approaches like pangenomics, which encompass the entire genomic diversity of a species, offer valuable insights into stress resistance and hold promise for sustainable crop improvement.

10.4.7 AI integration in traditional plant-breeding methods

AI-driven phenotyping significantly enhances the accuracy of trait selection in traditional plant breeding by integrating advanced ML techniques and high-throughput image analysis, which streamline the phenotyping process and improve predictive accuracy. Explainable AI (XAI) techniques, such as those applied in genotype-to-phenotype predictions, allow for the identification of key genetic factors influencing phenotypic traits, thereby supporting more informed breeding decisions and optimizing crop production (Novielli *et al.*, 2024). AI technologies, particularly in image analysis, facilitate the characterization of crops and the selection of optimal genotypes, eliminating manual methods and enabling the identification of subtle differences and complex QTL mapping that are challenging with conventional methods (Maraveas, 2024). The AMULET system serves as a prime example of how AI can anticipate morphological and physiological traits before they become visually apparent, thereby streamlining agricultural workflows and allowing for early interventions that support plant health and yield optimization (Zdrazil *et al.*, 2024). In the realm of plant breeding, AI plays a transformative role by enabling precise gene-editing approaches and forecasting the impacts of specific genetic variations. This capability helps bridge the gap between genotype and phenotype, supporting the creation of crop varieties adapted to evolving environmental conditions (Farooq *et al.*, 2024). Additionally, AI facilitates automated data collection and analysis of plant features, promoting sustainable farming practices. Its ability to enhance model accuracy and clarity is particularly valuable for developing resilient, high-performing crops (Gupta *et al.*, 2024).

10.5 Pangenomics: The Game-Changing Solution for Biotic Stress Resilience

The emergence of pangenomics has significantly advanced efforts to breed crops with enhanced resilience to biotic stresses. Unlike traditional reference genome approaches, pangenomics captures the full range of genetic diversity within a species, including structural variants (SVs) such as CNVs, PAVs, and other genomic rearrangements that are often overlooked (McHale *et al.*, 2012). This broader genomic perspective is transforming strategies to improve crop resistance to pests and pathogens.

A major strength of pangenomic analysis is its capacity to uncover diverse genetic variants linked to disease resistance. For example, in soybeans, resistance genes containing nucleotide-binding site (NBS) or NBS-leucine-rich repeat (LRR) domains frequently coincide with CNVs, suggesting their fundamental role in plant immune responses (Anderson *et al.*, 2014). Likewise, resistant soybean varieties have been shown to contain tandem duplications of genes associated with defense against the soybean cyst nematode (SCN), illustrating how SVs contribute to effective pathogen resistance (Cook *et al.*, 2012).

The influence of SVs on plant immunity is also evident in other crops. In maize, structural changes at the rp1 locus responsible for resistance to the common rust fungus led to variation in NBS-LRR gene copies, thereby producing differential resistance outcomes. Another example is a CNV within the ZmWAK gene, which provides resistance to head smut, highlighting the critical involvement of structural variants in mediating crop–pathogen dynamics (Richter *et al.*, 1995).

Pangenomics also enables the exploration of genetic resources in crop wild relatives (CWRs), which often possess traits for stress tolerance absent in cultivated lines (Petereit *et al.*, 2022). When coupled with modern breeding methods like marker-assisted and genomic selection, this knowledge supports the effective stacking of multiple resistance genes or QTLs to produce cultivars with durable disease resistance (Mekonnen *et al.*, 2017). Additionally, AI and ML tools streamline the interpretation of complex omics datasets, helping identify pivotal genetic markers involved in plant defense (Murmu *et al.*, 2024). This integrative approach not only accelerates the development of stress-resilient crops but also reinforces global food security amidst evolving environmental challenges (Soni *et al.*, 2024).

10.6 Constructing the Crop Pangenome: Blueprinting Success

The complete genomic landscape of crops requires an integrated approach leveraging three principal strategies: iterative mapping and assembly, de novo assembly, and graph-based assembly. These complementary techniques, powered by bioinformatics and computational methods, enable comprehensive pangenome construction (Fig. 10.1). By judiciously combining their strengths, scientists can assemble

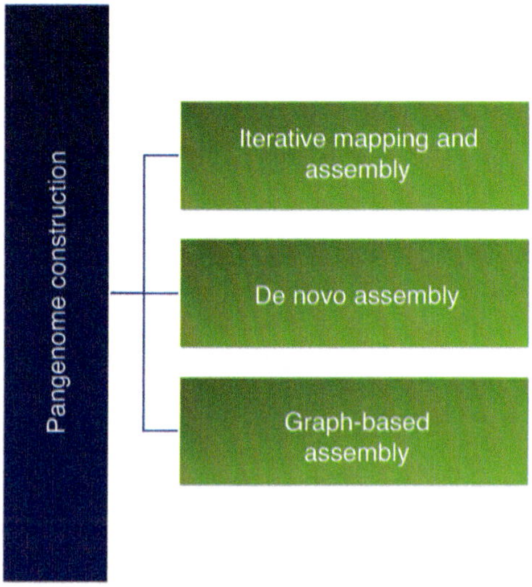

Fig. 10.1. Methods of pangenome construction. (Author's own image.)

a robust pangenomic blueprint, unlocking insights for crop improvement.

10.6.1 De novo assembly

10.6.1.1 Processing raw reads

In the process of handling raw reads, the initial genetic data, known as raw sequencing reads, can be sourced either locally or automatically retrieved from the European Nucleotide Archive (ENA) database. Subsequently, the acquired reads undergo a thorough clean-up procedure to eliminate low-quality segments and superfluous adapter sequences, thereby ensuring the data's integrity and suitability for further analysis. If the original data comprise paired-end reads, wherein genetic information is gleaned from both ends of a DNA fragment, these pairs are merged into longer single-end reads, enhancing the comprehensiveness of the sequences for subsequent analysis. Throughout these processes, rigorous quality checks are implemented to retain only reliable genetic information, thereby maintaining the integrity of the dataset. Moreover, the utilization of specialized software tools like Trimmomatic (Bolger *et al.*, 2014) and FLASH streamlines and standardizes the data preparation, enhancing efficiency in the workflow (Magoč and Salzberg, 2011).

10.6.1.2 Constructing contigs: forming continuous sequences from processed reads

To reconstruct longer sequences known as contigs from the processed reads of each low-confidence sample, one of three De Bruijn graph-based assemblers is typically employed: SPAdes (Bankevich *et al.*, 2012), MEGAHIT (Li *et al.*, 2015), or the GATB Minia pipeline. These tools utilize a multi-k approach, which involves analyzing the data using various k-mer lengths to capture a wide spectrum of sequence information. During the assembly phase, contigs that do not meet a user-defined minimum length are filtered out. This filtering step ensures that only high-quality and sufficiently long contigs are retained for subsequent analyses. By assembling short sequencing reads into longer, more continuous sequences, this method facilitates a more comprehensive understanding of the genetic material.

10.6.1.3 Refinement and scaffolding with references: improving and structuring sequences using reference data

During the reference-guided correction and scaffolding phase, assembled contigs are first aligned to a known reference genome using Minimap2 (Li, 2018). To address potential assembly errors and organize the corrected contigs into larger genomic structures known as pseudomolecules, the RagTag tool is utilized (Alonge *et al.*, 2019). Any contigs that do not align with the reference genome are grouped into a separate unit referred to as 'Chromosome0.' To assess the overall quality and completeness of the assembled genome, evaluation tools such as QUAST (Gurevich *et al.*, 2013) and BUSCO (Waterhouse *et al.*, 2018) are employed.

10.6.1.4 Genome annotation: identifying and labeling genes within the genome

Genome annotation for each LC sample involves two main steps. First, a process called 'lift-over' aligns reference genes to the LC genome assembly to utilize high-confidence information from the reference. This may result in slight differences due to genetic variations. Tools like GMAP (Wu and Watanabe, 2005) and GAWN are used for accurate alignment and gene model prediction. Predictions are refined by comparing protein sequences to a reference database and filtering for high-quality matches. Then, the genome is annotated using MAKER (Campbell *et al.*, 2014), which combines evidence from transcripts, proteins, and gene prediction tools. However, this automated process may generate errors, so unreliable gene models are filtered out based on parameters like annotation edit distance (AED) score. Additionally, overlapping gene models from the lift-over step are discarded, and the final genome annotation is produced by merging results from both annotation steps.

10.6.1.5 Simplification of gene variants: streamlining gene variations for analysis

In the removal of splice variants, only the longest transcript for genes with multiple variations is retained. This simplifies the analysis process and makes it easier to interpret the results. This practice is applied universally across all sample types, including reference, high-confidence, and LC samples. By focusing on the longest transcript, researchers can prioritize the most significant genetic information for further study and analysis.

10.6.1.6 Discovering orthologous connections: finding similar genes across different samples

Establishing orthology relationships requires the alignment of gene identifiers across various samples. To facilitate this, OrthoFinder2 clusters protein sequences into Orthogroups based on sequence similarity (Emms and Kelly, 2019). The process begins with the Markov Clustering algorithm, which groups similar sequences together. Subsequently, OrthoFinder2 uses phylogenomic methods to determine one-to-one and one-to-many orthologous relationships within these clusters. Since Orthogroups may also include paralogous genes originating from the same sample reflecting gene families rather than distinct genes further refinement is necessary. An extended version of the maximum weight orthogonal partitioning algorithm is employed to split these Ortho groups into smaller subclusters, each containing orthologous genes derived from different samples (Zheng *et al.*, 2011). These refined clusters, referred to as single pangenes, provide a more precise representation of gene relationships across multiple samples.

10.6.1.7 AI in de novo assembly

The use of advanced assembly methods and tools like the PanGenome Research Tool Kit (PGR-TK) has demonstrated the ability to resolve complex genomic variations, particularly in repetitive and clinically relevant genes, which are crucial for constructing comprehensive pangenomes (Chin *et al.*, 2023). The development of frameworks such as GNNome, which employs geometric deep learning for path identification in assembly graphs, further enhances the capability to reconstruct high-quality genomic sequences from complex regions, thereby facilitating the discovery of novel genomic features (Vrček *et al.*, 2024). Additionally, the de novo assembly of chicken genomes has revealed thousands of previously undetectable core genes, particularly in micro-chromosomes and sub-telomeric regions, underscoring the potential of de novo assembly in uncovering hidden genomic elements that are essential for understanding species diversity and evolution (Li *et al.*, 2022b). The integration of neural network models, such as the MPNN model for graph simplification, also supports the efficient assembly of genomes by simplifying complex graph structures, which is critical for identifying novel genomic features (Vrček *et al.*, 2020). Furthermore, the computational prediction of de novo emerged protein-coding genes highlights the role of de novo assembly in identifying genetic novelties that arise from noncoding sequences, contributing to the understanding of genetic diversity and evolution in eukaryotic genomes (Vakirlis and McLysaght, 2019). Collectively, these advancements illustrate the significant role of AI-driven de novo assembly in enhancing pangenome construction by identifying novel genomic features across diverse species.

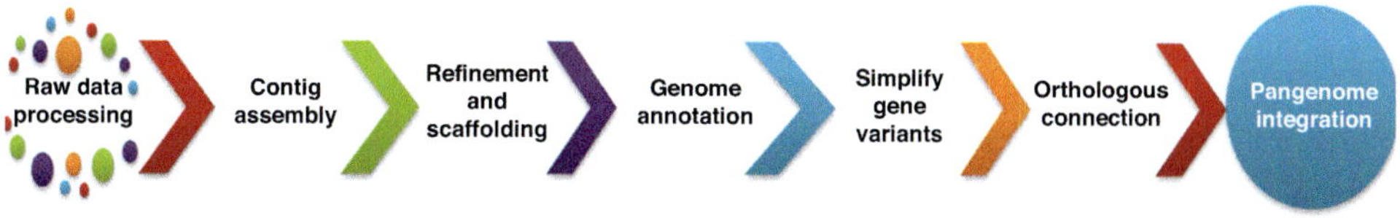

Fig. 10.2. Construction of pan genome using de novo assembly. (Author's own image.)

10.6.1.8 *Integration of pangenomic data: combining outcomes to form the complete pangenome*

In the final stage of the analysis pipeline, the results from earlier steps are consolidated to produce the definitive outputs. A PAV matrix is created based on the clustering information generated previously. For each pangene, a representative sequence is selected. If a reference gene is included in the cluster, its sequence and identifier are used; otherwise, the gene with the longest protein sequence is chosen as the representative. Following this selection, transcript and protein sequences of the representative genes are compiled into FASTA files for downstream analyses (Fig. 10.2).

10.6.2 Iterative mapping and assembly

10.6.2.1 *Data quality assurance*

Before initiating the mapping and assembly processes, it is essential to verify the quality of the sequencing data. Tools such as FastQC (Andrews, 2010) and MultiQC (Ewels *et al.*, 2016) are utilized to perform a thorough assessment of read quality and to provide a consolidated overview of any potential sequencing issues. Trimmomatic (Bolger *et al.*, 2014) is then employed to remove sequencing adapters and reads of low quality. Additionally, a custom Python script is used to filter out reads containing multiple ambiguous 'N' bases, ensuring the dataset is high-quality and suitable for downstream analyses.

10.6.2.2 *Mapping and extraction of unmapped reads*

Cleaned paired-end reads are aligned to a reference genome using Bowtie2 (Langmead and Salzberg, 2012). Post-alignment, SAMtools (Li *et al.*, 2009) is used to extract reads that do not map to the reference genome. These unmapped reads are then merged, converted back into FASTQ format, and sorted into paired and singleton categories. This step enables the recovery of sequences potentially representing novel genomic regions absent from the reference.

10.6.2.3 *De novo assembling unmapped reads*

Unmapped reads are subjected to de novo assembly using the MaSuRCA assembler (Zimin *et al.*, 2013) to reconstruct genomic regions not captured in the reference assembly. To minimize errors in assembly, contigs shorter than 200 base pairs are excluded. The reference genome is subsequently augmented with the newly assembled contigs, enabling iterative refinement and increased genome completeness.

10.6.2.4 *Contamination removal*

To maintain the integrity of the pangenome, assembled contigs are screened for contamination using NCBI-BLAST+ (Camacho *et al.*, 2009). Sequences that align significantly to non-plant organisms are considered contaminants and removed. Furthermore, redundant sequences are eliminated by aligning the new contigs against the reference genome, ensuring a non-redundant, plant-specific pangenome assembly.

10.6.2.5 *Gene prediction and functional annotation*

Gene prediction and annotation are performed using the MAKER pipeline, which integrates multiple lines of evidence to annotate genes across the updated pangenome. Redundant and contaminant-free contigs are incorporated into the reference assembly, contributing to the functional enrichment of the genome and supporting future analyses (Cantarel *et al.*, 2008).

10.6.2.6 Gene presence and absence calling

The finalized pangenome is indexed, and cleaned reads from each sample are mapped back to it using Bowtie2. The detection of gene presence and absence variations is carried out using SGSGeneLoss v0.1, allowing for the identification of genes that are variably present across different individuals. To interpret the functional relevance of these gene variations, Gene Ontology (GO) enrichment analyses are performed using tools like BLAST2GO (Conesa *et al.*, 2005) and the TopGO package (Alexa and Rahnenfuhrer, 2010).

This iterative strategy for pangenome construction offers a cost-effective alternative to full de novo genome assembly, as it enables the use of low-depth sequencing across hundreds of samples. However, it has limitations, particularly in resolving highly repetitive regions and detecting large structural variants (SVs) that are not spanned by short sequencing reads (Jiao and Schneeberger, 2017).

10.6.2.7 AI integration in iterative mapping and assembly

The integration of AI in iterative mapping and assembly significantly enhances the accuracy of pangenomic data by addressing the complexities and scale of genomic variations. The ability of AI to process vast amounts of sequencing data allows for a more comprehensive understanding of the genome, which is crucial for constructing accurate pangenomes that reflect the diversity within a species (Kalusivalingam *et al.*, 2021). The iterative mapping approach, as outlined by Fernandez, involves aligning and assembling data to create pangenomes, which are more representative than single reference genomes (Tay Fernandez, 2022). AI-driven tools, such as those discussed by Chandrashekar *et al.* (2024), automate and optimize the analysis pipeline, reducing computational time and minimizing human error, thus improving the quality of genomic research (Chandrashekar *et al.*, 2024). Furthermore, graph-based methods and multi-reference approaches, like those proposed by Anari *et al.* (2019), enhance the mapping accuracy by aligning reads to multiple genomes simultaneously, which is particularly effective for large collections of genomes (Anari *et al.*, 2019). These advancements are complemented by multiscale analysis techniques that enable the resolution of complex genomic variations, such as those in medically relevant genes, thereby improving the accuracy of variant calls in pangenomic studies (Chin *et al.*, 2023). Collectively, these AI-integrated methodologies facilitate a more precise and efficient analysis of pangenomic data, ultimately advancing our understanding of genomic diversity and its implications.

10.6.3 Graph-based assembly

10.6.3.1 Identification of genomic variations

To uncover genomic differences such as SNPs, insertions and deletions (InDels), and SVs, reference-based variant calling is conducted. Tools including MUMmer (Marçais *et al.*, 2018), minimap2 (Li, 2021), and PBSV are widely employed. MUMmer and minimap2 support intergenomic comparison and long-read alignments, while PBSV extracts structural variant data from these alignments, enhancing variant detection accuracy

10.6.3.2 Graph construction with variant information

Variant information obtained from VCF (variant call format) files is integrated with the reference genome to construct a graph-based pangenome. Tools such as vg (Garrison *et al.*, 2018) and Seven Bridges Graph Genome (Rakocevic *et al.*, 2019) facilitate this process. While effective, current approaches may face limitations in accurately representing complex or nested structural variants within the graph.

10.6.3.3 Improvement of long-read mapping tools

Improving tools like minimap2 is essential for increasing sensitivity to large InDels and SVs, particularly in repetitive plant genomic regions

(Li, 2021). Enhanced mappers contribute to more precise variant detection and better representation of genomic diversity.

10.6.3.4 Structural variation identification

Tools such as Sniffles (Sedlazeck *et al.*, 2018), SVIM (Heller and Vingron, 2020), and CuteSV (Jiang *et al.*, 2020) are utilized in this step to extract SV information from alignment results obtained from long-read mapping tools like minimap2. These tools help identify various types of structural variations present in the genome.

10.6.3.5 Alignment-based method

An alternative to variant-based graph construction involves direct alignment of entire genome assemblies to a reference genome. Tools like MGR and minigraph provide a fast and effective means to generate graph-based pangenomes without requiring extensive contig length, enabling more scalable analysis.

10.6.3.6 Graph-based pangenome construction pipelines

Comprehensive graph construction pipelines, such as PGGB and cactus, integrate various tools to support the end-to-end generation of graph-based pangenomes. These workflows typically involve pairwise sequence alignment, graph building, and normalization steps, ensuring structural accuracy and uniformity across the genome graph.

10.6.3.7 Acquisition and merging of PAVs

PAVs are extracted via comparative genomics or long-read mapping strategies using tools like MUMmer, minimap2, and PBSV. To reduce redundancy, overlapping PAVs are clustered and merged, resulting in a non-redundant PAV dataset that captures true variation across individuals.

10.6.3.8 Storage format

Graph-based pangenomes are stored using formats such as graphical fragment assembly (GFA) (Li, 2016), rGFA (Li *et al.*, 2020), or the vg format. Tools like vg and ODGI (Guarracino *et al.*, 2022) support conversion, compression, and querying of these formats, ensuring efficient data handling and retrieval.

10.6.3.9 Restoration of linear reference genome coordinates

Maintaining compatibility between the graph-based pangenome and the linear reference genome is vital for accurate annotation. Tools like vg allow for the conversion of coordinates and annotations from linear to graph formats, preserving the functional context of genomic features.

10.6.3.10 Visualization

To interpret the structural complexity of graph-based genomes, visualization tools such as Bandage, GfaViz (Gonnella *et al.*, 2019), vg viz, and Sequence Tube Map (Beyer *et al.*, 2019) are employed. These tools provide both high-level and nucleotide-resolution views, aiding in the exploration of genomic architecture and variation.

10.6.3.11 Annotation of the graph-based pangenome

Gene structures from the linear genome are projected onto the graph genome using tools like vg, facilitating the annotation of known genes. Furthermore, ongoing research focuses on developing algorithms to predict novel genes directly from the graph genome, thereby enriching our understanding of genomic organization and function within diverse populations.

10.6.3.12 AI-driven graph-based assembly

AI-driven graph-based assembly can indeed be used to identify novel genomic features in pangenome construction, as demonstrated by several studies. The construction of pangenome graphs allows for the identification of non-reference sequences (NRSs), which are crucial for understanding genetic diversity and functional genomics. For instance, in pigs, a pangenome graph constructed from 21 assemblies revealed 23,831 NRSs, highlighting their prevalence in genetically divergent breeds and their association with immune-related functions (Miao *et al.*, 2024). Similarly, in humans, a graph-based pangenome approach identified 5.1 million NRSs across five populations, with a significant portion being novel and associated with functional domains and phenotypic traits (Wu *et al.*, 2022). The use of tools like ggCaller further enhances this process by integrating gene prediction, annotation, and clustering, which improves the accuracy of identifying genomic variants and their functional implications, as seen in bacterial genomes (Horsfield *et al.*, 2023). Moreover, the development of algorithms like seqwish facilitates the scalable construction of pangenome graphs by transforming sequence alignments into variation graphs, thus enabling the exploration of large genomic datasets without the constraints of a single reference genome (Garrison and Guarracino, 2023). These methodologies collectively underscore the potential of graph-based assembly in uncovering novel genomic features, thereby advancing our understanding of genetic variation and its implications across different species.

10.7 Pangenomics in Action: Success Stories from the Field

The transformative impact of pangenomics has been clearly demonstrated across various crop species, enabling the discovery of novel genes and genetic architectures that underpin resistance to biotic stress (Table 10.1). One compelling example is the pangenome analysis of *Brassica oleracea*, which uncovered extensive gene diversity related to plant defense mechanisms. Notably, the study identified 439 putative resistance genes harboring LRR, toll/interleukin-1 receptor-like (TIR), and coiled-coil (CC) domains, well-known components of plant innate immunity (Golicz *et al.*, 2016). The observation of PAVs among these genes highlights their potential role in adaptive responses to diverse pathogenic threats. In rice (*Oryza sativa*), pangenome studies have distinguished between a core genome comprising 26,372 genes conserved across most accessions, and a dispensable genome containing 16,208 genes variably present among cultivars (Zhao *et al.*, 2018). Interestingly, genes within the dispensable set were found to be enriched in NBS-LRR and NB-ARC domains, which are closely associated with disease resistance and stress signaling. These insights facilitate the breeding of stress-tolerant rice varieties by targeting the most relevant genomic regions linked to pathogen defense.

Beyond discovery, pangenomics has profoundly accelerated marker-assisted selection in crop breeding programs. By integrating high-resolution genomic information, breeders can identify favorable alleles, stack resistance genes, and refine parental line selection with unprecedented precision. Successful applications of marker-assisted selection include the use of KASP markers in wheat, and trait-focused breeding in sunflower and tomato (Kucherenko *et al.*, 2022; Maji *et al.*, 2023; Nishio *et al.*, 2023; Orchard *et al.*, 2023). A landmark example is the introgression of bacterial blight resistance genes (Xa21, Xa13, Xa5) into the Jalmagna rice variety through marker-assisted backcrossing. This led to the development of pyramided lines capable of withstanding bacterial blight under deep-water cultivation systems (Pradhan *et al.*, 2015). Together, these advancements underscore the pivotal role of pangenomics in reshaping modern agriculture. By revealing genetic diversity and guiding targeted breeding, it offers a robust framework for enhancing biotic stress resilience and securing global food sustainability (Fig. 10.3).

10.8 Conclusion

The integration of pangenomics with AI marks a transformative advancement in modern crop improvement strategies. By unlocking the complete genetic repertoire of cultivated crops and their wild relatives, pangenomics provides an unprecedented resource for identifying novel resistance genes and structural variants essential for combating biotic stressors. AI-driven methodologies including machine learning and deep learning amplify the analytical power of this approach, enabling the processing and interpretation of vast, complex genomic datasets with remarkable accuracy and speed. These technologies facilitate the rapid discovery of key resistance traits and the optimization of

Table 10.1. Comprehensive overview of crop pangenome studies: approaches, metrics, and resources.

Genome assembly method	Sequencing	Species	Population size	Pangenome size	Reference
De novo assembly	Illumina	*Oryza sativa*	3	39,891 genes	Pagliarini *et al.*, 2012
	PacBio	*Medicago truncatula*	15	75,000 gene clusters	Juan *et al.*, 2017
	Illumina	*Glycine soya* (soybean)	7	59,080 gene clusters	Li *et al.*, 2014
	Illumina	*Brassica napus* (oilseed rape)	9	105,672 gene clusters	Kamburova *et al.*, 2021
	Illumina and PacBio	Barley	20	639 Mb PAV sequences	Jayakodi *et al.*, 2020
	Illumina	Juglans (walnut)	6	26,458 gene clusters	Martínez-Fortún *et al.*, 2022
	Illumina	*Brassica rapa*	3	41,858 genes	Kamburova *et al.*, 2021
	PacBio	*Setaria italica* (Foxtail millet)	110	51,401 pangenes	Martínez-Fortún *et al.*, 2022
	NA	*Oryza sativa* (indica/ japonica)	1483	NA	Singh *et al.*, 2015
	Illumina	*Zea mays* (maize)	503	41,903 transcripts	Darracq *et al.*, 2018
	Illumina	*Brachypodium distachyon* (stiff brome)	54	37,886 genes	Gordon *et al.*, 2017
	NA	*Sesamum indicum* (sesame)	5	26,472 gene clusters	Ahmad *et al.*, 2019
	Illumina	*Glycine max* (soybean)	29	57,492 gene clusters	Amaral *et al.*, 2019

Continued

Table 10.1. Continued

Genome assembly method	Sequencing	Species	Population size	Pangenome size	Reference
Iterative assembly	Illumina	*Helianthus annuus* (sunflower)	493	61,205 genes	Paknia *et al.*, 2020
	NA	*B. napus*	53	94,013 genes	Bridgen *et al.*, 2018
	Illumina	Amborella trichopoda	11	20.6 Mb PAVs and 30,189 pangenes	Figueroa *et al.*, 2013
	NA	Capsicum (pepper)	383	51,757 genes	Lal *et al.*, 2018
	Illumina	*Brassica oleracea*	10	61,379 genes	Golicz *et al.*, 2016
		O. sativa/Oryza rufipogon	67	42,580 genes	Zhao *et al.*, 2018
	Illumina	*Triticum aestivum* (bread wheat)	19	139,747 genes	Juan *et al.*, 2017
	Illumina	*Solanum lycopersicum* (tomato)	725	40,369 genes	Figueroa *et al.*, 2013
Map-to-pan	Illumina	Wheat	18	140 500 pangenes	Hickey *et al.*, 2017
	Illumina	Cotton	1807	Pangenome of 3388 Mb and 102 768 genes (*G. hirsutum*) Pangenome of 2575 Mb and 80 148 genes (G. barbadense)	Li, 2021
	Illumina	*O. sativa* (rice)	3010	48,098 genes	Wang *et al.*, 2024
	Illumina	Rice	3010	Pangenome of ~ 642 Mb	Wang *et al.*, 2024
	Illumina	Sorghum	176	Pangenome of 883.3 Mb and 35 719 pangenes	Rooney *et al.*, 2002

Continued

Table 10.1. Continued

Genome assembly method	Sequencing	Species	Population size	Pangenome size	Reference
Assembly-based graph genome	Illumina	Maize	721	Pangenome of ~ 6.71 Gb and 58 944 pangenes	Darracq *et al.*, 2018
	PacBio	Rice	31	66,636 pangenes	Wang *et al.*, 2024
	Nanopore	Rice	251	Pangenome of ~ 1.52 Gb and 51 359 pangenes	Wang *et al.*, 2024
	PacBio	Sorghum	16	Pangenome of 984.8 Mb and 44 079 pangenes	Rooney *et al.*, 2002
	PacBio	Cucumber	12	26,822 pangenes	Li *et al.*, 2022a
Genome assembly comparison	Illumina	Maize	6	34,352 core genes	Haroon *et al.*, 2020
	Nanopore	Rice	111	Pangenome of 1250 Mb and 75,305 pan-genes	Zhang *et al.*, 2021
	Illumina	Mung bean	217	Pangenome of ~ 762.9 Mb and 43 462 pangenes	Amaral *et al.*, 2019
	PacBio	Maize	26	103 000 pangenes	Hamadani *et al.*, 2021
	Illumina	Barley	20	638.6 Mb PAV sequences	Jayakodi *et al.*, 2020
Long-read mapping	Nanopore	Tomato	100	238,490 SVs	Alonge *et al.*, 2019
Read mapping (no assembly)	Illumina	Populus (poplar)	7	NA	Freschet *et al.*, 2018

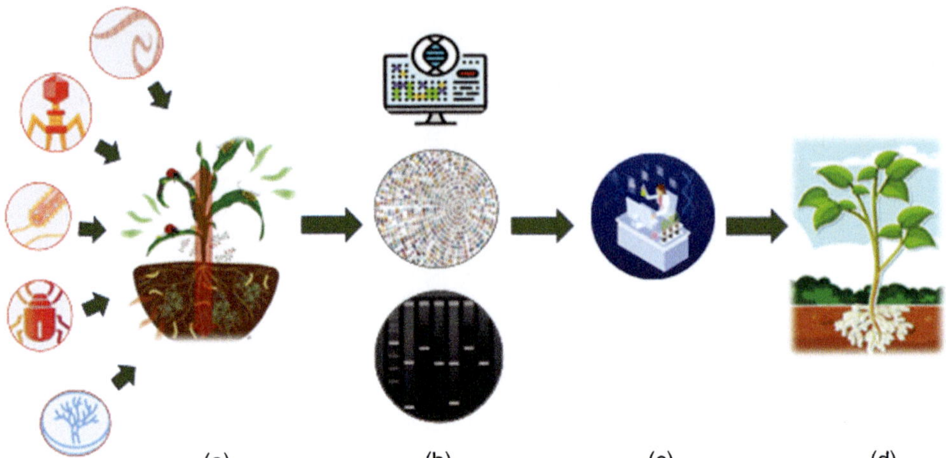

Fig. 10.3. Pangenomic approach for developing biotic stress-resilient crops. (Author's own image.) (a) Plant under biotic stress; (b) pangenome analysis; (c) researcher data analysis; and (d) healthy, resilient plant.

breeding pipelines. When synergistically applied alongside gene editing tools (e.g. CRISPR/Cas systems) and multi-omics integration (transcriptomics, proteomics, epigenomics), this model enables the precision engineering of robust crop varieties with tailored resistance profiles. Such innovation holds the key to fortifying global food systems in the face of climate change, emerging pathogens, and evolving pest pressures. By embracing this AI-augmented pangenomic framework, agricultural scientists and breeders can revolutionize breeding paradigms, improve resource-use efficiency, and secure sustainable crop production. Ultimately, this holistic approach is pivotal to ensuring future food security, empowering us to meet the challenges of a rapidly changing world while delivering resilient harvests for generations to come.

Acknowledgments

All the authors are thankful to their respective Universities and Institutions for their valuable support.

Data Availability

The authors confirm that the data supporting the findings of this study are available within the book/chapter.

References

Abadassi, J. (2014) Mass selection to improve grain yield in a traditional tropical maize population. *International Journal of Science and Technology* 4(3), 44–47.
Acquaah, G. (2009) *Principles of Plant Genetics and Breeding*. John Wiley & Sons.

Ahmad, B., Raina, A. and Khan, S. (2019) Impact of biotic and abiotic stresses on plants, and their responses. In: *Disease Resistance in Crop Plants: Molecular, Genetic and Genomic Perspectives*. pp. 1–19.

Alexa, A. and Rahnenfuhrer, J. (2010) topGO: Enrichment analysis for gene ontology. *R Package Version* 2.

Alonge, M., Soyk, S., Ramakrishnan, S., Wang, X., Goodwin, S. *et al.* (2019) RaGOO: Fast and accurate reference-guided scaffolding of draft genomes. *Genome Biology* 20(1), 224. DOI: 10.1186/s13059-019-1829-6.

Amaral, L.D.O., Bruzi, A.T., Resende, P.M.D. and Silva, K.B. (2019) Pure line selection in a heterogeneous soybean cultivar. *Crop Breeding and Applied Biotechnology* 19(3), 277–284. DOI: 10.1590/1984-70332019v19n3a39.

Anand, A., Subramanian, M. and Kar, D. (2023) Breeding techniques to dispense higher genetic gains. *Frontiers in Plant Science* 13, 1076094. DOI: 10.3389/fpls.2022.1076094.

Anari, S.S., de Ridder, D., Schranz, M.E. and Smit, S. (2019) Pangenomic read mapping. *Bioinformatics*. DOI: 10.1101/813634.

Anderson, J.E., Kantar, M.B., Kono, T.Y., Fu, F., Stec, A.O. *et al.* (2014) A roadmap for functional structural variants in the soybean genome. *G3: Genes, Genomes, Genetics* 4(7), 1307–1318. DOI: 10.1534/g3.114.011551.

Andrews, S. (2010) FastQC: A quality control tool for high throughput sequence data. Available at: https://www.bioinformatics.babraham.ac.uk/projects/fastqc/ (accessed 5 August 2025).

Ashok, P., Biswas, K.K., Shreenidhi, K.S., Anandaram, H., Karthikeyan, G. *et al.* (2025) Harnessing AI for climate resilience: Opportunities and challenges in combating climate change. In: *Text Mining and Sentiment Analysis in Climate Change and Environmental Sustainability*. pp. 141–170.

Azad, M.A.K., Yasmine, F., Kamruzzaman, M., Rani, M.H. and Begum, H.A. (2021) Development of climate-adaptable/resilient crop varieties through induced mutation. In: *Mutation Breeding, Genetic Diversity and Crop Adaptation to Climate Change*. CABI, Wallingford UK, pp. 157–171.

Baenziger, P.S., Elgin, J.H., Stavely, J.R. and Tomerlin, J.R. (1984) Breeding for disease resistance. In: *Applications of Genetic Engineering to Crop Improvement*. pp. 427–452.

Baktash, F.Y. and Hassan, L.K. (2015) Pure line selection from bread wheat for grain yield and it's components under different seeding rates. *Iraqi Journal of Agricultural Sciences* 46(5), 1132–1140.

Bankevich, A., Nurk, S., Antipov, D., Gurevich, A.A., Dvorkin, M. *et al.* (2012) SPAdes: A new genome assembly algorithm and its applications to single-cell sequencing. *Journal of Computational Biology* 19(5), 455–477. DOI: 10.1089/cmb.2012.0021.

Basandrai, A.K., Daisy Basandrai, D.B., Duraimurugan, P. and Srinivasan, T. (2011) Breeding for biotic stresses. In: *Biology and Breeding of Food Legumes*. CABI, Wallingford UK, pp. 220–240.

Bentivenga, G., Camerini, M., Belocchi, A., Fornara, M., Melloni, S. *et al.* (2016) Agronomic evaluation of fusarium head blight (FHB) resistance in Italian durum wheat cultivars and screening of advanced lines MAS selected for FHB resistance. *Plant Varieties Studying and Protection* 3(32), 30–41. DOI: 10.21498/2518-1017.3(32).2016.75978.

Bernardo, R. (2002) *Breeding for Quantitative Traits in Plants*, Vol. 1. Stemma Press, Woodbury, MN, p. 369.

Beyer, W., Novak, A.M., Hickey, G., Chan, J., Tan, V. *et al.* (2019) Sequence tube maps: Making graph genomes intuitive to commuters. *Bioinformatics* 35(24), 5318–5320. DOI: 10.1093/bioinformatics/btz597.

Bolger, A.M., Lohse, M. and Usadel, B. (2014) Trimmomatic: A flexible Trimmer for Illumina sequence data. *Bioinformatics* 30(15), 2114–2120. DOI: 10.1093/bioinformatics/btu170.

Boubakri, H. (2020) Induced resistance to biotic stress in plants by natural compounds: Possible mechanisms. In: *Priming-Mediated Stress and Cross-Stress Tolerance in Crop Plants*. Academic Press, pp. 79–99.

Bridgen, M.P., Van Houtven, W. and Eeckhaut, T. (2018) Plant tissue culture techniques for breeding. In: Van Huylenbroeck, J. (ed.) *Ornamental Crops*. Springer, Cham, Switzerland, pp. 127–144. DOI: 10.1007/9783319906980_6.

Brunner, S., Fengler, K., Morgante, M., Tingey, S. and Rafalski, A. (2005) Evolution of DNA sequence nonhomologies among maize inbreds. *The Plant Cell* 17(2), 343–360. DOI: 10.1105/tpc.104.025627.

Camacho, C., Coulouris, G., Avagyan, V., Ma, N., Papadopoulos, J. *et al.* (2009) BLAST+: Architecture and applications. *BMC Bioinformatics* 10, 1–9. DOI: 10.1186/1471-2105-10-421.

Campbell, M.S., Holt, C., Moore, B. and Yandell, M. (2014) Genome annotation and curation using MAKER and MAKER-P. *Current Protocols in Bioinformatics* 48(1), 4–11. DOI: 10.1002/0471250953.bi0411s48.

Canella Vieira, C., Zhou, J., Usovsky, M., Vuong, T., Howland, A.D. *et al.* (2022) Exploring machine learning algorithms to unveil genomic regions associated with resistance to southern root-knot nematode in soybeans. *Frontiers in Plant Science* 13, 883280. DOI: 10.3389/fpls.2022.883280.

Cantarel, B.L., Korf, I., Robb, S.M.C., Parra, G., Ross, E. *et al.* (2008) MAKER: An easy-to-use annotation pipeline designed for emerging model organism genomes. *Genome Research* 18(1), 188–196. DOI: 10.1101/gr.6743907.

Chandrashekar, K., Niranjan, V., Vishal, A. and Setlur, A.S. (2024) Integration of artificial intelligence, machine learning and deep learning techniques in genomics: Review on computational perspectives for NGS analysis of DNA and RNA seq data. *Current Bioinformatics* 19(9), 825–844. DOI: 10.2174/0115748936284044240108074937.

Chin, C.S., Behera, S., Khalak, A., Sedlazeck, F.J., Sudmant, P.H. *et al.* (2023) Multiscale analysis of pangenomes enables improved representation of genomic diversity for repetitive and clinically relevant genes. *Nature Methods* 20(8), 1213–1221. DOI: 10.1038/s41592-023-01914-y.

Conesa, A., Götz, S., García-Gómez, J.M., Terol, J., Talón, M. *et al.* (2005) Blast2GO: A universal tool for annotation, visualization and analysis in functional genomics research. *Bioinformatics* 21(18), 3674–3676. DOI: 10.1093/bioinformatics/bti610.

Cook, D.E., Lee, T.G., Guo, X., Melito, S., Wang, K. *et al.* (2012) Copy number variation of multiple genes at Rhg1 mediates nematode resistance in soybean. *Science* 338(6111), 1206–1209. DOI: 10.1126/science.1228746.

Cortés-Ortiz, L. (2017) Hybridization and hybrid zones. In: *The International Encyclopedia of Primatology*. John Wiley & Sons, Hoboken, NJ.

Danilevicz, M.F., Fernandez, C.G.T., Marsh, J.I., Bayer, P.E. and Edwards, D. (2020) Plant pangenomics: Approaches, applications and advancements. *Current Opinion in Plant Biology* 54, 18–25. DOI: 10.1016/j.pbi.2019.12.005.

Darracq, A., Vitte, C., Nicolas, S., Duarte, J., Pichon, J.-P. *et al.* (2018) Sequence analysis of European maize inbred line F2 provides new insights into molecular and chromosomal characteristics of presence/absence variants. *BMC Genomics* 19(1), 119. DOI: 10.1186/s12864-018-4490-7.

Das, A. (2020) Distant hybridization and introgression of alien genes for improving heterotic potential and resistance to various biotic stresses in vegetable and oilseed brassicas (*Brassica* Sp). *International Journal of Chemical Studies* 8(4), 1709–1717. DOI: 10.22271/chemi.2020.v8.i4q.9858.

Della Coletta, R., Qiu, Y., Ou, S., Hufford, M.B. and Hirsch, C.N. (2021) How the pan-genome is changing crop genomics and improvement. *Genome Biology* 22(1), 3. DOI: 10.1186/s13059-020-02224-8.

Emms, D.M. and Kelly, S. (2019) OrthoFinder: Phylogenetic orthology inference for comparative genomics. *Genome Biology* 20(1), 238. DOI: 10.1186/s13059-019-1832-y.

Ewels, P., Magnusson, M., Lundin, S. and Käller, M. (2016) MultiQC: Summarize analysis results for multiple tools and samples in a single report. *Bioinformatics* 32(19), 3047–3048. DOI: 10.1093/bioinformatics/btw354.

Falconer, D.S. (1996) *Introduction to Quantitative Genetics*. Pearson Education India.

Farooq, M.A., Gao, S., Hassan, M.A., Huang, Z., Rasheed, A. *et al.* (2024) Artificial intelligence in plant breeding. *Trends in Genetics* 40(10), 891–908. DOI: 10.1016/j.tig.2024.07.001.

Fichman, Y. and Mittler, R. (2020) Rapid systemic signaling during abiotic and biotic stresses: Is the ROS wave master of all trades?. *The Plant Journal* 102(5), 887–896. DOI: 10.1111/tpj.14685.

Figueroa, M., Alderman, S., Garvin, D.F. and Pfender, W.F. (2013) Infection of *Brachypodium distachyon* by formae speciales of *Puccinia graminis*: Early infection events and host-pathogen incompatibility. *PLoS ONE* 8(2), e56857. DOI: 10.1371/journal.pone.0056857.

Freschet, G.T., Violle, C., Bourget, M.Y., Scherer-Lorenzen, M. and Fort, F. (2018) Allocation, morphology, physiology, architecture: The multiple facets of plant above- and below-ground responses to resource stress. *The New Phytologist* 219(4), 1338–1352. DOI: 10.1111/nph.15225.

Garrison, E. and Guarracino, A. (2023) Unbiased pangenome graphs. *Bioinformatics* 39(1), btac743. DOI: 10.1093/bioinformatics/btac743.

Garrison, E., Sirén, J., Novak, A.M., Hickey, G., Eizenga, J.M. *et al.* (2018) Variation graph toolkit improves read mapping by representing genetic variation in the reference. *Nature Biotechnology* 36(9), 875–879. DOI: 10.1038/nbt.4227.

Golicz, A.A., Bayer, P.E., Barker, G.C., Edger, P.P., Kim, H. *et al.* (2016) The pangenome of an agronomically important crop plant Brassica oleracea. *Nature Communications* 7(1), 13390. DOI: 10.1038/ncomms13390.

Gonnella, G., Niehus, N. and Kurtz, S. (2019) GfaViz: Flexible and interactive visualization of GFA sequence graphs. *Bioinformatics* 35(16), 2853–2855. DOI: 10.1093/bioinformatics/bty1046.

Gordon, S.P., Contreras-Moreira, B., Woods, D.P., Marais, D.L., Burgess, D. *et al.* (2017) Extensive gene content variation in the *Brachypodium distachyon* pan-genome correlates with population structure. *Nature Communications* 8(1), 2184. DOI: 10.1038/s41467-017-02292-8.

Guarracino, A., Heumos, S., Nahnsen, S., Prins, P. and Garrison, E. (2022) ODGI: Understanding pangenome graphs. *Bioinformatics* 38(13), 3319–3326. DOI: 10.1093/bioinformatics/btac308.

Gul, D. and Banday, R.U.Z. (2024) Transforming crop management through advanced AI and machine learning: Insights into innovative strategies for sustainable agriculture. *AI, Computer Science and Robotics Technology*. DOI: 10.5772/acrt.20240030.

Gull, A., Lone, A.A. and Wani, N.U.I. (2019) Biotic and abiotic stresses in plants. In: Oliveira, A.B. (ed.) *Abiotic and Biotic Stress in Plants*. IntechOpen, London, pp. 1–19. DOI: 10.5772/intechopen.85832.

Gupta, D.K., Pagani, A., Zamboni, P. and Singh, A.K. (2024) AI-powered revolution in plant sciences: Advancements, applications, and challenges for sustainable agriculture and food security. *Exploration of Foods and Foodomics* 2(5), 443–459. DOI: 10.37349/eff.2024.00045.

Gurevich, A., Saveliev, V., Vyahhi, N. and Tesler, G. (2013) QUAST: Quality assessment tool for genome assemblies. *Bioinformatics* 29(8), 1072–1075. DOI: 10.1093/bioinformatics/btt086.

Hallauer, A.R., Carena, M.J. and Miranda Filho, J.D. (2010) *Quantitative Genetics in Maize Breeding*, Vol. 6. Springer Science & Business Media.

Hamadani, H., Rashid, S.M., Parrah, J.D., Khan, A.A., Dar, K.A. *et al.* (2021) Traditional farming practices and its consequences. In: *Microbiota and Biofertilizers: Ecofriendly Tools for Reclamation of Degraded Soil Environs*, Vol. 2. pp. 119–128.

Haroon, M., Zafar, M.M., Farooq, M.A., Afzal, R., Batool, M. *et al.* (2020) Conventional breeding, molecular breeding and speed breeding; brave approaches to Revamp the production of cereal crops. *Biology and Life Sciences*. DOI: 10.20944/preprints202011.0667.v1.

Heller, D. and Vingron, M. (2020) SVIM-asm: Structural variant detection from haploid and diploid genome assemblies. *Bioinformatics*. DOI: 10.1101/2020.10.27.356907.

Hickey, J.M., Chiurugwi, T., Mackay, I. and Powell, W. (2017) Genomic prediction unifies animal and plant breeding programs to form platforms for biological discovery. *Nature Genetics* 49(9), 1297–1303. DOI: 10.1038/ng.3920.

Horsfield, S.T., Tonkin-Hill, G., Croucher, N.J. and Lees, J.A. (2023) Accurate and fast graph-based pangenome annotation and clustering with ggcaller. *Genome Research* 33(9), 1622–1637. DOI: 10.1101/gr.277733.123.

Hussain, B. (2015) Modernization in plant breeding approaches for improving biotic stress resistance in crop plants. *Turkish Journal of Agriculture and Forestry* 39(4), 515–530. DOI: 10.3906/tar-1406-176.

Janaki Ramayya, P., Vinukonda, V.P., Singh, U.M., Alam, S., Venkateshwarlu, C. *et al.* (2021) Marker-assisted forward and backcross breeding for improvement of elite Indian rice variety Naveen for multiple biotic and abiotic stress tolerance. *PLoS ONE* 16(9), e0256721. DOI: 10.1371/journal.pone.0256721.

Jayakodi, M., Padmarasu, S., Haberer, G., Bonthala, V.S., Gundlach, H. *et al.* (2020) The barley pangenome reveals the hidden legacy of mutation breeding. *Nature* 588(7837), 284–289. DOI: 10.1038/s41586-020-2947-8.

Jiang, T., Liu, Y., Jiang, Y., Li, J., Gao, Y. *et al.* (2020) Long-read-based human genomic structural variation detection with cuteSV. *Genome Biology* 21(1), 1–24. DOI: 10.1186/s13059-020-02107-y.

Jiao, W.B. and Schneeberger, K. (2017) The impact of third generation genomic technologies on plant genome assembly. *Current Opinion in Plant Biology* 36, 64–70. DOI: 10.1016/j.pbi.2017.02.002.

Jones, J.D.G. and Dangl, J.L. (2006) The plant immune system. *Nature* 444(7117), 323–329. DOI: 10.1038/nature05286.

Juan, D., Agnieszka, A., Philipp, E. and Kenneth, C.K. (2017) Pangenome of hexaploid bread wheat. *Plant Journal*.

Kalusivalingam, A.K., Sharma, A., Patel, N. and Singh, V. (2021) Leveraging deep learning and random forest algorithms for AI-driven genomics in personalized medicine. *International Journal of AI and ML* (3).

Kamburova, V.S., Salakhutdinov, I.B., Shermatov, S.E. and Abdurakhmonov, I.Y. (2021) Using of genome editing methods in plant breeding. In: Abdurakhmonov, I.Y. (ed.) *Plant Breeding – Current and Future Views*. IntechOpen.

Kang, M.S., Subudhi, P.K., Baisakh, N. and Priyadarshan, P.M. (2007) Crop breeding methodologies: Classic and modern. In: Kang, M.S. and Priyadarshan, P.M. (eds) *Breeding Major Food Staples*. Blackwell Publishing, pp. 5–40. DOI: 10.1002/9780470376447.

Kochhar, S.L. and Gujral, S.K. (2020) Abiotic and biotic stress. In: *Plant Physiology: Theory and Applications*. Cambridge University Press, Chapter, Cambridge, pp. 545–589.

Kucherenko, Y., Zviahintseva, A., Kobyzeva, L., Kolomatska, V., Makliak, K. *et al.* (2022) Marker-assisted selection and use of molecular markers in sunflower breeding for resistance to diseases and parasites. *The Journal of VN Karazin Kharkiv National University, Series "Biology* 38(38), 14–29. DOI: 10.26565/2075-5457-2022-38-2.

Lal, R.K., Gupta, P., Chanotiya, C.S. and Sarkar, S. (2018) Traditional plant breeding in ocimum. In: Shasany, A.K. and Kole, C. (eds) *The Ocimum Genome*. Springer Nature, Cham, pp. 89–98. DOI: 10.1007/978-3-319-97430-9_7.

Langmead, B. and Salzberg, S.L. (2012) Fast gapped-read alignment with Bowtie 2. *Nature Methods* 9(4), 357–359. DOI: 10.1038/nmeth.1923.

Lei, L., Goltsman, E., Goodstein, D., Wu, G.A., Rokhsar, D.S. *et al.* (2021) Plant pan-genomics comes of age. *Annual Review of Plant Biology* 72(1), 411–435. DOI: 10.1146/annurev-arplant-080720-105454.

Liang, Y. and Liang, X. (2015) Pure line breeding method for increasing local chicken breed egg yield. Patent CN104604807A.

Li, H. (2016) Minimap and miniasm: Fast mapping and de novo assembly for noisy long sequences. *Bioinformatics* 32(14), 2103–2110. DOI: 10.1093/bioinformatics/btw152.

Li, H. (2018) Minimap2: Pairwise alignment for nucleotide sequences. *Bioinformatics* 34(18), 3094–3100. DOI: 10.1093/bioinformatics/bty191.

Li, H. (2021) New strategies to improve minimap2 alignment accuracy. *Bioinformatics* 37(23), 4572–4574. DOI: 10.1093/bioinformatics/btab705.

Li, D., Liu, C.M., Luo, R., Sadakane, K. and Lam, T.W. (2015) MEGAHIT: An ultra-fast single-node solution for large and complex metagenomics assembly via succinct de Bruijn graph. *Bioinformatics* 31(10), 1674–1676. DOI: 10.1093/bioinformatics/btv033.

Li, H., Handsaker, B., Wysoker, A., Fennell, T., Ruan, J. *et al.* (2009) The sequence alignment/map format and SAMtools. *Bioinformatics* 25(16), 2078–2079.

Li, H., Feng, X. and Chu, C. (2020) The design and construction of reference pangenome graphs with minigraph. *Genome Biology* 21(1), 1–19. DOI: 10.1186/s13059-020-02168-z.

Li, M., Sun, C., Xu, N., Bian, P., Tian, X. *et al.* (2022b) De novo assembly of 20 chicken genomes reveals the undetectable phenomenon for thousands of core genes on microchromosomes and subtelomeric regions. *Molecular Biology and Evolution* 39(4), msac066. DOI: 10.1093/molbev/msac066.

Li, W., Liu, J., Zhang, H., Liu, Z., Wang, Y. *et al.* (2022a) Plant pan-genomics: Recent advances, new challenges, and roads ahead. *Journal of Genetics and Genomics* 49(9), 833–846. DOI: 10.1016/j.jgg.2022.06.004.

Li, Y., Zhou, G., Ma, J., Jiang, W., Jin, L. *et al.* (2014) De novo assembly of soybean wild relatives for pan-genome analysis of diversity and agronomic traits. *Nature Biotechnology* 32(10), 1045–1052. DOI: 10.1038/nbt.2979.

Magoč, T. and Salzberg, S.L. (2011) FLASH: Fast length adjustment of short reads to improve genome assemblies. *Bioinformatics* 27(21), 2957–2963. DOI: 10.1093/bioinformatics/btr507.

Maji, A., Gorai, S., Hazra, S., Hasan, W., Parimala, G. *et al.* (2023) Marker-assisted breeding in vegetable crops. In: *Molecular Marker Techniques: A Potential Approach of Crop Improvement*. Springer Nature Singapore, Singapore, pp. 257–301.

Maraveas, C. (2024) Image analysis artificial intelligence technologies for plant phenotyping: Current state of the art. *AgriEngineering* 6(3), 3375–3407. DOI: 10.3390/agriengineering6030193.

Marçais, G., Delcher, A.L., Phillippy, A.M., Coston, R., Salzberg, S.L. *et al.* (2018) MUMmer4: A fast and versatile genome alignment system. *PLoS Computational Biology* 14(1), e1005944. DOI: 10.1371/journal.pcbi.1005944.

Martínez-Fortún, J., Phillips, D.W. and Jones, H.D. (2022) Natural and artificial sources of genetic variation used in crop breeding: A baseline comparator for genome editing. *Frontiers in Genome Editing* 4, 937853. DOI: 10.3389/fgeed.2022.937853.

McHale, L.K., Haun, W.J., Xu, W.W., Bhaskar, P.B., Anderson, J.E. *et al.* (2012) Structural variants in the soybean genome localize to clusters of biotic stress-response genes. *Plant Physiology* 159(4), 1295–1308. DOI: 10.1104/pp.112.194605.

Mekonnen, T., Haileselassie, T. and Tesfaye, K. (2017) Identification, mapping and pyramiding of genes/quantitative trait loci (qtls) for durable resistance of crops to biotic stresses. *Journal of Plant Pathology and Microbiology* 8(6). DOI: 10.4172/2157-7471.1000412.

Miao, J., Wei, X., Cao, C., Sun, J., Xu, Y. *et al.* (2024) Pig pangenome graph reveals functional features of non-reference sequences. *Journal of Animal Science and Biotechnology* 15(1), 32. DOI: 10.1186/s40104-023-00984-4.

Mitchell, N., Campbell, L.G., Ahern, J.R., Paine, K.C., Giroldo, A.B. *et al.* (2019) Correlates of hybridization in plants. *Evolution Letters* 3(6), 570–585. DOI: 10.1002/evl3.146.

Morgante, M., De Paoli, E. and Radovic, S. (2007) Transposable elements and the plant pan-genomes. *Current Opinion in Plant Biology* 10(2), 149–155. DOI: 10.1016/j.pbi.2007.02.001.

Murmu, S., Sinha, D., Chaurasia, H., Sharma, S., Das, R. *et al.* (2024) A review of artificial intelligence-assisted omics techniques in plant defense: Current trends and future directions. *Frontiers in Plant Science* 15, 1292054. DOI: 10.3389/fpls.2024.1292054.

Naithani, S., Deng, C.H., Sahu, S.K. and Jaiswal, P. (2023) Exploring pan-genomes: An overview of resources and tools for unraveling structure, function, and evolution of crop genes and genomes. *Biomolecules* 13(9), 1403. DOI: 10.3390/biom13091403.

Nazari, L., Aslan, M.F., Sabanci, K. and Ropelewska, E. (2023) Integrated transcriptomic meta-analysis and comparative artificial intelligence models in maize under biotic stress. *Scientific Reports* 13(1), 15899. DOI: 10.1038/s41598-023-42984-4.

Newman, M.A., Sundelin, T., Nielsen, J.T. and Erbs, G. (2013) MAMP (microbe-associated molecular pattern) triggered immunity in plants. *Frontiers in Plant Science* 4, 139. DOI: 10.3389/fpls.2013.00139.

Nishio, S., Moriya, S., Kunihisa, M., Takeuchi, Y., Imai, A. *et al.* (2023) Rapid and easy construction of a simplified amplicon sequencing (simplified AmpSeq) library for marker-assisted selection. *Scientific Reports* 13(1), 10575. DOI: 10.1038/s41598-023-37522-1.

Novielli, P., Romano, D., Pavan, S., Losciale, P., Stellacci, A.M. *et al.* (2024) Explainable artificial intelligence for genotype-to-phenotype prediction in plant breeding: A case study with a dataset from an almond germplasm collection. *Frontiers in Plant Science* 15, 1434229. DOI: 10.3389/fpls.2024.1434229.

Oh, I.S., Park, A.R., Bae, M.S., Kwon, S.J., Kim, Y.S. *et al.* (2005) Secretome analysis reveals an *Arabidopsis* lipase involved in defense against *Alternaria brassicicola*. *The Plant Cell* 17(10), 2832–2847. DOI: 10.1105/tpc.105.034819.

Orchard, C.J., Kressin, J., Chompookam, S., Chuapong, J., Onmanee, N. *et al.* (2023) Marker-assisted selection to combine alleles for four disease resistance genes of tomato collocated on chromosome 11. *HortScience* 58(5), 495–501. DOI: 10.21273/HORTSCI16982-22.

Pagliarini, M.S., Carneiro Vieira, M.L. and Borges do Valle, C. (2012) Meiotic behavior in intra-and interspecific sexual and somatic polyploid hybrids of some tropical species. In: Swan, A. (ed.) *Meiosis: Molecular Mechanisms and Cytogenetic Diversity*. IntechOpen, London, pp. 331–348. DOI: 10.5772/1232.

Paknia, R., Darvishzadeh, R., Shahriari, F., Malekzadeh, S. and Maleki, H.H. (2020) Analyses of genomic regions linked with resistance to basal stem rot in sunflower (*Helianthus annuus* L.) under field conditions. *Iranian Journal of Genetics and Plant Breeding* 9, 41–50. DOI: 10.30479/ijgpb.2020.13761.1277.

Pardal, S.J., Prihaningsih, A., Suharsono, S., Utari, R. and Sundasari, R. (2022) Backcrossing of soybean lines containing aluminium tolerance gene into superior soybean variety, Biosoy. In: *AIP Conference Proceedings*, Bogor, Indonesia (Vol. 2462, No. 1). DOI: 10.1063/5.0075187.

Patil, A. (2023) Use of artificial intelligence to hasten progress in plant genetics. *Biology and Life Sciences*. DOI: 10.20944/preprints202309.1067.v1.

Petereit, J., Bayer, P.E., Thomas, W.J.W., Tay Fernandez, C.G., Amas, J. *et al.* (2022) Pangenomics and crop genome adaptation in a changing climate. *Plants* 11, 15. DOI: 10.3390/plants11151949.

Pieterse, C.M.J., Zamioudis, C., Berendsen, R.L., Weller, D.M., Van Wees, S.C.M. *et al.* (2014) Induced systemic resistance by beneficial microbes. *Annual Review of Phytopathology* 52(1), 347–375. DOI: 10.1146/annurev-phyto-082712-102340.

Pradhan, S.K., Nayak, D.K., Mohanty, S., Behera, L., Barik, S.R. *et al.* (2015) Pyramiding of three bacterial blight resistance genes for broad-spectrum resistance in deepwater rice variety, Jalmagna. *Rice* 8(1), 51. DOI: 10.1186/s12284-015-0051-8.

Prajval, V., Krishnamoorthi, A., Thakur, R., Ruchitha, T., Kapoor, M. *et al.* (2024) From code to crop: How bioinformatics is transforming crop genomics in modern agriculture and bettering environment. *Journal of Advances in Biology & Biotechnology* 27(4), 27–49. DOI: 10.9734/jabb/2024/v27i4737.

Pratap, A., Gupta, S., Nair, R.M., Gupta, S.K., Schafleitner, R. *et al.* (2019) Using plant phenomics to exploit the gains of genomics. *Agronomy* 9(3), 126. DOI: 10.3390/agronomy9030126.

Prithiviraj, K., Suneetha, C., Lakshmi, T., Venkatachalam, K., Saikumar, P.J. *et al.* (2024) A novel approaches to crop variety development for sustainable agriculture using hybrid machine learning model. *Journal of Electrical Systems* 20(6s), 2811–2820. DOI: 10.52783/jes.3288.

Rakocevic, G., Semenyuk, V., Lee, W.-P., Spencer, J., Browning, J. *et al.* (2019) Fast and accurate genomic analyses using genome graphs. *Nature Genetics* 51(2), 354–362. DOI: 10.1038/s41588-018-0316-4.

Reddy, B.V.S., Mughogho, L.K., Narayana, Y.D., Nicodemus, K.D. and Stenhouse, J.W. (1992) Inheritance pattern of downy mildew resistance in advanced generations of sorghum. *Annals of Applied Biology* 121, 249–255. DOI: 10.1111/j.1744-7348.1992.tb03437.x.

Richter, T.E., Pryor, T.J., Bennetzen, J.L. and Hulbert, S.H. (1995) New rust resistance specificities associated with recombination in the Rp1 complex in maize. *Genetics* 141(1), 373–381. DOI: 10.1093/genetics/141.1.373.

Rooney, W.L., Collins, S.D., Klein, R.R., Mehta, P.J., Frederiksen, R.A. *et al.* (2002) Breeding sorghum for resistance to anthracnose, grain mold, downy mildew, and head smuts. *Sorghum and Millet Diseases* 47. DOI: 10.1002/9780470384923.ch47.

Saimon, A.S.M., Moniruzzaman, M., Islam, M.S., Ahmed, M. and Rahaman, M.M. (2023) Integrating genomic selection and machine learning: A data-driven approach to enhance corn yield resilience under climate change. *Journal of Environmental and Agricultural Studies* 4(2), 20–27. DOI: 10.32996/jeas.2023.4.2.6.

Scheben, A., Yuan, Y. and Edwards, D. (2016) Advances in genomics for adapting crops to climate change. *Current Plant Biology* 6, 2–10. DOI: 10.1016/j.cpb.2016.09.001.

Sedlazeck, F.J., Rescheneder, P., Smolka, M., Fang, H., Nattestad, M. *et al.* (2018) Accurate detection of complex structural variations using single-molecule sequencing. *Nature Methods* 15(6), 461–468. DOI: 10.1038/s41592-018-0001-7.

Serrano, M., Coluccia, F., Torres, M., L'Haridon, F. and Métraux, J.-P. (2014) The cuticle and plant defense to pathogens. *Frontiers in Plant Science* 5, 274. DOI: 10.3389/fpls.2014.00274.

Singh, R.P., Hodson, D.P., Jin, Y., Lagudah, E.S., Ayliffe, M.A. *et al.* (2015) Emergence and spread of new races of wheat stem rust fungus: Continued threat to food security and prospects of genetic control. *Phytopathology* 105(7), 872–884. DOI: 10.1094/PHYTO-01-15-0030-FI.

Soni, R., Ashaq, M., Manjunath, P., Pandey, S.K., Hosur, S. *et al.* (2024) Harnessing genomic resources and molecular breeding techniques for advancing crop resilience and productivity. *Plant Cell Biotechnology and Molecular Biology* 25(11–12), 124–136. DOI: 10.56557/pcbmb/2024/v25i11-128905.

Sterckeman, T., Cazes, Y. and Sirguey, C. (2019) Breeding the hyperaccumulator *Noccaea caerulescens* for trace metal phytoextraction: First results of a pure-line selection. *International Journal of Phytoremediation* 21(5), 448–455. DOI: 10.1080/15226514.2018.1537250.

Tay Fernandez, C. (2022) Making a pangenome using the iterative mapping approach. In: *Plant Bioinformatics: Methods and Protocols*. Springer US, New York, NY, pp. 259–271.

Tettelin, H., Masignani, V., Cieslewicz, M.J., Donati, C., Medini, D. *et al.* (2005) Genome analysis of multiple pathogenic isolates of *Streptococcus agalactiae*: Implications for the microbial "pangenome". *Proceedings of the National Academy of Sciences* 102(39), 13950–13955. DOI: 10.1073/pnas.0506758102.

Vakirlis, N. and McLysaght, A. (2019) Computational prediction of de novo emerged protein-coding genes. In: *Computational Methods in Protein Evolution*. pp. 63–81.

Vrček, L., Veličković, P. and Šikić, M. (2020) A step towards neural genome assembly *arXiv preprint arXiv*. DOI: 10.48550/arXiv.2011.05013.

Vrček, L., Bresson, X., Laurent, T., Schmitz, M., Kawaguchi, K. *et al.* (2024) Geometric deep learning framework for de novo genome assembly. *Genome Research* 35(4), 839–849. DOI: 10.1101/gr.279307.124.

Wang, J., Hu, H., Jiang, X., Zhang, S., Yang, W. *et al.* (2024) Pangenome-wide association study and transcriptome analysis reveal a Novel QTL and candidate genes controlling both panicle and leaf blast resistance in rice. *Rice* 17(1), 27. DOI: 10.1186/s12284-024-00707-x.

Waterhouse, R.M., Seppey, M., Simão, F.A., Manni, M., Ioannidis, P. *et al*. (2018) BUSCO applications from quality assessments to gene prediction and *phylogenomics*. *Molecular Biology and Evolution* 35(3), 543–548.

Wu, T.D. and Watanabe, C.K. (2005) GMAP: A genomic mapping and alignment program for mRNA and EST sequences. *Bioinformatics* 21(9), 1859–1875. DOI: 10.1093/bioinformatics/bti310.

Wu, Z., Li, T., Jiang, Z., Zheng, J., Liu, Y. *et al*. (2022) Graph pangenome reveals functional, evolutionary, and phenotypic significance of human nonreference sequences. *bioRxiv, 2022–09*.

Yan, G., Liu, H., Wang, H., Lu, Z., Wang, Y. *et al*. (2017) Accelerated generation of selfed pure line plants for gene identification and crop breeding. *Frontiers in Plant Science* 8, 1786. DOI: 10.3389/fpls.2017.01786.

Yaqoob, H., Tariq, A., Bhat, B.A., Bhat, K.A., Nehvi, I.B. *et al*. (2023) Integrating genomics and genome editing for orphan crop improvement: A bridge between orphan crops and modern agriculture system. *GM Crops & Food* 14(1), 1–20. DOI: 10.1080/21645698.2022.2146952.

Yu, J., Wang, J., Lin, W., Li, S., Li, H. *et al*. (2005) The genomes of Oryza sativa: A history of duplications. *PLoS Biology* 3(2), e38. DOI: 10.1371/journal.pbio.0030038.

Yuan, Y., Ton, B.L., Thomas, W.J.W., Batley, J. and Edwards, D. (2023) Supporting crop plant resilience during climate change. *Crop Science* 63(4), 1816–1828. DOI: 10.1002/csc2.21019.

Zdrazil, J., Kong, L., Klimeš, P., Jasso-Robles, F.I., Saiz-Fernández, I. *et al*. (2024) Decrypting the complex phenotyping traits of plants by machine learning. *Plant Biology*. DOI: 10.1101/2024.11.14.623623.

Zhang, H., Langham, D.R., Zhao, Y., Khalafalla, A. and Miao, H. (2021) Traditional breeding in sesame. In: Miao, H., Zhang, H. and Kole, C.Z. (eds) *The Sesame Genome (Compendium of Plant Genomes)*. Springer, Cham, pp. 145–158. DOI: 10.1007/978-3-319-98098-0_7.

Zhao, Q., Feng, Q., Lu, H., Li, Y., Wang, A. *et al*. (2018) Pan-genome analysis highlights the extent of genomic variation in cultivated and wild rice. *Nature Genetics* 50(2), 278–284. DOI: 10.1038/s41588-018-0041-z.

Zheng, C., Swenson, K., Lyons, E. and Sankoff, D. (2011) OMG! Orthologs in multiple genomes–competing graph-theoretical formulations. In: *International Workshop on Algorithms in Bioinformatics*, Springer Berlin Heidelberg, Berlin, Heidelberg, pp. 364–375. DOI: 10.1007/978-3-642-23038-7_30.

Zimin, A.V., Marçais, G., Puiu, D., Roberts, M., Salzberg, S.L. *et al*. (2013) The MaSuRCA genome assembler. *Bioinformatics* 29(21), 2669–2677. DOI: 10.1093/bioinformatics/btt476.

11 Machine Learning for Data Mining in Plant Research

Ishfaq Majid Hurrah[1], Loulaq Banday[2], Tabasum Mohi Ud Din[3], Iram Rasool[1] and Sayanti Mandal[4*]

[1]Plant Biotechnology Division, CSIR-Indian Institute of Integrative Medicine, Sanat Nagar, Srinagar, India; [2]Department of Biotechnology, University of Kashmir, Srinagar, India; [3]Department of Botany, Amar Singh College, Srinagar, India; [4]Department of Chemistry and Biochemistry, Sharda School of Engineering & Sciences, Sharda University, Greater Noida, India

Abstract

The integration of machine learning (ML) with data mining techniques is revolutionizing plant research, facilitating the uncovering of key insights from increasingly intricate and voluminous datasets. Here the foundational concepts of ML and their practical applications in analyzing diverse plant data types, including genomic, phenotypic, environmental, and sensor-derived information, are explored. Commonly used ML algorithms in the plant sciences and their value for objectives such as trait prediction, disease detection, yield estimation, and stress response analysis are highlighted. Drawing upon real-world case studies, the chapter illustrates how ML is being employed to enhance breeding efficiency, precision agriculture, and climate-resilient crop development. Fundamental tasks such as data heterogeneity, model interpretability, computational limitations, and ethical concerns are critically examined. Furthermore, the chapter outlines future directions and emerging trends, including explainable AI, multi-omics data integration, federated learning, and real-time decision support systems. Collectively, the chapter offers a broad overview of how ML is shaping the future of plant science and advancing sustainable agricultural innovation.

Keywords: machine learning, plant data mining, phenotyping and genomics, precision agriculture, artificial intelligence in agriculture, sustainable crop improvement

11.1 Introduction

Over the past few years, the merging of machine learning (ML) and plant science has unlocked novel frontiers for research, offering transformative tools to decode complex biological processes, enhance crop productivity, and ensure sustainable agriculture. With the arrival of high-throughput approaches like next-generation sequencing (NGS), advanced imaging systems, and remote sensing platforms, plant research is generating massive volumes of heterogeneous data. Uncovering important patterns within this massive data flow necessitates sophisticated analytical approaches, which is where ML plays a crucial part (Kumar *et al.*, 2020; Van Dijk *et al.*, 2021). ML is now recognized as a transformative force in plant science, significantly enhancing

*Corresponding author: mandalsayanti@gmail.com

DOI: 10.1079/9781800628649.0011

the ability of researchers to analyze complex biological data and derive meaningful insights. ML refers to computational methods that allow systems to recognize data patterns and deliver predictions or decisions autonomously, without detailed programming. In plant science, ML algorithms are increasingly utilized to analyze diverse datasets, including genomic information, phenotypic traits environmental factors, and agronomic practices. The integration of ML into plant science has enabled researchers to reveal concealed trends within massive datasets, facilitating advancements in plant breeding, genomics, phenotyping, and precision agriculture. Commonly employed ML techniques include random forests (RFs), support vector machines (SVMs), Bayesian networks, and deep learning neural networks (DLNNs). These algorithms are applied for roles such as classification, regression analysis, feature selection, anomaly detection, and image recognition (Niazian and Niedbała, 2020; Ahmad *et al.*, 2023; Ranjan *et al.*, 2024).

In plant breeding programs, ML methods improve the precision of predicating complex traits in hybrid breeding programs. By analyzing genomic data, ML models can identify genetic variations linked to vital traits like yield improvement and disease resistance. Additionally, ML-driven image recognition technologies have revolutionized plant phenotyping by automating the assessment and evaluation of traits such as leaf shape, size, color, and physiological responses in several environmental conditions (Rahaman *et al.*, 2019; Hatami *et al.*, 2024). Data mining (DM) is a perilous module of modern plant research that involves extracting valuable knowledge from large and complex datasets. It encompasses statistical methods and computational algorithms designed to discover patterns, relationships, and trends within data that might otherwise remain hidden. The application of DM techniques in plant science complements ML by providing robust frameworks for preprocessing data, identifying relevant features or traits, and validating predictive models (Swain *et al.*, 2020; Ali *et al.*, 2023). One significant area where DM contributes substantially is in high-throughput plant phenomics. Phenomics involves the systematic analysis of physical and biochemical characteristics (phenotypes) of plants at a large scale. High-throughput phenotyping spawns huge sums of image-based data that entails sophisticated analytical approaches. DM techniques such as linear discriminant analysis (LDA), RFs, and SVMs have been effectively used to classify plant status (e.g. stress vs non-stress conditions) based on extracted phenotypic features (Rahaman *et al.*, 2019; Sabat-Tomala *et al.*, 2020).

Furthermore, DM methods have proven effective in predicting crop yields by analyzing extensive datasets comprising agronomical parameters (such as planting density), genomic information (genetic markers), meteorological conditions (rainfall patterns), and soil properties. These predictive capabilities enable more informed decision-making processes for agricultural management practices. For example, advanced DM approaches have successfully predicted traits like flowering time, ear height, and yield performance in maize hybrids with greater accuracy than traditional prediction models (Ajayi *et al.*, 2024; Elbasi *et al.*, 2024). Additionally, DM facilitates soil analysis by predicting soil fertility traits based on climate conditions and soil ingredients. This allows farmers to select optimal crop types tailored specifically for particular soil conditions, thus maximizing agricultural productivity while promoting sustainable resource use. In summary, the combination of ML and DM methodologies represents a powerful approach to addressing complex challenges faced in contemporary plant research (Fig. 11.1). These technologies not only enhance our understanding of fundamental biological processes but also significantly improve agricultural productivity through optimized breeding strategies and precision farming practices (Mishra and Mishra, 2024; Titirmare *et al.*, 2024).

11.2 Overview of Data Mining in Plant Sciences

Plant sciences have entered a new era driven by data. Developments in high-throughput technologies like NGS, automated phenotyping, remote sensing, and sensor networks have directed to initiation of enormous and multifaceted datasets. These data hold the potential to unlock new biological insights, improve crop productivity, and address pressing challenges related to climate change, food security, and sustainable agriculture (Fu and Jiang, 2022; Munjal *et al.*, 2023). In plant sciences, DM helps researchers move from raw data to actionable knowledge, supporting

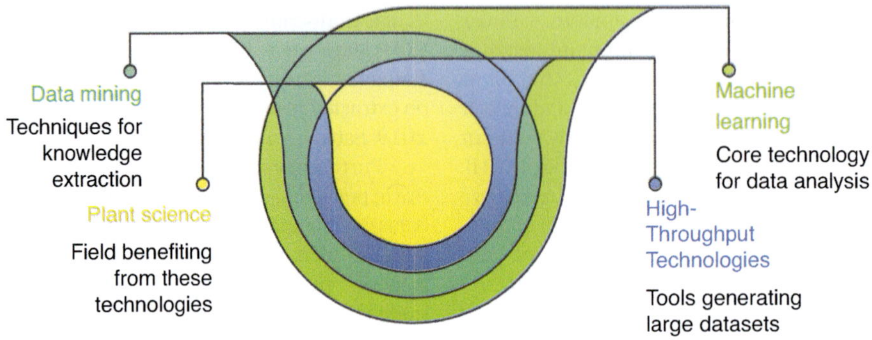

Fig. 11.1. Amalgamation of machine learning and plant science integration. (Author's own image.)

Table 11.1. Types of phenotypic data. (Author's own table.)

Morphological traits	Developmental traits	Physiological traits	Yield traits
Plant height	Flowering time	Chlorophyll content	Number of seeds
Leaf shape	Germination rate	Stomatal conductance	Fruit weight
Root architecture		Photosynthetic rate	Biomass

hypothesis generation, validation, and decision making. However, the sheer volume and complexity of these datasets require robust tools and methodologies for analysis, making DM indispensable.

11.2.1 Types and sources of plant data

The integration of ML into plant sciences is largely driven by the increasing availability of diverse and voluminous datasets. These datasets originate from a range of biological, environmental, and technological sources, each capturing different dimensions of plant behavior, growth, development, and interaction with their surroundings. Comprehension of the types and sources of these data is vital for designing effective DM pipelines and ML models.

11.2.1.1 Genomic and transcriptomic data

Genomic data represents the complete DNA blueprint of an organism, encompassing genes, regulatory sites and non-coding regions. In plant research, genomic data are pivotal for understanding heritable traits, evolutionary relationships, and gene functions. The sources include whole-genome sequencing (WGS), genotyping-by-sequencing (GBS), and single nucleotide polymorphism (SNP) arrays, while the public repositories are NCBI GenBank, Ensembl Plants, Phytozome. Transcriptomic data capture the gene expression profile under specific conditions, providing insights into the mechanism of plant response to biotic or abiotic stresses. The sources include RNA-seq data and microarrays, and the databases are gene expression omnibus (GEO) and ArrayExpress (Deng *et al.*, 2023; Chavhan *et al.*, 2024).

11.2.1.2 Phenotypic data

Phenotypic data encompass observable traits and characteristics of plants, from morphology to developmental stages and physiological responses. The sources include manual field observations and high-throughput phenotyping platforms (e.g. LemnaTec, PhenoBox), unmanned aerial vehicles (UAVs) and drones for aerial imaging, and image-based datasets using RGB, thermal, hyperspectral, or LiDAR sensors (Gano *et al.*, 2024; Angidi *et al.*, 2025) (Table 11.1).

11.2.1.3 Environmental data

Environmental data provides the context within which plant traits are expressed. These data are critical for understanding genotype ×

Table 11.2. Environmental variables. (Author's own table.)

Key environmental variables	References
Temperature, humidity, precipitation	Yadav *et al.*, 2020; Shree *et al.*, 2023
Soil type and composition	Yu *et al.*, 2020; Ahmad *et al.*, 2022
Solar radiation and light intensity	Li *et al.*, 2020; Ali *et al.*, 2023
Wind speed and direction	Ganti *et al.*, 2022; Xie *et al.*, 2022

Table 11.3. Sources of environmental data. (Author's own table.)

Sources	References
Remote sensing (satellite and aerial imagery)	El Mahrad *et al.*, 2020; Ouhami *et al.*, 2021
Ground-based weather stations	Kokhan *et al.*, 2020; Kępińska-Kasprzak and Struzik, 2023
IoT-enabled sensor networks (soil moisture sensors, weather nodes)	Sharma *et al.*, 2021; Ataei Kachouei *et al.*, 2023
Databases: NASA EarthData, WorldClim, Copernicus Climate Data Store	Sulova and Jokar Arsanjani, 2020; Chamorro-Padial *et al.*, 2024

environment (G×E) interactions (Costa-Neto *et al.*, 2021) (Tables 11.2 and 11.3).

11.2.1.4 Pathological and stress-related data

This type of data pertains to plant health, particularly in response to biotic (pathogens, pests) and abiotic (drought, salinity, heat) stress factors. The sources include imaging tools (e.g. spectral cameras, fluorescence sensors), laboratory assays and diagnostics, and annotated image databases for plant disease detection (e.g. PlantVillage) (Ambrosino *et al.*, 2020; Almeida-Silva and Venancio, 2022; Shakespear *et al.*, 2025) (Table 11.4).

Table 11.4. Examples of pathological and stress-related data. (Author's own table.)

Examples	References
Presence of disease symptoms (e.g. leaf spots, wilting, discoloration)	Dang *et al.*, 2024; Zhang *et al.*, 2024b
Pathogen infection levels (e.g. fungal spores, viral load)	Allwood *et al.*, 2021; Singh Saharan *et al.*, 2023
Stress-response gene expression profiles	Ambrosino *et al.*, 2020; Gupta *et al.*, 2023
Physiological indicators of stress (e.g. electrolyte leakage, ROS production)	Abdelaal *et al.*, 2021; Aina *et al.*, 2024

11.2.1.5 Agronomic and management data

Agronomic data reflect the influence of cultivation practices on plant performance, essential for precision agriculture and decision support. The sources include farm management software, field logs and extension services and Internet of Things (IoT) devices and automated farming systems (Ali *et al.*, 2023; Williamson *et al.*, 2023) (Table 11.5).

Table 11.5. Examples of agronomic management data. (Author's own table.)

Examples	References
Fertilizer application rates and schedules	Saiz-Rubio and Rovira-Más, 2020; Cravero *et al.*, 2022
Irrigation frequency and volume	Silva *et al.*, 2020; Torres-Sanchez *et al.*, 2020
Crop rotation and tillage practices	MacLaren *et al.*, 2021; Su *et al.*, 2021
Pesticide and herbicide use	Williamson *et al.*, 2023; Islam and Monjardino, 2025

11.2.1.6 Multi-omics and integrated data

Modern plant science increasingly employs multi-omics approaches that integrate data across multiple biological layers to gain a systems-level understanding. The sources include mass spectrometry, NMR spectroscopy, liquid chromatography (LC)–MS, and gas chromatography (GC)–MS platforms. The databases are plant metabolic network (PMN), PRIDE archive, and KEGG (Subramanian *et al.*, 2020; He *et al.*, 2022; Charagh *et al.*, 2024) (Table 11.6).

The diversity of plant data ranging from molecular to field-level information offers

Table 11.6. Examples of omics approaches. (Author's own table.)

Types	References
Proteomics (protein abundance and interactions)	Roychowdhury *et al.*, 2023; Sanches *et al.*, 2024
Metabolomics (metabolite profiles and pathways)	Eicher *et al.*, 2020; Shen *et al.*, 2023
Epigenomics (DNA methylation and histone modifications)	Wang *et al.*, 2023; Dao *et al.*, 2025

tremendous potential for DM and ML applications (Fig. 11.2). However, it also introduces challenges related to data integration, normalization, and interpretation. A robust understanding of these data types and their sources is foundational to developing successful computational models and unlocking meaningful biological insights in plant research.

11.3 Fundamentals of Machine Learning and Algorithms Commonly Used in Plant Research

ML has arisen as a transformative tool in plant research, allowing the discovery of complex designs in large, high-dimensional datasets which are often challenging to interpret with traditional statistical methods. ML algorithms do improve with data, advance over time, and make accurate predictions or classifications with minimal human intervention. This section outlines the fundamental concepts of ML and provides an overview of commonly used algorithms in plant science applications (Demidchik *et al.*, 2020; Cherian and Kumar, 2023). ML is a subfield of artificial intelligence (AI) focused on evolving algorithms that can automatically adapt and make decisions or predictions based on data. Rather than following explicit programming rules, ML models identify patterns within data and adapt their internal structure to improve performance on specific tasks. In plant research, ML is utilized to predict crop yield based on soil and weather conditions, identify plant diseases from leaf images, classify genotypes based on gene expression profiles, and model stress responses using environmental and physiological data (Van Dijk *et al.*, 2021; Chayalakshmi *et al.*, 2025).

11.3.1 Classifications of machine learning

ML techniques utilized in plant research are broadly classified into four core categories: supervised learning, unsupervised learning, semi-supervised learning, and reinforcement

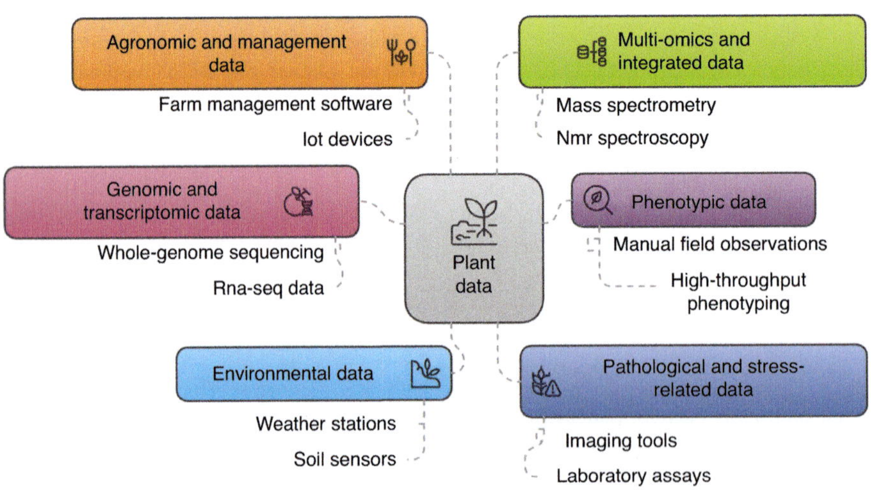

Fig. 11.2. Types and sources of plant data. (Author's own image.)

learning. Each class assists diverse analytical purposes subject to the nature of the data and the intended outcome.

11.3.1.1 Supervised learning

Supervised learning is the utmost class in plant research. It includes preparing a model on a labeled dataset, where each input data point is associated with a known output or label. The model learns the mapping between input features such as environmental variables, genetic markers, or image characteristics and corresponding target outcomes like plant height, disease presence, or yield. This procedure is extensively useful in trait prediction, disease classification, and yield estimation, providing accurate and actionable insights for plant breeders and agronomists (Su, 2020; Shetty *et al.*, 2022).

11.3.1.2 Unsupervised learning

Unsupervised learning, in contrast, deals with unlabeled data. The goal here is to uncover hidden patterns, relationships, or structures within the dataset without prior knowledge of output variables. In plant science, unsupervised learning is particularly useful for grouping plant varieties based on phenotypic similarities, detecting gene expression patterns, and clustering traits under environmental stress. This approach helps researchers explore complex datasets and generate hypotheses without predefined categories (Yan and Wang, 2022; Naeem *et al.*, 2023).

11.3.1.3 Semi-supervised learning

Semi-supervised learning bridges the gap between supervised and unsupervised learning. It influences a small set of labeled data along with a bulky volume of unlabelled data, making it very apt for scenarios where data annotation is costly, time-consuming, or limited. In plant research, this approach has found utility in classifying rare plant diseases from limited samples and in segmenting satellite or drone imagery for monitoring crop health or land use, especially in data-scarce regions (Riese and Keller, 2020; Zhou, 2021).

11.3.1.4 Reinforcement learning

Reinforcement learning is an emerging paradigm in plant sciences, where an agent interacts with an environment and learns to make optimal decisions through trial and error, guided by a system of rewards and penalties. Though still in its early stages of application within plant research, it shows promise in automated crop management systems, greenhouse climate control, and robotics-based weeding and harvesting. Its potential lies in developing intelligent, adaptive systems that can operate autonomously in real-time agricultural environments (Kelleher *et al.*, 2020; Gautron *et al.*, 2022).

11.3.2 Key machine learning algorithms in plant research with supervised learning algorithms

Several supervised ML algorithms have proven effective in plant research due to their robustness, interpretability, and performance across various biological and environmental datasets.

11.3.2.1 Linear and logistic regression

Linear and logistic regression are amongst the most modest yet most interpretable models. Linear regression is primarily used for predicting continuous variables, such as biomass yield based on nutrient inputs, while logistic regression is suited for binary classification roles, such as determining whether a plant is healthy or diseased. Despite their simplicity, these models are useful for quick, explainable insights and are often used as baselines in comparative studies (Sun *et al.*, 2020; Nagarhalli *et al.*, 2021).

11.3.2.2 Decision trees and random forests

These models offer a more flexible and powerful alternative. Decision trees split data based on feature thresholds to make predictions, whereas random forests build collective decision trees to improve generalizability and reduce overfitting. These models are commonly used for predicting crop yield, identifying important phenotypic or environmental predictors, and establishing genotype–phenotype associations (Kumar *et al.*, 2022; Salman *et al.*, 2024).

11.3.2.3 Support vector machines

SVMs are operational in high-dimensional feature spaces and are particularly well suited for classification problems. In plant research, they are used to classify plant species based on morphological traits or image features and to identify disease types based on gene expression profiles or leaf image analysis (Aliyu *et al.*, 2020; Montesinos López *et al.*, 2022).

11.3.2.4 k-Nearest neighbors (k-NN)

k-NN is a non-parametric algorithm that categorizes data points centered on the majority label of their nearest neighbors. It is particularly useful in plant variety identification and image-based phenotyping tasks, where similarities between samples in high-dimensional space can be leveraged for classification (Guo *et al.*, 2021; Mladenova and Valova, 2023).

11.3.2.5 Gradient boosting machines

Gradient boosting machines are advanced ensemble methods that build predictive models in a stage-wise fashion, correcting the errors of previous models at each step. These models are known for their high accuracy and have been widely applied in complex tasks such as yield prediction under variable stress conditions, modeling environmental responses, and trait prioritization in breeding programs (Aksoy and Genc, 2023; Kumar *et al.*, 2024).

11.3.3 Key machine learning algorithms in plant research with unsupervised learning algorithms

Unsupervised learning methods are essential tools for investigative analysis of data in plant research, uncovering latent patterns and relationships within datasets that are not immediately evident.

11.3.3.1 K-means clustering

This is a popular algorithm paneling datasets into a specified number of clusters based on feature similarity. In plant science, it is used to group genotypes based on phenotypic traits or stress responses, facilitating the identification of subpopulations with desirable characteristics (Singh *et al.*, 2021; Yan and Wang, 2022).

11.3.3.2 Principal component analysis (PCA)

PCA is a dimensionality reduction technique that transforms high-dimensional data into a lower-dimensional space, making it easier to visualize and interpret. PCA is frequently used in omics studies to identify key sources of variation, such as in transcriptomics or metabolomics data, and to visualize trait diversity among plant accessions (Lu *et al.*, 2023; Younes *et al.*, 2023; Raza *et al.*, 2025).

11.3.3.3 Hierarchical clustering

This method builds nested groupings of data points based on similarity or distance metrics and is especially useful in genomic and transcriptomic analyses. Applications include the construction of gene co-expression networks and varietal classification based on multi-trait or environmental profiles. Its visual output, often in the form of dendrograms, aids in understanding complex relationships among plant traits or genotypes (Siregar and Budiarto, 2021; García-Ariza *et al.*, 2022).

11.3.4 Deep learning algorithms

11.3.4.1 Deep learning (DL)

DL, a generalized subsection of ML, engages artificial neural networks (ANNs) with numerous layers to learn highly complex, non-linear relationships from large-scale and high-dimensional data. Its capacity to automatically extract relevant features makes it particularly advantageous in plant research where raw data such as images, time-series, or spatial data can be voluminous and intricate (Sujatha *et al.*, 2021; Van Dijk *et al.*, 2021).

11.3.4.2 Convolutional neural networks (CNNs)

CNNs are designed to process image and spatial data by learning hierarchical feature representations. They are widely used in plant disease detection, leaf shape classification, and object detection in drone or satellite imagery. Their success in image-based phenotyping has significantly advanced precision agriculture (Nazki *et al.*, 2020; Yan and Wang, 2022).

11.3.4.3 Recurrent neural networks (RNNs)

RNNs and their alternatives like long short-term memory (LSTM) networks are tailored for chronological data analysis. These models have been applied in modeling plant growth dynamics over time, forecasting environmental stress conditions, and analyzing time-series sensor data for smart irrigation systems or yield prediction (Gnanasaravanan *et al.*, 2021; Singh and Kumar, 2024).

11.3.4.4 Autoencoders

Autoencoders are unsupervised neural networks that acquire effectual coding of input data. They are mainly beneficial for feature extraction and dimensionality reduction in high-throughput phenotyping datasets, such as hyperspectral imaging or multi-modal sensor outputs. Autoencoders help compress data while retaining the most informative features for downstream tasks (Miranda *et al.*, 2022; Boukhris *et al.*, 2024).

Transformer models, initially developed for natural language processing, are now being explored in plant research for their ability to handle long-range dependencies in data. They are proving useful in remote sensing data analysis, spatio-temporal modeling of environmental variables, and fusing multimodal data sources such as genomic, phenotypic, and weather data to improve prediction accuracy in complex systems. Understanding the fundamentals of ML and the characteristics of commonly used algorithms provides a strong foundation for applying these tools in plant science. Whether for trait prediction, stress detection, or image-based phenotyping, ML offers powerful, scalable, and adaptable solutions for modern plant research challenges. As plant datasets endure to nurture in size and complexity, mastering these tools will be increasingly essential for the next generation of plant scientists (Huang *et al.*, 2023; Zhang *et al.*, 2023a, b).

11.3.5 Evaluation metrics for plant-related ML models

The performance of ML models in plant research is gauged using various evaluation metrics tailored to the type of problem being addressed—classification, regression, clustering, image analysis, or model robustness. These metrics help quantify model accuracy, reliability, and applicability to real-world plant science problems.

11.3.5.1 Classification metrics

Classification metrics are employed when the ML model categorizes plant data into discrete classes, such as plant species, disease states, or stress levels. One of the most commonly used metrics is accuracy, which is the ratio of appropriately forecast models to the total number of models. For example, accuracy is a straightforward metric for evaluating models that classify healthy vs diseased plants from leaf images. However, accuracy alone can be misleading, especially in imbalanced datasets. Therefore, precision, recall, and F1-score are critical metrics (Saleem *et al.*, 2020; Shrotriya *et al.*, 2024). Precision refers to the fraction of true positive predictions among all predicted positives, whereas recall (or sensitivity) processes the proportion of true positives among all actual positives. The F1-score, being the harmonic mean of precision and recall, is particularly valuable when class distribution is skewed, as in the detection of rare plant diseases like leaf blight. Moreover, the confusion matrix affords a comprehensive analysis of model predictions across multiple classes by tabulating true positives, true negatives, false positives, and false negatives, aiding in a deeper understanding of misclassifications. For binary classification problems, such as distinguishing weeds from crops, the receiver operating characteristic (ROC) curve and area under the curve (AUC) are effective tools. The ROC curve plots the true positive rate against the false positive rate, while the AUC quantifies the model's ability to distinguish between classes (Dadashzadeh *et al.*, 2020; Kamath *et al.*, 2020).

11.3.5.2 Regression metrics

Regression metrics are applied when ML models are used to envisage continuous variables, like crop yield, biomass, or chlorophyll content. The mean absolute error (MAE) computes the average absolute differences between predicted and actual values, rendering it a direct metric for tasks like estimating plant biomass using spectral data. Mean squared error (MSE) and root mean squared error (RMSE) are also extensively utilized, with MSE penalizing larger errors more significantly and RMSE providing interpretable results in the same units as the predicted variable. These are particularly relevant in tasks like forecasting yield or assessing drought tolerance. Another critical metric is R-squared (R^2), also identified as the coefficient of determination, which designates the section of variance in the target variable explained by the model. This is especially useful for assessing how well genotype and environmental data predict phenotypic traits (Adak *et al.*, 2021; Correndo *et al.*, 2022).

11.3.5.3 Clustering and unsupervised metrics

Clustering and unsupervised metrics are essential when models are applied to unlabeled data, such as grouping plant genotypes or clustering phenotypic traits. The Silhouette Score processes the fitting of data points in its allocated cluster compared to other clusters, making it useful for validating the clustering of plant cultivars. The Davies–Bouldin index evaluates intra-cluster similarity and inter-cluster differences, where lower values suggest better clustering performance useful in tasks such as grouping gene expression profiles. When some level of ground truth is available, the adjusted rand index (ARI) is also utilized to compare predicted cluster groupings to known classifications (Sun *et al.*, 2020; Benfenati *et al.*, 2023).

11.3.5.4 Image-based evaluation metrics

These are widely used in plant phenotyping applications involving computer vision, such as disease detection, leaf segmentation, and plant part counting. The intersection over union (IoU) metric assesses the intersection among predicted segmentation masks and ground truth regions, rendering appropriate for tasks like leaf segmentation from aerial or drone imagery. For object detection problems, the mean average precision (mAP) is a standard metric, that averages precision across classes and intersection thresholds, often used for detecting multiple plant structures like leaves, flowers, or fruits. The dice coefficient, akin to the F1 score but applied to pixel-level classification, is particularly effective for evaluating segmentation in plant disease detection tasks (Jiang and Li, 2020; Yang *et al.*, 2023a).

11.3.5.5 Model generalization and robustness metrics

The model generalization and robustness metrics ensure that models do not merely perform well on training data but also maintain accuracy across unseen datasets. Cross-validation scores, derived from techniques like k-fold validation, help assess performance consistency across different subsets of data, which is crucial in genotype–phenotype mapping. Additionally, overfitting indicators—such as a significant performance gap between training and testing datasets—can signal a lack of generalizability. Addressing overfitting by using appropriate regularization techniques and diverse data sources helps create more reliable and deployable ML models in plant research (Chai *et al.*, 2020; Classen *et al.*, 2020).

11.4 Applications of Machine Learning in Plant Research

ML has turned out to be a dynamic tool in plant research, revolutionizing how data are interpreted, patterns are discovered, and

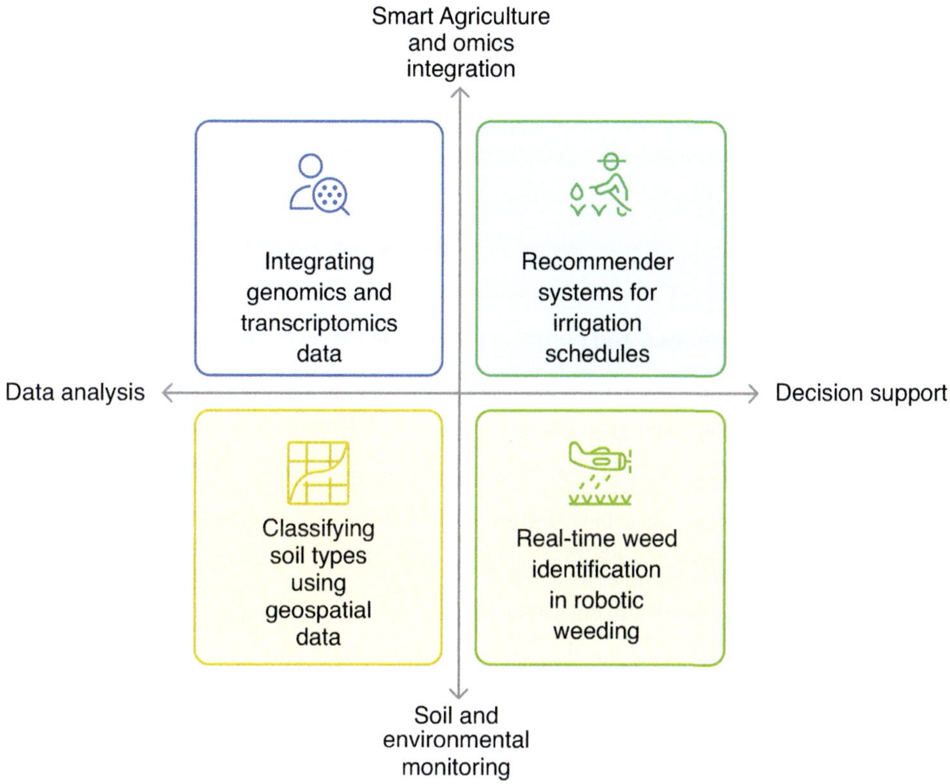

Fig. 11.3. Applications of machine learning in plant research. (Author's own image.)

decisions are made (Fig. 11.3). From basic plant biology to advanced crop improvement strategies, ML techniques enable researchers to analyze large-scale, complex datasets in ways that were previously impractical or impossible. This section discovers the wide-ranging uses of ML across several domains of plant science, highlighting how it contributes to innovations in agriculture, breeding, and sustainable plant management.

11.4.1 Plant phenotyping and image-based analysis

One of the most visible impacts of ML in plant research is in automated phenotyping, the process of extracting observable plant traits using imaging technologies and computer vision. It has wider applications such as detecting and quantifying morphological traits such as area, root length, stem thickness, and canopy cover; segmenting plant organs in RGB, hyperspectral, and thermal images using DL techniques such as CNNs; monitoring plant growth over time through time-series image analysis; and high-throughput phenotyping in field and greenhouse environments using drones, UAVs, and robotic platforms. One such example is that CNN-based models have been developed to detect and quantify drought stress in maize by analyzing changes in leaf wilting from multispectral images (Yang *et al.*, 2024; Murphy *et al.*, 2024).

11.4.2 Plant disease and pest detection

ML algorithms, especially those in the realm of DL and pattern recognition, have shown exceptional promise in detecting plant diseases and pest infestations at early stages. It has applications in the classification of plant leaf images to identify diseases (e.g. blight, rust, mildew), differentiation between biotic and abiotic stress symptoms using hyperspectral and fluorescence data, and real-time pest identification using edge devices (e.g. mobile phones, embedded sensors) (Liu and Wang, 2021; Wang *et al.*, 2025a).

11.4.3 Genomic selection and plant breeding

In modern plant breeding, ML supports the selection of superior genotypes by predicting phenotypic traits based on genomic data, a process known as genomic selection. It has applications in: predicting quantitative traits such as yield, biomass, or stress tolerance from SNPs and genotyping data; identifying candidate genes and genetic markers associated with desired traits using feature selection algorithms; and accelerating breeding cycles by prioritizing genotypes with high predicted performance. RFs and SVMs have been successfully used to envisage grain yield and flowering time in wheat and rice using genomic data, improving selection accuracy over traditional methods (Bayer *et al.*, 2021; Alemu *et al.*, 2024).

11.4.4 Prediction of crop yield

ML models are extensively utilized to estimate crop yield by integrating multidimensional datasets, including weather, soil, remote sensing, and management practices. It has applications in regional and plot-level yield prediction for precision agriculture, identifying environmental drivers of yield variability through model interpretation, and integrating time-series satellite data with weather data for forecasting purposes. Gradient boosting machines and LSTM networks have been engaged to forecast soybean and maize yields with high spatial and temporal resolution (Iniyan *et al.*, 2023; Morales and Villalobos, 2023).

11.4.5 Abiotic stress monitoring and prediction

ML enables the detection and forecasting of plant responses to abiotic stresses such as drought, salinity, temperature extremes, and nutrient deficiencies (Raza *et al.*, 2023). It has applications in: early stress detection through spectral signature analysis and time-series data; predicting stress tolerance from transcriptomic and metabolomic profiles; and modeling genotype × environment interactions to assess stress resilience. Unsupervised learning methods like clustering and PCA have been used to group cultivars based on their response patterns to salinity and heat stress (Gou *et al.*, 2024; Zhang *et al.*, 2024a).

11.4.6 Soil and environmental monitoring

ML is instrumental in analyzing complex environmental and soil datasets to inform precision farming decisions. It has applications in: classifying soil types and fertility zones using geospatial data; predicting soil nutrient levels, pH, and organic content; and mapping and forecasting microclimatic conditions using sensor networks and satellite data. SVMs and k-NN algorithms have been used to model soil organic carbon content from satellite imagery and *in-situ* measurements, aiding in soil health assessment (Padarian *et al.*, 2020; Diaz-Gonzalez *et al.*, 2022).

11.4.7 Weed detection and crop–weed differentiation

Automated weed detection is an essential section of site-specific weed management systems in sustainable agriculture. It has applications in: distinguishing crop plants from weeds using object detection algorithms in UAV imagery; real-time weed identification in robotic weeding platforms using CNNs; and reducing herbicide usage through targeted application based on ML classification. DL-based semantic segmentation models can identify and localize weed patches within crop rows, improving herbicide application efficiency (Hu *et al.*, 2024; Qu and Su, 2024).

11.4.8 Omics data integration and systems biology

Plant systems biology progressively depends on ML for multi-omics data integration, helping to comprehend the multifaceted networks that control plant functions. It has applications in integrating genomics, transcriptomics, proteomics, and metabolomics data to recognize regulatory pathways (Charagh *et al.*, 2024); inferring gene regulatory networks and identifying hub genes; and predicting metabolite concentrations and pathway fluxes under varying conditions. Bayesian networks and DL models have been used to infer gene interaction networks in Arabidopsis, giving insights into stress-response mechanisms (Jamil *et al.*, 2020; Krantz *et al.*, 2021).

11.4.9 Smart agriculture and decision support systems

ML is foundational to smart farming or Agriculture 4.0, where data-driven decision making optimizes resource use and crop productivity. It has applications in: recommender systems for irrigation and fertilization schedules; decision support tools for crop selection, rotation, and planting times; and integration with IoT devices for real-time monitoring and control of field conditions. Reinforcement learning algorithms have been projected to optimize irrigation schedules in real time by minimizing water use while maintaining crop health (Zhai *et al.*, 2020; Suresh and Manjupriya, 2022).

ML is driving a paradigm shift in plant research and agriculture by transforming data into actionable insights through an extensive series of applications from phenotyping and breeding to stress monitoring and smart farming. As more data become available through sensors, drones, and sequencing technologies, ML will continue to play a progressively critical role in addressing global agricultural challenges, improving crop productivity, and enhancing sustainability.

11.5 Case Studies and Real-World Applications

The applications of ML in plant research are no longer confined to theoretical or laboratory-based experiments. Today, it underpins a range of real-world agricultural technologies and scientific initiatives that have demonstrably enhanced crop productivity, disease management, and decision making. This section presents selected case studies and real-world implementations that exemplify how ML is being leveraged to solve complex plant science challenges and contribute to sustainable agriculture.

11.5.1 Deep learning for plant disease diagnosis (PlantVillage project)

Plant diseases are a major limitation in crop production, particularly in evolving regions where diagnostic expertise is scarce. The PlantVillage project, a collaboration between Penn State University and various international partners, developed a large, annotated dataset of plant leaf images across species and disease types. A deep CNN was competent on more than 50,000 labeled images spanning 38 classes (crop–disease pairs). The model achieved an accuracy exceeding 99% on training data and performed robustly on unseen datasets. The CNN model was deployed via a mobile application, enabling farmers to diagnose diseases in real time using smartphone cameras. It has been adopted in countries like Kenya, India, and Uganda, empowering smallholder farmers with accessible disease-detection tools (Mindhe *et al.*, 2020; Moupojou *et al.*, 2023).

11.5.2 Yield prediction in maize using machine learning

Accurate crop yield forecasting is vital for food security, especially in regions prone to climate variability. A collaborative project in sub-Saharan Africa explored ML models for yield prediction using environmental, management, and remote sensing data. RF, Gradient Boosting, and ANN models were trained on variables such as rainfall, temperature, NDVI (normalized difference vegetation index), and soil characteristics. The models predicted maize yield with high precision ($R^2 > 0.85$), outperforming traditional regression methods. These models were integrated into decision support systems to inform policy and resource allocation. Farmers received predictive insights weeks ahead of harvest, enabling better planning for logistics, storage, and market timing (Croci *et al.*, 2022; Kuradusenge *et al.*, 2023).

11.5.3 Genomic selection in wheat breeding (CIMMYT)

The International Maize and Wheat Improvement Center (CIMMYT) implemented ML techniques to enhance genomic selection in wheat breeding programs. SVMs, ridge regression, and gradient boosting models were used to predict yield, flowering time, and disease resistance based on genome-wide SNP markers. Feature selection methods helped identify informative markers, reducing computational load and improving accuracy. Accelerated selection cycles and improved identification of elite genotypes with desirable traits. Breeding efficiency improved by 20–30%, translating to faster development of high-yielding, disease-resistant wheat varieties (Juliana *et al.*, 2020; Dreisigacker *et al.*, 2021).

11.5.4 UAV-based weed detection in precision agriculture

Weed competition is a major yield reducer in many crops. A commercial farming operation in Australia deployed drone-based imaging systems for site-specific weed management. High-resolution aerial imagery was collected using drones. A DL model (YOLO; You Only Look Once) was trained to detect and localize weed species within crop fields. Semantic segmentation algorithms were used to generate weed maps, which enabled targeted herbicide application, reducing chemical usage by over 60%. The result was lower input costs and environmental impact, while maintaining crop health and yield (Phang *et al.*, 2023; Raptis *et al.*, 2023).

11.5.5 Early drought stress detection in soybean using multispectral data

Soybean is highly sensitive to drought during its reproductive stage. Early stress detection is crucial to mitigate yield losses. Multispectral imagery from UAVs and ground sensors was analyzed using PCA and k-means clustering. Supervised ML algorithms such as SVM and RF were trained to classify stress severity levels. Accurate drought-stress maps were generated before visual symptoms appeared. Farmers could prioritize irrigation and apply anti-stress agents selectively, enhancing water use efficiency (Zhang *et al.*, 2023c; Jones *et al.*, 2024).

11.5.6 Real-world deployment: John Deere's ML-powered crop intelligence platform

Agricultural equipment giant John Deere has incorporated ML into its precision farming tools. ML models process data from GPS-guided tractors, soil sensors, weather stations, and satellite imagery. They can be used for real-time decision making on seeding rates, pesticide application, and nutrient management, and improved field-level precision, resulting in better resource use and increased profitability. The platform supports millions of hectares globally, reflecting the scalability of ML in large-scale agriculture (Bogue, 2023; Kirpach and Riccoboni, 2023).

These case studies reflect the transformative potential of machine learning in real-world plant research and agricultural applications. Whether in disease detection, yield forecasting, breeding, or environmental monitoring, ML models are enabling data-driven decisions that enhance productivity, resilience, and sustainability. The growing availability of open datasets, improved computational power, and collaboration between plant scientists and data scientists will further accelerate the adoption and impact of ML in plant science.

11.6 Challenges and limitations

Despite the transformative potential of ML in plant research, several key challenges hinder its widespread adoption and effectiveness. One of the most significant obstacles is the handiness and class of data. Plant-related datasets are often limited in size, poorly annotated, and vary in format across different domains such as genomics, phenomics, and environmental sensing. Incomplete or noisy data can degrade model performance, while the heterogeneity of data sources complicates integration and standardization. Additionally, many ML models particularly DL approaches are inherently opaque, offering limited interpretability. This lack of transparency can lead to skepticism among end-users such as breeders and agronomists, who often need to understand the rationale behind predictions for informed decision making (Al-Badri *et al.*, 2022; Hesami *et al.*, 2022). Furthermore, the generalizability of models remains a pressing issue. Algorithms trained on specific crops, genotypes, or regions may perform poorly when applied to new conditions due to environmental variability and domain shift, necessitating constant model retraining and adaptation. Another significant challenge lies in the computational demands and technical infrastructure required to process and analyze large-scale plant data. High-performance computing resources are often inaccessible to smaller research institutions or those in developing regions, limiting the scalability of ML applications (Cravero *et al.*, 2022; Williamson *et al.*, 2023). The amalgamation of multimodal data such as combining molecular, imaging, and climate information is also complex due to differences in resolution, scale, and data structure, compounded by a lack of standardized protocols. Ethical and legal concerns further complicate the landscape, especially regarding data ownership, consent, and potential algorithmic bias that may disproportionately affect underrepresented crops or farming communities.

Finally, a shortage of interdisciplinary expertise continues to impede progress. Many plant scientists lack formal training in ML, while data scientists may not fully grasp biological complexities, leading to a gap in effective collaboration. Addressing these challenges through improved data practices, model interpretability, infrastructure investment, ethical frameworks, and cross-disciplinary training is necessary for harnessing the full potential of ML in evolving plant science and agriculture. Nevertheless, these challenges also present prospects for innovation and growth (Maganathan *et al.*, 2020; Soltis *et al.*, 2020). Advancements in open-access data platforms, interpretable AI, edge computing, and domain adaptation techniques can help overcome current barriers. Promoting cross-disciplinary education and creating robust frameworks for data governance and ethics can further strengthen the integration of ML into plant research. With continued efforts, the future holds significant promise for more transparent, scalable, and inclusive ML-driven solutions in agriculture and plant science.

11.7 Future directions and emerging trends

As the intersection of ML and plant research continues to evolve, new directions and technological advancements are opening up transformative possibilities for data mining in plant sciences (Fig. 11.4). These developments aim not only to address present restrictions but also to develop the possibility and depth of applications in agriculture, plant biology, and environmental sustainability. One of the most significant emerging trends is the integration of explainable AI (XAI) and interpretable ML models, aiming to elucidate the "black-box" nature of complex algorithms (Araújo *et al.*, 2023; Murmu *et al.*, 2024). By providing clearer insights into how predictions are made, these approaches foster greater trust and usability among plant scientists, agronomists, and policy makers. Coupled with this is the growing emphasis on federated learning and privacy-preserving ML, which allow model training across multiple decentralized datasets without compromising data privacy, an especially promising development for collaborative plant breeding and phenotyping projects involving multiple institutions

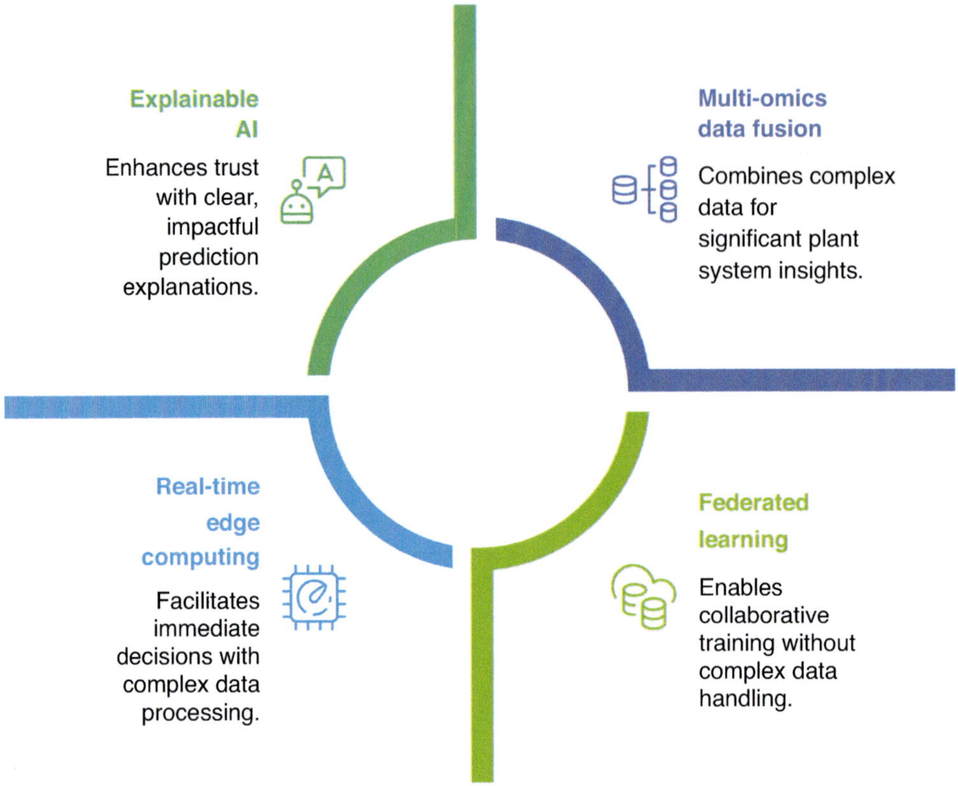

Fig. 11.4. Future directions of machine learning in plant research. (Author's own image.)

or countries (Prodhan *et al.*, 2022; Zhang and Wang, 2023).

Another key direction is the adoption of multi-omics and multimodal data fusion. By combining genomic, transcriptomic, metabolomic, phenotypic, and environmental data, researchers are gaining more holistic insights into plant systems biology. Advanced ML models, particularly graph neural networks and transformers, are being adapted to manage and learn from these complex, high-dimensional datasets. These tools are paving the way for discoveries in crop trait finding, gene function prediction, and plant–environment interaction analysis (Yang *et al.*, 2023b; Wang *et al.*, 2024). Additionally, real-time and edge computing capabilities are transforming how ML is applied in the field. Through the amalgamation of IoT devices, UAVs, and portable sensors, ML models can now process data at the source, enabling near-instantaneous decisions on irrigation, pest control, and crop health management. AutoML (automated ML) is also emerging as a powerful tool for non-experts, automating the process of model selection, tuning, and evaluation to democratize the use of ML in plant research. The rise of synthetic data generation using generative models like generative adversarial networks is helping address the problem of limited training data, particularly in plant disease diagnostics and rare phenotype classification. Meanwhile, domain adaptation and transfer learning are enhancing the ability of models to generalize across different crops, geographies, and growing conditions, making ML solutions more robust and scalable (Rizvi *et al.*, 2021; Rane *et al.*, 2024).

Furthermore, climate-smart agriculture is becoming a focal point, where ML models are used not just to optimize productivity but also to improve sustainability, carbon sequestration, and resilience to climate change. Predictive analytics for carbon fluxes, drought tolerance, and heat stress responses are gaining traction in breeding programs and policy frameworks. In terms of collaboration and dissemination, open science initiatives and data-sharing platforms like the Plant Phenome Journal, CyVerse, and the European Plant Phenotyping Network are promoting standardized protocols, reproducibility, and global cooperation, critical components for the future success of ML in plant research (Weil *et al.*, 2023; Wang *et al.*, 2025b). In conclusion, the future of ML in plant research lies in more integrative, transparent, and accessible systems. As computing power grows and interdisciplinary partnerships strengthen, ML will play an increasingly central role in addressing global food security, biodiversity conservation, and agricultural sustainability.

References

Abdelaal, K., AlKahtani, M., Attia, K., Hafez, Y., Király, L. *et al.* (2021) The role of plant growth-promoting bacteria in alleviating the adverse effects of drought on plants. *Biology* 10(6), 520. DOI: 10.3390/biology10060520.

Adak, A., Murray, S.C., Božinović, S., Lindsey, R., Nakasagga, S. *et al.* (2021) Temporal vegetation indices and plant height from remotely sensed imagery can predict grain yield and flowering time breeding value in maize via machine learning regression. *Remote Sensing* 13(11), 2141. DOI: 10.3390/rs13112141.

Ahmad, A., Saraswat, D. and El Gamal, A. (2023) A survey on using deep learning techniques for plant disease diagnosis and recommendations for development of appropriate tools. *Smart Agricultural Technology* 3, 100083. DOI: 10.1016/j.atech.2022.100083.

Ahmad, Z., Mulk Khan, S., Page, S., Alamri, S. and Hashem, M. (2022) Plants predict the mineral mines - A methodological approach to use indicator plant species for the discovery of mining sites. *Journal of Advanced Research* 39, 119–133. DOI: 10.1016/j.jare.2021.10.005.

Aina, O., Bakare, O.O., Fadaka, A.O., Keyster, M. and Klein, A. (2024) Plant biomarkers as early detection tools in stress management in food crops: A review. *Planta* 259(3), 60. DOI: 10.1007/s00425-024-04333-1.

Ajayi, O.O., Toromade, A.S. and Olagoke, A. (2024) Data-driven agropreneurship (DDA): Empowering farmers through predictive analytics. *Journal of Agribusiness and Technology*. DOI: 10.30574/msarr.2024.12.2.0177.

Aksoy, N. and Genc, I. (2023) Predictive models development using gradient boosting based methods for solar power plants. *Journal of Computational Science* 67, 101958. DOI: 10.1016/j.jocs.2023.101958.

Al-Badri, A.H., Ismail, N.A., Al-Dulaimi, K., Salman, G.A., Khan, A.R. *et al.* (2022) Classification of weed using machine learning techniques: A review—challenges, current and future potential techniques. *Journal of Plant Diseases and Protection* 129(4), 745–768. DOI: 10.1007/s41348-022-00612-9.

Alemu A., Åstrand, J., Montesinos-López, O.A., y Sánchez, J.I., Fernández-Gónzalez, J. *et al.* (2024) Genomic selection in plant breeding: Key factors shaping two decades of progress. *Molecular Plant* 17(4), 552–578. DOI: 10.1016/j.molp.2024.03.007.

Ali, A., Hussain, T., Tantashutikun, N., Hussain, N. and Cocetta, G. (2023) Application of smart techniques, internet of things and data mining for resource use efficient and sustainable crop production. *Agriculture* 13(2), 397. DOI: 10.3390/agriculture13020397.

Aliyu, M.A., Mokji, M.M.M. and Sheikh, U.U.U. (2020) Machine learning for plant disease detection: An investigative comparison between support vector machine and deep learning. *IAES International Journal of Artificial Intelligence* 9(4), 670. DOI: 10.11591/ijai.v9.i4.pp670-683.

Allwood, J.W., Williams, A., Uthe, H., van Dam, N.M., Mur, L.A.J. *et al.* (2021) Unravelling plant responses to stress-The importance of targeted and untargeted metabolomics. *Metabolites* 11(8), 558. DOI: 10.3390/metabo11080558.

Almeida-Silva, F. and Venancio, T.M. (2022) Pathogenesis-related protein 1 (PR-1) genes in soybean: Genome-wide identification, structural analysis and expression profiling under multiple biotic and abiotic stresses. *Gene* 809, 146013. DOI: 10.1016/j.gene.2021.146013.

Ambrosino, L., Colantuono, C., Diretto, G., Fiore, A. and Chiusano, M.L. (2020) Bioinformatics resources for plant abiotic stress responses: State of the art and opportunities in the fast evolving -omics era 9(5), 591. DOI: 10.3390/plants9050591.

Angidi, S., Madankar, K., Tehseen, M.M. and Bhatla, A. (2025) Advanced high-throughput phenotyping techniques for managing abiotic stress in agricultural crops—A comprehensive review. *Crops* 5(2), 8. DOI: 10.3390/crops5020008.

Araújo, S.O., Peres, R.S., Ramalho, J.C., Lidon, F. and Barata, J. (2023) Machine learning applications in agriculture: Current trends, challenges, and future perspectives. *Agronomy* 13(12), 2976. DOI: 10.3390/agronomy13122976.

Ataei Kachouei, M., Kaushik, A. and Ali, M.A. (2023) Internet of things-enabled food and plant sensors to empower sustainability. *Advanced Intelligent Systems* 5(12), 2300321. DOI: 10.1002/aisy.202300321.

Bayer, P.E., Petereit, J., Danilevicz, M.F., Anderson, R., Batley, J. *et al.* (2021) The application of pangenomics and machine learning in genomic selection in plants. *The Plant Genome* 14(3), e20112. DOI: 10.1002/tpg2.20112.

Benfenati, A., Causin, P., Oberti, R. and Stefanello, G. (2023) Unsupervised deep learning techniques for automatic detection of plant diseases: Reducing the need of manual labelling of plant images. *Journal of Mathematics in Industry* 13(1), 5. DOI: 10.1186/s13362-023-00133-6.

Bogue, R. (2023) The role of machine learning in robotics. *Industrial Robot: The International Journal of Robotics Research and Application* 50(2), 197–202. DOI: 10.1108/IR-11-2022-0279.

Boukhris, A., Jilali, A. and Asri, H. (2024) Deep learning and machine learning based method for crop disease detection and identification using autoencoder and neural network. *Revue d'Intelligence Artificielle* 38(2), 459–472. DOI: 10.18280/ria.380209.

Chai, S., Xu, Z., Jia, Y. and Wong, W.K. (2020) A robust spatiotemporal forecasting framework for photovoltaic generation. *IEEE Transactions on Smart Grid* 11(6), 5370–5382. DOI: 10.1109/TSG.2020.3006085.

Chamorro-Padial, J., García, R. and Gil, R. (2024) A systematic review of open data in agriculture. *Computers and Electronics in Agriculture* 219, 108775. DOI: 10.1016/j.compag.2024.108775.

Charagh, S., Wang, H., Wang, J., Raza, A., Hui, S. *et al.* (2024) Leveraging multi-omics tools to comprehend responses and tolerance mechanisms of heavy metals in crop plants. *Functional & Integrative Genomics* 24(6), 194. DOI: 10.1007/s10142-024-01481-1.

Chavhan, R.L., Hinge, V.R., Wankhade, D.J., Deshmukh, A.S., Mahajan, N. *et al.* (2024) Bioinformatics for molecular breeding and enhanced crop performance: Applications and perspectives. *Bioinformatics for Plant Research and Crop Breeding* 21–74.

Chayalakshmi, C.L., Pujar, R.S. and Mahabaleshwar, S.K. (2025) Machine learning fundamentals. In: *AI and Blockchain in Smart Grids*. Auerbach Publications, pp. 1–13.

Cherian, J.M. and Kumar, R. (2023) Fundamentals of machine learning. In: *A Guide to Applied Machine Learning for Biologists*. Springer International Publishing, Cham, pp. 147–174.

Classen, A., Eardley, C.D., Hemp, A., Peters, M.K., Peters, R.S. *et al.* (2020) Specialization of plant–pollinator interactions increases with temperature at Mt. Kilimanjaro. *Ecology and Evolution* 10(4), 2182–2195. DOI: 10.1002/ece3.6056.

Correndo, A.A., Rosso, L.H.M., Hernandez, C.H., Bastos, L.M., Nieto, L. *et al.* (2022) Metrica: An R package to evaluate prediction performance of regression and classification point-forecast models. *Journal of Open Source Software* 7(79), 4655. DOI: 10.21105/joss.04655.

Costa-Neto, G., Galli, G., Carvalho, H.F., Crossa, J. and Fritsche-Neto, R. (2021) EnvRtype: A software to interplay enviromics and quantitative genomics in agriculture. *G3 (Bethesda, Md.)* (4), jkab040. DOI: 10.1093/g3journal/jkab040.

Cravero, A., Pardo, S., Sepúlveda, S. and Muñoz, L. (2022) Challenges to use machine learning in agricultural big data: A systematic literature review. *Agronomy* 12(3), 748. DOI: 10.3390/agronomy12030748.

Croci, M., Impollonia, G., Meroni, M. and Amaducci, S. (2022) Dynamic maize yield predictions using machine learning on multi-source data. *Remote Sensing* 15(1), 100. DOI: 10.3390/rs15010100.

Dadashzadeh, M., Abbaspour-Gilandeh, Y., Mesri-Gundoshmian, T., Sabzi, S., Hernández-Hernández, J.L. *et al.* (2020) Weed classification for site-specific weed management using an automated stereo computer-vision machine-learning system in rice fields. *Plants (Basel, Switzerland)* 9(5), 559. DOI: 10.3390/plants9050559.

Dang, M., Wang, H., Li, Y., Nguyen, T.H., Tightiz, L. *et al.* (2024) Computer vision for plant disease recognition: A comprehensive review. *The Botanical Review* 90(3), 251–311. DOI: 10.1007/s12229-024-09299-z.

Dao, F., Xie, X., Zhang, H., Guan, Z., Wu, C. *et al.* (2025) PlantEMS: A comprehensive database of epigenetic modification sites across multiple plant species. *Plant Communications* 6(4), 101228. DOI: 10.1016/j.xplc.2024.101228.

Demidchik, V.V., Shashko, A.Y., Bandarenka, U.Y., Smolikova, G.N., Przhevalskaya, D.A. *et al.* (2020) Plant phenomics: Fundamental bases, software and hardware platforms, and machine learning. *Russian Journal of Plant Physiology* 67(3), 397–412. DOI: 10.1134/S1021443720030061.

Deng, C.H., Naithani, S., Kumari, S., Cobo-Simón, I., Quezada-Rodríguez, E.H. *et al.* (2023) Genotype and phenotype data standardization, utilization and integration in the big data era for agricultural sciences. *Database 2023* baad088. DOI: 10.1093/database/baad088.

Diaz-Gonzalez, F.A., Vuelvas, J., Correa, C.A., Vallejo, V.E. and Patino, D. (2022) Machine learning and remote sensing techniques applied to estimate soil indicators–review. *Ecological Indicators* 135, 108517. DOI: 10.1016/j.ecolind.2021.108517.

Dreisigacker, S., Crossa, J., Pérez-Rodríguez, P., Montesinos-López, O.A., Rosyara, U. *et al.* (2021) Implementation of genomic selection in the CIMMYT global wheat program, findings from the past 10 years. *Crop Breeding, Genetics and Genomics* 3(2). DOI: 10.20900/cbgg20210005.

Eicher, T., Kinnebrew, G., Patt, A., Spencer, K., Ying, K. *et al.* (2020) Metabolomics and multi-omics integration: A survey of computational methods and resources. *Metabolites* 10(5), 202. DOI: 10.3390/metabo10050202.

El Mahrad, B., Newton, A., Icely, J., Kacimi, I., Abalansa, S. *et al.* (2020) Contribution of remote sensing technologies to a holistic coastal and marine environmental management framework: A review. *Remote Sensing* 12(14), 2313. DOI: 10.3390/rs12142313.

Elbasi, E., Mostafa, N., Zaki, C., AlArnaout, Z., Topcu, A.E. *et al.* (2024) Optimizing agricultural data analysis techniques through ai-powered decision-making processes. *Applied Sciences* 14(17), 8018. DOI: 10.3390/app14178018.

Fu, X. and Jiang, D. (2022) High-throughput phenotyping: The latest research tool for sustainable crop production under global climate change scenarios. In: *Sustainable Crop Productivity and Quality Under Climate Change*. Academic Press, pp. 313–381.

Gano, B., Bhadra, S., Vilbig, J.M., Ahmed, N., Sagan, V. *et al.* (2024) Drone-based imaging sensors, techniques, and applications in plant phenotyping for crop breeding: A comprehensive review. *The Plant Phenome Journal* 7(1), e20100. DOI: 10.1002/ppj2.20100.

Ganti, P.K., Naik, H. and Barada, M.K. (2022) Environmental impact analysis and enhancement of factors affecting the photovoltaic (PV) energy utilization in mining industry by sparrow search optimization based gradient boosting decision tree approach. *Energy* 244, 122561. DOI: 10.1016/j.energy.2021.122561.

García-Ariza, J.V., Suarez-Barón, M.J., Junco-Orduz, E.A. and González-Sanabria, J.S. (2022) Application of unsupervised learning in the early detection of late blight in potato crops using image processing. *Inge CuC* 18(2), 89–100. DOI: 10.17981/ingecuc.18.2.2022.07.

Gautron, R., Maillard, O.A., Preux, P., Corbeels, M. and Sabbadin, R. (2022) Reinforcement learning for crop management support: Review, prospects and challenges. *Computers and Electronics in Agriculture* 200, 107182. DOI: 10.1016/j.compag.2022.107182.

Gnanasaravanan, S., Tharani, B. and Sahu, M. (2021) Long short-term memory recurrent neural networks for plant disease identification. *Plant Pathology* 3(2). DOI: 10.34256/ijcci2125.

Gou, C., Zafar, S., Fatima, N., Hasnain, Z., Aslam, N. *et al.* (2024) Machine and deep learning: Artificial intelligence application in biotic and abiotic stress management in plants. *Frontiers in Bioscience-Landmark* 29(1). DOI: 10.31083/j.fbl2901020.

Guo, Y., Du, C., Zhao, Y., Ting, T.F. and Rothfus, T.A. (2021) Two-level K-nearest neighbors approach for invasive plants detection and classification. *Applied Soft Computing* 108, 107523. DOI: 10.1016/j.asoc.2021.107523.

Gupta, S., Kaur, R., Sharma, T., Bhardwaj, A., Sharma, S. *et al.* (2023) Multi-omics approaches for understanding stressor-induced physiological changes in plants: An updated overview. *Physiological and Molecular Plant Pathology* 126, 102047. DOI: 10.1016/j.pmpp.2023.102047.

Hatami, S., Hatami Badrbani, M. and Kahrizi, D. (2024) Data mining approach in the agricultural industry, medicinal plants (case study); A Review. *Journal of Medicinal Plants and By-Products* 13(2), 247–256.

He, S., Yang, L., Ye, S., Lin, Y., Li, X. *et al.* (2022) MPOD: Applications of integrated multi-omics database for medicinal plants. *Plant Biotechnology Journal* 20(5), 797–799. DOI: 10.1111/pbi.13769.

Hesami, M., Alizadeh, M., Jones, A.M.P. and Torkamaneh, D. (2022) Machine learning: Its challenges and opportunities in plant system biology. *Applied Microbiology and Biotechnology* 106(9–10), 3507–3530. DOI: 10.1007/s00253-022-11963-6.

Huang, Y., Xu, J., Lai, J., Jiang, Z., Chen, T. *et al.* (2023) Advancing transformer architecture in long-context large language models: A comprehensive survey. *arXiv Preprint arXiv* 2311, 12351. DOI: 10.48550/arXiv.2311.12351.

Hu, K., Wang, Z., Coleman, G., Bender, A., Yao, T. *et al.* (2024) Deep learning techniques for in-crop weed recognition in large-scale grain production systems: A review. *Precision Agriculture* 25(1), 1–29. DOI: 10.1007/s11119-023-10073-1.

Iniyan, S., Akhil Varma, V. and Teja Naidu, C. (2023) Crop yield prediction using machine learning techniques. *Advances in Engineering Software* 175, 103326. DOI: 10.1016/j.advengsoft.2022.103326.

Islam, M.M. and Monjardino, M. (2025) The importance of developing an integrated data-driven modelling platform for herbicide resistance research: A review. *Heliyon* 11(4), e42564. DOI: 10.1016/j.heliyon.2025.e42564.

Jamil, I.N., Remali, J., Azizan, K.A., Nor Muhammad, N.A., Arita, M. *et al.* (2020) Systematic multi-omics integration (MOI) approach in plant systems biology. *Frontiers in Plant Science* 11, 944. DOI: 10.3389/fpls.2020.00944.

Jiang, Y. and Li, C. (2020) *Convolutional Neural Networks for Image-Based high-Throughput Plant Phenotyping: A Review*. Plant Phenomics.

Jones, S.E., Ayanlade, T.T., Fallen, B., Jubery, T.Z., Singh, A. *et al.* (2024) Multi-sensor and multi-temporal high-throughput phenotyping for monitoring and early detection of water-limiting stress in soybean. *The Plant Phenome Journal* 7(1), e70009. DOI: 10.1002/ppj2.70009.

Juliana, P., Singh, R.P., Braun, H.-J., Huerta-Espino, J., Crespo-Herrera, L. *et al.* (2020) Genomic selection for grain yield in the CIMMYT wheat breeding program-status and perspectives. *Frontiers in Plant Science* 11, 564183. DOI: 10.3389/fpls.2020.564183.

Kamath, R., Balachandra, M. and Prabhu, S. (2020) Paddy crop and weed discrimination: A multiple classifier system approach. *International Journal of Agronomy* 2020(1), 1–14. DOI: 10.1155/2020/6474536.

Kelleher, J.D., Mac Namee, B. and D'arcy, A. (2020) *Fundamentals of Machine Learning for Predictive Data Analytics: Algorithms, Worked Examples, and Case Studies*. MIT press.

Kępińska-Kasprzak, M. and Struzik, P. (2023) Monitoring of plant cultivation conditions using ground measurements and satellite products. *Water* 15(3), 449. DOI: 10.3390/w15030449.

Kirpach, M. and Riccoboni, A. (2023) AI in agriculture. In: *Engineering Mathematics and Artificial Intelligence*. CRC Press, pp. 331–364.

Kokhan, S., Dorozhynskyy, O., Burshtynska, K., Vostokov, A. and Drozdivskyi, O. (2020) Improved approach to the development of the crop monitoring system based on the use of multi-source spatial data. *Journal of Ecological Engineering* 21(7), 108–114. DOI: 10.12911/22998993/125442.

Krantz, M., Zimmer, D., Adler, S.O., Kitashova, A., Klipp, E. *et al.* (2021) Data management and modeling in plant biology. *Frontiers in Plant Science* 12, 717958. DOI: 10.3389/fpls.2021.717958.

Kumar, D., Kumar, Y., Gulati, A. and Kukreja, V. (2022) Wheat crop yield prediction using machine learning. In: *2022 International Conference on Data Analytics for Business and Industry (ICDABI)*, IEEE, Sakhir, Bahrain, pp. 433–437. DOI: 10.1109/ICDABI56818.2022.10041621.

Kumar, M., Kumar, A. and Palaparthy, V.S. (2020) Soil sensors-based prediction system for plant diseases using exploratory data analysis and machine learning. *IEEE Sensors Journal* 21(16), 17455–17468. DOI: 10.1109/JSEN.2020.3046295.

Kumar, S., Sohail, M.A., Jadhav, S., Gupta, R.K., AURO University, India *et al.* (2024) Light gradient boosting machine for optimizing crop maintence and yield prediction in agriculture. *ICTACT Journal on Soft Computing* 15(2), 3551–3555. DOI: 10.21917/ijsc.2024.0495.

Kuradusenge, M., Hitimana, E., Hanyurwimfura, D., Rukundo, P., Mtonga, K. *et al.* (2023) Crop yield prediction using machine learning models: Case of Irish potato and maize. *Agriculture* 13(1), 225. DOI: 10.3390/agriculture13010225.

Li, L., Chen, S., Yang, C., Meng, F. and Sigrimis, N. (2020) Prediction of plant transpiration from environmental parameters and relative leaf area index using the random forest regression algorithm. *Journal of Cleaner Production* 261, 121136. DOI: 10.1016/j.jclepro.2020.121136.

Liu, J. and Wang, X. (2021) Plant diseases and pests detection based on deep learning: A review. *Plant Methods* 17(1), 22. DOI: 10.1186/s13007-021-00722-9.

Lu, M., Gao, P., Hu, J., Hou, J. and Wang, D. (2023) A classification method of stress in plants using unsupervised learning algorithm and chlorophyll fluorescence technology. *Frontiers in Plant Science* 14, 1202092. DOI: 10.3389/fpls.2023.1202092.

MacLaren, C., Labuschagne, J. and Swanepoel, P.A. (2021) Tillage practices affect weeds differently in monoculture vs. crop rotation. *Soil and Tillage Research* 205, 104795. DOI: 10.1016/j.still.2020.104795.

Maganathan, T., Senthilkumar, S. and Balakrishnan, V. (2020) Machine learning and data analytics for environmental science: A review, prospects and challenges. In: *IOP Conference Series: Materials Science and Engineering*, p. 012107.

Mindhe, O., Kurkute, O., Naxikar, S. and Raje, N. (2020) Plant disease detection using deep learning. *International Research Journal of Engineering and Technology* 2497–2503.

Miranda, M., Zabawa, L., Kicherer, A., Strothmann, L., Rascher, U. *et al.* (2022) Detection of anomalous grapevine berries using variational autoencoders. *Frontiers in Plant Science* 13, 729097. DOI: 10.3389/fpls.2022.729097.

Mishra, H. and Mishra, D. (2024) AI for data-driven decision-making in smart agriculture: from field to farm management. In: *Artificial Intelligence Techniques in Smart Agriculture*. Springer Nature Singapore, Singapore, pp. 173–193.

Mladenova, T. and Valova, I. (2023) Classification with K-nearest neighbors algorithm: Comparative analysis between the manual and automatic methods for K-selection. *International Journal of Advanced Computer Science and Applications* 14(4), 396–404. DOI: 10.14569/IJACSA.2023.0140444.

Montesinos López, O.A., Montesinos López, A. and Crossa, J. (2022) Support vector machines and support vector regression. In: *Multivariate Statistical Machine Learning Methods for Genomic Prediction*. Springer International Publishing, Cham, pp. 337–378.

Morales, A. and Villalobos, F.J. (2023) Using machine learning for crop yield prediction in the past or the future. *Frontiers in Plant Science* 14, 1128388. DOI: 10.3389/fpls.2023.1128388.

Moupojou, E., Tagne, A., Retraint, F., Tadonkemwa, A., Wilfried, D. *et al.* (2023) FieldPlant: A dataset of field plant images for plant disease detection and classification with deep learning. *IEEE Access* 11, 35398–35410. DOI: 10.1109/ACCESS.2023.3263042.

Munjal, R., Benıwal, J., Dhundwal, A., Goyal, A., Kumarı, A. *et al.* (2023) Accelerating crop breeding in the 21st century: A comprehensive review of next generation phenotyping techniques and strategies. *Ekin Journal of Crop Breeding and Genetics* 9(2), 160–171.

Murmu, S., Sinha, D., Chaurasia, H., Sharma, S., Das, R. *et al.* (2024) A review of artificial intelligence-assisted omics techniques in plant defense: Current trends and future directions. *Frontiers in Plant Science* 15, 1292054. DOI: 10.3389/fpls.2024.1292054.

Murphy, K.M., Ludwig, E., Gutierrez, J. and Gehan, M.A. (2024) Deep learning in image-based plant phenotyping. *Annual Review of Plant Biology* 75(1), 771–795. DOI: 10.1146/annurev-arplant-070523-042828.

Naeem, S., Ali, A., Anam, S. and Ahmed, M.M. (2023) An unsupervised machine learning algorithms: Comprehensive review. *International Journal of Computing and Digital Systems* 13(1), 911–921. DOI: 10.12785/ijcds/130172.

Nagarhalli, T.P., Save, A.M. and Shekokar, N.M. (2021) Fundamental models in machine learning and deep learning. In: *Design of Intelligent Applications Using Machine Learning and Deep Learning Techniques*. Chapman and Hall/CRC, pp. 13–36.

Nazki, H., Yoon, S., Fuentes, A. and Park, D.S. (2020) Unsupervised image translation using adversarial networks for improved plant disease recognition. *Computers and Electronics in Agriculture* 168, 105117. DOI: 10.1016/j.compag.2019.105117.

Niazian, M. and Niedbała, G. (2020) Machine learning for plant breeding and biotechnology. *Agriculture* 10(10), 436. DOI: 10.3390/agriculture10100436.

Ouhami, M., Hafiane, A., Es-Saady, Y., El Hajji, M. and Canals, R. (2021) Computer vision, IoT and data fusion for crop disease detection using machine learning: A survey and ongoing research. *Remote Sensing* 13(13), 2486. DOI: 10.3390/rs13132486.

Padarian, J., Minasny, B. and McBratney, A.B. (2020) Machine learning and soil sciences: A review aided by machine learning tools. *SOIL* 6(1), 35–52. DOI: 10.5194/soil-6-35-2020.

Phang, S.K., Chiang, T.H.A., Happonen, A. and Chang, M.M.L. (2023) From satellite to UAV-based remote sensing: A review on precision agriculture. *IEEE Access* 11, 127057–127076. DOI: 10.1109/ACCESS.2023.3330886.

Prodhan, F.A., Zhang, J., Hasan, S.S., Pangali Sharma, T.P. and Mohana, H.P. (2022) A review of machine learning methods for drought hazard monitoring and forecasting: Current research trends, challenges, and future research directions. *Environmental Modelling & Software* 149, 105327. DOI: 10.1016/j.envsoft.2022.105327.

Qu, H.R. and Su, W.H. (2024) Deep learning-based weed–crop recognition for smart agricultural equipment: A review. *Agronomy* 14(2), 363. DOI: 10.3390/agronomy14020363.

Rahaman, M.M., Ahsan, M.A. and Chen, M. (2019) Data-mining techniques for image-based plant phenotypic traits identification and classification. *Scientific Reports* 9(1), 19526. DOI: 10.1038/s41598-019-55609-6.

Rane, N., Choudhary, S. and Rane, J. (2024) Machine learning and deep learning: A comprehensive review on methods, techniques, applications, challenges, and future directions. *SSRN*. DOI: 10.2139/ssrn.4850451.

Ranjan, A., Rajput, V.D., Prazdnova, E.V., Gurnani, M., Sharma, S. *et al.* (2024) Augmenting abiotic stress tolerance and root architecture: The function of phytohormone-producing PGPR and their interaction with nanoparticles. *South African Journal of Botany* 167, 612–629. DOI: 10.1016/j.sajb.2024.02.041.

Raptis, E.K., Krestenitis, M., Egglezos, K., Kypris, O., Ioannidis, K. *et al.* (2023) End-to-end precision agriculture UAV-based functionalities tailored to field characteristics. *Journal of Intelligent & Robotic Systems* 107(2), 23. DOI: 10.1007/s10846-022-01761-7.

Raza, A., Mubarik, M.S., Sharif, R., Habib, M., Jabeen, W. *et al.* (2023) Developing drought-smart, ready-to-grow future crops. *The Plant Genome* 16(1), e20279. DOI: 10.1002/tpg2.20279.

Raza, A., Anas, M., Bhardwaj, S., Mir, R.A., Charagh, S. *et al.* (2025) Harnessing metabolomics for enhanced crop drought tolerance. *The Crop Journal* 13(2), 311–327. DOI: 10.1016/j.cj.2025.01.001.

Riese, F.M. and Keller, S. (2020) Supervised, semi-supervised, and unsupervised learning for hyperspectral regression. In: *Hyperspectral Image Analysis: Advances in Machine Learning and Signal Processing*. pp. 187–232.

Rizvi, S.K.J., Azad, M.A. and Fraz, M.M. (2021) Spectrum of advancements and developments in multidisciplinary domains for Generative Adversarial Networks (gans). *Archives of Computational Methods in Engineering* 28(7), 4503–4521. DOI: 10.1007/s11831-021-09543-4.

Roychowdhury, R., Das, S.P., Gupta, A., Parihar, P., Chandrasekhar, K. *et al.* (2023) Multi-omics pipeline and omics-integration approach to decipher plant's abiotic stress tolerance responses. *Genes* 14(6), 1281. DOI: 10.3390/genes14061281.

Sabat-Tomala, A., Raczko, E. and Zagajewski, B. (2020) Comparison of support vector machine and random forest algorithms for invasive and expansive species classification using airborne hyperspectral data. *Remote Sensing* 12(3), 516. DOI: 10.3390/rs12030516.

Saiz-Rubio, V. and Rovira-Más, F. (2020) From smart farming towards agriculture 5.0: A review on crop data management. *Agronomy* 10(2), 207. DOI: 10.3390/agronomy10020207.

Saleem, M.H., Potgieter, J. and Arif, K.M. (2020) Plant disease classification: A comparative evaluation of convolutional neural networks and deep learning optimizers 9(10), 1319. DOI: 10.3390/plants9101319.

Salman, H.A., Kalakech, A. and Steiti, A. (2024) Random forest algorithm overview. *Babylonian Journal of Machine Learning* 69–79. DOI: 10.58496/BJML/2024/007.

Sanches, P.H.G., Melo, N.C., Porcari, A.M. and Carvalho, L.M. (2024) Integrating molecular perspectives: Strategies for comprehensive multi-omics integrative data analysis and machine learning applications in transcriptomics, proteomics, and metabolomics. *Biology* 13(11), 848. DOI: 10.3390/biology13110848.

Shakespear, S., Sivaji, M., Kumar, V., Arumugam Pillai, M., Wani, S.H. *et al.* (2025) Navigating through harsh conditions: Coordinated networks of plant adaptation to abiotic stress. *Journal of Plant Growth Regulation* 44(4), 1396–1414. DOI: 10.1007/s00344-023-11224-4.

Sharma, R.P., Ramesh, D., Pal, P., Tripathi, S. and Kumar, C. (2021) IoT-enabled IEEE 802.15.4 WSN monitoring infrastructure-driven fuzzy-logic-based crop pest prediction. *IEEE Internet of Things Journal* 9(4), 3037–3045. DOI: 10.1109/JIOT.2021.3094198.

Shen, S., Zhan, C., Yang, C., Fernie, A.R. and Luo, J. (2023) Metabolomics-centered mining of plant metabolic diversity and function: Past decade and future perspectives. *Molecular Plant* 16(1), 43–63. DOI: 10.1016/j.molp.2022.09.007.

Shetty, S.H., Shetty, S., Singh, C. and Rao, A. (2022) Supervised machine learning: algorithms and applications. In: *Fundamentals and Methods of Machine and Deep Learning: Algorithms, Tools and Applications*. pp. 1–16.

Shree, R., Gupta, R. and Srikanth, V. (2023) Forecasting diseases that affect plant leaves and moisture levels in the soil using a data mining approach. *Multidisciplinary Science Journal* 5, 2023ss0120. DOI: 10.31893/multiscience.2023ss0120.

Shrotriya, A., Sharma, A.K., Prabhu, S. and Bairwa, A.K. (2024) *An Approach Towards Classifying Plant-Leaf Diseases and Comparisons with the Conventional Classification*. IEEE Access.

Silva, A.O. da, Silva, B.A. da, Souza, C.F., Azevedo, B.M. de, Bassoi, L.H. *et al*. (2020) Irrigation in the age of agriculture 4.0: Management, monitoring and precision. *Revista Ciência Agronômica* 51(5), e20207695. DOI: 10.5935/1806-6690.20200090.

Singh, D. and Kumar, A. (2024) A deep recurrent neural network for plant disease classification. *SN Computer Science* 5(8), 1053. DOI: 10.1007/s42979-024-03400-4.

Singh, L., Huang, H., Bordoloi, S., Garg, A. and Jiang, M. (2021) Exploring simple K-means clustering algorithm for automating segregation of colors in leaf of *Axonopus compressus*: Towards maintenance of an urban landscape. *Journal of Intelligent & Fuzzy Systems* 40(1), 1219–1243. DOI: 10.3233/JIFS-201542.

Singh Saharan, G., Mehta, N.K. and Meena, P.D. (2023) Genomics of crucifer's host-pathosystem: Prologue. In: *Genomics of Crucifer's Host-Pathosystem*. Springer Nature, Singapore, pp. 1–126.

Siregar, J.J. and Budiarto, E. (2021) Comparison of unsupervised machine learning analysis using K-means, hierarchical clustering, and principal component analysis in horticulture agribusiness: Shallot and tomato productivity in Indonesia. In: *International Conference of Reliable Information and Communication Technology*, Springer International Publishing, pp. 270–282.

Soltis, P.S., Nelson, G., Zare, A. and Meineke, E.K. (2020) Plants meet machines: Prospects in machine learning for plant biology. *Applications in Plant Sciences* 8(6), e11371. DOI: 10.1002/aps3.11371.

Su, W.H. (2020) Advanced machine learning in point spectroscopy, RGB- and hyperspectral-imaging for automatic discriminations of crops and weeds: A review. *Smart Cities* 3(3), 767–792. DOI: 10.3390/smartcities3030039.

Su, Y., Gabrielle, B. and Makowski, D. (2021) A global dataset for crop production under conventional tillage and no tillage systems. *Scientific Data* 8(1), 33. DOI: 10.1038/s41597-021-00817-x.

Subramanian, I., Verma, S., Kumar, S., Jere, A. and Anamika, K. (2020) Multi-omics data integration, interpretation, and its application. *Bioinformatics and Biology Insights* 14, 1177932219899051. DOI: 10.1177/1177932219899051.

Sujatha, R., Chatterjee, J.M., Jhanjhi, N.Z. and Brohi, S.N. (2021) Performance of deep learning vs machine learning in plant leaf disease detection. *Microprocessors and Microsystems* 80, 103615. DOI: 10.1016/j.micpro.2020.103615.

Sulova, A. and Jokar Arsanjani, J. (2020) Exploratory analysis of driving force of wildfires in Australia: An application of machine learning within google earth engine. *Remote Sensing* 13(1), 10. DOI: 10.3390/rs13010010.

Sun, S., Wang, C., Ding, H. and Zou, Q. (2020) Machine learning and its applications in plant molecular studies. *Briefings in Functional Genomics* 19(1), 40–48. DOI: 10.1093/bfgp/elz036.

Suresh, M. and Manjupriya, S. (2022) An iot based intelligent decision support system for smart farming using machine learning techniques. *The Seybold Report 528–539*. DOI: 10.5281/zenodo.6622423.

Swain, M., Singh, R., Thakur, A.K. and Gehlot, A. (2020) A machine learning approach of data mining in agriculture 4.0. *International Journal Emerging Technologies* 11(1), 257–262.

Titirmare, S., Margal, P.B., Gupta, S. and Kumar, D. (2024) AI-powered predictive analytics for crop yield optimization. In: *Agriculture 4.0*. CRC Press, pp. 89–110.

Torres-Sanchez, R., Navarro-Hellin, H., Guillamon-Frutos, A., San-Segundo, R., Ruiz-Abellón, M.C. *et al*. (2020) A decision support system for irrigation management: Analysis and implementation of different learning techniques. *Water* 12(2), 548. DOI: 10.3390/w12020548.

Van Dijk, A.D.J., Kootstra, G., Kruijer, W. and Ridder, D. (2021) Machine learning in plant science and plant breeding. *Iscience* 24(1), 101890. DOI: 10.1016/j.isci.2020.101890.

Wang, M., Li, R. and Zhao, Q. (2023) Multi-omics techniques in genetic studies and breeding of forest plants. *Forests* 14(6), 1196. DOI: 10.3390/f14061196.

Wang, N., Zhang, G. and Li, M. (2024) Emerging trends in systems biology: Multi-omics integration and beyond. *Computational Molecular Biology* 14. DOI: 10.5376/cmb.2024.14.0024.

Wang, S., Xu, D., Liang, H., Bai, Y., Li, X. *et al*. (2025a) Advances in deep learning applications for plant disease and pest detection: A review. *Remote Sensing* 17(4), 698. DOI: 10.3390/rs17040698.

Wang, S., Huang, X., Zhang, M., Bao, S., Liu, L. *et al.* (2025b) Open science 2.0: Revolutionizing spatiotemporal data sharing and collaboration. *Computational Urban Science* 5(1), 4. DOI: 10.1007/s43762-025-00165-1.

Weil, H.L., Schneider, K., Tschöpe, M., Bauer, J., Maus, O. *et al.* (2023) PLANTdataHUB: A collaborative platform for continuous FAIR data sharing in plant research. *The Plant Journal* 116(4), 974–988. DOI: 10.1111/tpj.16474.

Williamson, H.F., Brettschneider, J., Caccamo, M., Davey, R.P., Goble, C. *et al.* (2023) Data management challenges for artificial intelligence in plant and agricultural research 324. DOI: 10.12688/f1000research.52204.2.

Xie, C., Guo, J., Yan, L., Jiang, R., Liang, A. *et al.* (2022) The influence of plant morphological structure characteristics on PM2.5 retention of leaves under different wind speeds. *Urban Forestry & Urban Greening* 71, 127556. DOI: 10.1016/j.ufug.2022.127556.

Yan, J. and Wang, X. (2022) Unsupervised and semi-supervised learning: The next frontier in machine learning for plant systems biology. *The Plant Journal* 111(6), 1527–1538. DOI: 10.1111/tpj.15905.

Yadav, S.A., Sahoo, B.M., Sharma, S. and Das, L. (2020) An analysis of data mining techniques to analyze the effect of weather on agriculture. In: *2020 International Conference on Intelligent Engineering and Management (ICIEM)*, IEEE, pp. 29–32. DOI: 10.1109/ICIEM48762.2020.9160110.

Yang, C., Baireddy, S., Méline, V., Cai, E., Caldwell, D. *et al.* (2023a) Image-based plant wilting estimation. *Plant Methods* 19(1), 52. DOI: 10.1186/s13007-023-01026-w.

Yang, C., Lei, J., Liu, Z., Xiong, S., Xi, L, *et al.* (2024) Estimation model of corn leaf area index based on improved CNN. *Agriculture* 15(5), 481. Available at: https://doi.org/10.3390/agriculture15050481

Yang, L., Yang, Y., Huang, L., Cui, X. and Liu, Y. (2023b) From single- to multi-omics: Future research trends in medicinal plants. *Briefings in Bioinformatics* 24(1), bbac485. DOI: 10.1093/bib/bbac485.

Younes, K., Kharboutly, Y., Antar, M., Chaouk, H., Obeid, E. *et al.* (2023) Application of unsupervised machine learning for the evaluation of aerogels' efficiency towards ion removal—A principal component analysis (PCA) approach. *Gels* 9(4), 304. DOI: 10.3390/gels9040304.

Yu, F., Lin, J., Xie, D., Yao, Y., Wang, X. *et al.* (2020) Soil properties and heavy metal concentrations affect the composition and diversity of the diazotrophs communities associated with different land use types in a mining area. *Applied Soil Ecology* 155, 103669. DOI: 10.1016/j.apsoil.2020.103669.

Zhai, Z., Martínez, J.F., Beltran, V. and Martínez, N.L. (2020) Decision support systems for agriculture 4.0: Survey and challenges. *Computers and Electronics in Agriculture* 170, 105256. DOI: 10.1016/j.compag.2020.105256.

Zhang, Y. and Wang, Y. (2023) Machine learning applications for multi-source data of edible crops: A review of current trends and future prospects. *Food Chemistry* 19, 100860. DOI: 10.1016/j.fochx.2023.100860.

Zhang, H., Wang, L., Jin, X., Bian, L. and Ge, Y. (2023a) High-throughput phenotyping of plant leaf morphological, physiological, and biochemical traits on multiple scales using optical sensing. *The Crop Journal* 11(5), 1303–1318. DOI: 10.1016/j.cj.2023.04.014.

Zhang, S., Fan, R., Liu, Y., Chen, S., Liu, Q. *et al.* (2023b) Applications of transformer-based language models in bioinformatics: A survey. *Bioinformatics Advances* 3(1), vbad001. DOI: 10.1093/bioadv/vbad001.

Zhang, T., Guan, H., Ma, X. and Shen, P. (2023c) Drought recognition based on feature extraction of multispectral images for the soybean canopy. *Ecological Informatics* 77, 102248. DOI: 10.1016/j.ecoinf.2023.102248.

Zhang, X., Ibrahim, Z., Khaskheli, M.B., Raza, H., Zhou, F. *et al.* (2024a) Integrative approaches to abiotic stress management in crops: Combining bioinformatics educational tools and artificial intelligence applications. *Sustainability* 16(17), 7651. DOI: 10.3390/su16177651.

Zhang, Z.Y., Yang, L. and Yang, W.C. (2024b) Research progress in biomarker-targeted optical biosensors for plant disease detection. *TrAC Trends in Analytical Chemistry* 178, 117841. DOI: 10.1016/j.trac.2024.117841.

Zhou, Z.H. (2021) Semi-supervised learning. In: Zhou, Z.H. (ed.) *Machine Learning*. Springer, Singapore, pp. 315–341.

12 Machine Learning-Assisted Plant Genome Editing

Sangeeta Sarma[1], Arpita Talukdar[2], Kamal Kashyap[3], Bhaswati Sarma[4], Maske Neha Chandrakant[5], Dhanawantari L. Singha[2]* and Salvinder Singh[5]*

[1]Department of Botany, Jagiroad College, Jagiroad, Morigaon, India; [2]Department of Botany, Rabindranath Tagore University, Hojai, India; [3]Department of Zoology, Rabindranath Tagore University, Hojai, India; [4]Department of Botany, ADP College, Haiborgaon, Nagaon, India; [5]Department of Agricultural Biotechnology, Assam Agricultural University, Jorhat, India

Abstract

The genome editing technique is a powerful tool for precision agriculture as it permits the modification of DNA sequences with the aim of crop improvemen—enhancing crop yield, improving stress tolerance, increasing resistance to pests and pathogens, and boosting nutritional qualities among others. Nevertheless, genome editing in plants is a critical and taxing operation as it requires precision in the way the enzymes work, how the editing process takes place, and the nature of the targets. With the advent of deep machine learning and big data, it is now possible to predict editing results and determine the parameters of genome editing. Machine learning has made genome editing easier and has led to the fine-tuning of the process. This chapter will primarily focus on the standardization of genome editing with the help of machine learning with a special focus on editing efficiency, editing specificity, and also prediction of genome editing by machine learning. The chapter will try to bridge the gap between plant breeding and genome editing through machine learning. Additionally, the chapter will also highlight the use of AI in genome editing. Further, it will try to analyze the opportunities and challenges of using machine learning in genome editing techniques. Combining genome editing with machine learning will definitely accelerate the progress of crop improvement.

Keywords: AI, crop improvement, genome editing, CRISPR/Cas, machine learning, precision agriculture

12.1 Introduction

Improving crop yield has become a challenge in today's world with the surging global population and increasing effects of climate change-related weather occurrences. In these testing times, a revolution in the field of agriculture is a must. Genome editing techniques have improved precision agriculture tremendously. It has led to climate-smart crops, disease-free crops and has also enhanced the nutritional qualities of crops. Rapid progress in machine learning and computational power has encouraged researchers to implement it in biological research for the improvement of various domains including agriculture. Machine learning (ML) can also be

*Corresponding authors: dhanawantarisingha@gmail.com and salvinder.singh@aau.ac.in

DOI: 10.1079/9781800628649.0012

employed in the calibration of several parameters of genome editing efficiency, specificity, and prediction of editing.

Here we provide a complete overview of the utilization of ML in genome editing. This chapter will evaluate the optimization of various parameters of genome editing using ML. It will also highlight the application of artificial intelligence (AI) technology in genome editing. Further, it will focus on the ways to connect plant breeding with genome editing via deep ML. Finally, it will also concentrate on the advantages and disadvantages of using ML in genome editing.

12.1.1 Genome editing techniques

Genome editing is a mechanism by virtue of which the organisms' genome can be manipulated to create heritable traits that can improve both quality and quantity. Genome editing includes insertion, deletion, single-nucleotide substitutions and also large fragment substitutions (Liu *et al.*, 2021). The study of plants has undergone a radical change since the advent of genome-editing technology such as clustered regularly interspaced short palindromic repeat/CRISPR-associated 9 (CRISPR/Cas9) based systems for site-specific mutagenesis (Petolino *et al.*, 2010; Noman *et al.*, 2016; Liu *et al.*, 2017a).

12.1.2 CRISPR/Cas-based genome editing

The CRISPR/Cas system serves as a powerful genome editing tool, building on earlier technologies like zinc finger nucleases (ZFNs) and transcription activator-like effector nucleases (TALENs). Unlike these methods, CRISPR uses a guide RNA (gRNA) to direct the Cas endonuclease to a specific DNA target site, enabling precise gene modifications (Fig. 12.1). This system has been widely adopted in crop science to enhance various agronomic traits such as yield, resistance to biotic and abiotic stresses, nutritional quality, and to reduce crop loss during pre- and post-harvest stages (Shi *et al.*, 2017; Lu *et al.*, 2018; Huang *et al.*, 2020; Sheng *et al.*, 2020; Xu *et al.*, 2021). Unlike conventional breeding, which relies on time-consuming processes like backcrossing and recombination, CRISPR-based genome editing enables the direct creation of elite plant varieties with desirable traits in a significantly shorter time (Wulff and Dhugga, 2018). Another notable advantage is multiplex editing, which allows simultaneous modification of multiple gene loci within a single generation, thereby accelerating the development of complex traits (Yang *et al.*, 2017). This precision, speed, and versatility make CRISPR/Cas a transformative tool in modern plant biotechnology, offering promising solutions for sustainable agriculture and food security in the face of climate change and a growing global population.

12.1.3 Fundamentals of machine learning

The CRISPR/Cas system is an advanced genome editing platform that has revolutionized plant biotechnology. Unlike earlier tools such as ZFNs and TALENs, CRISPR/Cas employs a guide RNA (gRNA) to direct the Cas endonuclease to a specific genomic location, enabling targeted and precise modifications. This simplicity and accuracy have led to its widespread use in plant science. In crop improvement, CRISPR/Cas has been successfully used to enhance several important traits, including higher yield, resistance to biotic and abiotic stresses, improved nutritional quality, and reduced pre- and post-harvest losses (Shi *et al.*, 2017; Lu *et al.*, 2018; Huang *et al.*, 2020; Sheng *et al.*, 2020; Xu *et al.*, 2021). These improvements are critical for addressing the demands of global food security and environmental sustainability. CRISPR also offers a significant advantage over conventional breeding methods, which are labor intensive and time consuming, often requiring multiple generations of backcrossing to achieve desired traits. In contrast, CRISPR enables the direct development of elite crop varieties with high precision and within a shorter timeframe (Wulff and Dhugga, 2018). Moreover, CRISPR supports multiplex genome editing, allowing multiple gene targets to be edited simultaneously in a single generation. This feature is especially valuable for engineering complex traits governed by multiple genes (Yang *et al.*, 2017). Therefore, the

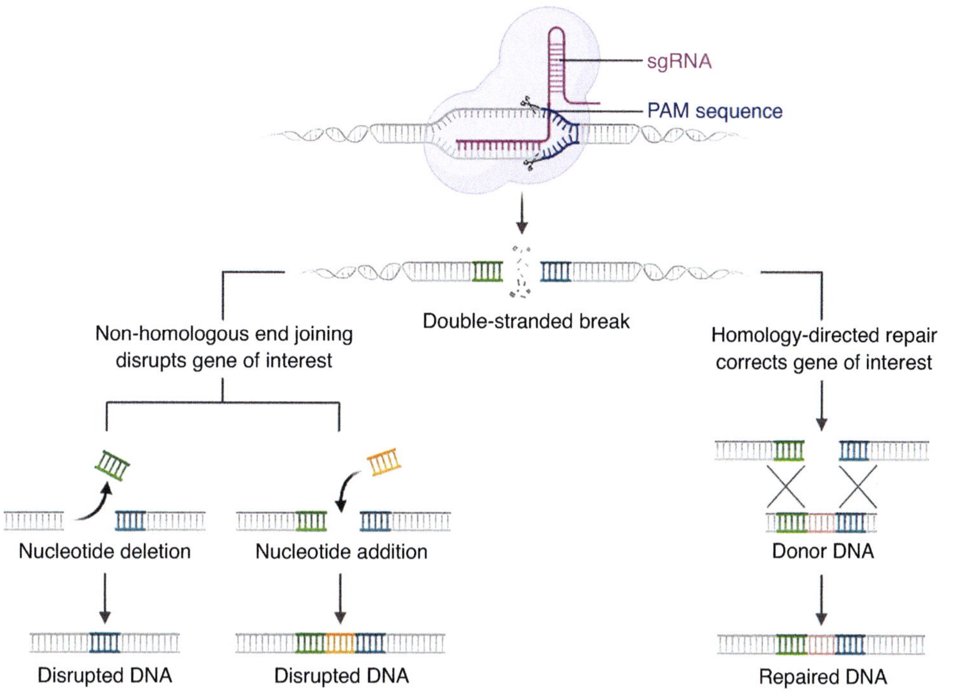

Fig. 12.1. CRISPR/Cas mediated genome editing mechanism (author's own image). The CRISPR/Cas system uses a single guide RNA (sgRNA) to direct the Cas nuclease to a specific DNA sequence, where it introduces a double-strand break (DSB). The cell repairs the DSB through two main pathways: non-homologous end joining (NHEJ) and homology-directed repair (HDR). HDR uses a donor DNA template to precisely repair the break, allowing for specific gene insertions or corrections. (Author's own image.)

CRISPR/Cas system represents a transformative approach in plant genome engineering, offering efficient, precise, and versatile solutions for crop improvement. It holds great promise for advancing sustainable agriculture, particularly in the context of climate change and increasing global food demand.

12.2 Standardization of CRISPR/Cas-Based Genome Editing Using Machine Learning

Genome editing has transformed life sciences and plant biotechnology by enabling precise manipulation of genetic material to improve traits and address biological challenges. Among these tools, CRISPR/Cas systems have emerged as the most powerful due to their efficiency and programmability. Beyond the widely used CRISPR/Cas9, other variants such as Cas12a (Type V) and endogenous Type I and III systems have been explored for plant genome editing (Liu *et al.*, 2020). These technologies have facilitated the development of disease-resistant crops, enhanced yield, and accelerated breeding. Parallel advancements in omics technologies have generated vast datasets related to plant genomics, proteomics, and metabolomics. To manage and interpret this data, ML has become increasingly valuable. ML helps breeders analyze complex datasets to identify key genetic elements for desirable traits (Najafabadi *et al.*, 2023). Despite the promise, standardization in CRISPR/Cas applications remains a critical need. Variations in editing efficiency across plant species, target genes, Cas enzymes, gRNA design, and delivery systems often lead to inconsistent outcomes. Without harmonized protocols for gRNA design, delivery methods, and off-target analysis, reproducibility and cross-study comparisons are challenging. Establishing centralized data repositories would enhance consistency, facilitate species-wide comparisons, and accelerate advancements in crop genome editing (Latif and Mustafa, 2025). ML algorithms can further aid by analyzing high-throughput CRISPR datasets to identify features like sequence motifs and chromatin states that influence editing efficiency, thereby guiding more precise and effective genome engineering strategies.

12.2.1 Editing efficiency optimization

The need for more accurate and efficient CRISPR editing is increasing as genome editing technology is being used more often in breeding. But there are still a number of issues that need to be resolved, including inconsistent CRISPR editing results and efficiency as well as enduring off-target consequences. The dependable and consistent use of CRISPR/Cas systems depends on optimizing editing efficiency. Numerous parameters, including gRNA design, target site accessibility, Cas-protein variations, delivery strategies, and cellular environment, affect the results of genome editing. Much research has been carried out to determine the elements that influence editing effectiveness, as the system of CRISPR/Cas editing efficiency varies greatly depending upon the target location. Prior research has demonstrated a correlation between editing efficiency and the properties of the gRNA and target site, including composition of the nucleotide, GC concentration, secondary structure of RNA, and target epigenetic statuses (Doench *et al.*, 2016).

12.2.1.1 Estimating the effect of different guide RNAs on editing efficiency

Doench *et al.* systematically optimized single guide RNA (sgRNA) design in human and mouse cells by generating multiple sgRNA libraries and developing ML models based on experimental data. Their first model, Rule Set 1, focused on sequence features. With expanded datasets and thermodynamic parameters, they developed Rule Set 2, which improved CRISPR/Cas9 on-target efficiency predictions (Doench *et al.*, 2014, 2016). Later, Rule Set 3 incorporated larger datasets and tracrRNA features, trained using a gradient boosting model, and showed superior sgRNA activity prediction (DeWeirdt *et al.*, 2022). These results highlight how expanding datasets and integrating new features enhance gRNA design. Liu *et al.* (2022) developed CRISPR/Cas9 vectors with single or dual sgRNAs and transformed them into rice cultivar Zhonghua11. Among 1283 plants analyzed, 371 showed successful edits. A or T insertions were most common. Dual sgRNA constructs exhibited higher editing efficiency than single sgRNA constructs. Editing efficiency declined with the deletion of one or two bases and dropped sharply with three-base deletions. Similarly, inserting 1–3 bases also reduced efficiency. In Arabidopsis, Mao *et al.* (2018) found that mutants with impaired RNA silencing had higher Cas9 and sgRNA transcript levels, resulting in increased mutagenesis. Introducing an AGO1-RNAi cassette enhanced gene editing efficiency by silencing *AGO1*, a key RNA-silencing component. These studies underscore the significance of guide RNA design, expression levels, and genetic background in optimizing genome editing outcomes.

12.2.1.2 Estimating the effect of different Cas proteins in editing efficiency

In addition to sgRNAs, the intrinsic nuclease activity also affects the effectiveness of genome editing. The most widely utilized protein, Cas9, may occasionally be restricted by off-target effects and the lack of a particular protospacer adjacent motif (PAM) sequence at the target location. Researchers have found numerous Cas-orthologs and created a number of variants of Cas with unique properties in contrast to conventional Cas proteins in order to achieve particular editing requirements (Chen *et al.*, 2024). 13 variants of Cas with good specificity and broad PAM compatibility were chosen by Kim *et al.* (2020), who then assessed the cleavage efficiency of these variations across 26,891 target sequences. They used the insertion and deletion (indel) frequency statistics obtained from target sequences that were compared with PAM sequences as training and testing datasets. The differences in the preferences of the target sequence between the various Cas-variants that lead to varying efficiency of editing with the same target region were revealed by their analysis. A convolutional neural network (CNN) deep learning (DL) model called DeepSpCas9 variants was created in order to precisely forecast the activity of several Cas variants at various locations. In a similar vein, Seo *et al.* (2023) created a DL-based predictive model named DeepSmallCas9 after methodically comparing the efficiency of 17 tiny Cas9 variants at various target sites. Researchers can better understand the traits of various Cas variants and forecast their effectiveness on target sequences by incorporating ML. This development increases the efficiency of genome editing by giving researchers the flexibility to choose a Cas that is designed for particular uses.

12.2.2 Editing specificity optimization

Precision is a key aspect of effective genome editing, especially in plant systems where off-target effects can lead to unforeseen phenotypic consequences or regulatory issues. ML offers a robust approach to improve the specificity of CRISPR/Cas-based editing by predicting and avoiding potential off-target interactions through advanced guide RNA design and editing protein modifications

12.2.2.1 Construction of high-specificity guide RNA

gRNAs are responsible for guiding Cas enzymes to the desired genomic locations for editing purposes. However, slight variations between the gRNA and the target DNA sequence can lead to off-target modifications. To address this problem, ML models have become valuable tools for designing gRNAs with improved specificity by analyzing a variety of sequence features and predicting their editing activity. These ML algorithms are trained on large datasets consisting of experimentally verified gRNAs, allowing them to recognize patterns and features that improve targeting specificity. Several such ML-driven tools have been developed:

1. **GuideScan2** is a faster and memory-saving software tool designed to help researchers create very accurate gRNAs for CRISPR genome editing. Unlike its predecessor, which required a lot of computational memory, GuideScan2 uses a faster and more efficient algorithm called the Burrows–Wheeler transform, making it run on ordinary laptops. It helps reduce off-target effects, which can cause inaccuracies in CRISPR experiments and allows researchers to design gRNAs specific for specific genetic variations, like those of hybrid mice. The tool also helps researchers design large gRNA libraries quickly and accurately, thus improving the results of gene-editing research. Overall, GuideScan2 improves the accuracy, consistency, and convenience of CRISPR experiments in multiple areas of study (Schmidt *et al.*, 2025).
2. **Launch-dCas9** is a ML system intended to improve the formulation of gRNAs for CRISPR interference (CRISPRi) purposes, aiming at gene control instead of gene editing. Through the use of data obtained from more than 1 million gRNAs, it predicts the effect of each gRNA on cell growth, gene expression, and wild-type cell abundance. The approach involves two strong models CNN (a DL method) and XGBoost (a gradient boosting library) and includes both sequence data and functional genomic data, including epigenetic modifications. Launch-dCas9 was found to have high accuracy and reliability across cell types, making it useful for a variety of biological studies. Additionally, it suggested that gRNAs affecting cell viability often influence gene expression, thus helping researchers choose more efficient gRNAs for gene regulation and potential therapeutic applications (Mu *et al.*, 2024).
3. **DeepCRISPR** is a unified DL system for obtaining on-target predictions of knockout efficiency and genome-wide off-target cleavage patterns of sgRNAs for CRISPR-Cas9 systems. It utilizes a hybrid neural network that consists of an unsupervised deep convolutional denoising autoencoder for pre-training and a supervised CNN for fine-tuning with labeled data. By combining large-scale unlabeled sgRNA sequences with epigenetic characteristics from various human cell types, DeepCRISPR learns high-level feature representations that enhance

prediction accuracy. The model overcomes major challenges like data sparsity, heterogeneity, and imbalance through data augmentation and bootstrapping methods. It performs better than current tools in both classification and regression tasks and generalizes well across various cell types, making it a potent tool for optimized and accurate sgRNA design (Chuai *et al.*, 2018).

4. **CRISPR-ONT** and **CRISPR-OFFT** are interpretable attention-based CNN models developed to accurately predict the on-target activity and off-target specificity of CRISPR/Cas9 sgRNAs. CRISPR-ONT uses attention mechanisms to highlight key nucleotide positions, particularly in the PAM-proximal region, that influence sgRNA efficiency, while CRISPR-OFFT evaluates sgRNA–DNA mismatches using bootstrapping and transfer learning to manage data imbalance in off-target prediction. Both models outperform existing tools like DeepCas9 and DeepHF in terms of accuracy and training efficiency and offer biological insights by identifying important sequence motifs and positional features. Their interpretable architecture and strong predictive performance make them valuable tools for designing highly specific and effective sgRNAs for genome editing.

12.2.2.2 Modification of editing protein

Apart from gRNA sequence optimization, editing specificity can also be increased by altering the Cas enzymes themselves. Native Cas9, for example, can accommodate a few mismatches, which results in off-target modifications. ML has enabled the engineering of Cas variants with changed DNA-binding characteristics and increased specificity. A few of those computational tools are:

1. **Protein2PAM:** It is a DL tool that makes predictions and designs PAM recognition in CRISPR-Cas systems to facilitate more extensive and specific genome editing. With training on more than 45,000 PAM sequences across various microbial genomes, it applies protein language models and a neural network for the prediction of 10 bp PAMs for Type I, II, and V Cas proteins with very high accuracy (≥86.8%). In contrast to earlier methods, Protein2PAM takes Cas protein sequences only and does not depend on structural information, thus being applicable to new and engineered variants. The method also allows for *in silico* mutational scanning to inform PAM reprogramming, as proven by the successful engineering of Nme1Cas9 variants with expanded PAM tolerance (Nayfach *et al.*, 2025).
2. **MODIFY:** To overcome the limitations of classical directed evolution, such as convergence to local optima and limited sequence diversity, Ding *et al.* (2024) introduced MODIFY (multi-objective directed evolution with fitness and diversity), an ML-based framework. MODIFY simultaneously optimizes enzyme activity and sequence diversity, balancing improved catalytic performance with a broader exploration of sequence space. It uses experimental sequence-fitness data to iteratively train predictive models and select new enzyme variants predicted to have high activity while maintaining sequence variation from prior candidates. This multi-objective approach enhances navigation of the protein fitness landscape, enabling the discovery of highly active and structurally diverse enzymes. In applications such as engineering *Rhodothermus marinus* cytochrome c (Rma cyt c) for abiological C–B and C–Si bond formation, MODIFY outperformed traditional methods by isolating superior and more diverse variants within a few evolutionary rounds. This method represents a breakthrough in enzyme engineering and holds significant promise for synthetic biology and biocatalyst discovery.
3. **ProtET:** It is a CLIP-inspired, multi-modal DL framework designed for controllable protein editing using natural language instructions (biotext). By aligning protein sequences with biotext representations through contrastive learning, ProtET enables context-aware protein function prediction and sequence optimization. It integrates a large protein encoder and a language model, trained on over 67 million protein–biotext pairs, and utilizes a FiLM-based fusion module for cross-modal conditioning. ProtET outperforms existing models in protein function

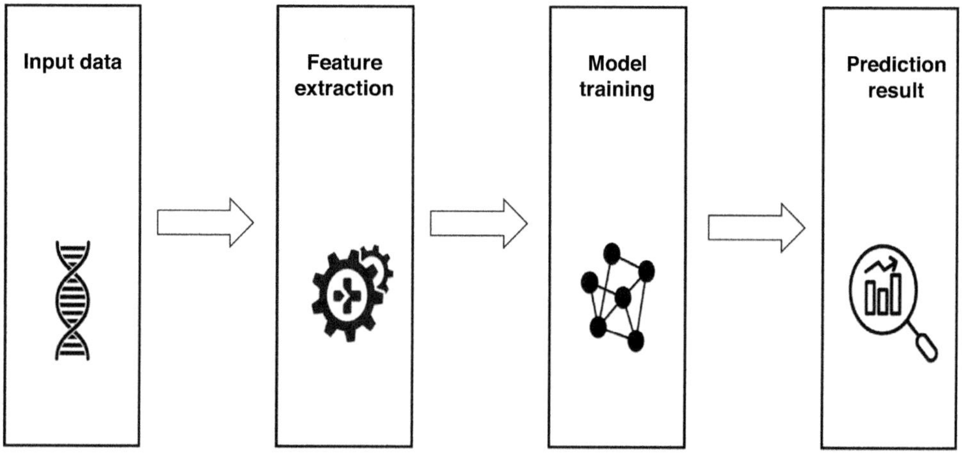

Fig. 12.2. ML pipeline for predicting genome editing outcome (author's own image). An MLpipeline for predicting genome editing outcomes begins with input data such as DNA sequences and target sites. Feature extraction includes nucleotide composition, PAM context, and secondary structure. The model is trained on known editing outcomes. It then predicts efficiency, off-target effects, or mutation patterns at specific loci. (Author's own image.)

classification, enzyme activity editing, and stability optimization, offering a novel and interpretable approach to ML-assisted protein engineering with applications in therapeutics, synthetic biology, and drug design.

12.2.3 Machine learning-based prediction of genome editing

ML is increasingly used to predict the outcomes of genome editing before experimental validation. These predictions include editing efficiency, specificity, and the likelihood of off-target events, providing a powerful decision-making tool for researchers (Fig. 12.2).

12.3 Plant Breeding with Genome Editing and Deep Machine Learning

Modern agriculture faces major challenges including a growing global population, shrinking arable land, and climate change, which disrupts precipitation and temperature patterns, thereby affecting crop yields. To meet rising food demands sustainably, advanced plant breeding techniques have become crucial (Razzaq *et al.*, 2021). Traditional breeding relies on naturally occurring mutations and is time consuming and less precise. In contrast, genome editing offers a targeted, efficient approach to trait improvement. It enables the rapid development of superior crop varieties by eliminating the need for lengthy backcrossing, saving two to 3 years compared to conventional methods (Wulff and Dhugga, 2018). Plant breeding has evolved through four key stages: conventional breeding, genetic breeding, molecular breeding, and emerging intelligent breeding. Conventional breeding depends on natural variation and is slow and unpredictable. Genetic breeding, rooted in 19th-century discoveries, improved outcomes but remained time consuming and costly. Molecular breeding introduced precise tools such as genetic markers and gene editing, offering greater efficiency but with added complexity and cost. Intelligent breeding integrates genetic engineering with AI, especially DL, to optimize breeding outcomes (Wang *et al.*, 2023). AI, particularly ML and DL, has further revolutionized plant breeding. These tools analyze large, complex datasets to enhance prediction accuracy and trait selection, supporting climate-resilient and high-yield crop development (van Dijk *et al.*, 2020). DL is especially effective in decoding the complex relationships between genotype and phenotype, a fundamental concept in plant biology (Mochida *et al.*, 2018; Niazian and Niedbała, 2020; Montesinos-López *et al.*, 2024; Kajrolkar, 2025). Together, genome editing and AI-driven tools represent a transformative shift in sustainable agriculture, offering precise, fast, and scalable solutions for modern crop improvement.

12.3.1 Gene discovery

Traditional statistical methods, such as univariate and multivariate analyses, are widely used in plant breeding. However, due to the nonlinear and nondeterministic nature of plant traits influenced by genotype, phenotype, and environmental interactions, these methods often fall short. ML, DL in particular, offers advanced alternatives by handling complex, large-scale genomic data efficiently. DL models can uncover hidden patterns in data, enabling accurate genomic predictions and the discovery of novel genes related to desirable traits (Liu *et al.*, 2020; Niazian and Niedbała, 2020). DL is increasingly applied in plant breeding to identify gene functions, predict regulatory elements like promoters and enhancers, and suggest effective breeding strategies. By analyzing RNA sequencing data, DL can detect genes and gene fragments, offering deeper insights into genome structure and regulation. These models are also adept at integrating diverse data sources to model complex gene interactions and understand phenotypic outcomes. For genomic prediction, DL helps breeders make informed cross-selection decisions and introduce beneficial alleles into elite germplasm. However, the accuracy of DL models depends heavily on the availability of large, high-quality datasets—a key limitation in some breeding programs. Techniques such as CNNs are used for automated analysis of phenotypic images, while multimodal deep learning (MMDL) integrates genomic and environmental data for better predictions. Other architectures like recurrent neural networks (RNNs), residual networks (ResNet), transformers, and graph neural networks (GNNs) are increasingly adopted for genomic prediction tasks. These tools accelerate breeding by supporting the development of high-yielding, climate-resilient crop varieties with improved traits (Montesinos-López *et al.*, 2024).

12.3.2 Identification of cis-regulatory elements and key nucleotide sequences

Cis-regulatory elements (CREs), including promoters and enhancers, play a key role in controlling gene expression. Promoters initiate transcription, while enhancers modulate their rate. These CREs contain specific nucleotide sequences recognized by transcription factors (TFs), which bind to various regulatory sequences to control gene activity (Worsley-Hunt *et al.*, 2011). Genome editing techniques allow breeders to modify CREs for trait improvement (Zhang *et al.*, 2018; Zhou *et al.*, 2023). Identifying CREs and TF binding sites is essential for understanding gene regulation. DL methods, especially CNNs, are increasingly applied in plant breeding to accurately identify CREs from DNA sequences. These models are effective because they learn complex sequence patterns indicative of regulatory activity. Tools like iCREPCP and PlantDeepSEA's Sequence Profiler use CNNs to predict CREs in species like Arabidopsis, sorghum, tomato, and maize, using chromatin accessibility data for improved accuracy (Peleke *et al.*, 2024). However, selecting appropriate DL tools remains challenging without systematic evaluation. Model effectiveness depends on data size, type, and quality. Ongoing research focuses on improving model performance and developing adaptable frameworks for use across various plant species and environmental conditions (Zhang *et al.*, 2022; Xu and Liu, 2024).

12.4 Applications of AI in Plant Genome Editing

Recent progress in plant genome editing technologies, particularly the CRISPR/Cas systems, has significantly advanced agricultural biotechnology by facilitating targeted and efficient genetic alterations aimed at improving crop yield, abiotic and biotic stress resilience, nutritional quality, and resistance to pathogens. However, challenges such as polyploidy and complex gene regulation hinder editing efficiency (Yoosefzadeh Najafabadi *et al.*, 2023). AI, through ML and DL, addresses these challenges by analyzing genomic and phenotypic data, optimizing gRNA design, and predicting editing outcomes and trait performance (Dixit *et al.*, 2024). Accurate and annotated genome sequences are essential for CRISPR applications, especially in polyploid species, as they enable precise target identification and functional gene studies. AI and genome editing together can accelerate crop improvement toward sustainable agriculture (Ma *et al.*, 2024). AI-driven tools are employed at multiple stages of the editing pipeline, including both pre-editing and post-editing processes, to enhance the efficiency, accuracy, and predictive capabilities of genome modification strategies in plants (Table 12.1).

12.4.1 AI tools and computational algorithms for the design of guide RNAs in plant systems

CRISPR/Cas-based genome editing in plants begins with designing precise sgRNAs that guide Cas nucleases to specific genomic loci. These nucleases induce double-strand breaks (DSBs), which are repaired via non-homologous end joining (NHEJ) or homologous recombination (HR), enabling targeted gene modifications. The success of genome editing largely depends on the specificity and efficiency of sgRNAs (Hwang *et al.*, 2021a). Advances in genome sequencing, CRISPR tools, and computational methods have led to the development of various platforms for sgRNA design tailored to plant genomes. Web tools such as CHOPCHOP (v3) support multiple nucleases and editing functions across 75 plant genomes (Labun *et al.*, 2019), while CRISPR enables sgRNA design for over 219 genomes with on-/off-target scoring and batch processing (Concordet and Haeussler, 2018). CCTop, CRISPRdirect, and CRISPR-GE allow customized sgRNA selection with various parameters, species support, and integrated modules for primer design, mutation detection, and sequence retrieval (Xie *et al.*, 2017; Saha *et al.*, 2025). Plant-specific tools include CRISPR-P 2.0, supporting 82 genomes with detailed annotations (Liu *et al.*, 2017b); CRISPR-Local, which allows genotype-specific design from non-reference genomes (Sun *et al.*, 2019); and CRISPR-Cereal, which integrates gene expression and chromatin data for cereals like rice and maize (He *et al.*, 2021). CRISPR-PLANT focuses on model plants, classifying sgRNAs based on seed mismatch for improved specificity (Minkenberg *et al.*, 2019). The RGEN tools suite (Cas-Designer, PE-Designer, BE-Designer) enables standard, base, and prime editing across 231 plant genomes with fine control over spacer length and PAMs (Jeongbin *et al.*, 2015; Hwang *et al.*, 2021b). Additional platforms like CRISPRscan and E-CRISP offer scoring and genomic context analysis for key crops (Heigwer *et al.*, 2014; Moreno-Mateos *et al.*, 2015). Therefore, these tools form a robust ecosystem for accurate, species-specific plant genome editing and crop improvement.

12.4.2 AI tools for the selection, design, and visualization of CRISPR/Cas vectors

A variety of platforms support plant genome editing through customizable and efficient vector design. Addgene, a major plasmid repository, provides access to more than 132,000 plasmids, including more than 500 vectors for CRISPR-Cas-based plant editing, categorized by editing function and target organism. Vector Bee and its VectorBuilder module enable custom vector design with editable components, codon optimization, and online ordering. The GoldenBraid (GB) CRISPR tool offers modular cloning for synthetic plant biology, allowing the assembly of complex constructs using standardized parts for both monocots and dicots. GenSmart™ Design by GenScript facilitates plasmid construction with tools for codon optimization, error checking, and vector annotation. Lastly, SnapGene provides an intuitive interface for CRISPR vector design, supporting various cloning methods and features such as sgRNA design, homology-directed repair (HDR) template selection, and mutation prediction. Together, these

Table 12.1. The various web tools used for plant genome editing (Author's own table.)

Sr.NO	Applications/tools	Supports	Features	URL
A	Guide RNA design tools			
1	CHOPCHOP (v3)	Cas9 Cpf1 TALEN	Multifunctional tool supporting multiple nucleases and 75 plant genomes.	https://chopchop.cbu.uib.no/
2	CRISPOR	Cas9 Cpf1	Web and command-line tool for sequences < 2300 bp, supporting 219 plant genomes and batch sgRNA selection.	https://crispor.gi.ucsc.edu/
3	CCTop	Cas9	Customizable tool for sgRNA design across 52 plant genomes with mismatch and PAM flexibility.	https://cctop.cos.uni-heidelberg.de/
4	CRISPRdirect	Cas9	An easy-to-use tool with 98 plant genomes, providing detailed sgRNA characteristics.	http://crispr.dbcls.jp/
5	CRISPR-GE	Cas9 Cpf1	Comprehensive suite supporting various PAMs, off-target analysis, and integrated modules for plant editing.	http://skl.scau.edu.cn/targetdesign/
6	CRISPR-P 2.0	Cas9 Cpf1	Produces a set of sgRNA sequences annotated with genomic coordinates and essential parameters such as on-target efficacy, genomic context, GC content, and predicted off-target sites.	http://crispr.hzau.edu.cn/
7	CRISPR-Local	Cas9	Standalone or graphical user interface tool for sgRNA design from non-reference or 72 pre-annotated plant genomes.	http://crispr.hzau.edu.cn/CRISPR-Local/
8	CRISPR-Cereal	Cas9 Cpf1	sgRNA design platform that integrates regulome data and genomic variation for key crops such as wheat, maize, and rice.	http://crispr.hzau.edu.cn/CRISPR-Cereal/
9	CRISPR-PLANT	Cas9	Supports seven model species with high-specificity spacer classification.	http://omap.org/crispr2/
10	CRISPR RGEN tools	Cas9 Prime editing, Base editing.	Includes Cas-Designer, PE-Designer, and BE-Designer for versatile editing with 231 plant genomes.	http://www.rgenome.net/cas-designer/
11	CRISPRscan	Cas9 Cpf1	Supports five plant species with ranking based on efficiency and off-target potential.	https://www.crisprscan.org/
12	E-CRISP	Cas9	Versatile tool analyzing genomic context with customizable sgRNA parameters.	http://www.e-crisp.org/
B	Vector designing tools			
1	Addgene		Over 500 plant genome editing vectors available (e.g. DSBs, nicks, base editing, prime editing). Vectors categorized by editing mechanism, function, or target organism.	https://www.addgene.org/
2	VectorBuilder		Allows modification of vector components (e.g. promoter, open reading frame, selectable markers). eatures sequence editing, annotation, and codon optimization.	https://en.vectorbuilder.com/

Continued

Table 12.1. Continued

Sr.NO	Applications/tools	Supports	Features	URL
3	GoldenBraid (GB)		Enables modular assembly of large constructs (15–20 Kb with 4–6 transcriptional units), Monocot/dicot-specific promoter and target selection,	https://goldenbraidpro.com/
4	GenSmart™ Design		Auto-annotation of functional plasmid elements, Built-in codon optimization for rapid sequence refinement.	https://www.genscript.com/
5	SnapGene		Includes 2700 + annotated plasmids, including plant CRISPR vectors. Supports multiple cloning methods: Gateway, Gibson, In-Fusion, Golden Gate. Enables repair template selection for HDR using ssODN or synthetic DNA.	https://www.snapgene.com/
C	Prediction and analysis			
1	ICE Synthego	Cas9 Cpf1	A freely accessible web-based tool designed for the analysis of Sanger sequencing data. It necessitates the input of gRNA sequences, nuclease type, and .ab1 files.	https://www.synthego.com/products/bioinformatics/crispr-analysis
2	TIDE	CRISPR/Cas9, TALENs and ZFNs	Indel frequency is determined through Sanger sequencing and gRNA sequence analysis. In "batch mode," TIDE enables the simultaneous analysis of multiple data files.	https://tide.nki.nl/
3	Cas-Analyzer	ZFNs, TALENs, Cas9 nickases, and dCas9-FokI nucleases, Cpf1, etc	The outcomes are classified according to mutation type and organized based on indel frequency, with additional information on their locations and sizes.	http://www.rgenome.net/cas-analyzer/
4	CRISPR-detector	Cas9, Cpf1, Base editor, Prime editor.	Web-based platform for analyzing CRISPR-induced edits from sequencing data, designed for fast and accurate identification of genome editing outcomes.	https://db.cngb.org/crispr-detector/
5	CRISPRcloud2	Cas9 Knockout	Web tool optimized for analyzing NGS data from CRISPR-edited samples. Offers cloud-based, user-friendly interface for mutation detection.	https://crispr.nrihub.org/
6	CRISPResso2	Cas9, Base editor, Prime editor	Available as a web tool and standalone software. Performs read quality filtering, adapter trimming, and reference alignment.	http://crispresso2.pinellolab.org/

See text for abbreviations.

tools streamline the development of plant gene editing vectors for diverse applications (Saha *et al.*, 2025).

12.4.3 AI tools for CRISPR editing prediction and analysis

In the post-editing phase of plant genome modification, CRISPR/Cas-induced changes are identified using Sanger sequencing or next-generation sequencing (NGS). Inference of CRISPR Edits (ICE) by Synthego and Tracking of Indels by Decomposition (TIDE) both analyze Sanger data to quantify indels with 1–5% sensitivity using sequence trace decomposition (Eva *et al.*, 2014; Conant *et al.*, 2022). CRISPResso2 processes NGS data to assess NHEJ, HR, and base editing outcomes. It supports batch processing, allele-specific quantification, biologically informed alignments, and rapid data analysis (Clement *et al.*, 2019). Cas-Analyzer quantifies mutations from uploaded NGS data by identifying flanking markers around cleavage sites, selecting valid reads with up to one mismatch, filtering by frequency, and classifying sequences into insertions, deletions, or wild-type/substitutions (Park *et al.*, 2017). CRISPR-detector enhances analysis with whole genome sequencing (WGS) support, haplotype-based variant calling, structural variant detection, and background noise filtering using paired samples. It was validated on simulated data with indels, substitutions, and sequencing errors (Huang *et al.*, 2023). CRISPRcloud2 is a secure, cloud-based tool for pooled NGS data analysis, offering statistical tools, customizable outputs, and user-friendly visualization via a web interface (Hyun-Hwan *et al.*, 2017). The Plant Genome Editing Database (PGED), by the Boyce Thompson Institute, documents CRISPR-edited plant lines with economic traits, detailing target genes, gRNAs, transformation methods, and phenotypes—searchable by species, gene ID, or gRNA (Zheng *et al.*, 2019). The Global Gene Editing Regulation Tracker, maintained by the Genetic Literacy Project, provides country-wise updates on gene editing regulations across agriculture, health, and livestock (Saha *et al.*, 2025). Together, these tools and databases form a comprehensive ecosystem for post-editing validation, analysis, and regulatory insight in plant genome editing.

12.5 Future Prospects of Machine Learning-Assisted Plant Genome Editing

ML is rapidly emerging as a powerful tool in the field of genome editing, offering enhanced efficiency and precision. Its integration into plant genome editing is expected to revolutionize the development of crops with improved traits and stress resilience.

12.5.1 Role in epigenetic regulatory genome editing

ML is poised to play a transformative role in targeting genes involved in epigenetic regulation. By predicting and designing highly specific sgRNAs, ML algorithms can assist in editing regions of DNA that control gene expression through modifications of DNA or chromatin structure. This approach enhances the accuracy and efficacy of genome editing, particularly for complex traits governed by epigenetic mechanisms.

12.5.2 Role in single-cell technology

An emerging application of ML in genome editing is within single-cell sequencing technologies. Single-cell CRISPR screening has already shown success in animal systems and is now being adapted for plant cells, despite challenges such as rigid cell walls and heterogeneity in cell size. The combination of single-cell technology and ML holds great promise for unraveling plant cell-specific responses and guiding precise genome edits at the individual cell level.

12.5.3 Applications in orphan crop improvement

ML also holds immense potential for the improvement of orphan crops—underutilized species like sorghum that are primarily cultivated in developing regions and often exhibit tolerance to extreme environmental stresses. By analyzing genomic data using ML models, such as artificial neural networks, researchers can identify unique genes conferring resilience to biotic and abiotic stress. These genes can be targeted for genome editing or transferred to staple crops like rice, wheat, and maize to enhance global food security. Genomic databases such as the African Orphan Crops Consortium (AOCC) and Crops for the Future (CFF) provide valuable resources for this endeavor.

12.6 Conclusions

Crop genome editing is currently one of the most powerful and cutting-edge technologies in agricultural biotechnology. It has demonstrated significant potential in improving various traits, developing climate-resilient crops, and enhancing overall crop productivity. Due to fewer regulatory restrictions compared to traditional genetic modification, genome editing is also more feasible for practical applications in the near future.

The integration of ML further strengthens this approach. ML can enhance the precision, speed, and efficiency of genome editing by automating key processes such as target identification, sgRNA design, and off-target prediction. This synergy not only maximizes the full potential of genome editing but also opens up new frontiers in crop research and development. As a result, ML-driven genome editing is expected to transform plant breeding and agricultural innovation on a larger scale in the coming years.

References

Chen, L., Liu, G. and Zhang, T. (2024) Integrating machine learning and genome editing for crop improvement. *aBIOTECH* 5(2), 262–277. DOI: 10.1007/s42994-023-00133-5.

Chuai, G., Ma, H., Yan, J., Chen, M., Hong, N. *et al.* (2018) DeepCRISPR: Optimized CRISPR guide RNA design by deep learning. *Genome Biology* 19(1), 80. DOI: 10.1186/s13059-018-1459-4.

Clement, K., Rees, H., Canver, M.C., Gehrke, J.M., Farouni, R. *et al.* (2019) CRISPResso2 provides accurate and rapid genome editing sequence analysis. *Nature Biotechnology* 37(3), 224–226. DOI: 10.1038/s41587-019-0032-3.

Conant, D., Hsiau, T., Rossi, N., Oki, J., Maures, T. *et al.* (2022) Inference of CRISPR edits from sanger trace data. *The CRISPR Journal* 5(1), 123–130. DOI: 10.1089/crispr.2021.0113.

Concordet, J.P. and Haeussler, M. (2018) CRISPOR: Intuitive guide selection for CRISPR/Cas9 genome editing experiments and screens. *Nucleic Acids Research* 46(W1), W242–W245. DOI: 10.1093/nar/gky354.

DeWeirdt, P.C., McGee, A.V., Zheng, F., Nwolah, I., Hegde, M. *et al.* (2022) Accounting for small variations in the tracrRNA sequence improves sgRNA activity predictions for CRISPR screening. *Nature Communications* 13(1), 5255. DOI: 10.1038/s41467-022-33024-2.

Ding, K., Chin, M., Zhao, Y., Huang, W., Mai, B.K. *et al.* (2024) Machine learning-guided co-optimization of fitness and diversity facilitates combinatorial library design in enzyme engineering. *Nature Communications* 15(1), 6392. DOI: 10.1038/s41467-024-50698-y.

Dixit, S., Kumar, A., Srinivasan, K., Vincent, P.M.D.R. and Ramu Krishnan, N. (2024) Advancing genome editing with artificial intelligence: Opportunities, challenges, and future directions. *Frontiers in Bioengineering and Biotechnology* 11, 1335901. DOI: 10.3389/fbioe.2023.1335901.

Doench, J.G., Hartenian, E., Graham, D.B., Tothova, Z., Hegde, M. *et al.* (2014) Rational design of highly active sgRNAs for CRISPR-Cas9-mediated gene inactivation. *Nature Biotechnology* 32(12), 1262–1267. DOI: 10.1038/nbt.3026.

Doench, J.G., Fusi, N., Sullender, M., Hegde, M., Vaimberg, E.W. *et al.* (2016) Optimized sgRNA design to maximize activity and minimize off-target effects of CRISPR-Cas9. *Nature Biotechnology* 34(2), 184–191. DOI: 10.1038/nbt.3437.

Eva, K.B., Tao, C., Mario, A. and Bas, V. (2014) Easy quantitative assessment of genome editing by sequence trace decomposition. *Nucleic Acids Research* 42(22), e168–e168. DOI: 10.1093/nar/gku936.

He, C., Liu, H., Chen, D., Xie, W.Z., Wang, M. *et al.* (2021) CRISPR-cereal: A guide RNA design tool integrating regulome and genomic variation for wheat, maize and rice. *Plant Biotechnology Journal* 19(11), 2141–2143. DOI: 10.1111/pbi.13675.

Heigwer, F., Kerr, G. and Boutros, M. (2014) E-CRISP: Fast CRISPR target site identification. *Nature Methods* 11(2), 122–123. DOI: 10.1038/nmeth.2812.

Huang, L., Li, Q., Zhang, C., Chu, R., Gu, Z. *et al.* (2020) Creating novel Wx alleles with fine-tuned amylose levels and improved grain quality in rice by promoter editing using CRISPR/Cas9 system. *Plant Biotechnology Journal* 18(11), 2164–2166. DOI: 10.1111/pbi.13391.

Huang, L., Wang, D., Chen, H., Hu, J., Dai, X. *et al.* (2023) CRISPR-detector: Fast and accurate detection, visualization, and annotation of genome-wide mutations induced by genome editing events. *Journal of Genetics and Genomics* 50(8), 563–572. DOI: 10.1016/j.jgg.2023.03.010.

Hwang, G., Song, B. and Bae, S. (2021a) Current widely used web-based tools for CRISPR nucleases. *Gene and Genome Editing* 1, 100004. DOI: 10.1016/J.GGEDIT.2021.100004.

Hwang, G.H., Jeong, Y.K., Habib, O., Hong, S.A., Lim, K. *et al.* (2021b) PE-designer and PE-analyzer: Web-based design and analysis tools for CRISPR prime editing. *Nucleic Acids Research* 49(W1), W499–W504. DOI: 10.1093/nar/gkab319.

Hyun-Hwan, H.-H., Kim, S.Y., Rousseaux, M.W.C., Zoghbi, H.Y. and Liu, Z. (2017) CRISPRcloud: A secure cloud-based pipeline for CRISPR pooled screen deconvolution. *Bioinformatics* 33(18), 2963–2965. DOI: 10.1093/bioinformatics/btx335.

Jeongbin, J., Sangsu, B. and Jin-Soo, K. (2015) Cas-designer: A web-based tool for choice of CRISPR-Cas9 target sites. *Bioinformatics* 31(24), 4014–4016. DOI: 10.1093/bioinformatics/btv537.

Kajrolkar, A. (2025) Deep learning applications in plant biology: Bridging genotype and phenotype. *Premier Journal of Plant Biology* 3, 100011. DOI: 10.70389/PJPB.100011.

Kim, N., Kim, H.K., Lee, S., Seo, J.H., Choi, J.W. *et al.* (2020) Prediction of the sequence-specific cleavage activity of Cas9 variants. *Nature Biotechnology* 38(11), 1328–1336. DOI: 10.1038/s41587-020-0537-9.

Labun, K., Montague, T.G., Krause, M., Torres Cleuren, Y.N., Tjeldnes, H. *et al.* (2019) CHOPCHOP v3: Expanding the CRISPR web toolbox beyond genome editing. *Nucleic Acids Research* 47(W1), W171–W174. DOI: 10.1093/nar/gkz365.

Latif, M. and Mustafa, M. (2025) Crispr/Cas-mediated biofortification of crop plants: Advancing nutritional enhancement through precision genome editing. *Insights-Journal of Life and Social Sciences*. DOI: 10.71000/jyf5gh19.

Liu, H., Ding, Y., Zhou, Y., Jin, W., Xie, K. *et al.* (2017b) CRISPR-p 2.0: An improved CRISPR-Cas9 tool for genome editing in plants. *Molecular Plant* 10(3), 530–532. DOI: 10.1016/j.molp.2017.01.003.

Liu, J., Li, J., Wang, H. and Yan, J. (2020) Application of deep learning in genomics. *Science China Life Sciences* 63(12), 1860–1878. DOI: 10.1007/s11427-020-1804-5.

Liu, Q., Yang, F., Zhang, J., Liu, H., Rahman, S. *et al.* (2021) Application of CRISPR/Cas9 in crop quality improvement. *International Journal of Molecular Sciences* 22(8), 4206. DOI: 10.3390/ijms22084206.

Liu, X., Wu, S., Xu, J., Sui, C. and Wei, J. (2017a) Application of CRISPR/Cas9 in plant biology. *Acta Pharmaceutica Sinica B* 7(3), 292–302. DOI: 10.1016/j.apsb.2017.01.002.

Liu, X., Yang, J., Song, Y., Zhang, X., Wang, X. *et al.* (2022) Effects of sgRNA length and number on gene editing efficiency and predicted mutations generated in rice. *The Crop Journal* 10(2), 577–581. DOI: 10.1016/j.cj.2021.05.015.

Lu, H.-P., Luo, T., Fu, H.-W., Wang, L., Tan, Y.-Y. *et al.* (2018) Resistance of rice to insect pests mediated by suppression of serotonin biosynthesis. *Nature Plants* 4(6), 338–344. DOI: 10.1038/s41477-018-0152-7.

Ma, S., Qu, Z., Li, X. and Tang, Y. (2024) The innovative application and prospects of AI and gene editing technology in high-yield crop cultivation. *Geographical Research Bulletin* 3, 219–222. DOI: 10.50908/grb.3.0_219.

Mao, Y., Yang, X., Zhou, Y., Zhang, Z., Botella, J.R. *et al.* (2018) Manipulating plant RNA-silencing pathways to improve the gene editing efficiency of CRISPR/Cas9 systems. *Springer Nature* 19(1). DOI: 10.1186/s13059-018-1529-7.

Minkenberg, B., Zhang, J., Xie, K. and Yang, Y. (2019) CRISPR-PLANT v2: An online resource for highly specific guide RNA spacers based on improved off-target analysis. *Plant Biotechnology Journal* 17(1), 5–8. DOI: 10.1111/pbi.13025.

Mochida, K., Koda, S., Inoue, K., Hirayama, T., Tanaka, S. *et al.* (2018) Computer vision-based phenotyping for improvement of plant productivity: A machine learning perspective. *GigaScience* 8(1), 1–12. DOI: 10.1093/gigascience/giy153.

Montesinos-López, O.A., Chavira-Flores, M., Kiasmiantini, C.-H.L., Crespo-Herrera, L., Piere, C. *et al.* (2024) A review of multimodal deep learning methods for genomic-enabled prediction in plant breeding. *Genetics* 228(4), iyae161. DOI: 10.1093/genetics/iyae161.

Moreno-Mateos, M.A., Vejnar, C.E., Beaudoin, J.D., Fernandez, J.P., Mis, E.K. *et al.* (2015) CRISPRscan: Designing highly efficient sgRNAs for CRISPR-Cas9 targeting *in vivo*. *Nature Methods* 12(10), 982–988. DOI: 10.1038/nmeth.3543.

Mu, W., Luo, T., Barrera, A., Bounds, L.R., Klann, T.S. *et al.* (2024) Machine learning methods for predicting guide RNA effects in CRISPR epigenome editing experiments. *Biorxiv*. DOI: 10.1101/2024.04.18.590188.

Najafabadi, M.Y., Hesami, M. and Eskandari, M. (2023) Machine learning-assisted approaches in modernized plant breeding programs. *Genes* 14(4), 777. DOI: 10.3390/genes14040777.

Nayfach, S., Bhatnagar, A., Novichkov, A., Estevam, G.O., Kim, N. *et al.* (2025) Engineering of CRISPR-Cas PAM recognition using deep learning of vast evolutionary data. *bioRxiv*. DOI: 10.1101/2025.01.06.631536.

Niazian, M. and Niedbała, G. (2020) Machine learning for plant breeding and biotechnology. *Agriculture* 10(10), 436. DOI: 10.3390/agriculture10100436.

Noman, A., Aqeel, M. and He, S. (2016) CRISPR-Cas9: Tool for qualitative and quantitative plant genome editing. *Frontiers in Plant Science* 7, 1740. DOI: 10.3389/fpls.2016.01740.

Park, J., Lim, K., Kim, J.S. and Bae, S. (2017) Cas-analyzer: An online tool for assessing genome editing results using NGS data. *Bioinformatics (Oxford, England)* 33(2), 286–288. DOI: 10.1093/bioinformatics/btw561.

Peleke, F.F., Zumkeller, S.M., Gültas, M., Schmitt, A. and Szymański, J. (2024) Deep learning the cis-regulatory code for gene expression in selected model plants. *Nature Communications* 15(1), 3488. DOI: 10.1038/s41467-024-47744-0.

Petolino, J.F., Worden, A., Curlee, K., Connell, J., Strange Moynahan, T.L. *et al.* (2010) Zinc finger nuclease-mediated transgene deletion. *Plant Molecular Biology* 73(6), 617–628. DOI: 10.1007/s11103-010-9641-4.

Razzaq, A., Kaur, P., Akhter, N., Wani, S.H. and Saleem, F. (2021) Next-generation breeding strategies for climate-ready crops. *Frontiers in Plant Science* 12, 620420. DOI: 10.3389/fpls.2021.620420.

Saha, D., Panda, A.K. and Datta, S. (2025) Critical considerations and computational tools in plant genome editing. *Heliyon* 11(1), e41135. DOI: 10.1016/j.heliyon.2024.e41135.

Schmidt, H., Zhang, M., Chakarov, D., Bansal, V., Mourelatos, H. *et al.* (2025) Genome-wide CRISPR guide RNA design and specificity analysis with GuideScan2. *Genome Biology* 26(1), 41. DOI: 10.1186/s13059-025-03488-8.

Seo, S.Y., Min, S., Lee, S., Seo, J.H., Park, J. *et al.* (2023) Massively parallel evaluation and computational prediction of the activities and specificities of 17 small Cas9s. *Nature Methods* 20(7), 999–1009. DOI: 10.1038/s41592-023-01875-2.

Sheng, X., Sun, Z., Wang, X., Tan, Y., Yu, D. *et al.* (2020) Improvement of the rice “Easy-to-Shatter” trait via CRISPR/Cas9-mediated mutagenesis of the qSH1 gene. *Frontiers in Plant Science* 11, 619. DOI: 10.3389/fpls.2020.00619.

Shi, J., Gao, H., Wang, H., Lafitte, H.R., Archibald, R.L. *et al.* (2017) ARGOS8 variants generated by CRISPR-Cas9 improve maize grain yield under field drought stress conditions. *Plant Biotechnology Journal* 15(2), 207–216. DOI: 10.1111/pbi.12603.

Sun, J., Liu, H., Liu, J., Cheng, S., Peng, Y. *et al.* (2019) CRISPR-Local: A local single-guide RNA (sgRNA) design tool for non-reference plant genomes 35(14), 2501–2503. DOI: 10.1093/bioinformatics/bty970.

van Dijk, A.D.J., Kootstra, G., Kruijer, W. and Ridder, D. (2020) Machine learning in plant science and plant breeding. *Science* 24(1), 101890. DOI: 10.1016/j.isci.2020.101890.

Wang, X., Zeng, H., Lin, L., Huang, Y., Lin, H. *et al.* (2023) Deep learning-empowered crop breeding: Intelligent, efficient and promising. *Frontiers in Plant Science* 14, 1260089. DOI: 10.3389/fpls.2023.1260089.

Worsley-Hunt, R., Bernard, V. and Wasserman, W.W. (2011) Identification of cis-regulatory sequence variations in individual genome sequences. *Genome Medicine* 3(10), 65. DOI: 10.1186/gm281.

Wulff, B.B.H. and Dhugga, K.S. (2018) Wheat-the cereal abandoned by GM. *Science* 361(6401), 451–452. DOI: 10.1126/science.aat5119.

Xie, X., Ma, X., Zhu, Q., Zeng, D., Li, G. *et al*. (2017) CRISPR-GE: A convenient software toolkit for CRISPR-based genome editing. *Molecular Plant* 10(9), 1246–1249. DOI: 10.1016/j.molp.2017.06.004.

Xu, L. and Liu, Y. (2024) Identification, design, and application of noncoding cis-regulatory elements. *Biomolecules* 14(8), 945. DOI: 10.3390/biom14080945.

Xu, X., Xu, Z., Li, Z., Zakria, M., Zou, L. *et al*. (2021) Increasing resistance to bacterial leaf streak in rice by editing the promoter of susceptibility gene OsSULRT3;6. *Plant Biotechnology Journal* 19(6), 1101–1103. DOI: 10.1111/pbi.13602.

Yang, H., Wu, J.-J., Tang, T., Liu, K.-D. and Dai, C. (2017) CRISPR/Cas9-mediated genome editing efficiently creates specific mutations at multiple loci using one sgRNA in *Brassica napus*. *Scientific Reports* 7(1), 7489. DOI: 10.1038/s41598-017-07871-9.

Yoosefzadeh Najafabadi, M., Hesami, M. and Eskandari, M. (2023) Machine learning-assisted approaches in modernized plant breeding programs. *Genes* 14(4), 777. DOI: 10.3390/genes14040777.

Zhang, H., Si, X., Ji, X., Fan, R., Liu, J. *et al*. (2018) Genome editing of upstream open reading frames enables translational control in plants. *Nature Biotechnology* 36(9), 894–898. DOI: 10.1038/nbt.4202.

Zhang, S., Ma, A., Zhao, J., Xu, D., Ma, Q. *et al*. (2022) Assessing deep learning methods in cis -regulatory motif finding based on genomic sequencing data. *Briefings in Bioinformatics* 23(1), 1–10. DOI: 10.1093/bib/bbab374.

Zheng, Y., Zhang, N., Martin, G.B. and Fei, Z. (2019) Plant genome editing database (PGED): A call for submission of information about genome-edited plant mutants. *Molecular Plant* 12(2), 127–129. DOI: 10.1016/j.molp.2019.01.001.

Zhou, J., Liu, G., Zhao, Y., Zhang, R., Tang, X. *et al*. (2023) CRISPR-Cas12a promoter editing system for crop improvement. *Nature Plants* 9(4), 588–604. DOI: 10.1038/s41477-023-01384-2.

13 Machine Learning-Assisted Genome-Wide Association Study (GWAS) in Plants

Jemaa Essemine[1,2]*, Zhan Xu[3], Mokhtar Guerfel[4], Yogesh K. Ahlawat[5], Jen-Tsung Chen[6]* and Mingnan Qu[1]*

[1]Jiangsu Key Laboratory of Crop Genetics and Physiology, Agricultural College of Yangzhou University, Yangzhou, China; [2]Bioinformatics Section, Department of Informatics, University of Quebec at Montreal, Pavillon President-Kennedy, Montreal, Canada; [3]Yazhouwan National Laboratory, Yazhou District, Sanya City, China; [4]Department of Biological Sciences, College of Education Afif, Shaqra University, Ar-Riyad, Saudi Arabia; [5]Centre of Research Impact and Outcome, Chitkara University Institute of Engineering and Technology, Rajpura, India; [6]Department of Life Sciences, National University of Kaohsiung, Taiwan

Abstract

Genome-wide association studies (GWAS) have become a pivotal tool for dissecting the genetic basis of complex traits in plants. However, traditional statistical methods often face limitations in addressing the intricate genetic architectures and environmental interactions that govern many agronomically important phenotypes. The integration of machine learning (ML) algorithms into GWAS, termed ML–GWAS, offers a transformative approach to overcome these challenges. This chapter explores the revolutionary impact of ML on plant genomics, highlighting its capacity to enhance predictive accuracy for complex traits, uncover novel genetic associations including epistatic interactions and small-effect loci, and improve the efficiency of breeding programs. We discuss how ML techniques, such as deep learning and ensemble methods, can model non-linear relationships, perform feature selection, and ultimately provide a more comprehensive understanding of the genotype–phenotype map in plants. Furthermore, we touch upon the crucial advancements in the interpretability of ML models in the context of GWAS. By leveraging the power of ML, researchers are gaining unprecedented insights into the genetic control of complex traits, paving the way for the development of superior crop varieties with enhanced yield, nutritional quality, and resilience to environmental stresses. This chapter provides an overview of the current state and future perspectives of ML-assisted GWAS in plant genomics, emphasizing its potential to accelerate crop improvement for global food security.

Keywords: ML, GWAS, complex traits, genomic prediction, epistasis, gene–environment interaction (G × E), feature selection, crop improvement, breeding

*Corresponding authors: jentsung@nuk.edu.tw, qmn@yzu.edu.cn and jemaa.essemine@gmail.com

DOI: 10.1079/9781800628649.0013

13.1 Introduction

The imperative to enhance crop productivity and resilience in the face of a burgeoning global population and escalating environmental challenges necessitates a deeper understanding of the genetic underpinnings of complex agronomic traits. Genome-wide association studies (GWAS) have emerged as a powerful tool and cornerstone for dissecting the genetic architecture of these traits in plant populations (Bernardo, 2020). By leveraging extensive genetic variation and high-density molecular markers, GWAS aims to identify significant statistical associations between genetic variants, predominantly single nucleotide polymorphisms (SNPs), and observed phenotypic diversity.

However, conventional GWAS approaches in plants grapple with inherent limitations. The intricate genomic landscape of many plant species, often characterized by extensive linkage disequilibrium (LD), polyploidy, and historical admixture, can impede the precise mapping of causal variants (Flint-Garcia and Thornsberry, 2003). Furthermore, numerous agronomically important traits exhibit a polygenic architecture, influenced by a multitude of genes with individually small effects, posing a significant challenge for detection using traditional linear statistical models in GWAS (Mackay *et al.*, 2009). The complex interplay of gene–gene (epistasis) and gene–environment (G × E) interactions, which substantially shape phenotypic expression, further complicates the analysis and interpretation of GWAS results using standard methodologies (Cuevas *et al.*, 2023).

The integration of machine learning (ML) methodologies presents a transformative opportunity to address these limitations and enhance the power, resolution, and interpretability of GWAS in plants. ML, a subfield of artificial intelligence (AI) focused on enabling systems to learn from data without explicit programming, has demonstrated remarkable success in analyzing complex, high-dimensional biological datasets, including human genomics (Angermueller *et al.*, 2016; Wang *et al.*, 2023). The application of ML to plant GWAS holds the potential to overcome several challenges inherent in traditional statistical approaches.

Specifically, ML offers several advantages for plant GWAS. First, ML algorithms can model intricate non-linear relationships and complex interactions between genetic variants and phenotypes, moving beyond the additive assumptions of linear models typically employed in GWAS (Gianola *et al.*, 2018). Techniques such as random forests (RF) (Breiman, 2001), support vector machines (SVM) (Cortes and Vapnik, 1995), and increasingly, deep neural networks (DNN) (LeCun *et al.*, 2015) are capable of capturing epistatic effects and G × E interactions that might be overlooked by linear analyses. ML can facilitate feature selection by identifying the most informative genetic variants associated with a trait of interest from a vast pool of SNPs. Feature importance measures derived from ML models can aid in reducing noise and prioritizing genomic regions with the strongest predictive power, thereby enhancing statistical power and potentially improving the resolution of causal variant identification (Guyon and Elisseeff, 2003).

While distinct from traditional GWAS, ML is central to genomic selection, where whole-genome marker data are used to predict the breeding value of individuals (Meuwissen *et al.*, 2001). Insights gained from ML models trained on GWAS datasets can contribute to a more nuanced understanding of the genetic architecture of complex traits and the identification of synergistic combinations of genetic variants influencing phenotype. Eventually, a critical area of ongoing research focuses on enhancing the interpretability of ML models applied to GWAS. Understanding the biological rationale behind model predictions is crucial for translating statistical associations into biological insights and identifying potential causal mechanisms (Cho and Kang, 2020; Sidak *et al.*, 2022). Explainable AI (XAI) techniques are increasingly being explored to demystify complex ML models and identify the key genetic features driving phenotypic variation.

This chapter delves into the diverse applications of ML techniques (or algorithms) to advance GWAS in plants. We explore the strengths and limitations of various ML approaches, illustrate their utility with examples from recent literature (to the best of our current knowledge), and discuss future directions in this rapidly evolving field. The overarching goal is to provide a comprehensive overview of how ML is empowering researchers to unravel the genetic complexity of crucial agronomic traits in plants and accelerate the development of improved crop varieties for future food security and sustainability. This piece of work aims to bring together (or collect) the recent advances in ML-assisted GWAS topics in plants and the impact of implementing this integrative technique (ML–GWAS) on fostering plant breeding strategies.

13.2 ML–GWAS Revolution in Plant Genomics

The advent of ML algorithms has ushered in a transformative era for GWAS, fundamentally reshaping the landscape of plant genomics. While traditional GWAS methodologies have provided invaluable insights into the genetic architecture of numerous plant traits (Flint-Garcia and Thornsberry, 2003; Bernardo, 2020), they often encounter significant hurdles in dissecting the complexities underlying economically and biologically important phenotypes. These restrictions (or limitations) frequently stem from factors such as constrained statistical power, particularly when dealing with traits governed by numerous genes with small individual effects (Mackay *et al.*, 2009), and the pervasive presence of confounding variables like population structure and cryptic (or hidden) relatedness that can lead to spurious associations (Price *et al.*, 2006; Kang *et al.*, 2008). In contrast, ML–GWAS approaches harness the sophisticated pattern recognition and predictive capabilities of advanced algorithms, including deep learning (DL) architectures (Enoma *et al.*, 2022; Lakiotaki *et al.*, 2023; Yu *et al.*, 2024; Sigala *et al.*, 2024) and ensemble methods like RF (Breiman, 2001), to overcome these challenges and unlock a more comprehensive understanding of the genotype–phenotype (G × P) map in plants. This paradigm shift enables researchers to move beyond the inherent constraints of linear models and explore the intricate, non-linear relationships that often characterize the genetic control of complex traits (Gianola *et al.*, 2018). By leveraging the ability of ML to handle high-dimensional data (or big data), capture subtle interactions, and make robust predictions, the ML–GWAS approach paves the way for more accurate gene discovery, enhanced breeding strategies, and ultimately, the development of superior crop varieties with improved yield, nutritional quality, and resilience to environmental stresses (biotic and abiotic).

13.2.1 Enhance predictive accuracy

By discerning intricate patterns and interplays among genetic variations, ML–GWAS offers a substantial enhancement in the prediction of complex traits such as yield potential, resistance to diseases, and tolerance to abiotic stresses (Jannink *et al.*, 2010; Crossa *et al.*, 2017). For instance, ML models have demonstrated considerable promise and a beneficial influence in elevating the precision of genomic selection for complex traits in plants (Montesinos-López *et al.*, 2018), potentially outperforming conventional linear models through their capacity to capture non-additive genetic effects and G × E) interactions. This improved predictive capability allows breeders to make more informed decisions earlier in the breeding cycle, potentially reducing the need for extensive and resource-intensive field trials. Furthermore, the ability of ML models to learn from large and diverse datasets can lead to more robust and generalizable predictions across different genetic backgrounds and environmental conditions.

13.2.2 Uncover novel genetic associations

These methods can detect hidden (or subtle) genetic effects and epistatic interactions that might be missed by traditional approaches (including GWAS), leading to the discovery of novel genes and pathways underlying important agronomic traits (Huang and Mackay, 2016; Bajad *et al.*, 2024). By employing algorithms capable of modeling non-linear relationships, ML–GWAS can dissect complex genetic architectures and identify previously undetected associations (Michael and VanBuren, 2015). For example, ML approaches have been shown to enhance the detection of quantitative trait loci (QTLs) with smaller effects that might be overlooked in standard linear GWAS models (Szymczak *et al.*, 2009; Yoosefzadeh-Najafabadi *et al.*, 2022). Furthermore, the ability of ML to handle high-dimensional data (big data) and capture complex interactions allows for the exploration of epistatic effects contributing, thus, to phenotypic variation, potentially revealing novel genetic combinations and influencing agronomic performance.

13.2.3 Improve the efficiency of breeding programs

By precisely predicting the phenotypic outcomes of different breeding lines, ML–GWAS may hasten (or accelerate) the development and creation of improved crop varieties with enhanced productivity and resilience (Heffner *et al.*, 2009). By offering more accurate predictions, ML helps breeders make superior selection decisions, minimizing thus the number of extensive and time-consuming field trials required. For instance, integrating ML-derived insights from GWAS can improve the selection of superior parental lines with favorable combinations of alleles for complex traits (Bernardo, 2020). Furthermore, ML models can be used to predict the performance of novel genotypes in diverse environments, facilitating the development of broadly adapted or locally optimized varieties more efficiently.

13.3 ML-Enhanced GWAS: Advancing Plant Genetic Understanding

The integration of ML within plant GWAS signifies a notable leap forward in our comprehension of plant heredity. While classical GWAS approaches have provided valuable insights into specific genetic regions linked to certain characteristics (or traits), they frequently fall short of fully elucidating the intricate and multi-layered control of quantitative traits. These inherent restrictions often stem from challenges such as insufficient statistical capacity to detect minor genetic influences, the obscuring presence of interacting factors, and the complexities involved in deciphering intricate gene–gene relationships (epistasis; Mackay *et al.*, 2009). By leveraging the capabilities of advanced computational algorithms, including DL frameworks and RF methodologies (Fig. 13.1), scientists can navigate beyond these constraints (Breiman, 2001; LeCun *et al.*, 2015). These sophisticated tools demonstrate exceptional proficiency in examining large (or big) and intricate datasets, discerning non-linear correlations, and identifying accurate (or subtle) patterns and interplays among genetic variations that might be undetectable by traditional means (Bishop, 2006) such GWAS solely. This augmented analytical capacity empowers ML–GWAS to uncover previously unknown genetic linkages and significantly enhance the precision of predicting observable characteristics, laying the groundwork for more efficient breeding efforts.

13.4 Boosting Plant GWAS: The Power of ML

The integration of sophisticated ML algorithms into plant GWAS underscores its catalytic influence in supercharging this field. While traditional robust computational tools, including GWAS, offer foundational insights, their capacity to fully unravel the intricate web of biological systems often encounters obstacles. Elements such as restricted statistical capacity, interacting variables, and the existence of complex gene–gene relationships (epistasis) can impede the precise pinpointing of genetic regions linked to crucial agricultural characteristics (Phillips, 2008). ML algorithms, endowed with the capability to process extensive datasets and discern non-linear correlations (Hastie *et al.*, 2009; Chafai *et al.*, 2023), present a potent strategy for navigating big raw data and addressing the shortcomings of traditional GWAS approaches. Through strategically employing techniques like DL, RF, and SVM (Fig. 13.1), scientists can amplify the power and accuracy of GWAS (Cortes and Vapnik, 1995; Breiman, 2001; Goodfellow *et al.*, 2016). This enhancement encompasses improved precision in forecasting complex traits (Crossa *et al.*, 2017, 2025), the discovery of previously unknown genetic linkages, and a more profound comprehension of the genetic framework underpinning key agricultural attributes. Finally, the integration of ML with plant GWAS

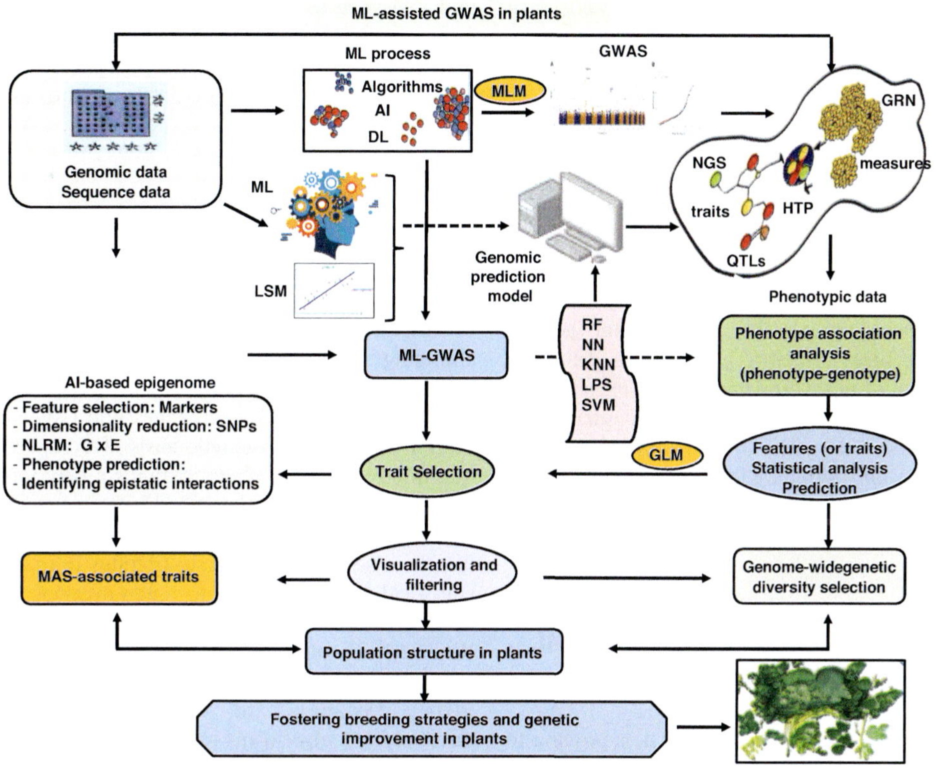

Fig. 13.1. Diagram illustrates how genomic and phenotypic data are analyzed using machine learning (ML) within the framework of GWAS studies. This approach aims to identify the associations between genes and traits, which subsequently allows for the improved selection of desirable traits and genetic diversity for plant breeding and crop improvement (Author's own image). AI, artificial intelligence; ML, machine learning; DL, deep learning; GWAS, Genome-wide association study; GxE: gene–environment interactions; MAS, marker-assisted selection; SNPs, single nucleotide polymorphisms; QTLs, quantitative trait loci; LSM, linear statistical method; GLM, Gaussian linear model; MLM, mixed linear model; RF, random forest; NN, neuronal network; KNN, K-nearest neighbor; PLS, partial least squares; SVM, support vector machine; GRN, gene regulatory network; NGS, next-generation sequencing; HTP, high-throughput. (Author's own image.)

represents a pivotal advancement, pushing the field closer to developing the next generation of superior crop varieties. This progress is consistent with the goals highlighted by the genomic selection approaches (Crossa *et al.*, 2017; Alemu *et al.*, 2024).

13.5 Decoding Plant Genomes: ML's Integral Role in GWAS

The transformative potential of ML-assisted GWAS highlights the indispensable function of advanced computational methods in elucidating the complex hereditary instructions embedded within plant genomes (Kersey, 2019). Although traditional GWAS has offered valuable insights, their capacity to fully resolve the multifaceted nature of biological systems is often constrained, particularly in the detection of precise genetic influences, the dissection of gene-gene (G × G) interactions, and the comprehensive assessment of environmental impacts on gene expression (G × E) (Manolio, 2010). To truly grasp these complexities, we need to move beyond the limitations of standard GWAS approaches. Thus,

ML algorithms, by virtue of their proficiency in processing extensive data, discerning non-linear correlations, and revealing previously obscured patterns, furnish a robust toolbox for surmounting these analytical barriers (Eraslan *et al.*, 2019). Through the synergistic integration of ML techniques with GWAS, scientists can more effectively pinpoint genetic variations linked to critical features such as yield potential, resistance to pests, and tolerance to environmental stresses (Hu *et al.*, 2020). This enhanced understanding of the fundamental genetic architecture of plants promises to drive the creation of novel crop varieties, foster advancements in plant development, and contribute to amplified yields, thereby bolstering global food security and flattening the avenue for a resilient and sustainable agricultural future (Singh *et al.*, 2020).

13.6 Unveiling Plant Secrets: ML and GWAS Breakthrough

The synergistic integration of ML is proving to be a catalytic force in advancing plant genetics, opening (or offering) novel avenues to decipher the complexities underlying economically significant traits (Michael and VanBuren, 2015). Despite their role as fundamental analytical instruments, conventional GWAS frequently exhibit limitations in thoroughly elucidating the genetic regulation of intricate plant phenotypes (Tam *et al.*, 2019). These inherent restrictions frequently stem from challenges in resolving accurate genetic signals due to restricted statistical power, the obscuring effects of confounding factors, and the inherent complexity arising from the interplay of multiple genes (Flint and Mackay, 2009). However, the application of sophisticated ML algorithms, including DL frameworks and ensemble methods like RFs, offers a paradigm shift in our ability to interrogate plant genomes (Singh *et al.*, 2023). By harnessing ML proficiency in processing extensive datasets, discerning intricate non-linear associations, and revealing previously cryptic patterns, researchers can achieve a more profound and accurate understanding of the genetic architecture governing complex traits such as yield potential, resistance to diseases, and tolerance to environmental stresses. This fundamental breakthrough in analytical capability transforms our approach to genetic discovery, moving beyond the inherent limitations of traditional methods.

13.7 ML–GWAS Advantage: Precision Plant Breeding

Precision plant breeding represents a game-changing advancement towards refined accuracy in crop improvement, driven by the potent synergy of ML approaches and GWAS. Fully elucidating the genetic basis of complex quantitative traits in plants remains a defy (or challenge), even with the valuable frameworks provided by conventional GWAS. This difficulty largely arises from intrinsic restrictions in their statistical sensitivity, the often-overlooked influence of environmental variation, and the intricate multi-gene interactions known as epistasis (Buckler *et al.*, 2009; Gupta *et al.*, 2014). Decisively, the implementation of advanced ML algorithms exhibits (or presents) a robust strategy to overcome these limitations by facilitating the detection of nuanced genetic signals and the construction of sophisticated, non-linear models that more accurately detect (or capture) the intricate relationship between genetic makeup (fine-tuning) and observable traits (de los Campos *et al.*, 2009; van Dijk *et al.*, 2021).

Cutting-edge ML methodologies, encompassing gradient boosting machines, artificial neural networks, and kernel-based regression, exhibit a heightened capacity to process intricate, high-dimensional genomic datasets and yield more reliable forecasts of complex phenotypic expressions. This enhanced predictive capability enables a more precise pinpointing of the underlying causal genetic variations (Zou *et al.*, 2010; de Los Campos *et al.*, 2013; Pérez-Harguindeguy *et al.*, 2013; Ma *et al.*, 2017), directly empowering more insightful and efficient decision making within breeding initiatives (He and Li, 2020). Consequently, ML–GWAS substantially (or significantly) speeds up (or accelerates) the formation (generation or creation) of enhanced crop varieties endowed with (or displaying) superior yield potential, improved nutritional profiles, and augmented resistance to both biological and environmental constraints, fostering, thereby, more sustainable and reliable food systems.

13.8 Pioneering Plant GWAS: Leveraging Advanced ML Methodologies

The advent of sophisticated ML methodologies in plant GWAS marks substantial progress in the discipline. This evolution transcends conventional statistical frameworks by harnessing the intricate capabilities of cutting-edge ML techniques to achieve enhanced insights in plant genomics. A complete understanding of traits influenced by multiple genes often eludes classical GWAS, despite its foundational role in mapping genotype-phenotype correlations. This shackle stems from its often-simplistic models, which predominantly assume linear genetic effects and purely additive contributions (Holland, 2007). In contrast, advanced ML algorithms, encompassing DL architectures, SVMs, and ensemble learning paradigms such as RF and gradient boosting, offer transformative approaches to address these limitations. ML approaches (or ML algorithms) achieve this by effectively modeling non-linear associations, detecting complex epistatic effects, and managing big-data or high-dimensional genomic datasets with greater proficiency or dexterity (Breiman, 2001; Hastie *et al.*, 2009; Goodfellow *et al.*, 2016).

The incorporation of state-of-the-art ML techniques is fundamentally reshaping plant GWAS, facilitating, therefore, a more granular comprehension of the genotype-phenotype landscape. For example, convolutional neural networks (CNNs) are being deployed on genomic sequence information to discern (or distinguish) predictive patterns related to various plant characteristics. Moreover, Bayesian ML methodologies provide a probabilistic structure for GWAS, enabling the assessment of uncertainty in gene-trait linkages and potentially fostering the reliability of research outcomes (Stephens, 2013). These progressive methodologies hold the power (or potential) to identify previously undiscovered genetic markers and refine the accuracy (or precision) of genomic selection, leading to more streamlined and targeted breeding strategies aimed at improving crop performance and adaptability to challenging environmental conditions. Furthermore, integrated assays of GWAS, multi-omics colocalization, and ML-associated calcium signaling genes with oilseed rape resistance to *Sclerotinia sclerotiorum* (Wang *et al.*, 2024a).

13.9 Streamlining Plant GWAS: The Imperative of ML Integration

The imperative of streamlining plant GWAS through the strategic integration of ML algorithms within established analytical workflows is becoming increasingly apparent in plant breeding (Hu *et al.*, 2020). This necessitates a well-planned and efficient incorporation of ML across the different phases of the analytical process. Besides, this includes the application of advanced data preprocessing techniques designed to manage the intricate nature (or aspect) of genomic and phenotypic datasets (Märtens *et al.*, 2016), the implementation of intelligent feature selection approaches to pinpoint the most relevant genetic variations (Sghaier *et al.*, 2022; Yu *et al.*, 2022), the development of resilient model training protocols specifically adapted to the unique characteristics of plant genetic structures, and the establishment of effective frameworks for the interpretation and biological validation of ML-generated findings in the context of plant traits. By achieving a fluid and comprehensive integration of ML methodologies into GWAS pipeline, researchers can fully exploit the advanced analytical power of these techniques to substantially improve both the precision and the efficiency of identifying (or pinpointing) genetic variants associated with crucial agronomic characteristics (Jannink *et al.*, 2010; Martini *et al.*, 2022), leading, thus, to a faster and more accurate generation of enhanced crop varieties that can significantly contribute to progress in plant breeding and long-term agricultural viability (Kaushik, 2025).

13.10 ML for Complex Trait Analysis in Plant GWAS

The application of ML for the analysis of complex traits within plant GWAS underscores its crucial contribution to significantly deepening our understanding of the intricate genetic frameworks that underpin multifaceted characteristics in plants. Hence, unraveling the complex genetic relationships that control economically important traits, such as yield potential and abiotic stress tolerance, often requires sophisticated tools (e.g. GWAS). These advanced methodologies enable researchers to pinpoint the genetic factors underlying these crucial agricultural characteristics. The above-mentioned traits are typically governed by numerous genes engaging in sophisticated interactions, making their dissection challenging (difficult or tricky) via traditional statistical methods that frequently operate under the assumption of linear and additive effects (Nishio and Satoh, 2015). However, ML algorithms, leveraging their inherent ability to process and interpret large quantities of data (or big datasets), identify non-linear correlations, and uncover latent patterns, provide a powerful toolkit (suite of tools) for overcoming these analytical hitches (Montesinos-López *et al.*, 2024b). By incorporating ML approaches, researchers can effectively model intricate genetic interactions, determine (or pinpoint) subtle yet impactful genetic influences, and enhance the accuracy of forecasting phenotypic expressions (Montesinos-López *et al.*, 2024a; Kumar *et al.*, 2024). This enhanced insight into the genetic underpinnings of complex traits holds the potential to substantially hasten (or accelerate) the breeding of improved crop varieties and contribute to a more sustainable and resilient agricultural future, better positioned to address global food security needs in the context of environmental change (Wang *et al.*, 2024b).

13.11 Ushering in a New Era of Crop Improvement: ML's Synergistic Role in GWAS

Revolutionizing plant breeding via ML-powered GWAS stresses the profound and transformative influence of merging ML with GWAS, triggering (or signaling) a new epoch in the science of plant breeding (Schluttenhofer and Yuan, 2017). Although modern GWAS approaches have yielded significant knowledge regarding the associations (or links) between genetic makeup and observable traits, their efficacy in fully dissecting the complex genetic regulation of multifaceted characteristics is often restricted by factors such as constrained statistical sensitivity and the challenge of comprehensively detecting non-additive genetic influences, e.g. epistasis, and the interplay between genes and environment (Viana and Garcia, 2022). The emergence of ML-enhanced GWAS presents a powerful solution to these shortcomings, empowering researchers to use advanced computational algorithms adept at pinpointing accurate (tiny or infinitesimal) QTLs, elucidating complex epistatic interactions (Stolyarova *et al.*, 2022), and attaining more precise predictions of phenotypic expressions based on genomic information (Sinha *et al.*, 2023). This improved clarity and predictive capacity equips plant breeders with the suitable and robust tools (or instruments) to carry out more strategic selection choices, considerably speeding up the creation of elite crop varieties distinguished by a high yield potential, enhanced resistance to insect infestations, and greater robustness against unfavorable environmental stresses (biotic and abiotic), contributing, thereby, to a more ecologically sound and food-secure future for an expanding global populace.

13.12 Realizing Full Spectrum of Plant Breeding: ML's Empowering Role in GWAS

Unlocking plant potential through implementing ML in GWAS emphasizes the revolutionary capacity of merging ML to completely actualize the latent possibilities within plant breeding endeavors. Although the modern GWAS tool has established a bedrock comprehension of plant heredity, it frequently encounters difficulties in fully grasping the intricate and multifaceted essence of vital plant characteristics. Through the integration of advanced ML algorithms, scientists can overcome (or surmount) these constraints (or hurdles) and access the concealed genetic intelligence residing within plant genomes (Szymczak *et al.*, 2009). Notably, ML methodologies demonstrate exceptional proficiency in scrutinizing complex data arrays, discerning delicate genetic influences that might elude standard statistical analyses, and revealing complex and interconnected gene–gene (epistasis) and gene–environment interactions that substantially contribute to observable trait diversity (Wang *et al.*, 1998; Moore, 2003; Mackay, 2014). This amplified aptitude (or capability) for interpreting the genetic foundations (or bases) of plant traits will enable plant breeders to cultivate enhanced crop varieties exhibiting greater yield capabilities, enriched nutritional composition, and heightened resilience to diverse surrounding pressures (Varshney, 2021).

13.13 Charting a New Course for Crop Improvement: ML's Integrative Role in GWAS

The future of plant breeding strategies can be shaped by the huge progress made in prediction algorithms and smart fine technologies. Thus, ML-assisted GWAS paints a vision of a fundamentally reshaped future for crop improvement, propelled by the collaborative fusion of ML and GWAS. While traditional plant breeding approaches including but not limited to GWAS have generated considerable advancements, they often encounter bottlenecks (or challenges) inefficiently dissecting the intricate genetic control of key agronomic traits. By harnessing the computational power and pattern recognition abilities inherent in ML algorithms, researchers (or breeders) can transcend these constraints and investigate innovative pathways for enhancing crops, particularly in forecasting (or predicting) complex phenotypic features (Chen *et al.*, 2024). ML-assisted GWAS streamlines the detection of genes with smaller individual effects and the clarification of intricate epistatic interactions, leading to more accurate predictive frameworks for plant performance across assorted (or varied) environmental conditions (Bernardo, 2020). This sophisticated analytical structure offers the potential to considerably promote crop improvement for enhanced varieties displaying superior yield capacity, strong resistance to common pathogens threats, and amplified tolerance to environmental hurdles (Hu *et al.*, 2020; Niazian and Niedbała, 2020).

13.14 Untangling the Intricacies of Plant Genomics: ML's Essential Contribution to GWAS

ML for plant GWAS corroborates the vital function of ML in elucidating the inherent intricacies present within plant GWAS. The manifestation of plant characteristics frequently arises from elaborate interplays among a multitude of genes, diverse environmental factors, and fundamental complex genetic organizations that can prove difficult to analyze and decipher. Though current genomic investigations have certainly offered valuable initial insights, they frequently fall short of comprehensively unraveling the intricate web of these complex connections. ML algorithms, with their sophisticated ability to process and analyze large-scale datasets, identify non-linear associations between genotypes and phenotypes, and uncover latent patterns, offer a powerful suite of tools to deal with this complexity (Wang *et al.*, 1998). Thus, through integrating ML methods, researchers can more effectively characterize precise genetic effects that might be missed by using linear models, unravel intricate gene–gene and gene–environment interactions (Yoosefzadeh-Najafabadi *et al.*, 2022), and significantly ameliorate (or improve) the accuracy of predicting phenotypic outcomes based on genomic embedded information (Romano *et al.*, 2022; Wójcik-Gront *et al.*, 2024). This enhanced capability to navigate the complexities of plant genomes promises to accelerate the development of improved crop varieties with desirable traits and contribute significantly to progress in agricultural productivity and sustainability.

13.15 Confronting the Complexities of Plant Genomes: The ML Imperative in GWAS

Unraveling plant genomes acknowledges the considerable impediments faced in reaching a thorough comprehension of plant genomes through the use of standard GWAS methodologies. These difficulties include the intrinsic intricacy of genetic organizations that dictate numerous characteristics, the existence of diverse confounding ecological and organic elements, interconnected convoluted metabolic pathways, and the restrictions of established statistical techniques in discerning faint genetic indicators and dissecting elaborate gene–gene interactions (Hill, 2010). The crux of the ML–GWAS challenge resides in the efficacious deployment of ML algorithms to overcome these stumbling blocks. By formulating and implementing refined ML models adept at managing extensive genomic datasets, identifying non-linear correlations, and executing resilient feature selection, scientists endeavor to decipher the complexities of plant genomes, pinpoint novel and potentially unexpected genetic links, and ultimately hasten the creation of enhanced crop varieties with improved output, nutritional value, and stress resilience to satisfy the needs of a growing global populace for food and feeds (Edwards and Batley, 2010; Ballard *et al.*, 2024).

13.16 Taming the Labyrinth of Plant Genomes: ML's Pivotal Role in GWAS

Investigating the genetic effects, within traditional genetic approaches, confronts intricacies and ambiguity due to the lack of sophisticated tools and suitable effective predictive models. To disentangle the interconnected genetic functions (or effects), scientists resorted to the integration of a subfield of AI tools, which is ML, with the traditional screening methods including, but not limited, to GWAS. In this regard, ML in plant GWAS recognizes the considerable difficulties that the intrinsic intricacy of plant genomes poses for conventional GWAS investigations. The elaborate interconnectedness of numerous genes, varied ecological impacts, and sophisticated gene–gene interactions frequently conceal (hide) the exact genetic foundation of multifaceted features (Esposito *et al.*, 2019). ML algorithms furnish a potent change in perspective to address this complexity. By capitalizing on ML's aptitude for examining expansive and high-dimensional data collections, discerning non-linear associations between genetic variations and observable traits, and revealing subtle, hidden regularities, scientists can reach more profound and nuanced understandings of the genetic organization of complex traits (Ma *et al.*, 2017). This sophisticated analytical capacity permits the identification of delicate genetic impacts that linear models might overlook, the detection of elaborate epistatic interactions, and the more precise forecasting (or prediction) of phenotypic expressions, facilitating, eventually, the creation of more efficacious and focused plant breeding approaches aimed at improving vital agricultural characteristics (Khan *et al.*, 2022; Gou *et al.*, 2024).

13.17 Refining Crop Improvement: Harnessing ML and GWAS for Smart Agriculture Practices

The fusion of ML algorithms with GWAS signifies a revolutionary advancement toward highly accurate plant breeding. Conventional crop improvement, frequently depending on observable characteristic selection across extensive plant populations, can be protracted and may lack the necessary clarity to dissect the intricate genetic regulation of significant attributes (Moose and Mumm, 2008). Conversely, ML-augmented GWAS presents a more refined and precise methodology for interpreting the genetic framework of these attributes. The capacity of ML techniques to examine multifaceted genomic and phenotypic datasets enables researchers to identify non-linear correlations and complex interplays, such as epistasis and gene–environment interactions, that impact characteristics like drought resistance and nutrient utilization (Danilevicz *et al.*, 2022; Cuevas *et al.*, 2023). This enhanced comprehension facilitates the exact localization of QTLs linked to sought-after characteristics, including augmented photosynthetic performance (Heckmann *et al.*, 2017; Liu and Xu, 2023; Khatibi and Ali, 2024).

However, plant breeders may use this detailed genetic intelligence to implement more effective marker-assisted selection (MAS) and genomic selection (GS) approaches, leading to accelerated genetic advancement and the generation of elite varieties endowed with improved performance and growth consistency. This shift toward precision breeding not only optimizes the crop improvement procedure but also aids in the development of crops better suited (or adaptive) to specific environmental circumstances, promoting thus more ecologically sound and robust agricultural and economic systems.

13.18 Decoding Genomes for Tailored Farming Practices: ML's Role in GWAS-Driven Precision Agriculture

Towards modern and sustainable agriculture, researchers look to integrate robust and subtle predictive models to improve the detection of critical agronomic and economic traits. ML-assisted plant GWAS underlines the pivotal function of ML-driven GWAS in propelling accurate and smart modern agriculture, a domain focused on refining farming practices by adapting interventions to particular field circumstances and crop requirements. This refinement can be attained through a more profound grasp of the intricate relationship between hereditary elements and environmental impacts on crop productivity (Cheng and Wang, 2024; Farooq *et al.*, 2024; Al-Mamun *et al.*, 2025). The incorporation of ML algorithms within GWAS allows (or empowers) researchers to dissect the complex genetic organization of traits, even within varied and changing environments, thereby facilitating the creation of location-specific management approaches. These approaches include variable rate nutrient application, optimized water delivery, and focused pest management, all guided by the identification of genetic variations related (or linked) to specific environmental reactions (Liu *et al.*, 2024). Moreover, ML assists in the enhancement of crop characteristics through improved accuracy in gene-editing technologies, permitting the development of crop varieties with heightened adaptation to the prevailing local conditions (Farooq *et al.*, 2024). The utilization of ML in GWAS not only sharpens our comprehension of genotype–phenotype correlations but also harbors the potential to considerably enhance resource-use efficiency, lessen environmental consequences, and bolster agricultural sustainability (Ferguson *et al.*, 2021; Bai *et al.*, 2024). Accordingly, this amplified understanding and accuracy steer the progression toward a more precise and sustainable agricultural future, optimizing crop output while minimizing ecological impacts (Ferguson *et al.*, 2021).

13.19 Enhancing Crop Improvement Strategies: ML's Role in Precise GWAS Application

ML in GWAS underlines the vital contribution of ML in refining plant breeding approaches by amplifying the precision and effectiveness of GWAS (Gill *et al.*, 2022; Cheng and Wang, 2024; Farooq *et al.*, 2024). Conventional crop improvement often employs large-scale selection techniques, which can be lengthy and may not consistently yield the most favorable results, particularly when addressing complex traits (Yoosefzadeh Najafabadi *et al.*, 2023). Through the incorporation (or integration) of ML algorithms, breeders can achieve a more detailed comprehension of the genetic framework underlying complex features, dissecting the intricate connections between genetic makeup and observable characteristics (Miao *et al.*, 2024; Al-Mamun *et al.*, 2025). ML methodologies can pinpoint subtle hereditary influences, identify complex epistasic (gene–gene) interactions, and enhance the precision of predicting phenotypic expressions, leading to more strategically informed breeding choices (Yoosefzadeh-Najafabadi *et al.*, 2021; Gill *et al.*, 2022). This improved understanding provides breeders with the ability to implement more strategic selections, refine crop development programs, and swiftly produce elite plant types that offer greater yields, stronger immunity to diseases, and increased abiotic stress tolerance. Consequently, this accelerates genetic advancement and boosts overall breeding effectiveness (Tekeu *et al.*, 2023).

13.20 Conclusions and General Remarks

This chapter has traversed the transformative landscape of ML-assisted GWAS in plant genomics, illustrating its profound impact on our ability to unravel the genetic complexities underlying crucial agronomic traits. We have moved from the inherent limitations of traditional GWAS methodologies to the exciting potential unlocked by the integration of sophisticated ML algorithms. The discussion has highlighted how ML empowers researchers to confront the challenges posed by intricate genetic architectures, gene–environment interactions, and the elusive nature of small-effect loci. The ML–GWAS revolution is evident in its capacity to enhance predictive accuracy for complex traits, enabling more effective genomic selection and a deeper understanding of the genotype–phenotype map. Furthermore, the ability of ML methods to uncover novel genetic associations, including epistatic interactions often overlooked by linear models, pledges to accelerate gene discovery and provide a more holistic view of the genetic control of important phenotypes. This enhanced understanding directly translates to the ML-GWAS advantage in precision plant breeding, allowing for more informed selection decisions, optimized breeding strategies, and ultimately, the faster development of superior crop varieties.

The exploration of advanced ML methods for plant GWAS showcased the cutting-edge techniques being employed to model non-linear relationships and handle high-dimensional genomic data (big data) with greater efficacy. The emphasis on integrating ML into plant GWAS pipelines underscores the importance of seamless and strategic incorporation of these powerful tools into standard analytical workflows. The chapter, further, delved into the critical role of ML for complex trait analysis, demonstrating its ability to dissect the intricate genetic networks that govern essential agronomic characteristics. The narrative progressed through the revolutionizing plant breeding and unlocking plant potential afforded by ML-powered GWAS, emphasizing the enhanced resolution and predictive capabilities that empower breeders to make more targeted interventions. The vision for the future of plant breeding, ML-assisted GWAS paints a picture of accelerated breeding cycles and the development of crops better adapted to diverse and changing environments. The debate on navigating complexity and confronting the ML–GWAS challenge highlighted the indispensable role of ML in deciphering the intricate interplay of genetic and environmental factors. In addition, this chapter illustrated the practical implications of ML–GWAS in fine-tuning plant breeding for enhanced precision and efficiency, and its crucial contribution towards precision and smart agriculture by enabling the development of site-specific management strategies and crop varieties optimized for local conditions.

In conclusion, the integration of ML into GWAS is a monumental leap forward in plant genomics and breeding. By providing powerful tools to analyze complex datasets (genomics, transcriptomics, etc.), uncover hidden patterns, and make more accurate predictions, ML–GWAS is not just an incremental improvement but a fundamental shift towards a more precise, efficient, and ultimately, more successful era of crop improvement. The continued advancement and application of these techniques hold the key to unlocking the full potential of plant genomes, ensuring food security, and promoting sustainable agricultural practices for a growing global population facing unprecedented environmental challenges.

References

Alemu, A., Åstrand, J., Montesinos-López, O.A., y Sánchez, J.I., Fernández-Gónzalez, J. *et al.* (2024) Genomic selection in plant breeding: Key factors shaping two decades of progress. *Molecular Plant* 17(4), 552–578. DOI: 10.1016/j.molp.2024.03.007.

Al-Mamun, H.A., Danilevicz, M.F., Marsh, J.I., Gondro, C. and Edwards, D. (2025) Exploring genomic feature selection: A comparative analysis of GWAS and machine learning algorithms in a large-scale soybean dataset. *The Plant Genome* 18(1), e20503. DOI: 10.1002/tpg2.20503.

Angermueller, C., Pärnamaa, T., Parts, L. and Stegle, O. (2016) Deep learning for computational biology. *Molecular Systems Biology* 12(7), 878. DOI: 10.15252/msb.20156651.

Bai, W., Li, C., Li, W., Wang, H., Han, X. *et al.* (2024) Machine learning assists prediction of genes responsible for plant specialized metabolite biosynthesis by integrating multi-omics data. *BMC Genomics* 25(1), 418. DOI: 10.1186/s12864-024-10258-6.

Bajad, S., Anil Kumar, A., Kumar, K. and Sharma, S. (2024) Genomic selection in plant breeding: A new era for crop improvement. *Agriculture Food: E-Newsletter* 6(11), 579–582.

Ballard, J.L., Wang, Z., Li, W., Shen, L. and Long, Q. (2024) Deep learning-based approaches for multi-omics data integration and analysis. *BioData Mining* 17(1), 38. DOI: 10.1186/s13040-024-00391-z.

Bernardo, R. (2020) Reinventing quantitative genetics for plant breeding: Something old, something new, something borrowed, something BLUE. *Heredity (Edinb)* 125(6), 375–385. DOI: 10.1038/s41437-020-0312-1.

Bishop, C.M. (2006) *Pattern Recognition and Machine Learning*. Internetarchivebooks, Springer, New York. Available at: ark:/13960/s2xgzs87wvn

Breiman, L. (2001) Random forests. *Machine Learning* 45(1), 5–32. DOI: 10.1023/A:1010933404324.

Buckler, E.S., Holland, J.B., Bradbury, P.J., Acharya, C.B., Brown, P.J. *et al.* (2009) The genetic architecture of maize flowering time. *Science* 325(5941), 714–718. DOI: 10.1126/science.1174276.

Chafai, N., Hayah, I., Houaga, I. and Badaoui, B. (2023) A review of machine learning models applied to genomic prediction in animal breeding. *Frontiers in Genetics* 14, 1150596. DOI: 10.3389/fgene.2023.1150596.

Chen, L., Liu, G. and Zhang, T. (2024) Integrating machine learning and genome editing for crop improvement. *aBIOTECH* 5(2), 262–277. DOI: 10.1007/s42994-023-00133-5.

Cheng, Q. and Wang, X. (2024) Machine learning for AI breeding in plants. *Genomics, Proteomics & Bioinformatics* 22(4), qzae051. DOI: 10.1093/gpbjnl/qzae051.

Cho, Y.R. and Kang, M. (2020) Interpretable machine learning in bioinformatics. *Methods* 179, 1–2. DOI: 10.1016/j.ymeth.2020.05.024.

Cortes, C. and Vapnik, V. (1995) Support-vector networks. *Machine Learning* 20(3), 273–297. DOI: 10.1023/A:1022627411411.

Crossa, J., Pérez-Rodríguez, P., Cuevas, J., Montesinos-López, O., Jarquín, D. *et al.* (2017) Genomic selection in plant breeding: Methods, models, and perspectives. *Trends in Plant Science* 22(11), 961–975. DOI: 10.1016/j.tplants.2017.08.011.

Crossa, J., Montesinos-Lopez, O.A., Costa-Neto, G., Vitale, P., Martini, J.W.R. *et al.* (2025) Machine learning algorithms translate big data into predictive breeding accuracy. *Trends in Plant Science* 30(2), 167–184. DOI: 10.1016/j.tplants.2024.09.011.

Cuevas, J., Reslow, F., Crossa, J. and Ortiz, R. (2023) Modeling genotype × environment interaction for single and multitrait genomic prediction in potato (*Solanum tuberosum* L.). *G3 (Bethesda)* 13(2), jkac322. DOI: 10.1093/g3journal/jkac322.

Danilevicz, M.F., Gill, M., Anderson, R., Batley, J., Bennamoun, M. *et al.* (2022) Plant genotype to phenotype prediction using machine learning. *Frontiers in Genetics* 13, 822173. DOI: 10.3389/fgene.2022.822173.

de los Campos, G., Naya, H., Gianola, D., Crossa, J., Legarra, A. *et al.* (2009) Predicting quantitative traits with regression models for dense molecular markers and pedigree. *Genetics* 182(1), 375–385. DOI: 10.1534/genetics.109.101501.

de Los Campos, G., Vazquez, A.I., Fernando, R., Klimentidis, Y.C. and Sorensen, D. (2013) Prediction of complex human traits using the genomic best linear unbiased predictor. *PLoS Genetics* 9(7), e1003608. DOI: 10.1371/journal.pgen.1003608.

Edwards, D. and Batley, J. (2010) Plant genome sequencing: Applications for crop improvement. *Plant Biotechnology Journal* 8(1), 2–9. DOI: 10.1111/j.1467-7652.2009.00459.x.

Enoma, D.O., Bishung, J., Abiodun, T., Ogunlana, O. and Osamor, V.C. (2022) Machine learning approaches to genome-wide association studies. *Journal of King Saud University - Science* 34(4), 101847. DOI: 10.1016/j.jksus.2022.101847.

Eraslan, G., Avsec, Ž., Gagneur, J. and Theis, F.J. (2019) Deep learning: New computational modelling techniques for genomics. *Nature Reviews. Genetics* 20(7), 389–403. DOI: 10.1038/s41576-019-0122-6.

Esposito, S., Carputo, D., Cardi, T. and Tripodi, P. (2019) Applications and trends of machine learning in genomics and phenomics for next-generation breeding. *Plants* 9(1), 34. DOI: 10.3390/plants9010034.

Farooq, M.A., Gao, S., Hassan, M.A., Huang, Z., Rasheed, A. *et al.* (2024) Artificial intelligence in plant breeding. *Trends in Genetics* 40(10), 891–908. DOI: 10.1016/j.tig.2024.07.001.

Ferguson, J.N., Fernandes, S.B., Monier, B., Miller, N.D., Allen, D. *et al.* (2021) Machine learning-enabled phenotyping for GWAS and TWAS of WUE traits in 869 field-grown sorghum accessions. *Plant Physiology* 187(3), 1481–1500. DOI: 10.1093/plphys/kiab346.

Flint, J. and Mackay, T.F.C. (2009) Genetic architecture of quantitative traits in mice, flies, and humans. *Genome Research* 19(5), 723–733. DOI: 10.1101/gr.086660.108.

Flint-Garcia, S.A. and Thornsberry, J.M. (2003) Buckler ES 4th. structure of linkage disequilibrium in plants. *Annual Review of Plant Biology* 54, 357–374. DOI: 10.1146/annurev.arplant.54.031902.134907.

Gianola, D., Campos, G., González-Recio, O., Long, N., Okut, H. *et al.* (2018) Statistical learning methods for genome-based analysis of quantitative traits. Available at: https://www.researchgate.net/publication/327318405_Statistical_Learning_Methods_For_Genome-based_Analysis_Of_Quantitative_Traits (accessed 15 August 2025).

Gill, M., Anderson, R., Hu, H., Bennamoun, M., Petereit, J. *et al.* (2022) Machine learning models outperform deep learning models, provide interpretation and facilitate feature selection for soybean trait prediction. *BMC Plant Biology* 22(1), 180. DOI: 10.1186/s12870-022-03559-z.

Goodfellow, I., Bengio, Y. and Courville, A. (2016) *Deep Learning*. MIT press. Available at: http://www.deeplearningbook.org (accessed August 2025).

Gou, C., Zafar, S., Fatima, N., Hasnain, Z., Aslam, N. *et al.* (2024) Machine and deep learning: Artificial intelligence application in biotic and abiotic stress management in plants. *Frontiers in Bioscience-Landmark* 29(1), 20. DOI: 10.31083/j.fbl2901020.

Gupta, P.K., Kulwal, P.L. and Jaiswal, V. (2014) Association mapping in crop plants: Opportunities and challenges. *Advances in Genetics* 85, 109–147. DOI: 10.1016/B978-0-12-800271-1.00002-0.

Guyon, I. and Elisseeff, A. (2003) An introduction to variable and feature selection. *The Journal of Machine Learning Research* 3, 1157–1182.

Hastie, T., Tibshirani, R. and Friedman, J. (2009) *The Elements of Statistical Learning: Data Mining, Inference, and Prediction*, 2nd edn. Springer Series in Statistics, Springer New York. DOI: 10.1007/978-0-387-84858-7.

He, T. and Li, C. (2020) Harness the power of genomic selection and the potential of germplasm in crop breeding for global food security in the era with rapid climate change. *The Crop Journal* 8(5), 688–700. DOI: 10.1016/j.cj.2020.04.005.

Heckmann, D., Schlüter, U. and Weber, A.P.M. (2017) Machine learning techniques for predicting crop photosynthetic capacity from leaf reflectance spectra. *Molecular Plant* 10(6), 878–890. DOI: 10.1016/j.molp.2017.04.009.

Heffner, E.L., Sorrells, M.E. and Jannink, J.L. (2009) Genomic selection for crop improvement. *Crop Science* 49(1), 1–12. DOI: 10.2135/cropsci2008.08.0512.

Hill, WG. (2010) Understanding and using quantitative genetic variation. *Philosophical Transactions of the Royal Society of London. Series B, Biological Sciences* 365(1537), 73–85. DOI: 10.1098/rstb.2009.0203.

Holland, J.B. (2007) Genetic architecture of complex traits in plants. *Current Opinion in Plant Biology* 10(2), 156–161. DOI: 10.1016/j.pbi.2007.01.003.

Hu, T., Darabos, C. and Urbanowicz, R. (2020) Editorial: Machine learning in genome-wide association studies. *Frontiers in Genetics* 11, 593958. DOI: 10.3389/fgene.2020.593958.

Jannink, J.L., Lorenz, A.J. and Iwata, H. (2010) Genomic selection in plant breeding: From theory to practice. *Briefings in Functional Genomics* 9(2), 166–177. DOI: 10.1093/bfgp/elq001.

Huang, W. and Mackay, T.F.C. (2016) The genetic architecture of quantitative traits cannot be inferred from variance component analysis. *PLoS Genetics* 12(11), e1006421. DOI: 10.1371/journal.pgen.1006421.

Kang, H.M., Zaitlen, N.A., Wade, C.M., Kirby, A., Heckerman, D. *et al.* (2008) Efficient control of population structure in model organism association mapping. *Genetics* 178(3), 1709–1723. DOI: 10.1534/genetics.107.080101.

Kaushik, P. (2025) AI-driven breeding enhances stress tolerance in high-elevation extremophytes: A proof-of-concept study with cross-component validation. *Bioinformatics*. DOI: 10.1101/2025.02.21.639605.

Kersey, P.J.. (2019) Plant genome sequences: Past, present, future. *Current Opinion in Plant Biology* 48, 1–8. DOI: 10.1016/j.pbi.2018.11.001.

Khan, M.H.U., Wang, S., Wang, J., Ahmar, S., Saeed, S. *et al.* (2022) Applications of artificial intelligence in climate-resilient smart-crop breeding. *International Journal of Molecular Sciences* 23(19), 11156. DOI: 10.3390/ijms231911156.

Khatibi, S.M.H. and Ali, J. (2024) Harnessing the power of machine learning for crop improvement and sustainable production. *Frontiers in Plant Science* 15, 1417912. DOI: 10.3389/fpls.2024.1417912.

Kumar, R., Das, S.P., Choudhury, B.U., Kumar, A., Prakash, N.R. *et al.* (2024) Advances in genomic tools for plant breeding: Harnessing DNA molecular markers, genomic selection, and genome editing. *Biological Research* 57(1), 80. DOI: 10.1186/s40659-024-00562-6.

Lakiotaki, K., Papadovasilakis, Z., Lagani, V., Fafalios, S., Charonyktakis, P. *et al.* (2023) Automated machine learning for genome wide association studies. *Bioinformatics* 39(9), btad545. DOI: 10.1093/bioinformatics/btad545.

LeCun, Y., Bengio, Y. and Hinton, G. (2015) Deep learning. *Nature* 521(7553), 436–444. DOI: 10.1038/nature14539.

Liu, H. and Xu, Z. (2023) Editorial: Machine vision and machine learning for plant phenotyping and precision agriculture. *Frontiers in Plant Science* 14, 1331918. DOI: 10.3389/fpls.2023.1331918.

Liu, Q., Zuo, S., Peng, S., Zhang, H., Peng, Y. *et al.* (2024) Development of machine learning methods for accurate prediction of plant disease resistance. *Engineering* 40, 100–110. DOI: 10.1016/j.eng.2024.03.014.

Martini, J.W.R., Hearne, S.J., Gardunia, B., Wimmer, V. and Toledo, F.H. (2022) Editorial: Genomic selection: Lessons learned and perspectives. *Frontiers in Plant Science* 13, 890434. DOI: 10.3389/fpls.2022.890434.

Ma, W., Qiu, Z., Song, J., Cheng, Q. and Ma, C. (2017) DeepGS: Predicting phenotypes from genotypes using deep learning. *Bioinformatics*. DOI: 10.1101/241414.

Mackay, T.F.C. (2014) Epistasis and quantitative traits: Using model organisms to study gene-gene interactions. *Nature Reviews Genetics* 15(1), 22–33. DOI: 10.1038/nrg3627.

Mackay, T.F.C., Stone, E.A. and Ayroles, J.F. (2009) The genetics of quantitative traits: Challenges and prospects. *Nature Reviews. Genetics* 10(8), 565–577. DOI: 10.1038/nrg2612.

Manolio, T.A. (2010) Genomewide association studies and assessment of the risk of disease. *New England Journal of Medicine* 363(2), 166–176. DOI: 10.1056/NEJMra0905980.

Märtens, K., Hallin, J., Warringer, J., Liti, G. and Parts, L. (2016) Predicting quantitative traits from genome and phenome with near perfect accuracy. *Nature Communications* 7, 11512. DOI: 10.1038/ncomms11512.

Meuwissen, T.H., Hayes, B.J. and Goddard, M.E. (2001) Prediction of total genetic value using genome-wide dense marker maps. *Genetics* 157(4), 1819–1829. DOI: 10.1093/genetics/157.4.1819.

Miao, J., Wu, Y., Sun, Z., Miao, X., Lu, T. *et al.* (2024) Valid inference for machine learning-assisted genome-wide association studies. *Nature Genetics* 56(11), 2361–2369. DOI: 10.1038/s41588-024-01934-0.

Michael, T.P. and VanBuren, R. (2015) Progress, challenges and the future of crop genomes. *Current Opinion in Plant Biology* 24, 71–81. DOI: 10.1016/j.pbi.2015.02.002.

Montesinos-López, A., Montesinos-López, O.A., Gianola, D., Crossa, J. and Hernández-Suárez, C.M. (2018) Multi-environment genomic prediction of plant traits using deep learners with dense architecture. *G3 (Bethesda)* 8(12), 3813–3828. DOI: 10.1534/g3.118.200740.

Montesinos-López, A., Crespo-Herrera, L., Dreisigacker, S., Gerard, G., Vitale, P. *et al.* (2024a) Deep learning methods improve genomic prediction of wheat breeding. *Frontiers in Plant Science* 15, 1324090. DOI: 10.3389/fpls.2024.1324090.

Montesinos-López, O.A., Chavira-Flores, M., Kismiantini, Crespo-Herrera, L., Saint Piere, C. *et al.* (2024b) A review of multimodal deep learning methods for genomic-enabled prediction in plant breeding. *Genetics* 228(4), iyae161. DOI: 10.1093/genetics/iyae161.

Moore, J.H. (2003) The ubiquitous nature of epistasis in determining susceptibility to common human diseases. *Human Heredity* 56(1–3), 73–82. DOI: 10.1159/000073735.

Moose, S.P. and Mumm, R.H. (2008) Molecular plant breeding as the foundation for 21st century crop improvement. *Plant Physiology* 147(3), 969–977. DOI: 10.1104/pp.108.118232.

Niazian, M. and Niedbała, G. (2020) Machine learning for plant breeding and biotechnology. *Agriculture* 10(10), 436. DOI: 10.3390/agriculture10100436.

Nishio, M. and Satoh, M. (2015) Genomic best linear unbiased prediction method reflecting the degree of linkage disequilibrium. *Journal of Animal Breeding and Genetics* 132(5), 357–365. DOI: 10.1111/jbg.12162.

Pérez-Harguindeguy, N., Díaz, S., Garnier, E., Lavorel, S., Poorter, H. *et al.* (2013) New handbook for standardised measurement of plant functional traits worldwide. *Australian Journal of Botany* 61(3), 167. DOI: 10.1071/BT12225.

Phillips, P.C. (2008) Epistasis--the essential role of gene interactions in the structure and evolution of genetic systems. *Nature Reviews. Genetics* 9(11), 855–867. DOI: 10.1038/nrg2452.

Price, A.L., Patterson, N.J., Plenge, R.M., Weinblatt, M.E., Shadick, N.A. *et al.* (2006) Principal components analysis corrects for stratification in genome-wide association studies. *Nature Genetics* 38(8), 904–909. DOI: 10.1038/ng1847.

Romano, L.E., Iovane, M., Izzo, L.G. and Aronne, G. (2022) A machine-learning method to assess growth patterns in plants of the family *lemnaceae*. *Plants* 11, 15. DOI: 10.3390/plants11151910.

Schluttenhofer, C. and Yuan, L. (2017) Challenges towards revitalizing hemp: A multifaceted crop. *Trends in Plant Science* 22(11), 917–929. DOI: 10.1016/j.tplants.2017.08.004.

Sghaier, N., Essemine, J., Ayed, R.B., Gorai, M., Ben Marzoug, R. *et al.* (2022) An evidence theory and fuzzy lLogic combined approach for the prediction of potential ARF-regulated genes in Quinoa. *Plants* 12(1), 71. DOI: 10.3390/plants12010071.

Sidak, D., Schwarzerová, J., Weckwerth, W. and Waldherr, S. (2022) Interpretable machine learning methods for predictions in systems biology from omics data. *Frontiers in Molecular Biosciences* 9, 926623. DOI: 10.3389/fmolb.2022.926623.

Sigala, R.E., Lagou, V., Shmeliov, A., Atito, S., Kouchaki, S. *et al.* (2024) Machine learning to advance human genome-wide association studies. *Genes* 15(1), 34. DOI: 10.3390/genes15010034.

Singh, A., Ganapathysubramanian, B., Singh, A.K. and Sarkar, S. (2020) Machine learning for high-throughput stress phenotyping in plants. *Trends Plant Science* 21(2), 110–124. DOI: 10.1016/j.tplants.2015.10.015.

Singh, B., Kumar, S., Elangovan, A., Vasht, D., Arya, S. *et al.* (2023) Phenomics based prediction of plant biomass and leaf area in wheat using machine learning approaches. *Frontiers in Plant Science* 14, 1214801. DOI: 10.33898/fpls.2023.12144801.

Sinha, D., Maurya, A.K., Abdi, G., Majeed, M., Agarwal, R. *et al.* (2023) Integrated genomic selection for accelerating breeding programs of climate-smart cereals. *Genes* 14(7), 1484. DOI: 10.3390/genes14071484.

Stephens, M. (2013) A unified framework for association analysis with multiple related phenotypes. *PLoS ONE* 8(7), e65245. DOI: 10.1371/journal.pone.0065245.

Stolyarova, A.V., Neretina, T.V., Zvyagina, E.A., Fedotova, A.V., Kondrashov, A.S. *et al.* (2022) Complex fitness landscape shapes variation in a hyperpolymorphic species. *eLife* 11, e76073. DOI: 10.7554/eLife.76073.

Szymczak, S., Biernacka, J.M., Cordell, H.J., González-Recio, O., König, I.R. *et al.* (2009) Machine learning in genome-wide association studies. *Genetic Epidemiology* 33(S1), S51–S57. DOI: 10.1002/gepi.20473.

Tam, V., Patel, N., Turcotte, M., Bossé, Y., Paré, G. *et al.* (2019) Benefits and limitations of genome-wide association studies. *Nature Reviews Genetics* 20(8), 467–484. DOI: 10.1038/s41576-019-0127-1.

Tekeu, H., Jean, M., Ngonkeu, E.L.M. and Belzile, F. (2023) Machine learning-GWAS reveals the role of WSD1 gene for cuticular wax ester biosynthesis and key genomic regions controlling early maturity in bread wheat. *Genomics*. DOI: 10.1101/2023.11.03.565125.

van Dijk, A.D.J., Kootstra, G., Kruijer, W. and Ridder, D. (2021) Machine learning in plant science and plant breeding. *iScience* 24(1), 101890. DOI: 10.1016/j.isci.2020.101890.

Varshney, R.K. (2021) The plant genome special issue: Advances in genomic selection and application of machine learning in genomic prediction for crop improvement. *The Plant Genome* 14(3), e20178. DOI: 10.1002/tpg2.20178.

Viana, J.M.S. and Garcia, A.A.F. (2022) Significance of linkage disequilibrium and epistasis on genetic variances in noninbred and inbred populations. *BMC Genomics* 23(1), 286. DOI: 10.1186/s12864-022-08335-9.

Wang, M.L., Shao, W., Hao, X.K. and Zhang, D.Q. (2023) Machine learning for brain imaging genomics methods: A review. *Machine Intelligence Research* 20(1), 57–78. DOI: 10.1007/s11633-022-1361-0.

Wang, P., Lehti-Shiu, M.D., Lotreck, S., Segura Abá, K., Krysan, P.J. *et al.* (2024b) Prediction of plant complex traits via integration of multi-omics data. *Nature Communications* 15(1), 6856. DOI: 10.1038/s41467-024-50701-6.

Wang, T., Fernando, R.L. and Grossman, M. (1998) Genetic evaluation by best linear unbiased prediction using marker and trait information in a multibreed population. *Genetics* 148(1), 507–515. DOI: 10.1093/genetics/148.1.507.

Wang, X.-Y., Ren, C.-X., Fan, Q.-W., Xu, Y.-P., Wang, L.-W. *et al.* (2024a) Integrated assays of genome-wide association study, multi-omics co-localization, and machine learning associated calcium signaling genes with oilseed rape resistance to *Sclerotinia sclerotiorum*. *International Journal of Molecular Sciences* 25(13), 6932. DOI: 10.3390/ijms25136932.

Wójcik-Gront, E., Zieniuk, B. and Pawełkowicz, M. (2024) Harnessing AI-powered genomic research for sustainable crop improvement. *Agriculture* 14(12), 2299. DOI: 10.3390/agriculture14122299.

Yoosefzadeh Najafabadi, M., Hesami, M. and Eskandari, M. (2023) Machine learning-assisted approaches in modernized plant breeding programs. *Genes* 14(4), 1–22. DOI: 10.3390/genes14040777.

Yoosefzadeh-Najafabadi, M., Torabi, S., Tulpan, D., Rajcan, I. and Eskandari, M. (2021) Genome-wide association studies of soybean yield-related hyperspectral reflectance bands using machine learning-mediated data integration methods. *Frontiers in Plant Science* 12, 777028. DOI: 10.3389/fpls.2021.777028.

Yoosefzadeh-Najafabadi, M., Eskandari, M., Torabi, S., Torkamaneh, D., Tulpan, D. *et al.* (2022) Machine-learning-based genome-wide association studies for uncovering QTL underlying soybean yield and its components. *International Journal of Molecular Sciences* 23(10), 5538. DOI: 10.3390/ijms23105538.

Yu, C., Dong, Z., Essemine, E., Zhu, Z., Mo, R. *et al.* (2022) A feature selection approach guided an early prediction of anthocyanin accumulation using massive untargeted metabolomics data in mulberry. *Plant Cell Physiology* 63(5), 671–682. DOI: 10.1093/pcp/pcac010.

Yu, S., Wu, J., Shao, Y., Qiu, D., Qin, Z.S. *et al.* (2024) A novel classification framework for genome-wide association study of whole brain MRI images using deep learning. *PLoS Computational Biology* 20(10), e1012527. DOI: 10.1371/journal.pcbi.1012527.

Zou, F., Huang, H., Lee, S. and Hoeschele, I. (2010) Nonparametric *Bayesian* variable selection with applications to multiple quantitative trait loci mapping with epistasis and gene-environment interaction. *Genetics* 186(1), 385–394. DOI: 10.1534/genetics.110.113103.

14 Predicting Gene Functions Using Machine Learning Techniques

Suman Dutta[1]*, Subham Ghosh[2], Abhik Sarkar[2] and Dipro Sinha[2]

[1]School of Agriculture and Rural Development, Ramakrishna Mission Vivekananda Educational and Research Institute, Belur Math, Howrah, India; [2]Discipline of Bioinformatics, ICAR-Indian Agricultural Statistics Research Institute, Pusa, India

Abstract

Predicting gene function is a central challenge in genomics, essential for understanding biological processes, identifying regulatory mechanisms, and advancing applications in medicine and agriculture. Machine learning (ML) techniques have emerged as powerful tools for automating and improving the accuracy of gene function prediction. This chapter provides a comprehensive overview of ML-based approaches, beginning with sequence-based features such as nucleotide composition, codon usage bias, motifs, protein domains, and k-mer frequencies that serve as inputs for predictive models. It explores both supervised and unsupervised learning models, including clustering algorithms, dimensionality reduction techniques, and autoencoders, highlighting their roles in functional classification. A detailed workflow for implementing ML-based gene function prediction is outlined, encompassing data preprocessing, feature extraction, model training, and validation. The chapter further delves into the integration of gene expression analysis and regulatory network inference, emphasizing the role of unsupervised and supervised ML techniques in understanding gene interactions and constructing gene regulatory networks. Advances in functional genomics are discussed through multi-omics data integration using ML and deep learning frameworks. Additionally, the application of ML for functional annotation of uncharacterized genes is examined, along with associated challenges such as data sparsity, model interpretability, and biological validation. The chapter also covers phenotype prediction and trait association, as well as emerging applications of natural language processing in extracting gene function information from biomedical literature. Collectively, this work underscores the transformative potential of ML in decoding gene functions and shaping the future of genomics research.

Keywords: Gene function prediction, machine learning, sequence-based features, supervised learning, unsupervised learning, gene expression analysis, multi-omics integration, gene regulatory networks

14.1 Introduction

The functional characterization of genes is a cornerstone of modern biology, providing critical insights into the molecular mechanisms governing growth, development, and adaptation in all living organisms (Dutta *et al.*, 2024a; Kumar *et al.*, 2024). Despite the advances in sequencing technologies that have exponentially increased the availability of genomic data, the pace of functional annotation of genes has lagged behind (Roy *et al.*, 2024). A significant proportion of genes in even well-studied organisms remain uncharacterized, posing a major

*Corresponding author: duttasumansiliguri@gmail.com

 AI-Enhanced Plant Omics (ed. Jen-Tsung Chen)
DOI: 10.1079/9781800628649.0014

challenge for systems biology, functional genomics, and precision breeding (Auffray *et al.*, 2003). Traditional experimental methods for gene function prediction, such as mutagenesis, gene knockouts, and biochemical assays, are often time-consuming, labor intensive, and expensive (Dutta *et al.*, 2024b). This growing gap between gene discovery and functional annotation has necessitated the development of computational approaches to accelerate functional prediction. Machine learning (ML), a subset of artificial intelligence (AI), has emerged as a powerful tool for analyzing large-scale biological data and uncovering complex patterns that are not readily discernible through conventional methods (Dutta *et al.*, 2024a; Raghunandan *et al.*, 2025). ML algorithms can learn from diverse biological datasets—such as transcriptomics, proteomics, metabolomics, and epigenomics—to predict gene functions by identifying associations, patterns, and correlations (Perakakis *et al.*, 2018). Techniques including supervised learning, unsupervised learning, semi-supervised learning, and deep learning (DL) have shown significant promise in functionally classifying genes based on various features such as gene expression profiles, sequence homology, protein–protein interactions, and evolutionary conservation (Nguyen and Ho, 2012). Supervised learning models like support vector machines (SVMs), random forests (RFs), and neural networks have been widely applied to assign gene functions based on annotated training datasets (Boateng *et al.*, 2020; Dutta *et al.*, 2023). Unsupervised methods, including clustering algorithms, facilitate the identification of gene modules with shared functions or biological pathways. More recently, DL architectures, such as convolutional neural networks (CNNs) and recurrent neural networks (RNNs), have enabled more nuanced functional predictions by capturing hierarchical and temporal patterns in genomic data (Zafar *et al.*, 2023). Incorporating ML into gene function prediction not only enhances annotation accuracy but also enables high-throughput hypothesis generation for experimental validation. As integrative multi-omics approaches gain momentum, ML models trained on heterogeneous data types hold promise for uncovering novel gene functions and regulatory networks (Mukherjee *et al.*, 2025). However, challenges such as data quality, feature selection, model interpretability, and biological validation remain central concerns (Cembrowska-Lech *et al.*, 2023; Maji *et al.*, 2025). Addressing these limitations through improved algorithms, curated datasets, and domain-specific customization will further unlock the potential of ML in functional genomics. This chapter explores the principles, methodologies, and applications of ML techniques in gene function prediction, highlighting recent advancements, case studies, and future directions in this rapidly evolving field.

14.2 Overview of Gene Function Prediction

Gene function prediction involves assigning biological roles to genes, such as their involvement in specific cellular processes, molecular functions, or participation in biological pathways (Dutta *et al.*, 2025). Traditionally, this has been achieved through experimental methods like knockout studies, co-expression analyses, or phylogenetic profiling. However, these methods are time consuming and costly. As high-throughput sequencing technologies have become ubiquitous, computational approaches have gained prominence, offering faster and more scalable alternatives to experimental methods. ML utilizes both labeled (supervised learning) and unlabeled (unsupervised learning) datasets to predict gene functions (Yan and Wang, 2022). By analyzing sequence-based features, such as motifs, codon usage, and protein domains, these models can generalize patterns that correlate with specific functions.

14.2.1 Sequence-based features for gene function prediction

ML relies on informative features that effectively represent input data, and for gene function prediction, sequence-based features offer valuable insights into a gene's potential biological roles. Key features include nucleotide composition, which involves the frequency of nucleotides (A, T, G, C) in a gene sequence and can indicate regulatory functions or coding potential; codon usage bias, where codon usage patterns reflect evolutionary pressures and functional constraints; and motifs and regulatory elements, which are short conserved sequences serving as transcription factor binding sites that reveal regulatory mechanisms (Behura and Severson, 2013; Dutta *et al.*, 2024c). Additionally, protein domains predicted from gene sequences often correlate with specific biological functions, providing structural and functional context. K-mer frequencies, which capture the occurrence of fixed-length subsequences within the genome, are also informative of sequence-specific functions (Karollus *et al.*, 2024). These diverse features are extracted from raw genomic sequences and transformed into numerical representations, making them suitable inputs for training ML models aimed at accurate gene function prediction.

14.2.2 Supervised learning models for gene function prediction

Supervised learning encompasses various algorithms that can be adapted to predict gene functions. Below are some commonly used models.

14.2.2.1 Decision trees and random forests

Decision trees partition data based on feature thresholds, creating a hierarchy of decision rules. RFs aggregate multiple decision trees for robust predictions (Fig. 14.1). An extremely versatile and user-friendly supervised ML technique, RF generates competitive predictions of continuous, binary, categorical, and count outcomes with no effort (Montesinos López *et al.*, 2022). These models handle non-linear relationships well and are interpretable, making them suitable for feature importance analysis in gene function prediction.

14.2.2.2 Support vector machines

SVMs aim to find the optimal hyperplane that separates different classes in a high-dimensional feature space (Fig. 14.2). SVMs are one of the most powerful and robust

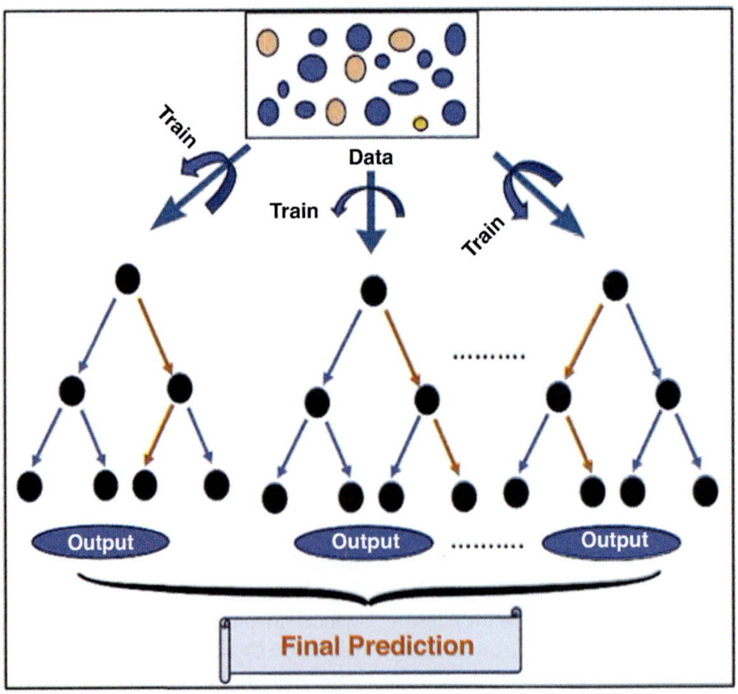

Fig. 14.1. A schematic illustration of random forest consisting of decision trees. (Author's own image.)

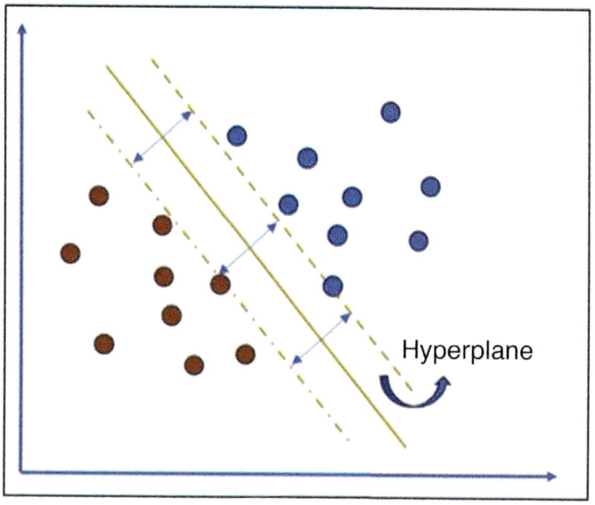

Fig. 14.2. Classification of data by support vector machine (SVM). (Author's own image.)

classification and regression algorithms in multiple fields of application (Cervantes *et al.*, 2020). They perform well in high-dimensional datasets with small sample sizes, which are common in genomics.

14.2.2.3 Neural networks and deep learning

Neural networks, particularly DL models, consist of multiple layers that capture complex, non-linear patterns in data (Fig. 14.3). CNNs and RNNs are widely used for sequence data. CNNs identify spatial motifs, while RNNs capture sequential dependencies. DL modeling can handle massive quantities of characteristics to create an efficient data-driven model, which is very helpful when working with large volumes of data (Sarker, 2021).

14.2.2.4 k-Nearest neighbors (k-NN)

This non-parametric method predicts gene function by finding the closest neighbors in the feature space. While simple, it is effective when the feature space is well-represented. The choice of k-NN is motivated by its simplicity, flexibility to incorporate different data types and adaptability to irregular feature spaces (Yao and Ruzzo, 2006).

14.2.3 Unsupervised learning models for gene function prediction

Unsupervised learning identifies patterns in unlabeled data, making it particularly useful when labeled datasets are scarce. Common unsupervised techniques are described below.

14.2.3.1 Clustering algorithms

Clustering methods, such as k-means and hierarchical clustering, group genes based on similarity in sequence-based features. Genes with similar sequence characteristics are often grouped together, revealing potential functional similarities (Wei *et al.*, 2012).

14.2.3.2 Dimensionality reduction techniques

Methods like principal component analysis (PCA) and t-SNE reduce the dimensionality of data while preserving important variance. Dimensionality reduction aids in visualizing gene relationships and identifying clusters corresponding to functional groups. In micro-array data analysis, dimensionality reduction is used to improve prediction quality, cut down on computation time, and build stronger

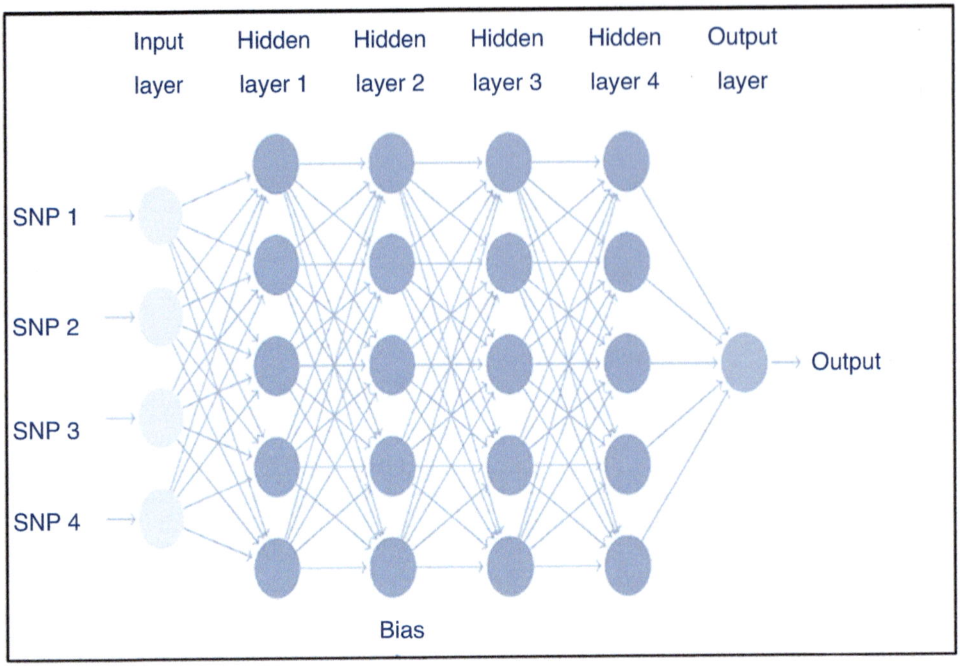

Fig. 14.3. Multi-layer perceptron (MLP) diagram with four hidden layers. SNP, single nucleotide polymorphism. (Author's own image.)

models. Furthermore, compared to the classes (samples), the algorithm learning performance involves an expressive number of characteristics (genes) (Taveira De Souza *et al.*, 2019).

14.2.3.3 Autoencoders

Autoencoders are neural networks that learn efficient encodings of data by compressing and reconstructing input features. They can uncover latent features in sequence data, which may correlate with gene functions (Sheena and Nair, 2024).

14.2.4 Workflow for machine learning-based gene function prediction

1. *Data collection:* Datasets containing labeled gene sequences with annotated functions are collected from databases such as GenBank, UniProt, and GO (Gene Ontology). For unsupervised learning, unlabeled sequences are sufficient (Mahood *et al.*, 2020).
2. *Feature extraction:* Sequence-based features are computationally derived using bioinformatics tools and encoded into numerical formats.
3. *Data preprocessing:* Steps include normalization, handling missing values, and addressing class imbalances through techniques like oversampling or SMOTE (synthetic minority oversampling technique).
4. *Model training and evaluation:* Supervised learning algorithms are trained on labeled datasets, while unsupervised methods analyze data without labels. Metrics such as silhouette score or explained variance are used for unsupervised methods, alongside traditional metrics for supervised models.
5. *Prediction and validation:* The trained model predicts the functions of unlabeled genes, and experimental validation is conducted to confirm predictions.

14.2.5 Challenges and considerations

While ML has shown great promise, several challenges persist:

1. *Data quality and quantity:* Labeled datasets for gene functions are often incomplete or biased, limiting model generalization (Mahood *et al.*, 2020).
2. *Feature selection:* Identifying the most informative features without introducing noise is critical for robust predictions.
3. *Class imbalance:* Some gene functions are underrepresented, making it difficult to train balanced models.
4. *Model interpretability:* Understanding how ML models arrive at predictions is essential for biological insights but remains challenging for complex models like DL.
5. *Scalability:* Genomic datasets are large and high-dimensional, necessitating computationally efficient algorithms.

14.3 Gene Expression Analysis and Regulatory Networks

Although RNA-seq generates complex, high-dimensional data, it measures gene expression throughout the transcriptome. By finding patterns, reducing dimensionality (e.g. PCA, t-SNE), choosing significant features (e.g. LASSO, RFs), and creating predictive models (e.g. SVMs, neural networks), ML aids in the analysis of this data (Wu and Zhang, 2020). ML also facilitates multi-omics integration, noise correction, and clustering. In order to better understand gene regulation, find biomarkers, and advance research and clinical applications, RNA-seq and ML work together to provide strong tools (Borah *et al.*, 2024).

14.3.1 Unsupervised ML in gene expression analysis

By identifying hidden patterns, classifying genes according to their functions, and displaying intricate relationships, unsupervised ML aids in the analysis of gene expression data. New techniques are being created to make these models simpler to understand and more beneficial for biological insights as DL becomes more and more popular in this field. These techniques can be divided into two categories: strategies to create naturally interpretable models using biologically constrained architectures and post hoc evaluations of black-box models using feature attribution techniques (Janizek *et al.*, 2023). The main techniques applied are **clustering algorithms**. Clustering is the process of grouping similar data points (genes or samples) together, revealing patterns of gene expression or cellular behavior. Key clustering methods used for RNA-seq data analysis include:

1. *K-means clustering:* Principle—this algorithm partitions the data into a pre-specified number of clusters (k), minimizing the variance within each cluster. It iteratively assigns genes or samples to clusters by updating centroids. A common clustering technique for assembling genes or samples with comparable expression patterns is K-means (Gasch and Eisen, 2002). Based on gene expression, it aids in the identification of patterns such as cancer subtypes. In molecular biology, clustering algorithms such as K-means have shown promise in the analysis of gene expression data. Genes are grouped according to how similar their expression profiles are using clustering methods like K-means, hierarchical clustering, SOM, etc. Genes that are functionally linked are so found (Lu *et al.*, 2004). Choosing the optimal value for k can be tricky, and the method assumes clusters to be spherical and of equal size, which may not always hold true in biological data.
2. *Hierarchical clustering:* Principle—hierarchical clustering creates a tree-like structure (dendrogram) by recursively merging or splitting clusters based on similarity. There are two approaches: agglomerative (bottom-up) and divisive (top-down). It is useful for exploring gene relationships, as it produces a dendrogram that can visually show how genes or samples are related. It is particularly helpful when the number of clusters is unknown. Hierarchical clustering methods

are useful for analyzing gene expression data as well as many data in other contexts (Yin *et al.*, 2006). It can be computationally expensive and sensitive to noise, making it less efficient for large datasets.

3. *DBSCAN (density-based spatial clustering of applications with noise):* Principle—unlike K-means, DBSCAN does not require specifying the number of clusters (Bikku *et al.*, 2023). It groups genes or samples based on density, identifying regions of high data point concentration and marking isolated points as noise. DBSCAN is effective in identifying gene expression patterns where some genes are not well-defined in terms of clusters. It is useful for detecting rare patterns or subgroups in gene expression data. It may not perform well when clusters are of varying densities or in high-dimensional data.
4. *Dimensionality reduction:* Dimensionality reduction methods aim to reduce the complexity of high-dimensional gene expression data while preserving the most relevant information (Taveira De Souza *et al.*, 2019). This helps in visualizing and interpreting large datasets, as well as improving the performance of downstream analyses like clustering and classification.
5. *Principal component analysis (PCA):* Principle—PCA identifies the directions (principal components) in which the data varies the most, reducing the number of variables (genes) while retaining the most significant patterns of variation (Yeung and Ruzzo, 2001). PCA is used for visualizing gene expression data, typically by plotting the first two or three principal components, helping to discern global patterns such as clustering of samples based on condition or disease state. PCA is a linear method and may not capture complex, non-linear relationships in gene expression data.
6. *t-SNE (t-distributed stochastic neighbor embedding):* Principle—t-SNE is a non-linear technique that reduces dimensionality by mapping high-dimensional data to two or three dimensions while preserving local neighborhood relationships (Kobak and Berens, 2019). t-SNE is particularly useful for visualizing gene expression data in single-cell RNA-seq experiments, where subtle relationships between cells or genes can be better understood. t-SNE is sensitive to hyperparameters like perplexity and can be computationally expensive for large datasets.
7. *UMAP (uniform manifold approximation and projection):* Principle—UMAP is a non-linear dimensionality reduction method similar to t-SNE but is faster and more scalable. It attempts to preserve both local and global structures in the data. UMAP is often used for visualizing high-dimensional RNA-seq data, especially in single-cell RNA-seq, revealing clusters or patterns that are biologically meaningful (Dorrity *et al.*, 2020). While faster than t-SNE, UMAP still requires careful tuning of parameters to get the best results.
8. *Gene co-expression networks:* Co-expression networks identify groups of genes that exhibit similar expression patterns across various samples or conditions, suggesting they are regulated together or participate in the same biological processes. Unsupervised clustering methods are used to group genes with highly correlated expression profiles, forming gene modules (Serin *et al.*, 2016). These modules are thought to represent functional gene groups that are co-regulated. Co-expression networks can reveal gene modules that are involved in specific biological pathways or processes, such as immune response or cancer progression. Modules may not always correspond to known pathways, and some genes may belong to multiple modules, complicating the interpretation.
9. *Pathway enrichment analysis:* Method—after identifying gene modules, pathway enrichment analysis (e.g. gene ontology or KEGG pathways) is performed to understand which biological processes are overrepresented in each module (Zhou *et al.*, 2017). This provides insight into the functional roles of co-expressed genes, linking gene expression patterns to cellular processes. False-positive results can occur if the gene list used for enrichment is too broad or incomplete.

14.3.2 Supervised machine learning for gene expression analysis

Supervised ML is widely used in gene expression analysis to predict outcomes like disease status or classify samples into different biological groups (Pirooznia *et al.*, 2008). These models learn from labeled data and are effective in high-dimensional settings like microarray or RNA-seq datasets.

1. *Support vector machines (SVM):* Suitable for analyzing high-dimensional gene expression data; effective in classifying biological conditions or sample types.
2. *Random forest (RF):* A powerful ensemble method that can rank important genes, handle noisy data, and perform classification or regression tasks.
3. *Artificial neural networks (ANNs):* Able to learn complex relationships in gene expression data; useful for identifying patterns across various biological conditions.
4. *Logistic regression:* A simple, interpretable model often used for binary classification tasks, such as distinguishing between two biological states.
5. *K-nearest neighbors (k-NN):* Intuitive and easy to implement; works well when data is properly preprocessed and dimensionality is reduced.
6. *Gradient boosting (e.g. XGBoost):* The highly accurate method used for classification, regression, and gene ranking based on their contribution to prediction.
7. *Naive Bayes:* Fast and efficient, especially with smaller datasets; performs well despite its assumption of feature independence.

14.3.3 Unsupervised ML in regulatory network inference

Unsupervised ML also plays a crucial role in the reconstruction of gene regulatory networks, where the goal is to identify interactions between genes and their regulators, such as transcription factors (Maetschke *et al.*, 2014). Regulatory networks depict the interactions between genes and transcription factors (TFs) that control gene expression. Unsupervised ML methods help infer these networks from gene expression data.

1. *Correlation-based analysis:* Genes that show strong co-expression may be regulated by the same transcription factor or participate in the same regulatory pathway. Gene expression correlation matrices can be used to infer potential regulatory relationships between genes. Correlation alone is not sufficient to establish causality or regulatory relationships.
2. *Mutual information (MI) methods:* MI measures the dependence between two variables (genes), capturing non-linear relationships that correlation-based methods may miss. MI methods are used to infer direct or indirect regulatory relationships by evaluating the statistical dependence between genes and their potential regulators. MI can sometimes overestimate dependencies, particularly in noisy data.
3. *Graph-based network inference:* Gene regulatory networks are often represented as graphs, where nodes represent genes and edges represent regulatory interactions. Several graph-based methods help infer these networks from gene expression data.
4. *Markov random fields (MRF):* Principle—MRF is a probabilistic graphical model that captures the relationships between genes by modeling the conditional dependencies between them. MRFs are used to construct gene regulatory networks by incorporating gene expression data to identify direct interactions between genes and transcription factors. Because it directly interprets edges with the conditional dependence between nodes, MRF, an undirected graphical model, is a popular method for describing the network structure of a collection of genetic variables (Yin *et al.*, 2006). The complexity of the model can lead to overfitting, and computational demands may increase with large datasets.
5. *Network clustering:* Principle—network clustering algorithms, like Louvain or Leiden, group genes based on shared regulatory features and expression patterns. These methods help identify gene modules that may be co-regulated by the same transcription factors. Inaccurate clustering due to

noisy or incomplete data can lead to incorrect network predictions.

6. *Transcription factor prediction:* Unsupervised ML methods help predict transcription factors (TFs) that regulate gene clusters by analyzing gene co-expression networks.
7. *Co-Expression Network Analysis:* Principle—co-expression network analysis links genes that are co-expressed with potential transcription factors, providing insights into gene regulation. By identifying TFs that are highly correlated with gene expression modules, researchers can predict regulatory interactions. Some co-expressed genes may not necessarily be regulated by the same TFs, complicating predictions.
8. *ChIP-seq Data Integration:* Principle—chromatin immunoprecipitation sequencing (ChIP-seq) data can be used to identify binding sites of TFs, which can then be integrated with RNA-seq data. ChIP-seq data enhances the prediction of TF–gene interactions by confirming the physical binding of transcription factors to gene promoters. ChIP-seq data are not always available for all genes or transcription factors. Inferring gene regulatory networks (GRNs) and revealing hidden gene expression patterns are made easier by advanced ML, particularly unsupervised approaches. These techniques are essential for comprehending how transcription factors control genes, which is critical for researching cellular processes, disease progression, and development. The identification of novel therapeutic targets, disease diagnosis, and personalized medicine are all aided by ML-based GRN inference.

14.3.4 Gene regulatory networks and their importance

TFs regulate gene expression in systems known as gene regulatory networks (GRNs). Diseases may result from disturbances in GRNs. Although experimental techniques such as ChIP-seq are expensive, ML reveals how genes are regulated in cells and helps predict GRNs from gene expression data. Analysis results can provide insight into the mechanisms underlying life and illness by integrating experimental data and biological knowledge with careful computational modeling to unravel regulatory networks (Karlebach and Shamir, 2008). The use of ML for GRN inference has led to several key advancements in biology:

1. *Cancer research:* ML models have been used to predict the regulatory networks underlying cancerous gene expression profiles, helping identify potential biomarkers and therapeutic targets.
2. *Drug discovery:* By understanding the regulatory networks involved in diseases, ML models can help predict how drugs may affect gene expression, guiding the development of more effective treatments.
3. *Developmental biology:* ML models have been applied to study how gene regulation changes during development, providing insights into processes such as differentiation and tissue formation.
4. *Single-cell RNA-seq analysis:* ML techniques, especially DL, are increasingly used to analyze single-cell RNA-seq data, helping to uncover gene regulatory networks at the single-cell level and understand cellular heterogeneity.

14.4 Functional Genomics: Multi-Omics Integration

The study of functional genomics looks at how genes and their byproducts affect phenotypes and biological processes. It focuses on the roles of genes in processes like stress response and development, in contrast to traditional genomics. Deep insights are made possible by technologies like mass spectrometry and next-generation sequencing, and multi-omics integration aids in addressing biological complexity for a more thorough understanding. While genome-wide association studies (GWAS) can identify genetic variants associated with abdominal fat deposition, the precise roles and mechanisms of these variants remain largely unclear (Shen *et al.*, 2024). While sequencing the genome has been a monumental achievement, understanding the functional implications of genetic information is essential to address many unanswered questions in biology. Functional genomics provides this understanding by focusing on:

1. *Gene expression and regulation:* Investigating when, where, and how genes are turned on or off in response to developmental cues or environmental changes.
2. *Protein function:* Analyzing the roles of proteins, which are the primary effectors of cellular processes, and understanding how their structure and interactions contribute to their functions.
3. *Molecular pathways:* Mapping the interactions between genes, RNA, proteins, and metabolites to understand how cellular networks maintain homeostasis or respond to disruptions.

Despite its successes, functional genomics alone cannot capture the full complexity of biological systems. Living organisms operate through interconnected layers of molecular interactions—DNA, RNA, proteins, and metabolites—that influence each other dynamically. Studying any single layer in isolation provides an incomplete picture. The rise of multi-omics integration addresses this limitation by combining data from multiple omics layers, including genomics (understanding the blueprint encoded in DNA), transcriptomics (capturing gene expression profiles to study RNA dynamics), proteomics (analyzing protein abundance, modifications, and interactions), metabolomics (profiling metabolites to explore metabolic pathways), and epigenomics (investigating changes in chromatin and DNA modifications that regulate gene expression) (Subramanian *et al.*, 2020). Multi-omics approaches are increasingly adopted because of the following reasons:

1. *Provide holistic insights:* By linking genetic mutations to alterations in RNA, protein, and metabolites in diseases, for example, multi-omics integration provides a systems-level understanding of how various molecular layers interact to shape phenotypes.
2. *Enhance precision medicine:* Particularly in oncology, where it is essential to comprehend tumor diversity and forecast treatment response, multi-omics aids in the identification of intricate biomarkers for individualized treatment.
3. *Reveal hidden mechanisms:* Integration across layers can uncover regulatory mechanisms and pathways that are invisible to single-omics approaches. For example, integrating epigenomics with transcriptomics has revealed how chromatin modifications control gene expression.

By connecting genomics, transcriptomics, proteomics, and metabolomics, this method aids in the deciphering of intricate molecular networks and advances biological and medical discoveries. To help with the integrated analysis of metabolomics data with various omics, a wide range of computational tools have been created (Jendoubi, 2021).

14.4.1 Applications of machine learning and deep learning in multi-omics data integration

ML and DL approaches are transforming biological research by providing insights into gene function, regulatory networks, and disease mechanisms. In the biomedical domain, multi-omics data includes molecular (genomics, transcriptomics, proteomics, epigenomics, metabolomics, etc.) and imaging (radiomics, pathomics) modalities that, when combined, have the potential to improve performance on prediction, classification, clustering, and other tasks (Ballard *et al.*, 2024). Multi-omics data from genomics representing molecular, metabolomic, transcriptomic, proteomic, or interatomic cell measurements constitute fertile ground for a combination of computational ML with biomedical applications to analyze, diagnose and treat common and chronic ailments using cutting-edge technologies (El-Hassan, 2024). Below are key applications where these techniques have made significant contributions:

1. *Gene function prediction:* ML/DL models integrate multi-omics data to identify genes linked to biological processes, pathways, or diseases.
2. *Disease biomarker discovery:* ML helps discover disease-related biomarkers by analyzing genomic, transcriptomic, and metabolomic features.
3. *Precision medicine:* ML enables personalized treatments by predicting individual responses based on their multi-omics profiles.

14.5 Functional Annotations of Uncharacterized Genes

Annotation Click or tap here to enter text. (Mount, 2000) can be defined as the process of identifying the locations of genes and all of the coding regions in a genome and determining the activities of a gene. Functional annotation (Carnielli *et al.*, 2015) relies on the assignment of functional information to the identified genes or regions, such as gene ontology terms, enzyme codes, or metabolic pathway associations. However, some genes that lack functional annotations or known roles in biological processes are termed uncharacterized genes (Eisen, 1998). There are main approaches that are being used to functionally annotate uncharacterized genes. There are several approaches by which functional annotations of uncharacterized genes can be done as shown in Fig. 14.4 Some important approaches are:

1. *Sequence homology:* By using BLAST or PSI-BLAST to compare sequences against comprehensive databases (e.g. NCBI, UniProt).
2. *Gene ontology annotation:* Application of eggNOG-mapper, Blast2GO to assign functional information based on known orthologs.
3. *Expression analysis:* Perform RNA-seq to check if the gene is being expressed under certain conditions or not. Correlate expression patterns with known pathways or conditions.
4. *Protein structure prediction:* Use of protein structure prediction tools like AlphaFold, I-TASSER, or Phyre2 to infer possible functions.
5. *Phylogenetic analysis:* Build phylogenetic trees to identify evolutionary relationships and infer potential functions.

14.5.1 Machine learning in functional annotation of uncharacterized genes

ML algorithms like k-NN, RF, and others can predict gene function, protein family, or pathway membership based on input features derived from sequence and other biological data (Andorf

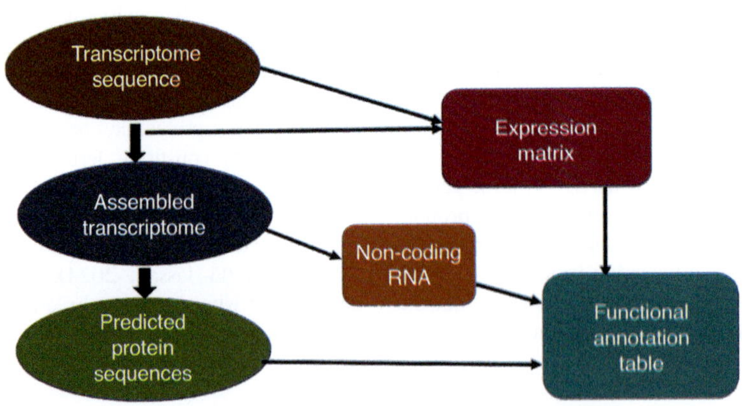

Fig. 14.4. Different approaches of functional annotation. (Author's own image.)

Table 14.1. Machine learning algorithms for functional annotation.

Serial No	Algorithm	Advantages	Considerations
1	k-Nearest neighbors (k-NN) (Guo *et al.*, 2003)	Assigns function based on the majority of nearest neighbors	Requires well-clustered feature space
2	Random forest (RF) (Breiman, 2001)	Provides feature importance	May overfit with too many trees
3	Support vector machines (SVM) (Cortes *et al.*, 1995)	Works well with a clear margin of separation	Less effective for large datasets
4	Neural networks (Rosenblatt, 1958)	Captures high-order dependencies	Prone to overfitting without regularization
5	Gradient boosting (e.g. XGBoost, LightGBM) (Bentéjac *et al.*, 2021)	Efficient with noisy features	Requires hyperparameter tuning
6	Clustering algorithms (e.g. k-means, DBSCAN) (Rodriguez *et al.*, 2019)	Groups genes with similar features	Results depend on the choice of distance metrics and parameters

et al., 2007). The workflow for functional annotation using ML is given below:

1. *Data preparation:* Data used to get prepared by extracting various features like sequence composition like k-mers, GC content, di-/tri-nucleotide frequencies, BLAST scores, orthologous groups, or phylogenetic profiles, secondary structure, domain architecture and operon structure, co-expression networks.
2. *Data processing:* Datasets are used to be processed through normalizing/scaling numerical features, encoding categorical features, and handling missing values by imputation or removal.
3. *Data labeling:* Using known gene functions derived from gene ontology or KEGG as data labels for model development.
4. *Algorithm selection:* Different ML algorithms can be used for annotation of uncharacterized genes, which are shown in Table 14.1.
5. *Model training and evaluation:* Data is split into training and testing or k-fold cross-validation. Performance evaluation metrics like accuracy, precision, recall, and F1-score are mainly used for classification tasks. Hyperparameter tuning using Grid Search or Bayesian Optimization can also improve the model.
6. *Prediction and functional assignment:* Then the trained model can be used to predict functions for uncharacterized genes. Integration with biological knowledge (e.g.pathway databases) for validation is more efficient.

14.5.2 Challenges and considerations of machine learning approaches

Apart from holding a great potential for functional annotation of genes, several challenges of the ML approach must be addressed to ensure accurate and reliable predictions (Leung *et al.*, 2015). Crucial challenges and their solutions are:

1. *Data quality and quantity:* Many genes remain uncharacterized, creating a bottleneck for supervised learning. Furthermore, noisy and incomplete datasets can reduce model performance.
2. *Feature engineering:* Extracting meaningful features from sequence data is complex and requires domain expertise. Features like sequence composition, structural motifs, or evolutionary information must capture relevant biological insights while avoiding redundancy.

3. *Imbalanced data:* Functional categories often exhibit imbalances, with some being over-represented and others scarcely annotated. This can lead to biased models favoring dominant classes.
4. *Interpretability:* Biologists need interpretable models to trust and validate predictions. Algorithms like RF offer feature importance, but DL models are often seen as "black boxes."
5. *Validation:* Experimental validation is critical but time-consuming and costly. False positives can lead to misleading biological conclusions.

14.6 Phenotype Prediction and Trait Association

The visible physical, physiological, and behavioral characteristics of plants that arise from the interplay between their genetic composition (genotype) and environmental factors are referred to as plant phenotypes (Kothari and Schweiger, 2022). Phenotypic traits include a wide range of characteristics such as plant height, leaf shape, flower color, seed yield, drought tolerance, and resistance to pests. ML has become a key tool for predicting plant phenotypic traits from genomic, transcriptomic, proteomic, and environmental data. Different ML algorithms, including gradient boosting and DL, offer unique strengths for modeling the complex relationships between genotypes, environmental factors, and phenotypes. The workflow for functional annotation using ML is:

1. *Data preparation:* Data preparation starts with the compilation of datasets with features like genetic markers (SNPs), transcriptomic data, or environmental factors.
2. *Feature engineering:* Extract meaningful features like k-mer frequencies, gene expression levels, and quantitative trait loci (QTL) regions are being extracted from the dataset. Reduction of dimensionality is very much needed by using methods like PCA or feature selection.
3. *Model training:* Different ML models can be implemented in the context. Some of the significant models are shown in Table 14.2.
4. *Evaluation:* Metrics like R^2, MAE, and RMSE for regression and accuracy, F1-score, and ROC-AUC for classification are used for evaluation of model performance along with k-fold cross-validation to ensure robustness.
5. *Interpretation and validation:* SHAP (SHapley Additive exPlanations) (Antwarg *et al.*, 2021) or feature importance plots are widely used for interpretability. Validation of predictions needs experiments in controlled or field conditions.

Table 14.2. Machine learning algorithms for phenotypic trait prediction.

Serial No	Algorithm	Advantages	Challenges
1	Gradient Boosting (e.g. XGBoost, LightGBM) (Bentéjac *et al.*, 2021)	Handles high-dimensional data and noise well	Computationally intensive for large datasets
2	Random forest (Breiman, 2001)	Robust to overfitting;handles nonlinear relationships	Less precise than gradient boosting for certain problems
3	Deep learning (CNNs, RNNs) (Chauhan and Singh, 2019)	Effective for large, high-dimensional datasets	Requires significant computational resources and data
4	k-Nearest neighbors (k-NN) (Guo *et al.*, 2003)	Effective for small datasets with clear clusters	Sensitive to irrelevant features and feature scaling
5	Support vector machines (SVMs) (Cortes *et al.*, 1995)	Effective for small to medium datasets	Computationally intensive for large datasets
6	Linear models (e.g. Ridge, LASSO) (Ogutu *et al.*, 2012)	Effective for linear relationships between genotype and phenotype	Limited in capturing complex feature interactions

14.6.1 Challenges and considerations of machine learning approaches

Instead of having more accuracy and efficiency ML approaches have some shortcomings in phenotype prediction and trait association. Some of its challenges and their solutions are:

1. *Data quality and availability:* High-quality, labeled datasets are essential for training effective models, but phenotypic data often have missing, noisy, or inconsistent values.
2. *High-dimensional data:* Genomic and phenotypic datasets often have a large number of features (e.g. SNPs) compared to the number of samples, leading to the "curse of dimensionality."
3. *Complex trait architecture:* Many traits are polygenic and influenced by environmental factors, making it difficult to model the genotype–phenotype relationship.
4. *Overfitting:* Overfitting is common, especially with small datasets and complex models like DL. The employment of techniques such as cross-validation, regularization, dropout, and early stopping can be done to prevent overfitting.
5. *Interpretability:* Many ML models, especially DL, are often considered "black boxes," which limits their utility in biological research. Use of interpretable models (e.g. RF, linear models) or tools like SHAP and LIME to explain predictions are recommended.
6. *Genotype-by-environment interactions (G x E):* Phenotypic expression depends on environmental conditions, which are challenging to account for in predictive models.
7. *Computational costs:* ML methods, particularly DL, require significant computational resources and time for large-scale data.
8. *Generalizability:* Models trained on specific datasets or species may not perform well on new datasets or related species.
9. *Experimental validation:* Validating ML predictions experimentally is time-consuming and expensive.

By addressing these challenges, ML approaches can be more effectively applied to predict phenotypic traits and accelerate advancements in plant breeding and biology.

14.7 Applications of Natural Language Processing in Gene Function Prediction

Natural language processing (NLP) (Chowdhary, 2020) is a subfield of AI that enables machines to understand, interpret, and generate human language. Combining computational linguistics and ML, NLP bridges the gap between human communication and computer understanding, making it pivotal in a wide range of applications. In biomedical research, NLP automates tasks such as extracting information about genes, proteins, and diseases, facilitating the annotation of gene functions (Zhao *et al.*, 2020), drug discovery (Drews, 2000), and personalized medicine. For example, NLP can analyze abstracts from PubMed to identify novel gene-disease relationships or classify genes (Kalfoglou and Schorlemmer, 2003) into functional categories using textual evidence. NLP faces several challenges in the domain of gene function prediction due to the complexity of biological data. One of the primary obstacles is domain-specific language. Biomedical texts often include specialized terminology, acronyms, and technical jargon that standard NLP models struggle to interpret accurately. Additionally, ambiguity presents a significant challenge, as biological terms can have multiple meanings depending on context. For instance, a single gene might be associated with varying functions under different conditions, making it difficult for traditional NLP models to disambiguate and extract precise meaning from the text. Another challenge is the lack of labelled data for gene function prediction. Moreover, the heterogeneity of data sources complicates predictions, as gene functions are often described across various forms, such as scientific literature, protein interaction databases, and experimental datasets. Finally, the integration of diverse data types, including genomic sequences and experimental results, into NLP models remains a challenging task for accurate gene function prediction (Miller *et al.*, 2022).

14.7.1 Considerations for the challenges of NLP in gene function prediction

Some relevant solutions for those above-mentioned challenges are mentioned in the following-

1. *Domain-specific language and terminology:* One effective solution to handling domain-specific language in gene function prediction is the use of domain-specific NLP models. Pre-trained models like BioBERT or SciBERT, which are fine-tuned in biomedical literature, can understand the specialized terminology found in genetic data.
2. *Addressing ambiguity:* Disambiguation techniques can be used to resolve multiple meanings of biological terms. One approach is contextual word embeddings, such as those generated by BioBERT or ELMo, which consider the surrounding words and sentences to determine the most likely meaning of a term. Another solution involves relationship extraction models that analyze the interaction between words to understand context.
3. *Lack of labeled data:* The scarcity of annotated datasets can be addressed using semi-supervised learning and transfer learning techniques. In semi-supervised learning, models can be trained on a smaller set of labeled data and leverage a large amount of unlabeled data. Transfer learning involves pre-training NLP models on large, general datasets and then fine-tuning them on smaller, domain-specific datasets.
4. *Handling heterogeneous data sources:* To address the challenge of heterogeneity in data sources, the integration of multi-source data is essential. Using multi-modal DL models that can combine textual data with structured data (e.g. gene expression profiles, and protein interaction data) enables a more comprehensive understanding of gene function.
5. *Integration of diverse data types:* For integrating diverse data types, the use of multi-modal DL techniques proves effective. These models are capable of handling both unstructured text (such as literature) and structured data (like genomic sequences or protein structures).

Above all, addressing the challenges of NLP in gene function prediction requires a multi-faceted approach that includes domain-specific models, advanced ML techniques like transfer learning and semi-supervised learning, and the integration of heterogeneous data sources. By leveraging state-of-the-art models like BioBERT, combining diverse data types, and ensuring ethical considerations, NLP can become a powerful tool for predicting gene functions with high accuracy and reliability.

14.8 Conclusion

In conclusion, ML offers a powerful and efficient approach to predicting gene functions, helping to bridge the gap between gene discovery and functional annotation. By leveraging informative sequence-based features—such as nucleotide composition, codon usage, motifs, protein domains, and k-mer frequencies—ML models can uncover complex patterns and relationships that are difficult to detect through traditional methods. These models not only enhance the accuracy of functional predictions but also facilitate high-throughput analysis across diverse genomes. As genomic data continue to grow, integrating multi-omics information and refining ML algorithms will further improve predictive performance and biological relevance. Despite challenges like data quality and model interpretability, ML remains a promising tool for accelerating functional genomics research and advancing our understanding of gene roles in health, development, and evolution. Continued advancements in this field are expected to drive innovations in areas such as crop improvement, disease research, and personalized medicine.

References

Andorf, C., Dobbs, D. and Honavar, V. (2007) Exploring inconsistencies in genome-wide protein function annotations: A machine learning approach. *BMC Bioinformatics* 8, 1–12. DOI: 10.1186/1471-2105-8-284.

Antwarg, L., Miller, R.M., Shapira, B. and Rokach, L. (2021) Explaining anomalies detected by autoencoders using shapley additive explanations. *Expert Systems with Applications* 186, 115736. DOI: 10.1016/j.eswa.2021.115736.

Auffray, C., Imbeaud, S., Roux-Rouquié, M. and Hood, L. (2003) From functional genomics to systems biology: Concepts and practices. *Comptes Rendus Biologies* 326(10–11), 879–892. DOI: 10.1016/j.crvi.2003.09.033.

Ballard, J.L., Wang, Z., Li, W., Shen, L. and Long, Q. (2024) Deep learning-based approaches for multi-omics data integration and analysis. *BioData Mining* 17(1), 38. DOI: 10.1186/s13040-024-00391-z.

Behura, S.K. and Severson, D.W. (2013) Codon usage bias: Causative factors, quantification methods and genome-wide patterns: With emphasis on insect genomes. *Biological Reviews of the Cambridge Philosophical Society* 88(1), 49–61. DOI: 10.1111/j.1469-185X.2012.00242.x.

Bentéjac, C., Csörgő, A. and Martínez-Muñoz, G. (2021) A comparative analysis of gradient boosting algorithms. *Artificial Intelligence Review* 54(3), 1937–1967. DOI: 10.1007/s10462-020-09896-5.

Bikku, T., Ramu, J., Sekhar, J.C., Pratap, V.K. and Pujari, J.J. (2023) Optimizing gene expression analysis using clustering algorithms. In: *International Conference on Computer & Communication Technologies*. Springer Nature Singapore, pp. 163–171.

Boateng, E.Y., Otoo, J. and Abaye, D.A. (2020) Basic tenets of classification algorithms K-nearest-neighbor, support vector machine, random forest and neural network: A review. *Journal of Data Analysis and Information Processing* 08(4), 341–357. DOI: 10.4236/jdaip.2020.84020.

Borah, K., Das, H.S., Seth, S., Mallick, K., Rahaman, Z. *et al.* (2024) A review on advancements in feature selection and feature extraction for high-dimensional NGS data analysis. *Functional & Integrative Genomics* 24(5), 139. DOI: 10.1007/s10142-024-01415-x.

Breiman, L. (2001) Random forests. *Machine Learning* 45(1), 5–32. DOI: 10.1023/A:1010933404324.

Carnielli, C.M., Winck, F.V. and Paes Leme, A.F. (2015) Functional annotation and biological interpretation of proteomics data. *Biochimica et Biophysica Acta (BBA) - Proteins and Proteomics* 1854(1), 46–54. DOI: 10.1016/j.bbapap.2014.10.019.

Cembrowska-Lech, D., Krzemińska, A., Miller, T., Nowakowska, A., Adamski, C. *et al.* (2023) An integrated multi-omics and artificial intelligence framework for advance plant phenotyping in horticulture. *Biology* 12(10), 1298. DOI: 10.3390/biology12101298.

Cervantes, J., Garcia-Lamont, F., Rodríguez-Mazahua, L. and Lopez, A. (2020) A comprehensive survey on support vector machine classification: Applications, challenges and trends. *Neurocomputing* 408, 189–215. DOI: 10.1016/j.neucom.2019.10.118.

Chauhan, N.K. and Singh, K. (2019) A review on conventional machine learning vs deep learning. In: *2018 International Conference on Computing, Power and Communication Technologies (GUCON)*, Greater Noida, Uttar Pradesh, India, pp. 347–352. DOI: 10.1109/GUCON.2018.8675097.

Chowdhary, K.R. (2020) Natural language processing. *Fundamentals of Artificial Intelligence* 603–649. DOI: 10.1007/978-81-322-3972-7_19.

Cortes, C., Vapnik, V. and Saitta, L. (1995) Support-vector networks. *Machine Learning* 20(3), 273–297. DOI: 10.1023/A:1022627411411.

Dorrity, M.W., Saunders, L.M., Queitsch, C., Fields, S. and Trapnell, C. (2020) Dimensionality reduction by UMAP to visualize physical and genetic interactions. *Nature Communications* 11(1), 1537. DOI: 10.1038/s41467-020-15351-4.

Drews, J. (2000) Drug discovery: A historical perspective. *Science (New York, N.Y.)* 287(5460), 1960–1964. DOI: 10.1126/science.287.5460.1960.

Dutta, S., Zunjare, R.U., Muthusamy, V. and Hossain, F. (2023) Prediction of CENH3 protein in maize using machine learning techniques. *The Pharma Innovation* 12(7S), 01–06. DOI: 10.22271/tpi.2023.v12.i7Sa.21185.

Dutta, S., Chhabra, R., Muthusamy, V., Gain, N., Subramani, R. *et al.* (2024a) Allelic variation and haplotype diversity of matrilineal (MTL) gene governing *in vivo* maternal haploid induction in maize. *Physiology and Molecular Biology of Plants* 30(5), 823–838. DOI: 10.1007/s12298-024-01456-3.

Dutta, S., Paul, D., Das, A., Kumar, S., Roy, A. *et al.* (2024b) Directed mutagenesis in plants through genome editing using guide RNA library. *The Nucleus* 67(1), 61–73. DOI: 10.1007/s13237-024-00481-7.

Dutta, S., Chhabra, R., Muthusamy, V., Gain, N., Subramani, R. *et al.* (2024c) Allelic variation and haplotype diversity of matrilineal (MTL) gene governing in vivo maternal haploid induction in maize. *Physiology and Molecular Biology of Plants* 30(5), 823–838. DOI: 10.1007/s12298-024-01456-3.

Dutta, S., Islam, Z., Das, S., Barman, A., Chowdhury, M. *et al.* (2025) Harmonizing plant resilience: Unveiling the symphony of membrane lipid dynamics in response to abiotic stresses: A review. *Discover Plants* 2(1), 61. DOI: 10.1007/s44372-025-00152-0.

Eisen, J.A. (1998) Phylogenomics: Improving functional predictions for uncharacterized genes by evolutionary analysis. *Genome Research* 8(3), 163–167. DOI: 10.1101/gr.8.3.163.

El-Hassan, A. (2024) Machine learning from multi-omics: Applications and data integration. In: *Machine Learning Methods for Multi-Omics Data Integration*. Springer International Publishing, pp. 13–21. DOI: 10.1007/978-3-031-36502-7_2.

Gasch, A.P. and Eisen, M.B. (2002) Exploring the conditional coregulation of yeast gene expression through fuzzy K-means clustering. *Genome Biology* 3(11), 1–22. DOI: 10.1186/gb-2002-3-11-research0059.

Guo, G., Wang, H., Bell, D., Bi, Y. and Greer, K. (2003) KNN model-based approach in classification. *Lecture Notes in Computer Science (Including Subseries Lecture Notes in Artificial Intelligence and Lecture Notes in Bioinformatics)* 2888, 986–996. DOI: 10.1007/978-3-540-39964-3_62.

Janizek, J.D., Spiro, A., Celik, S., Blue, B.W., Russell, J.C. *et al.* (2023) PAUSE: Principled feature attribution for unsupervised gene expression analysis. *Genome Biology* 24(1), 81. DOI: 10.1186/s13059-023-02901-4.

Jendoubi, T. (2021) Approaches to integrating metabolomics and multi-omics data: A primer. *Metabolites* 11(3), 184. DOI: 10.3390/metabo11030184.

Kalfoglou, Y. and Schorlemmer, M. (2003) Ontology mapping: The state of the art. *The Knowledge Engineering Review* 18(1), 1–31. DOI: 10.1017/S0269888903000651.

Karlebach, G. and Shamir, R. (2008) Modelling and analysis of gene regulatory networks. *Nature Reviews. Molecular Cell Biology* 9(10), 770–780. DOI: 10.1038/nrm2503.

Karollus, A., Hingerl, J., Gankin, D., Grosshauser, M., Klemon, K. *et al.* (2024) Species-aware DNA language models capture regulatory elements and their evolution. *Genome Biology* 25(1), 83. DOI: 10.1186/s13059-024-03221-x.

Kobak, D. and Berens, P. (2019) The art of using T-SNE for single-cell transcriptomics. *Nature Communications* 10(1), 5416. DOI: 10.1038/s41467-019-13056-x.

Kothari, S. and Schweiger, A.K. (2022) Plant spectra as integrative measures of plant phenotypes. *Journal of Ecology* 110(11), 2536–2554. DOI: 10.1111/1365-2745.13972.

Kumar, P., Paul, D., Jhajhriya, S., Kumar, R., Dutta, S. *et al.* (2024) Understanding heat-shock proteins' abundance and pivotal function under multiple abiotic stresses. *Journal of Plant Biochemistry and Biotechnology* 1–22.

Leung, M.K.K., Delong, A., Alipanahi, B. and Frey, B.J. (2015) Machine learning in genomic medicine: A review of computational problems and data sets. *Proceedings of the IEEE* 104(1), 176–197. DOI: 10.1109/JPROC.2015.2494198.

Lu, Y., Lu, S., Fotouhi, F., Deng, Y. and Brown, S.J. (2004) Incremental genetic K-means algorithm and its application in gene expression data analysis. *BMC Bioinformatics* 5(1), 172. DOI: 10.1186/1471-2105-5-172.

Maetschke, S.R., Madhamshettiwar, P.B., Davis, M.J. and Ragan, M.A. (2014) Supervised, semi-supervised and unsupervised inference of gene regulatory networks. *Briefings in Bioinformatics* 15(2), 195–211. DOI: 10.1093/bib/bbt034.

Mahood, E.H., Kruse, L.H. and Moghe, G.D. (2020) Machine learning: A powerful tool for gene function prediction in plants. *Applications in Plant Sciences* 8(7), e11376. DOI: 10.1002/aps3.11376.

Maji, A.K., Das, S., Marwaha, S., Kumar, S., Dutta, S. *et al.* (2025) Intelligent decision support for drought stress (IDSDS): An integrated remote sensing and artificial intelligence-based pipeline for quantifying drought stress in plants. *Computers and Electronics in Agriculture* 236, 110477. DOI: 10.1016/j.compag.2025.110477.

Miller, D., Stern, A. and Burstein, D. (2022) Deciphering microbial gene function using natural language processing. *Nature Communications* 13(1), 5731. DOI: 10.1038/s41467-022-33397-4.

Montesinos López, O.A., Montesinos López, A. and Crossa, J. (2022) Random forest for genomic prediction. In: *In Multivariate Statistical Machine Learning Methods for Genomic Prediction*. Springer International Publishing, pp. 633–681. DOI: 10.1007/978-3-030-89010-0_15.

Mount, S.M. (2000) Genomic sequence, splicing, and gene annotation. *American Journal of Human Genetics* 67(4), 788–792. DOI: 10.1086/303098.

Mukherjee, A., Abraham, S., Singh, A., Balaji, S. and Mukunthan, K.S. (2025) From data to cure: A comprehensive exploration of multi-omics data analysis for targeted therapies. *Molecular Biotechnology* 67(4), 1269–1289. DOI: 10.1007/s12033-024-01133-6.

Nguyen, T.P. and Ho, T.B. (2012) Detecting disease genes based on semi-supervised learning and protein-protein interaction networks. *Artificial Intelligence in Medicine* 54(1), 63–71. DOI: 10.1016/j.artmed.2011.09.003.

Ogutu, J.O., Schulz-Streeck, T. and Piepho, H.P. (2012) Genomic selection using regularized linear regression models: Ridge regression, lasso, elastic net and their extensions. *BMC Proceedings* 6(Suppl 2), 1–6. DOI: 10.1186/1753-6561-6-S2-S10.

Perakakis, N., Yazdani, A., Karniadakis, G.E. and Mantzoros, C. (2018) Omics, big data and machine learning as tools to propel understanding of biological mechanisms and to discover novel diagnostics and therapeutics. *Metabolism* 87, A1–A9. DOI: 10.1016/j.metabol.2018.08.002.

Pirooznia, M., Yang, J.Y., Yang, M.Q. and Deng, Y. (2008) A comparative study of different machine learning methods on microarray gene expression data. *BMC Genomics* 9 Suppl 1(Suppl 1), 1–13. DOI: 10.1186/1471-2164-9-S1-S13.

Raghunandan, K., Dutta, S., Thribhuvan, R., Bhowmick, R., Chourasia, K.N. *et al.* (2025) Breeding minor pulses for climate resilience in the era of genomics: Opportunities and prospects. In: *Breeding Climate Resilient and Future Ready Pulse Crops*. Springer Nature Singapore, Singapore, pp. 351–372.

Rodriguez, M.Z., Comin, C.H., Casanova, D., Bruno, O.M., Amancio, D.R. *et al.* (2019) Clustering algorithms: A comparative approach. *PLOS ONE* 14(1), e0210236. DOI: 10.1371/journal.pone.0210236.

Rosenblatt, F. (1958) The perceptron: A probabilistic model for information storage and organization in the brain. *Psychological Review* 65(6), 386–408. DOI: 10.1037/h0042519.

Roy, A., Dutta, S., Das, S. and Choudhury, M.R. (2024) Next-generation sequencing in the development of climate-resilient and stress-responsive crops–A review. *The Open Biotechnology Journal* 18(1). DOI: 10.2174/0118740707301657240517063244.

Sarker, I.H. (2021) Deep learning: A comprehensive overview on techniques, taxonomy, applications and research directions. *SN Computer Science* 2(6), 420. DOI: 10.1007/s42979-021-00815-1.

Serin, E.A.R., Nijveen, H., Hilhorst, H.W.M. and Ligterink, W. (2016) Learning from co-expression networks: Possibilities and challenges. *Frontiers in Plant Science* 7, 444. DOI: 10.3389/fpls.2016.00444.

Sheena, K.S. and Nair, M.S. (2024) Gencoder: A novel convolutional neural network based autoencoder for genomic sequence data compression. *IEEE/ACM Transactions on Computational Biology and Bioinformatics* 21(3), 405–415. DOI: 10.1109/TCBB.2024.3366240.

Shen, L., Bai, X., Zhao, L., Zhou, J., Chang, C. *et al.* (2024) Integrative 3D genomics with multi-omics analysis and functional validation of genetic regulatory mechanisms of abdominal fat deposition in chickens. *Nature Communications* 15(1), 9274. DOI: 10.1038/s41467-024-53692-6.

Subramanian, I., Verma, S., Kumar, S., Jere, A. and Anamika, K. (2020) Multi-omics data integration, interpretation, and its application. *Bioinformatics and Biology Insights* 14, 1177932219899051. DOI: 10.1177/1177932219899051.

Taveira De Souza, J., Carlos De Francisco, A. and Macedo, D.C.D. (2019) Dimensionality reduction in gene expression data sets. *IEEE Access* 7, 61136–61144. DOI: 10.1109/ACCESS.2019.2915519.

Wei, D., Jiang, Q., Wei, Y. and Wang, S. (2012) A novel hierarchical clustering algorithm for gene sequences. *BMC Bioinformatics* 13, 1–15. DOI: 10.1186/1471-2105-13-174.

Wu, Y. and Zhang, K. (2020) Tools for the analysis of high-dimensional single-cell RNA sequencing data. *Nature Reviews. Nephrology* 16(7), 408–421. DOI: 10.1038/s41581-020-0262-0.

Yan, J. and Wang, X. (2022) Unsupervised and semi-supervised learning: The next frontier in machine learning for plant systems biology. *The Plant Journal* 111(6), 1527–1538. DOI: 10.1111/tpj.15905.

Yao, Z. and Ruzzo, W.L. (2006) A regression-based K nearest neighbor algorithm for gene function prediction from heterogeneous data. *BMC Bioinformatics* 7(S1), S11. DOI: 10.1186/1471-2105-7-S1-S11.

Yeung, K.Y. and Ruzzo, W.L. (2001) Principal component analysis for clustering gene expression data. *Bioinformatics (Oxford, England)* 17(9), 763–774. DOI: 10.1093/bioinformatics/17.9.763.

Yin, L., Huang, C.-H. and Ni, J. (2006) Clustering of gene expression data: Performance and similarity analysis. *BMC Bioinformatics* 7(S4), S19. DOI: 10.1186/1471-2105-7-S4-S19.

Zafar, I., Anwar, S., kanwal, F., Yousaf, W., Un Nisa, F. *et al.* (2023) Reviewing methods of deep learning for intelligent healthcare systems in genomics and biomedicine. *Biomedical Signal Processing and Control* 86, 105263. DOI: 10.1016/j.bspc.2023.105263.

Zhao, Y., Wang, J., Chen, J., Zhang, X., Guo, M. *et al.* (2020) A literature review of gene function prediction by modeling gene ontology. *Frontiers in Genetics* 11, 524174. DOI: 10.3389/fgene.2020.00400.

Zhou, T., Yao, J. and Liu, Z. (2017) Gene ontology, enrichment analysis, and pathway analysis. In: *Bioinformatics in Aquaculture: Principles and Methods*. pp. 150–168.

15 Data-Driven Healing: Integrating Machine Learning with Traditional Chinese Medicinal Plants

Elshan Musazade[1], Isack Ibrahim Mrisho[2] and Xianzhong Feng[1]*
[1]Key Laboratory of Soybean Molecular Design Breeding, Northeast Institute of Geography and Agroecology, Chinese Academy of Sciences, Changchun, PR China; [2]College of Chinese Medicinal Material, Jilin Agricultural University, Changchun, PR China

Abstract
Traditional Chinese Medicine (TCM) has long utilized medicinal plants for therapeutic purposes; however, scientific validation and standardization remain challenging due to their complex compositions and multi-target interactions. The integration of artificial intelligence (AI), particularly machine learning (ML) and deep learning (DL), has transformed herbal medicine research by enhancing the identification of medicinal plants, bioactive compound discovery, and pharmacological analysis. AI-powered approaches, including cheminformatics, network pharmacology, and spectral analysis, have streamlined quality control and prescription optimization processes. However, challenges such as data scarcity, a lack of standardized datasets, and the interpretability of AI-driven models hinder their widespread adoption. Addressing these limitations requires interdisciplinary collaboration, explainable AI models, and regulatory advancement. By bridging traditional wisdom with computational intelligence, AI can modernize TCM and improve its scientific credibility and integration into global healthcare.

Keywords: Artificial intelligence, traditional chinese medicine, machine learning, herbal drug discovery, pharmacological mechanisms

15.1 Introduction

Traditional Chinese Medicine (TCM) has been a cornerstone of healthcare for millennia, emphasizing holistic, plant-based therapies rooted in balance and synergy. Medicinal plants contain bioactive compounds that enhance efficacy and minimize toxicity (Wang *et al.*, 2021a; Wang *et al.*, 2021b). Notably, artemisinin from *Artemisia annua* has revolutionized malaria treatment (Zhai *et al.*, 2016), and TCM-based prescriptions have played a vital role in managing COVID-19 (An *et al.*, 2021; Wang *et al.*, 2021c). Despite its potential, TCM faces challenges in standardization, authentication, and mechanistic elucidation, necessitating advanced methodologies to decode its complex multi-component and multi-target mechanisms (Liu and Guo, 2020; Wei *et al.*, 2023).

Accurate medicinal plant identification is crucial for ensuring the efficacy and safety of TCM. Traditional methods, including

*Corresponding author: fengxianzhong@iga.ac.cn

DOI: 10.1079/9781800628649.0015

morphological, anatomical, and chemical analyses, are often subjective, time consuming, and prone to misidentifications. Variability in plant subtypes and inconsistencies in quality complicate authentication (Chen *et al.*, 2021a; Chen and He, 2022). Expert-reliant conventional approaches can be labor intensive and error prone, potentially affecting clinical outcomes (Zhou *et al.*, 2021a; Chu *et al.*, 2022). To address these challenges, researchers have increasingly employed artificial intelligence (AI) and computational technologies to improve TCM standardization and applications (Li *et al.*, 2022a).

AI has transformed healthcare, pharmacology, and bioinformatics by enabling complex data analyses, pattern recognition, and precise predictions. In TCM, AI-powered techniques significantly enhance plant authentication, ensuring quality control and safety of herbal products. Convolutional neural networks (CNNs) and DNA barcoding with machine learning (ML) automate species recognition. In contrast, spectral imaging and metabolomics-based AI models improve the detection of adulterated or substandard herbal materials (Kattenborn *et al.*, 2019; Picon *et al.*, 2019).

Automated classification has gained traction in TCM research. Wu *et al.* (2016) used an electronic nose and computer vision to analyze herbal odors and colors through backpropagation (BP) and support vector machines (SVM) (Wu *et al.*, 2016). Hussein *et al.* (2020) employed discriminant analysis, random forest (RF), and SVM for plant identification, although early pixel-based models were vulnerable to environmental variations (Hussein *et al.*, 2020). Advances in deep learning (DL) have significantly improved classification accuracy, with CNNs extracting hierarchical features that surpass traditional manual methods in terms of reliability and precision (Kattenborn *et al.*, 2019; Picon *et al.*, 2019). CNN-based approaches, such as those by Lee *et al.* (2015) and Hu *et al.* (2018), have demonstrated superior performance in plant identification, while Park *et al.* (2021) introduced a multi-rate 3D CNN with a joint fusion classifier, enhancing medicinal plant recognition in TCM (Lee *et al.*, 2015; Hu *et al.*, 2018; Park *et al.*, 2021).

Beyond plant identification, AI accelerates bioactive compound discovery and pharmacological analyses. Traditional methods, such as nuclear magnetic resonance (NMR) spectroscopy and high-performance liquid chromatography (HPLC), are labor intensive. In contrast, AI-driven cheminformatics and quantitative structure–activity relationship (QSAR) modeling analyze extensive phytochemical databases, predict molecular properties, and identify pharmacologically relevant compounds. AI-assisted mass spectrometry and metabolomics streamline bioactive molecule screening and can reduce time and cost (Yang *et al.*, 2019a ; Tripathi *et al.*, 2021).

Unlike single-target Western drugs, TCM formulations act via multi-component and multi-target interactions. AI-driven network pharmacology integrates multi-omics data (transcriptomics, proteomics, and metabolomics) to predict the effects of compounds. DL enhances virtual screening, molecular docking, absorption, distribution, metabolism, excretion, and toxicity (ADMET) predictions, thereby optimizing drug discovery (Rajpurkar *et al.*, 2022). Computational approaches, such as virtual screening and computer-aided drug design, further accelerate TCM drug development. AI frameworks, such as DeepDDI, predict drug–drug interactions, thereby improving safety (Ryu *et al.*, 2018).

AI is revolutionizing TCM by modernizing traditional practices and strengthening scientific validation. ML and DL enhance medicinal plant identification, bioactive compound discovery, and pharmacological analysis, thereby supporting quality control and prescription refinement (Fig. 15.1). AI-driven methodologies, such as cheminformatics, network pharmacology, and spectral analysis, facilitate multi-target interaction analysis. Despite these advancements, challenges such as data scarcity, lack of standardized datasets, and model interpretability hinder widespread adoption. This review explores AI-driven advancements in TCM, demonstrating its potential to bridge ancient wisdom with modern technology, drive innovation, and improve the therapeutic outcomes of TCM.

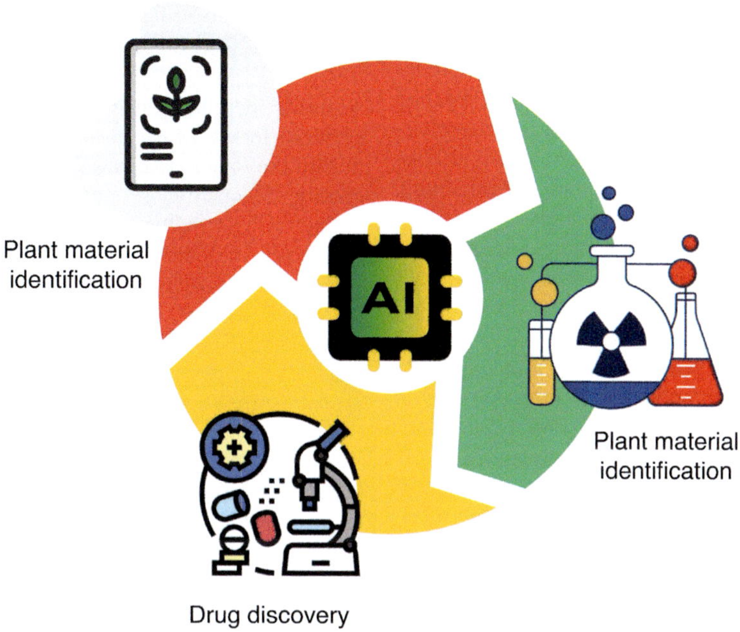

Fig. 15.1. AI applications in TCM plant materials, compound identification, and drug discovery. This process focuses on identifying plant materials, profiling bioactive compounds, and developing plant-based drugs, highlighting the role of AI in advancing TCM-related pharmaceutical research. (Author's own image.)

15.2 AI in Traditional Chinese Medicine

AI has transformed life sciences and medicine through advances in big data analytics, computational hardware, and algorithms. It encompasses ML, natural language processing (NLP), robotics, and computer vision, enabling human-like cognitive functions, such as learning, reasoning, and decision making. In TCM, AI applications, such as ML and NLP, are modernizing research, improving diagnostics, and enhancing pharmacological understanding (Zeng and Jia, 2024).

ML, a core AI branch, analyzes large datasets to identify patterns, predicts outcomes, and refines diagnostics through data preprocessing, model selection, and evaluation (Tian *et al.*, 2024). It aids molecular studies, meridian classification, and TCM component analysis (Wang *et al.*, 2019). DL models such as AlphaFold have revolutionized structural biology by predicting protein structures, thereby advancing herbal pharmacology. By autonomously processing complex data, ML accelerates drug discovery and reduces the reliance on manual feature extraction (Jumper *et al.*, 2021).

NLP enhances medical literature analysis and interpretation of patient records. Transformer-based models, such as generative pre-trained transformer 4 (GPT-4) and bidirectional encoder representations from transformers (BERT), improve machine translation and text comprehension and benefit molecular research (Zhang *et al.*, 2024a). In TCM, NLP standardizes symptom classification and improves data accuracy and terminology consistency (Zhou *et al.*, 2022b). AI-driven diagnostic systems using bidirectional long short-term memory (BiLSTM) and CNNs have achieved high accuracy in disease diagnosis and syndrome classification from electronic health records (Zhang *et al.*, 2020a), promoting TCM standardization.

AI-driven quantitative analysis advances TCM by systematically evaluating the physicochemical properties, chemical compositions, and biopharmaceutical behaviors of herbal components. Traditional methods are often inefficient, whereas AI enables high-throughput screening and precise pharmacokinetic modeling (Zeng and Jia, 2024). AI frameworks optimize compound screening, refine herbal formulations, and enhance personalized medicine applications. Additionally, AI integrates multi-omics data and provides insights into molecular mechanisms (Li *et al.*, 2022b).

Initially focused on Western medicine, AI progressed from image recognition and disease diagnosis to robotic surgery and personalized medicine (Staub and Sadrameli, 2019; Lee *et al.*, 2023). Network pharmacology advances TCM by supporting tongue and pulse diagnosis, syndrome differentiation, and herbal compatibility analysis (Zhang *et al.*, 2020a; Yue *et al.*, 2021). Algorithms, such as SVM, RF, and DL, improve quality control, bioactive ingredient discovery, and formulation analysis (Xu *et al.*, 2021a). However, large-scale validation and rigorous clinical trials remain essential for the integration of AI into TCM to address standardization and reproducibility challenges (Normile, 2003).

Beyond diagnosis and drug discovery, AI aids in the preservation and interpretation of classical medical texts. DL facilitates digitalization, resolves ambiguities in syndrome differentiation, and improves accessibility. By bridging tradition with modern validation, AI propels TCM toward standardization and evidence-based practices. As AI advances, its role in TCM will expand, driving innovations in herbal medicine, personalized treatments, and holistic healthcare (Li *et al.*, 2022b).

In summary, AI is revolutionizing TCM by enhancing diagnostics, accelerating drug discovery, and standardizing traditional practices in TCM. ML and NLP process vast datasets, refine disease classification and advance pharmacological research. AI-driven quantitative analysis modernizes TCM by addressing the inefficiencies of traditional methods. As AI continues to evolve, its synergy with TCM will drive scientific progress, strengthening the integration of traditional healing with modern medicine.

15.3 Medicinal Plant Identification and Quality Assurance

Accurate identification of medicinal plants is essential for ensuring the efficacy and safety of Chinese Patent Medicines (CPMs). However, quality inconsistencies arise from species variations, geographical origin, harvest timing, and processing methods, all of which affect chemical composition and potency (Bai *et al.*, 2020; He *et al.*, 2021; Chen and He, 2022; Xue *et al.*, 2023). Although effective, traditional identification relies on empirical methods that are subjective, time-consuming, and error-prone (Wang *et al.*, 2020; Zhou *et al.*, 2021a). These challenges necessitate the development of intelligent and high-precision identification methods to improve TCM quality control and standardization (Chu *et al.*, 2022).

AI and DL are transforming medicinal plant identification by analyzing large image datasets, improving recognition accuracy, and reducing the workload (Xu *et al.*, 2021b) (Fig. 15.2). Traditional methods struggle with high-level semantic representations (Zhao *et al.*, 2022), and DL integrates low- and high-level features, enabling precise species and sub-variety identification (Han *et al.*, 2022; Wang *et al.*, 2023a). Given the complex classification of TCM, AI models enhance both accuracy and efficiency (Hao *et al.*, 2022).

Quality control remains a major challenge due to adulteration, heavy metal contamination, and artificial coloring, which compromise safety and efficacy (Tu *et al.*, 2021; Xu *et al.*, 2021b). High-quality raw materials are crucial; however, inconsistent cultivation practices and inadequate seed screening exacerbate this variability (Wakchaure *et al.*, 2023). AI-driven databases integrate species classification, origin tracking, chemical composition analysis, pharmacology, and data processing (Wongchai *et al.*, 2022). By leveraging big data analytics, AI facilitates authenticity verification, origin tracking, and quality prediction, thereby strengthening standardization and quality assurance (Guo *et al.*, 2020a).

AI also optimizes TCM processing, which is a critical factor for product safety and efficacy (Zhou *et al.*, 2024). Processing methods influence chemical composition and therapeutic

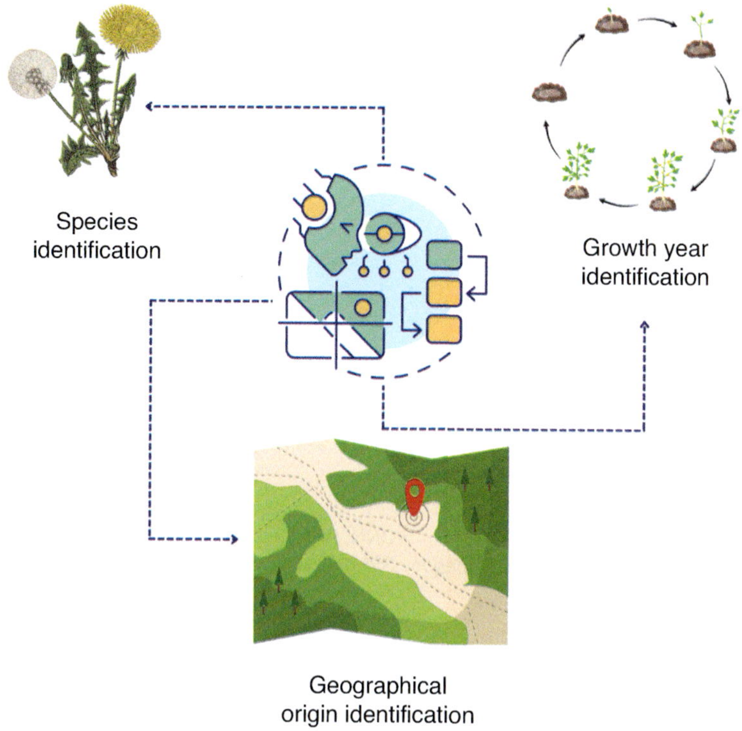

Fig. 15.2. AI has applications in TCM plant material identification, including species identification through morphological and visual analysis, growth year estimation using lifecycle modeling, and geographical origin tracing based on environmental, geospatial, and compound profiling data. (Author's own image.)

properties; however, inefficiencies, poor standardization, and environmental contamination pose challenges (Jia *et al.*, 2020; Li *et al.*, 2020b; Shi *et al.*, 2022; Zhang *et al.*, 2023a; Pan *et al.*, 2024). AI enhances processing conditions, ensures consistency, and preserves therapeutic efficacy.

In summary, AI-driven DL and big data analytics are revolutionizing the identification, classification, and quality control of TCM. Various ML techniques have been applied to identify herb-material sources with high accuracy, enhancing species differentiation, authenticity verification, and processing standardization. Table 15.1 summarizes key applications, including spectral analysis, image-based classification, and chemical fingerprinting. By improving efficiency, accuracy, and consistency, AI ensures the safety and efficacy of TCM in modern healthcare.

15.3.1 Medicinal material sources and species verification

AI is transforming the identification and quality control of medicinal materials in TCM. Technologies such as electronic noses and computer vision extract odor and color features, whereas ML models such as BP and SVM enhance classification accuracy (Wu *et al.*, 2016; Hussein *et al.*, 2020). However, traditional image processing relies on shallow pixel-based features, which makes it vulnerable to environmental variations and limits its reliability (Sun and Qian, 2016). These challenges highlight the need for advanced AI-driven identification methods to address these issues.

DL, particularly CNNs, has significantly improved TCM authentication by autonomously extracting hierarchical features to enhance

Table 15.1. Applications of machine learning in herb-material source identification. (Author's own table.)

Category	Purpose	Data used	AI techniques	Key findings	Reference
Identification					
Herbal Variety Recognition	Classify herbal plants using images	Plant images	CNN	DL models achieved high classification accuracy; ConvNeXt network with Stacked ACMix module enhanced feature extraction.	Miao *et al.*, 2023
Detection of *Atractylodis Rhizome* Adulteration	Identify adulterated samples	HSI and NIR spectral data	PLSR	Image-spectroscopy fusion improved prediction accuracy, which was further enhanced with CARS data.	Jiang *et al.*, 2023
Rapid Detection of Adulterated Saffron	Identify adulterated saffron samples	Plant and staining images	CNN, RF, SVM	CNN outperformed other ML models, achieving 99.5% accuracy, with ResNet18 and DenseNet121 yielding superior results.	Husaini *et al.*, 2022
Detection of Adulterated *Ziziphi Spinosae Semen*	Identify and quantify adulteration levels	FT–NIR spectral data	PLSR, SVM, KNN, ANN	Accuracy improved from 77.06 to 97.58% after applying pre-processing techniques.	Li *et al.*, 2023a
Geographical origin					
Tracking *Gentiana rigescens* origins	Determine the geographical origin of medicinal plant	HPLC–MS, FTIR, 2D correlation spectra	PLS-DA, PLSR	Sequential pre-processing and orthogonalization enhanced accuracy by 12.8%.	Liu *et al.*, 2021a
Identifying *Radix Paeoniae Alba* source	Classify dried plant slices by origin	HSI images	CNN, KNN, LR, SVM, RF, XGBoost	CNN with attention mechanism achieved 93.64–95.27% accuracy, surpassing conventional ML models (72.00%–90.55%).	Cai *et al.*, 2023
Fritillaria thunbergii variety identification	Differentiate plant varieties	HSI images	CNN, PLS-DA, SVM	CNN yielded the highest accuracy (98.88% training, 88.89% test), followed by PLS-DA (92.59%, 81.94%) and SVM (99.65%, 79.17%).	Kabir *et al.*, 2022
Cultivation and processing					

Continued

Table 15.1. Continued

Category	Purpose	Data used	AI techniques	Key findings	Reference
Astragalus membranaceus seed classification	Differentiate Astragalus seeds from similar species	HSI and general images of seeds	SVM, PLS-DA, ANN	ML models accurately classified seeds, with SVM-based models reaching 100% validation accuracy.	Xu *et al.*, 2022
Coix seed storage year classification	Determine storage duration of Coix seeds	HSI images of seeds	ResNet, SVM, KNN, RF, XGBoost	DL models exceeded 85% accuracy, with the CARS-ResNet model improving classification by over 10%.	Bai *et al.*, 2024
Ginger processing stage recognition	Classify different levels of processed ginger	Volatile compound data (GC-ion mobility spectrometry)	SVM, KNN, XGBoost, LR, RF, LDA, QDA	SVM (linear kernel), LR, and QDA models outperformed others, improving accuracy by 2.94–14.61%.	Liu *et al.*, 2023a

See text for additional abbreviations.

accuracy (Kattenborn *et al.*, 2019; Picon *et al.*, 2019). CNNs outperform traditional handcrafted features in plant identification, as demonstrated by deconvolution networks (DN) and multi-scale fusion CNNs, which achieve a high classification accuracy (Lee *et al.*, 2015; Hu *et al.*, 2018). Advanced CNN architectures, such as segmented deep networks, multirate 3D CNNs, and hierarchical models, have further improved plant recognition (Zhu *et al.*, 2018; Kim *et al.*, 2019; Park *et al.*, 2021). Notably, Tan *et al.* (2020) achieved 99.35% classification accuracy for *Zanthoxyli Pericarpium* species using ResNet101, VGG16, DenseNet121, and InceptionV4, with ResNet101 demonstrating superior speed and precision (Tan *et al.*, 2020; Liu *et al.*, 2022a).

AI, combined with DNA barcoding, enhances species verification by analyzing genetic data for precise differentiation (Zhu *et al.*, 2022a). AI-assisted spectral techniques, such as Raman and near-infrared spectroscopy (NIRS), further improve authentication and quality control by analyzing the morphology and chemical composition and preventing adulteration (Yan *et al.*, 2020; Azadnia *et al.*, 2022; Tao *et al.*, 2022; Shi *et al.*, 2024).

Species identification remains challenging when morphologically similar plants share a genus, which complicates their differentiation (Wang *et al.*, 2022; Zhang and Wang, 2023). Traditional visual inspection methods are often inadequate (Li *et al.*, 2016; Fan *et al.*, 2021). AI-driven spectral analysis offers a solution, as demonstrated by CNN-optimized hyperspectral imaging (HSI) in distinguishing *Fritillaria thunbergii* varieties (Kabir *et al.*, 2022) and partial least squares discriminant analysis (PLS-DA)/SVM models that classify cinnamon species with over 90% accuracy (Cruz-Tirado *et al.*, 2023).

Adulteration remains a significant issue as it compromises the therapeutic efficacy of TCM and introduces toxic components (Zhang *et al.*, 2024b). The similarity in taste and appearance between authentic herbs and adulterants makes visual inspection insufficient, thereby increasing error rates (Zhang and Wang, 2023). ML enhances adulteration detection, as seen in *Atractylodis rhizoma* analysis using partial least squares regression (PLSR) with NIRS and HSI (Jiang *et al.*, 2023) and saffron adulteration detection using CNN models combined with Foldscope technology (Husaini *et al.*, 2022). Similarly, SVM-based HSI effectively differentiates *Astragalus membranaceus* from *Melilotus officinalis* (Xu *et al.*, 2022), thereby demonstrating the role of AI in safeguarding TCM authenticity (Amirvaresi *et al.*, 2021; Xing *et al.*, 2021; Zhang *et al.*, 2023b).

Standardized identification is crucial, as many adulterants morphologically resemble genuine herbs or share vernacular names, complicating their differentiation (Zhu *et al.*, 2022a). For example, *Ziziphi Spinosae Semen*, valued for its sedative properties, is often adulterated with *Ziziphi Mauritianae Semen* and *Hovenia Acerba Semen* due to their similar appearance (Ren *et al.*, 2023). To address this, Fourier transform near-infrared (FT–NIR) spectroscopy combined with multivariate algorithms was used to distinguish authentic samples effectively (Li *et al.*, 2023a). Additionally, DL models integrating the ConvNeXt and ACMix networks have achieved high classification accuracy for herbal categorization (Miao *et al.*, 2023).

AI is revolutionizing TCM by improving the identification, processing, and quality control of TCM formulations through DL, spectral analysis, and DNA barcoding. These advancements enhance the precision, efficiency, and automation of medicinal material authentication, strengthening the reliability of TCM. However, challenges in standardization and data integration still persist. Future efforts should focus on refining methodologies and integrating AI with expert knowledge to advance AI-enhanced plant omics and ensure more consistent and accessible TCM practices.

15.3.2 Geographical origin and cultivation assessment

The geographical origin of TCM plants significantly affects their chemical composition, influencing their therapeutic efficacy and quality. For example, *Gentiana rigescens* exhibit compositional variations based on production area (Li *et al.*, 2021), highlighting the need for precise geographic authentication. However, integrating multi-source data for this purpose remains challenging because of the complexity of data-fusion strategies (Cheng *et al.*, 2022). To address this, researchers have employed ML and spectral analysis for accurate origin identification.

Liu *et al.* (2021a) developed a sequential pre-processing through orthogonalization (SPORT)-based PLS-DA model using HPLC, improving *G. rigescens* authentication accuracy by 48.5% over conventional spectral models. Similarly, CNNs applied to spectral images of *Radix Paeoniae Alba* have enabled precise geographic classification (Cai *et al.*, 2023). Wang *et al.* (2023b) combined HSI with an attention-enhanced CNN and LSTM (ACLSTM) model to predict *Coix* seed origin, improving wavelength selection and interpretability. These advancements establish AI-driven spectral analysis as a rapid and non-destructive tool for TCM authentication and standardization.

In addition to origin authentication, AI is transforming TCM plant cultivation. ML optimizes growth conditions by analyzing environmental factors, such as temperature, humidity, and light, thereby enhancing yield quality and sustainability (Wakchaure *et al.*, 2023). The integration of ML with HSI improves quality control by addressing issues such as mixed germplasm and inconsistent seed quality (Wang *et al.*, 2021d). Bai *et al.* (2024) identified the CARS-ResNet model as the most effective for determining the *Coix* seed storage age, advancing seed purity, and cultivation methods (Bai *et al.*, 2024).

An accurate assessment of the growth period is crucial for preserving medicinal efficacy. DL techniques, such as two-dimensional correlation spectroscopy (2DCOS) with attenuated total reflection FT infrared spectroscopy (ATR–FTIR) and residual networks (ResNet), have achieved 100% classification accuracy in determining the growth years of *Paris polyphylla* var. *yunnanensis* (Yue *et al.*, 2021, 2022). These findings highlight the potential of AI in optimizing harvesting periods and refining TCM authentication.

By integrating AI with traditional agricultural knowledge, researchers have enhanced TCM plant cultivation, authentication, and harvesting. These innovations improve the quality, sustainability, and consistency of TCM, ensuring its continued reliability in modern healthcare.

15.4 Processing Stage and Quality Control

Accurate processing is crucial for maintaining the quality and efficacy of TCM, as preparation techniques influence clinical effectiveness by preserving the active compounds and eliminating harmful substances (Zhou *et al.*, 2022a; Tsai *et al.*, 2023). AI-driven technologies, particularly DL and neural networks, are transforming TCM processing by automating and standardizing preparation and ensuring consistency and stability (Xu *et al.*, 2021c).

AI has notably improved the stir-frying stage identification, which is a key step in herbal preparation. Zhang *et al.* (2022) applied DL and transfer learning to classify the *Gardeniae Fructus Praeparatus* (GFP) stir-frying stages using image analysis. Pre-trained models, including GoogLeNet, VGG16, and ResNet34, achieved over 95.82% accuracy, with MobileNetV3 excelling at 98.77% while maintaining efficiency (Zhang *et al.*, 2022). This highlights the potential of DL in optimizing TCM standardization and quality control.

A decoction is essential for maximizing the therapeutic effects and benefits of AI-powered systems that integrate real-time monitoring and computer-controlled equipment, ensuring consistency and large-scale customization (Peng *et al.*, 2022; Wang *et al.*, 2023c). AI also enhances quality control across cultivation, processing, and component analysis by establishing quality markers (Q-markers) and refining the preparation of medicinal plants. However, inconsistent AI methodologies have hindered the comparison of results and data sharing (Zhou *et al.*, 2024).

To enhance TCM processing, researchers have integrated ML with Raman spectroscopy and NIRS for real-time production monitoring (Yan *et al.*, 2020; Tao *et al.*, 2022; Shi *et al.*, 2024). Liu *et al.* (2023a) used ML models with nine key indicator compounds to accurately differentiate ginger processing levels, demonstrating a fast, cost-effective approach.

AI-driven DL models automate herbal slice identification, improving precision and standardization (Xu *et al.*, 2021c). However, AI must be integrated with expert knowledge to refine the methodologies, establish standardized protocols, and ensure high-quality TCM development. As AI advances, its synergy with traditional expertise will further modernize TCM, enhancing therapeutic efficacy and reliability.

15.5 AI-Driven Compound Characterization in TCM

Quantitative analysis is crucial for evaluating the physicochemical properties of TCM, chemical composition, and biopharmaceutical behavior of TCMs, forming the foundation for mechanistic studies and clinical applications. However, traditional methods are inefficient and subjective and lack automation, which slows TCM modernization (Zeng and Jia, 2024). AI integration enhances the efficiency and precision of chemical component identification, offering a transformative solution.

AI-driven high-throughput screening and automated analysis have significantly advanced the characterization of the complex chemical composition of TCM, which is essential for therapeutic efficacy (Fig. 15.3) (McGibbon *et al.*, 2024). ML models, particularly graph neural networks (GNNs), predict compound properties directly from molecular structures, eliminating the reliance on traditional molecular descriptors. This improves the efficiency and accuracy of chemical analysis, accelerating drug discovery and bioactive compound evaluation in TCM (Yang *et al.*, 2021).

15.5.1 Compound identification and classification

AI-driven methodologies have revolutionized the identification of bioactive compounds in TCM, enhancing the precision and efficiency of chemical profiling. As the therapeutic efficacy of TCM depends on diverse plant constituents, accurate identification is crucial for quality evaluation and standardization (Jing *et al.*, 2023). ML optimizes spectroscopic and chromatographic techniques by integrating AI with HSI and ultra-high-performance liquid chromatography quadrupole time-of-flight mass spectrometry (UHPLC–QTOF/MS) for refined chemical analyses (Wan *et al.*, 2023; Xu *et al.*, 2023; Zhang and Wang, 2023).

AI synergizes with analytical methods to improve the detection of bioactive compounds. Zhu *et al.* (2024) applied AI-driven models, such as Savitzky-Golay (SG)–Competitive Adaptive Reweighted Sampling (CARS)–Multiple Linear

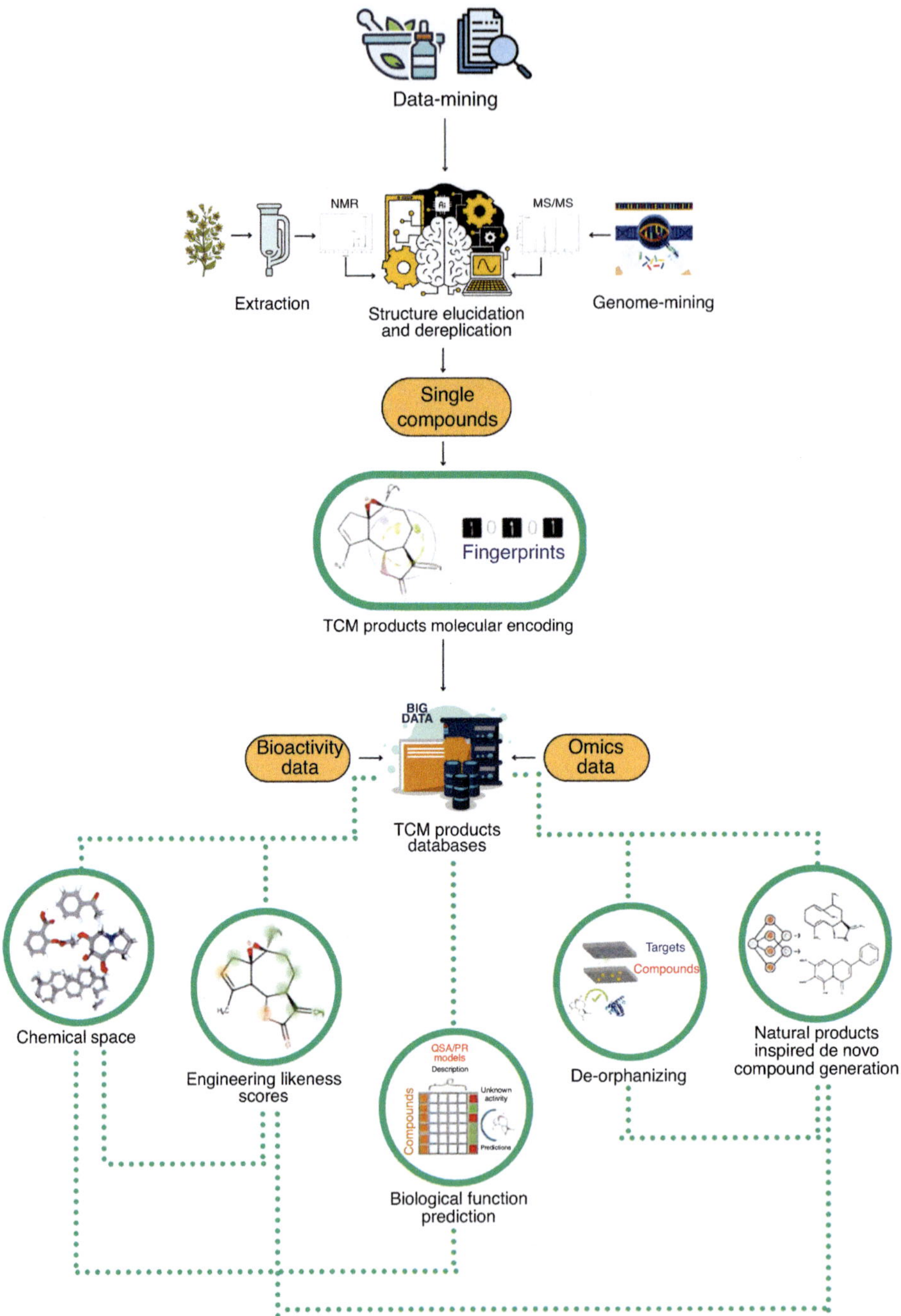

Fig. 15.3. Overview of AI/ML algorithms in the TCM product discovery pipeline. The pipeline consists of two sections: (i) computer-assisted discovery of TCM products, including data mining from traditional medicines and literature, genome mining, and structural elucidation/dereplication; and (ii) ML applications to TCM products, including molecular encoding, molecular descriptors, engineering likeness scores, chemical space mapping, biological function prediction, de-orphanizing, and *de novo* TCM-inspired compound generation. (Author's own image.)

Regression (MLR), to flaxseed nutrition analysis, achieving high-accuracy predictions for protein, oil, linoleic acid, and lignin (Zhu *et al.*, 2024). Similarly, CNNs combined with visible-near-infrared (Vis-NIR) spectroscopy enhanced the analysis of *Gastrodia elata Blume* (Ma *et al.*, 2023). The AI-integrated HSI also improved the evaluation of *Puerariae Thomsonii Radix* content (Hu *et al.*, 2023). Additionally, AI-driven spectral analysis refines complex chemical structure identification (Wang *et al.*, 2021e), while Zuo *et al.* (2021) leveraged AI-assisted LC-MS for advanced herbal component characterization (Zuo *et al.*, 2021).

The integration of AI with analytical chemistry has significantly advanced TCM research. Liu *et al.* (2024) identified 138 compounds in Kangshuai Lao Pian using UPLC/Q--Orbitrap-MS/MS, including phenylethanol glycosides, ginsenosides, iridoids, ionones, and alkaloids (Liu *et al.*, 2024). AI-driven analysis of third-order chemical data, such as fluorescence excitation-emission matrices, further enhances the identification of bioactive compounds (Lu *et al.*, 2020). These developments demonstrate the ability of AI to process multi-dimensional chemical datasets, accelerating TCM standardization and bridging traditional herbal knowledge with computational techniques.

TCM classification, which is essential for understanding therapeutic applications, is often hindered by the overlapping bioactive properties of the multi-component formulations. AI, particularly ML and DL, improves classification accuracy by systematically categorizing medicinal herbs and their active compounds. AI-driven chemical fingerprint analysis distinguishes structurally similar compounds, identifies unique chemical markers, and enhances classification reliability (Bui *et al.*, 2023). ML models trained on physical and chemical properties provided more profound insights into the therapeutic potential of herbs (Yang *et al.*, 2019b). At the same time, DL and data mining uncovered intricate patterns beyond human detection, improving predictive capabilities in herbal classification.

Network pharmacology, an AI-driven approach, has revolutionized the classification of TCM by elucidating the multi-target and multi-pathway actions of herbal formulations. AI-assisted analysis has revealed the synergistic effects of herbal compounds, thereby enhancing our understanding of their therapeutic mechanisms (Bi *et al.*, 2021; Chai *et al.*, 2023). AI also assesses herb–herb interactions at the molecular level, offering insights into collective therapeutic effects, particularly for complex diseases (Ren *et al.*, 2019; Zhang *et al.*, 2020b). As AI evolves, its role in TCM research extends beyond chemical profiling to uncover complex herbal interactions, optimize therapeutic applications, and drive novel drug discovery (Chung *et al.*, 2024).

15.5.2 Chemical interactions and mechanism of action

Understanding compound–target interactions is essential for elucidating therapeutic mechanisms of TCM. AI- and network-based approaches have significantly improved the quantification of these complex interactions. Wang *et al.* (2021b) constructed a protein–protein interaction (PPI) network of 349,197 herb pairs, showing that frequently used pairs have shorter network distances, suggesting a more significant influence on adjacent proteins (Wang *et al.*, 2021f). Their findings highlight the central role of key components in herbal interactions, providing a data-driven foundation for optimizing TCM formulations.

AI-driven computational models further enhance the identification of synergistic herbal relationships through network topology analysis, refining our understanding of TCM mechanisms. These advancements provide a systematic framework for optimizing herbal combinations and integrating AI into herbal medicine research and modernization (Zeng and Jia, 2024).

AI-powered omics technologies have revolutionized the identification of bioactive compounds and the elucidation of their mechanisms in TCM. AI models rapidly analyze natural product databases to identify therapeutic molecules and accelerate drug development (Mullowney *et al.*, 2023). Computational frameworks can predict molecular interactions with target proteins, thereby providing deeper insights into the efficacy of TCM formulations (Mastropietro *et al.*, 2023).

SAR models identify pharmacologically active components, whereas QSAR models predict compound toxicity, aiding pharmacotoxicological assessments (He *et al.*, 2022a). These AI-driven models improve bioactive compound screening efficiency, reducing time and cost compared to traditional methods.

Zhang *et al.* (2019) developed a computational pipeline for *de novo* protein target identification in TCM by integrating pharmacophore comparison with reverse ligand–protein docking. Their method identified known targets and uncovered novel molecular structures, aiding in mechanism elucidation and bioactive compound discovery (Zhang *et al.*, 2019). These advancements reinforce the role of AI in bridging traditional herbal medicine with modern pharmaceutical research, driving data-driven innovation in TCM.

15.5.3 Pharmacological activity and functional analysis

AI and systems pharmacology have revolutionized TCM research by analyzing the gene expression profiles induced by herbal compounds and elucidating their molecular mechanisms. Lee *et al.* (2022) demonstrated AI-driven systems pharmacology's ability to integrate complex biological data, enhancing the understanding of the multi-target efficacy of TCM in treating complex diseases (Lee *et al.*, 2022). By constructing interaction networks of herbal compounds, AI improves drug–gene interaction predictions, thereby modernizing and optimizing herbal treatments.

AI has also advanced SAR analysis by establishing links between chemical structures and biological activities, thereby refining TCM pharmacology (Gao *et al.*, 2020). Predictive modeling and data mining have uncovered novel associations between chemical components and pharmacological effects, accelerating drug discovery and improving insights into the therapeutic potential of TCM (Gong *et al.*, 2019). ML and DL enhance chemical data processing through pre-processing, feature extraction, and predictive modeling, thereby increasing the efficiency and accuracy of pharmacological research (Lin *et al.*, 2019).

AI-driven clustering identifies structural similarities, predicts biological activity, and identifies drug candidates (Chen *et al.*, 2019). DL, particularly CNNs, advances chemical structure analysis by detecting intricate patterns and improving drug discovery efficiency (Tripathi *et al.*, 2021; Raza *et al.*, 2023). NLP models, including GPT, extract chemical and pharmacological data from TCM literature and enhance compound analysis (Walters and Barzilay, 2021).

AI-powered functional analysis facilitates biomarker identification, therapeutic target discovery, and pathway mapping in TCM. AI-integrated omics technologies, including genomics, transcriptomics, and metabolomics, enhance the molecular-level interpretation of bioactive compounds, promoting a systematic approach to studying synergistic herbal effects. Herbal medicine databases preserve key data on medicinal materials, chemical components, and therapeutic efficacies. However, their complexity necessitates AI-driven data extraction for meaningful insights (Yang *et al.*, 2019b; Fang *et al.*, 2021).

Despite these advancements, AI-driven herbal analyses have faced challenges in terms of data quality, model accuracy, and interpretability (Mater and Coote, 2019). Overcoming these limitations requires improved AI models, enhanced data processing, and interdisciplinary collaboration between AI researchers and TCM experts (Lipinski *et al.*, 2019; Isert *et al.*, 2023).

As AI evolves, its role in TCM research will expand, bridging the gap between traditional and modern medicine. AI-driven analysis unlocks ancient medicinal wisdom using contemporary science, driving innovations in global healthcare. Advancements in AI algorithms and computational platforms will accelerate drug discovery and optimize TCM formulations, ushering in a new era of precision herbal medicines.

15.6 AI in TCM Drug Discovery and Development

TCM drug discovery is complex and resource intensive, requiring the identification of bioactive compounds in a vast chemical space. Approximately 90% of natural compounds fail

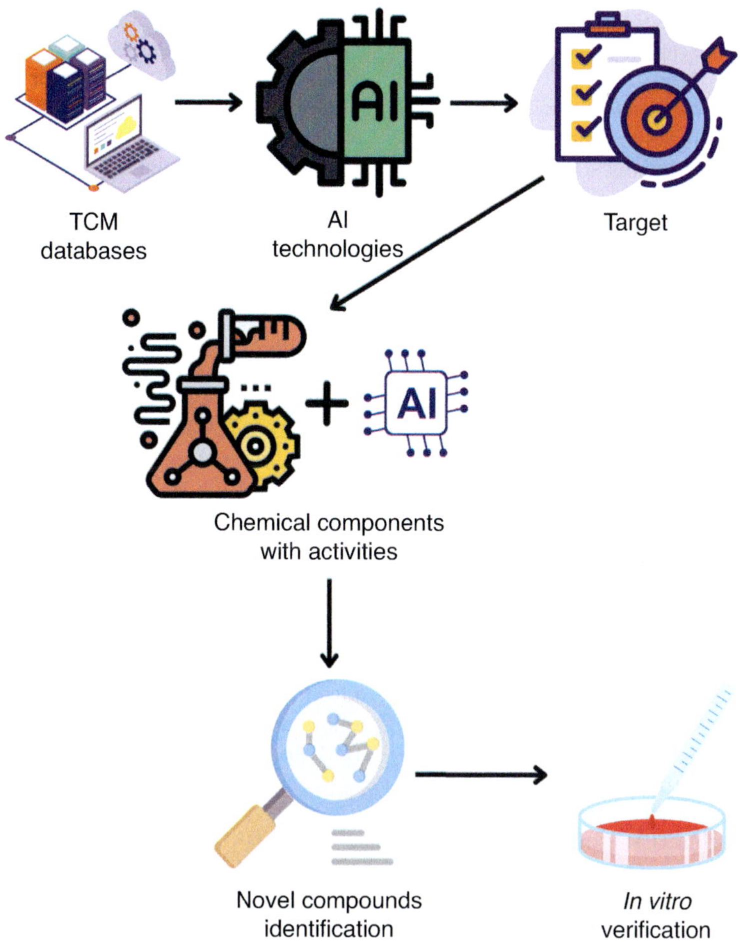

Fig. 15.4. AI-driven TCM drug discovery workflow. The process involves utilizing AI technologies to perform virtual screening of TCM databases to identify targets and active chemical components. This is followed by novel compound identification and *in vitro* validation to confirm biological activity. (Author's own image.)

Phase II trials, primarily because of unclear mechanisms and multi-target interactions (Matias *et al.*, 2021). Traditionally, discovery has relied on empirical prescriptions and human expertise, leading to inefficiency and high resource consumption (Zhou and Wang, 2014; Xu *et al.*, 2019).

AI and ML have transformed TCM drug discovery by improving lead compound identification, ADMET prediction, and small-molecule screening techniques (Fig. 15.4). Since the 1990s, ML has enhanced early-stage drug development, whereas DL has revolutionized the field by extracting biological patterns from large datasets (Gupta *et al.*, 2021). AI-driven multimodal multi-omics analysis enables the precise identification of drug targets, active compounds, synergistic effects, and regulatory mechanisms in complex TCM formulations (Li *et al.*, 2023b).

TCM evaluation and quality control remain challenging because traditional methods focus on single herbs and lack a universal framework for multi-component formulations. AI-assisted quality control, which leverages big data and network analysis, enhances the consistency and efficacy of TCM preparations (Song *et al.*, 2024).

However, these techniques require further refinement to develop standardized AI-based quality control methodologies.

The regulatory guidelines for botanical drugs differ from those for synthetic pharmaceuticals. The US FDA's Botanical Drug Development Guidance for Industry provides flexibility in the initial clinical trial requirements (IND stage) owing to historical use. However, it mandates stringent safety and efficacy standards for new drug applications (Guo *et al.*, 2020b). AI-driven approaches facilitate regulatory compliance by analyzing clinical data, identifying key TCM components, and optimizing the formulations.

With the increasing demand for high-quality TCM, AI, big data, and network pharmacology are crucial for modernizing drug research and development. AI-powered biological networks enhance phenotype–drug interaction analysis, precise screening, and pharmaceutical innovation. Table 15.2 summarizes the key applications of ML in understanding TCM mechanisms across three major research areas: TCM active ingredient mechanisms, pharmacological effects of TCM, and TCM compatibility studies. These approaches help uncover bioactive compounds, elucidate their mechanisms of action, predict therapeutic effects, and analyze complex interactions within TCM formulations at molecular, cellular, and systemic levels. Establishing a standardized AI framework for TCM quality control and drug discovery will ensure the safety and efficacy of herbal medicines and promote data-driven advancement in the field.

15.6.1 Target identification and validation

Big data and AI have revolutionized bioactive compound discovery in TCM, particularly in the identification of targets. ML mines biomedical databases to identify and validate disease-associated target proteins, accelerate therapeutic target discovery, and optimize drug screening (Liu *et al.*, 2022b).

AI-driven models have demonstrated a strong potential for TCM target identification. Liu *et al.* developed a multitarget-based polypharmacology prediction model (mTPP) using support vector regression (SVR), multi-layer perceptron (MLP), gradient boost regression (GBR), and decision tree regressor (DTR). This model identified 20 candidate compounds for drug-induced liver injury (DILI), with two experimentally validated compounds, highlighting their efficacy in linking multi-target effects to drug outcomes (Liu *et al.*, 2022b). Similarly, He *et al.* (2020) applied a QSAR model integrating multiple ML algorithms to screen TCM compounds for binding to Alzheimer's disease-related proteins (He *et al.*, 2020).

AI-powered platforms extend target discovery beyond traditional ML models. Pun *et al.* (2022) used PandaOmics, an AI-driven platform, to analyze CNS expression profiles and iPSC-derived motor neurons (diMNs) from Answer ALS and identified 28 potential amyotrophic lateral sclerosis (ALS) targets, including eight novel genes (Pun *et al.*, 2022).

By integrating AI-driven methodologies with the multi-target approach of TCM, researchers have enhanced the precision, efficiency, and scalability of pharmacological research. These advancements accelerate target discovery and refine drug screening, thereby modernizing TCM-based drug development.

15.6.2 Virtual screening and ADMET prediction

AI-driven virtual screening has significantly advanced bioactive compound discovery in TCM, accelerating drug identification while reducing costs and resource consumption. Virtual screening employs molecular docking to predict ligand–target interactions, thereby enhancing lead compound selection through receptor- and ligand-based screening (Yang *et al.*, 2019c; Yang *et al.*, 2020a). The ability of AI to identify multi-target compounds aligns with the holistic approach of TCM, facilitating the rapid discovery of novel compounds and their mechanisms of action (Thomford *et al.*, 2018; He *et al.*, 2020).

AI-driven virtual screening is instrumental in antiviral research, osteoporosis treatment, and oncology. During the 2009 H1N1 pandemic, Chang *et al.* (2011) employed computational models to screen TCM compounds targeting H1 and N1 proteins, expediting antiviral drug discovery. AI-based models have also identified

Table 15.2. Applications of machine learning in analyzing mechanisms of Traditional Chinese Medicine (TCM) at various levels. (Author's own table.)

Research area	Objective	Dataset used	AI techniques	Key findings	Reference
TCM active ingredient mechanisms	Identify platelet-derived growth factor receptor α (PDGFRA) inhibitors enhancing radioiodine sensitivity in thyroid cancer	PDGFRA inhibitors dataset	RF, SVM	Identified 4',5,7-trimethoxyflavone as a potential inhibitor	Yu *et al*., 2022
	Predict the hepatoprotective potential of TCM compounds	Hepato-protective TCM dataset	RF, Naïve Bayes, AdaBoost, K-star	Predicted 12 hepatoprotective TCMs, 9 validated by literature	He *et al*., 2022a
	Assess potential hepatoprotective ingredients in TCM	Binding scores from molecular docking	SVR, GBR, decision tree regressor, ANN	Chelerythrine and Biochanin A predicted to improve APAP-injured cells	Liu *et al*., 2022b
	Screen feature genes in SDECGs	Gene expression data related to Tongmai Jiangtang treatment	RF, SVM, Generalized linear, XGBoost	Identified nine feature genes ranked by importance	(Liu *et al*., 2023b)
Pharmacological effects of TCM	Investigate herb-disease correlation for treatment discovery	A herb-disease network including compounds and targets	CNN	Explored applications of Chuan Xiong and *Astragalus membranaceus*	Hu *et al*., 2024
	Analyze licorice pharmacological mechanisms in COVID-19 treatment	SMILES strings, amino acid sequences, and binding affinity values	Graph convolutional network (GCN), CNN	Licorice is proposed as COVID-19 treatment via PTEN, MAP3K8, Mpro, etc.	Fu *et al*., 2023
	Explore bioactive compounds and targets in coronary heart disease (CHD) formulas	Network clustering of bioactive compounds and CHD targets	Hierarchical clustering, random walk with restart	Developed ranking algorithm integrating network topology and function	Yang *et al*., 2020b

Continued

Table 15.2. Continued

Research area	Objective	Dataset used	AI techniques	Key findings	Reference
TCM compatibility studies	Evaluate the safety of TCM prescriptions recommended for COVID-19	Dataset of TCM prescription side effects	ANN	Identified 10 safety indicators (values > 0.8) in 38 prescriptions	Wang *et al*. 2021h
	Develop an herbal recommendation system using TCM knowledge	Publicly available TCM prescriptions	GNN, GCN, LSTM	Developed MGAT method for herb recommendations	Jin *et al*., 2023
	Predict tonic formulae based on herbal properties	Classic tonic formulae from ancient TCM books	GCN, SVM, KNN, LR, Naïve Bayes	HPE-GCN model outperformed others in precision and recall	Liu *et al*., 2022c

See text for additional abbreviations.

potent anti-osteoporosis compounds using self-attentive neural networks (SANs) (Liu *et al.*, 2021b). In oncology, Pan *et al.* (2019) applied virtual screening for cancer target identification, whereas DrugCell, a DL model, mapped drug responses in human cancer cells, offering insights into TCM-based cancer treatments (Pan *et al.*, 2019; Kuenzi *et al.*, 2020).

Beyond infectious diseases and cancer, AI-assisted virtual screening has contributed to neurodegenerative, cardiovascular, and metabolic research. DL and RF models have predicted optimal TCM formulas for Alzheimer's disease with high accuracy (Chen and He, 2022), while AI-driven similarity network fusion identified intersecting omics biomarkers for prostate cancer (Wang *et al.*, 2021g). AI further refines drug screening by identifying hit compounds and optimizing them into lead candidates with a more substantial therapeutic potential (Atanasov *et al.*, 2021; Jiménez-Luna *et al.*, 2021).

DL extends virtual screening beyond traditional molecular docking by analyzing large compound libraries and predicting their pharmacological effects. Tang *et al.* (2023) used Deep Docking to screen over three million compounds and identified two novel adenosine receptor antagonists validated *in vitro* (Tang *et al.*, 2023). Similarly, Lin *et al.* (2022) applied structure-based virtual screening to discover three renal glutaminase inhibitors confirmed via enzymatic assays (Lin *et al.*, 2022). These studies highlight the ability of AI to accelerate the discovery of TCM bioactive compounds and precision medicine (Amendola and Cosconati, 2021; Zhang *et al.*, 2021).

Following the virtual screening, rigorous ADMET prediction is crucial to assess drug viability. AI-driven tools, such as SwissADME, efficiently predict pharmacokinetics and drug-likeness, improve lead optimization, and reduce failure rates (Daina *et al.*, 2017; Ajala *et al.*, 2023). However, AI-based ADMET modeling faces challenges in TCM owing to multi-component formulations, unpredictable interactions, and lack of standardized databases, limiting predictive accuracy and clinical translation (Guo *et al.*, 2020b; Mu *et al.*, 2023).

Enhancing AI-driven ADMET predictions requires improving the quality and standardization of TCM databases. Establishing a unified classification system and implementing stringent quality controls will provide deeper insights into the molecular mechanisms of TCM (Wang *et al.*, 2023d). While AI optimizes *in silico* screening, experimental validation remains essential for confirming the efficacy and safety of the compounds. The integration of AI with high-throughput experimental techniques and improved data curation will create more reliable TCM drug discovery pipelines in the future. As AI continues to evolve, its synergy with plant omics is expected to refine TCM-based drug discovery, driving data-driven advancements in modern medicine.

15.6.3 Network pharmacology and multi-target strategies

TCM follows a holistic, multi-target approach, making its pharmacological mechanisms complex and challenging to elucidate. AI-integrated network pharmacology leverages large-scale biomedical data, computational modeling, and biological network analysis to decode these complex interactions. AI-driven methodologies enable rapid identification of bioactive compounds, target prediction, and pharmacological pathway analysis with unprecedented accuracy (Guo *et al.*, 2019). DL and association rule mining (ARM) models have validated the efficacy of *Atractylodis Macrocephalae Rhizoma* and *Astragali Radix* for gastric cancer through *in vivo* and *in vitro* studies (Xu *et al.*, 2021a). Similarly, unsupervised ML models have identified functional similarities between compounds, aiding in the understanding of TCM mechanisms (Guo *et al.*, 2020a).

AI-driven network pharmacology and molecular docking provide insights into the complex TCM formulations. The Shenyan Kangfu tablet (SYKFT) was systematically analyzed through computational prediction and experimental validation, strengthening its clinical basis (Kim *et al.*, 2021). Bayesian ML models have identified neuroprotective compounds in Xiaoxuming Decoction, demonstrating the role of AI in elucidating TCM mechanisms (Yang *et al.*, 2019c). AI also enhances safety evaluations, as observed in predictive models for herb-induced liver injury (HILI), which assesses hepatotoxicity risks (Wu *et al.*, 2019).

By integrating AI with network pharmacology, researchers have accelerated compound discovery and optimization. AI-driven screening has identified bioactive compounds such as ginsenoside cepharanthine, Rg3, and β-elemene, which show significant therapeutic potential (Tao *et al.*, 2014; Chen *et al.*, 2021b; Zhu *et al.*, 2022b). Multi-view DL models predict activation and inhibition mechanisms, uncovering novel compound-target interactions (Zhang *et al.*, 2023c). ML and network pharmacology have also optimized herbal formulations, such as by identifying 11 key compounds in *Cocos Poria* and the active ingredients in Wuwei Musk Pill for rheumatoid arthritis (Li *et al.*, 2022c; He *et al.*, 2022b).

AI-driven virtual bioequivalence (VBE) prediction modernizes TCM by identifying bioequivalent substitutes for rare or costly ingredients while preserving their efficacy. AI models assess molecular structures, chemical compositions, and bioactive targets to identify viable alternatives, as demonstrated by studies on *Scutellariae Radix* (Liang *et al.*, 2019) and the Da-Cheng-Qi Decoction (Li *et al.*, 2020a).

AI can also predict drug interactions, synergies, and incompatibilities in multi-component TCM formulations. DL models, such as DeepSynergy, integrate chemical descriptors and genomic data to predict synergistic drug interactions and enhance formulation optimization (Preuer *et al.*, 2018). Tools such as SynergyFinder 3.0 use network-based modeling to quantify multi-drug synergies, aiding clinical translation (Ianevski *et al.*, 2022). However, predicting drug–drug interactions (DDIs) remains challenging because of the holistic nature of TCM formulations. AI models such as XGBoost and artificial neural networks (ANNs) improve DDI prediction accuracy and refine drug compatibility and safety assessments (Romm and Tsigelny, 2020; Ke *et al.*, 2024).

AI-based network pharmacology has demonstrated high accuracy in predicting herb-disease associations. Hu *et al.* (2024) developed a dual-channel hypergraph convolutional network to forecast TCM indications, particularly for *Chuanxiong Rhizome* and *Astragalus membranaceus*. Similarly, DL-based pharmacology has identified bioactive compounds in licorice for COVID-19 treatment (Fu *et al.*, 2023). Knowledge graph embedding methods, such as HTINet2, have predicted herbal targets for *A. annua* and *Coptis chinensis*, thereby advancing personalized medicine research (Duan *et al.*, 2024).

Despite these advancements, challenges remain, including the need for high-quality multiherb interaction datasets and improved AI model interpretability. Expanding standardized databases, integrating DL with network pharmacology, and enhancing model transparency will improve predictive accuracy and facilitate regulatory compliance (Ke *et al.*, 2024).

AI is revolutionizing TCM by elucidating its mechanisms of action, discovering bioactive compounds, and optimizing multi-herb formulations. By integrating ML, DL, and network pharmacology, AI enhances ingredient screening, pharmacological analysis, and compatibility prediction. As AI advances, its integration with multi-omics analyses will further bridge traditional herbal medicine with modern pharmaceutical sciences, paving the way for precision medicine and personalized herbal therapies.

15.6.4 Safety, toxicity, and regulatory considerations

Ensuring the safety of TCM formulations is critical because multi-component prescriptions can lead to adverse interactions and toxicities. AI-driven models that integrate ML, DL, molecular docking, and network pharmacology have become powerful tools for predicting toxicity, adverse drug reactions (ADRs), and herb–drug interactions. These models enhance safety assessments, optimize formulations, and mitigate risks in the development of TCM drugs.

AI-based toxicity prediction analyzes chemical structures, biological pathways, and clinical data to identify potential toxic effects before clinical trials are conducted (Tran *et al.*, 2023). For example, rhubarb (Dahuang) exhibits hepatoprotective and hepatotoxic properties, which are mitigated by glycyrrhizin in licorice via the Nrf2 pathway (Ni *et al.*, 2022). AI models can predict these interactions, thereby improving formulation safety.

Advanced ML models such as DeepHit, FP-ADMET, and ResNet18DNN predict DILI and other toxicities by assessing endpoints such as hERG inhibition, LD50, Ames mutagenesis, and carcinogenesis (Tran *et al.*, 2023). AI also enhances cross-system safety evaluations. Chen *et al.* (2022) developed an ML model integrating SVMs, neural networks, decision trees, and RF to detect hepatotoxic compounds in both TCM and Western medicine (Chen *et al.*, 2022).

AI can be used to predict drug–food interactions (DFIs), which can alter the pharmacokinetics and efficacy of TCM formulations. Kha *et al.* (2023) applied XGBoost- and MLP-based classifiers trained on DrugBank datasets to predict DFIs, and ensemble learning techniques improved accuracy (Kha *et al.*, 2023). Regular updates with new toxicology data ensure AI model refinement and reliability.

In addition to toxicity prediction, AI optimizes formulations to minimize risks. Lowering the herb dosage while maintaining efficacy through strategic combinations reduces toxicity (Xie *et al.*, 2021). Liu *et al.*, (2021b) developed an AI framework to assess toxicity risks in TCM formulations for COVID-19 treatment, ensuring safer prescriptions (Liu *et al.*, 2021b).

AI also aids in the screening of safer bioactive compounds. He *et al.* (2020) used molecular docking and DL to identify multi-target inhibitors of Alzheimer's disease, integrating ligand-based feature analysis with 3D-QSAR modeling (He *et al.*, 2020). Similarly, Li *et al.* (2019) combined network pharmacology, ML, and molecular dynamics simulations to identify promising lead compounds (Li *et al.*, 2019). Gong *et al.* (2021) applied ML regression models, including SVMs and RF, to predict DPP-4-targeting compounds for diabetes and experimentally validated the role of AI in rational drug design (Gong *et al.*, 2021).

AI has advanced pharmacokinetics and metabolite identification in TCM drug safety. Feng *et al.* (2019) used UPLC–MS/MS to detect active metabolites in biological samples and refine pharmacodynamic predictions (Feng *et al.*, 2019). Reverse target identification and predictive modeling further improve safety assessments (Hassanzadeh *et al.*, 2019).

Despite the transformative impact of AI, challenges remain, including data quality, model interpretability, and interdisciplinary collaboration. Standardizing TCM data and aligning AI models with global pharmacovigilance frameworks will enhance their predictive power and regulatory acceptance.

AI is revolutionizing TCM drug safety by predicting toxicological risks, optimizing formulations, and refining pharmacokinetic profiles. By integrating ML, DL, molecular docking, and network pharmacology, AI reduces toxicity risks and improves clinical success rates. As AI continues to evolve, its role in toxicity modeling and regulatory compliance will be crucial for modernizing TCM and ensuring its global integration into contemporary healthcare.

15.7 Challenges and Outlook

The integration of AI with TCM has led to significant advancements in quality control, prescription optimization, and drug discovery. However, to fully harness the potential of AI, it is essential to address challenges such as data limitations, model interpretability, collaboration gaps, and regulatory constraints (Ke *et al.*, 2024). Overcoming these obstacles requires strategic interdisciplinary solutions to ensure that AI-driven TCM research is reliable and effective.

A major challenge is the lack of high-quality, standardized datasets for plant identification, compound profiling, and pharmacological validation. Many TCM databases are incomplete or poorly structured, limiting the performance of AI models. Additionally, plant authentication is complicated by morphological similarities, adulteration, and environmental variations, which reduce the accuracy of AI-based identification and raise concerns regarding reproducibility (Zhou *et al.*, 2020; Xu *et al.*, 2021c).

AI model interpretability remains a critical issue, particularly in complex biological systems. Many DL algorithms function as "black-box" models, making their predictions challenging to explain. This lack of transparency hinders clinical adoption and regulatory approval because evidence-based validation is essential for acceptance. Biases in the training datasets further affect model generalizability, limiting their ability to adapt to real-world variations in plant species, compounds, and patient responses (Vo *et al*., 2022; Nguyen *et al*., 2023).

Regulatory and ethical considerations further impact the adoption of AI in TCM. AI-driven drug discovery and herbal research must comply with global safety standards, intellectual property regulations, and ethical guidelines for data use. Collaboration among AI researchers, pharmacologists, botanists, and regulators is essential for developing frameworks that balance innovation with public health protection.

Despite these challenges, AI has the potential to transform TCM by enhancing research precision, improving quality control, and advancing personalized herbal medicines. Addressing data standardization, interpretability, and regulatory alignment is crucial for integrating AI into modern TCM practice, ensuring safe and effective applications in global healthcare.

15.7.1 Data challenges and standardization

The integration of AI into TCM is hindered by data scarcity, quality issues, and the complexity of TCM practices. Despite rich empirical knowledge of TCM, the lack of structured and standardized datasets limits the effectiveness of AI. Addressing these challenges is crucial for advancing AI-driven TCM research and its applications.

A major obstacle is the scarcity of data. Traditional medicinal records are often unstructured, scattered, or undocumented, complicating the creation of comprehensive datasets. Clinical studies on TCM are constrained by small sample sizes, inconsistent standards, and limited digitization, which reduce the reliability of AI models. Efforts to enhance herb cultivation and quality control in processing have led to more structured medicinal plant databases; however, significant gaps still remain (Ke *et al*., 2024).

Even with the available data, ensuring accuracy and reliability is challenging. TCM formulations involve multiple herbs and complex interactions, leading to variability in the datasets. Incomplete or biased data can undermine AI predictions, thereby highlighting the need for validated pharmacological and clinical databases to enhance AI applications (Xu *et al*., 2021c).

The lack of standardization in data collection and classification further impedes AI integration. Unlike modern medicine, TCM diagnostic methods and treatment principles vary widely, resulting in heterogeneous datasets that reduce the effectiveness of AI models. Standardized data protocols are essential to improve the reliability of AI-driven TCM applications (Zhou *et al*., 2020).

Beyond technical challenges, the holistic approach of TCM adds complexity to AI models. Treatments aim to restore balance rather than target single diseases, requiring AI to analyze multi-component interactions, herbal synergies, and combined therapeutic effects (Wang *et al*., 2021b; Zhao *et al*., 2023). The "black-box" nature of AI models complicates the interpretation of multi-dimensional principles of TCM, necessitating interdisciplinary efforts to refine AI-driven insights.

Addressing data scarcity, quality, standardization, and complexity is key to maximizing the role of AI in TCM modernization. Improving data collection, ensuring accuracy, standardizing databases, and developing AI models that accommodate the holistic nature of TCM will unlock the potential of AI in drug discovery, herbal formulation optimization, and global TCM research.

15.7.2 Interdisciplinary collaboration

The integration of AI with TCM requires collaboration across multiple disciplines. A key challenge is the shortage of professionals proficient in both AI techniques and TCM principles. Bridging this expertise gap is essential for advancing AI-driven TCM research, necessitating cooperation among TCM practitioners, chemists, biologists, statisticians, and AI specialists (Ke *et al*., 2024).

In TCM quality control and pharmacological research, advanced separation techniques such as UPLC-Q/TOF and UHPLC-Q-TOF/MS, combined with AI-driven data analysis, enhance precision and reliability. Collaboration between analytical chemists and AI experts can improve data interpretation, model development, and validation, thereby ensuring accurate and reproducible results (Ke *et al.*, 2024).

Strengthening interdisciplinary partnerships will accelerate the application of AI in herbal identification, quality control, drug discovery, and clinical decision support. By fostering synergy between traditional expertise and computational advancements, collaboration is key to modernizing TCM and facilitating its global acceptance in the future.

15.7.3 Model interpretability and trustworthiness

DL models often function as opaque "black-box" systems, which makes their decision-making processes challenging to interpret (Vo *et al.*, 2022; Nguyen *et al.*, 2023). Although highly effective, this lack of transparency complicates trust and regulatory acceptance of TCM applications. Developing explainable AI models is essential to improve confidence in AI-driven insights (Ke *et al.*, 2024).

To enhance interpretability, researchers have integrated knowledge graphs and attention mechanisms. The structure of a knowledge graph is domain-specific data, improving contextual understanding, while attention mechanisms highlight key features, clarifying AI-driven decisions. These approaches have increased the transparency, credibility, and usability of TCM research (Ke *et al.*, 2024).

DL has significantly advanced TCM identification by offering more objective and efficient methods. Integrating traditional knowledge with modern AI-driven solutions will further strengthen trust in AI applications, supporting the modernization and global acceptance of TCM.

15.8 Perspectives and Future Prospects

AI is transforming TCM by improving diagnostic precision, personalizing treatments, and elucidating pharmacological mechanisms (Song *et al.*, 2024). ML and large language models (LLMs) enhance data analysis and accelerate the processing of large datasets (Shi *et al.*, 2024). AI-driven image recognition, spectral analysis, and genomic sequencing refine medicinal plant identification, reducing misclassification and ensuring herbal authenticity. These innovations are advancing TCM toward scientific validation and global acceptance.

AI plays a key role in optimizing herb combinations and treatment strategies (Dou *et al.*, 2021; Gan *et al.*, 2023). Network pharmacology and AI algorithms map the interactions between herbal compounds and therapeutic pathways, validating TCM formulations (Bi *et al.*, 2021; He *et al.*, 2022b). AI-driven systems biology approaches have revealed multi-target mechanisms that support precision medicine. The integration of multi-omics data with AI facilitates the discovery of bioactive compounds, leading to safer and more effective therapeutics. Explainable AI (XAI) enhances transparency by addressing the "black-box" issue and improving trust among researchers and regulators.

Future AI-driven TCM research should focus on developing models that capture the holistic nature of TCM (Cheng *et al.*, 2021; Zhou *et al.*, 2022b). AI will also bridge TCM with Western medicine, promoting an integrative healthcare approach (Wang *et al.*, 2021a; Xiong *et al.*, 2021). Expanding AI applications in preventive medicine, health monitoring, and patient education will further enhance personalized care. Additionally, open-access AI-compatible TCM databases can improve the reliability and reproducibility of ML.

Collaboration among academia, industry, and regulatory bodies is crucial for standardizing methodologies, establishing regulations, and integrating TCM with modern pharmaceuticals. As AI evolves, it will modernize TCM while preserving its fundamental principles. Future research should align AI models with the holistic approach of TCM, ensuring its evolution alongside modern science (Lai *et al.*, 2020; Zhou *et al.*, 2021b). Addressing the current challenges and leveraging AI advancements will enhance the efficacy, precision, and accessibility of AI, revolutionizing drug discovery, healthcare, and personalized medicine.

15.9 Conclusion

The integration of AI and ML into TCM marks a significant step toward modernization. AI enhances plant authentication, bioactive compound discovery, drug development, and quality control. Through data mining and computational modeling, literature analysis, drug synergy predictions, and efficacy assessments can be facilitated while optimizing resource management and cloud-based platforms for TCM applications.

Various AI algorithms, including SVM, RF, and Bayesian networks, have been applied to quality evaluation, pharmacological studies, and virtual screening. DL models, such as CNNs and ANNs, further improve predictive accuracy and offer deeper insights into the therapeutic mechanisms of TCM. However, the complexity of these models, often seen as "black-box" systems, limits their interpretability. Explainable AI techniques, such as surrogate models and local interpretable model-agnostic explanations (LIME), provide potential solutions to this problem.

AI has revolutionized pharmacological research by decoding the intricate interactions between TCM formulations. AI accelerates drug discovery through applications in network pharmacology, virtual screening, and toxicity assessments, while component-based structural theories and *in vitro–in vivo* correlation (IVIVC) models enhance the understanding of the pharmacological foundation of TCM. However, despite these advancements, significant challenges remain, including the lack of standardized TCM databases, inconsistencies in pharmacological evaluations, and regulatory constraints. These gaps hinder the seamless integration of AI into TCM research and drug development. Addressing these issues requires interdisciplinary collaboration, rigorous AI model validation, and standardized data collection methods to ensure reliability, reproducibility, and compliance with regulations.

Future advancements in AI, combined with experimental techniques such as HSI and sensor-based quality detection, will refine TCM research and improve quality control. Aligning AI methodologies with international standards is essential for enhancing credibility and facilitating the integration of TCM into modern healthcare. Although challenges in data standardization and model reliability persist, AI offers unprecedented opportunities to bridge traditional knowledge with computational intelligence. By refining AI-driven methodologies to align with the holistic principles of TCM, this field can evolve into a more evidence-based and globally recognized discipline. Ultimately, AI strengthens the scientific foundation of TCM, expanding its role in precision and integrative healthcare.

Acknowledgments

We thank the editors and reviewers for their critical comments on the manuscript.

Funding

This research was supported by the National Natural Science Foundation of China (U21A20215 and 32488102) and the Biological Breeding-National Science and Technology Major Project of China (2023ZD0403203).

Conflicts of Interest

The authors declare no conflicts of interest.

Author Contributions

All authors have read and agreed to the published version of the manuscript.

References

Ajala, A., Uzairu, A., Shallangwa, G.A. and Abechi, S.E. (2023) Virtual screening, molecular docking simulation and ADMET prediction of some selected natural products as potential inhibitors of NLRP3 inflammasomes as drug candidates for alzheimer disease. *Biocatalysis and Agricultural Biotechnology* 48, 102615. DOI: 10.1016/j.bcab.2023.102615.

Amendola, G. and Cosconati, S. (2021) PyRMD: A new fully automated AI-powered ligand-based virtual screening tool. *Journal of Chemical Information and Modeling* 61(8), 3835–3845. DOI: 10.1021/acs.jcim.1c00653.

Amirvaresi, A., Nikounezhad, N., Amirahmadi, M., Daraei, B. and Parastar, H. (2021) Comparison of near-infrared (NIR) and mid-infrared (MIR) spectroscopy based on chemometrics for saffron authentication and adulteration detection. *Food Chemistry* 344, 128647. DOI: 10.1016/j.foodchem.2020.128647.

An, X., Zhang, Y., Duan, L., Jin, D., Zhao, S. *et al.* (2021) The direct evidence and mechanism of traditional Chinese medicine treatment of COVID-19. *Biomedicine & Pharmacotherapy* 137, 111267. DOI: 10.1016/j.biopha.2021.111267.

Atanasov, A.G., Zotchev, S.B., Dirsch, V.M., Supuran, C.T. and International Natural Product Sciences Taskforce (2021) Natural products in drug discovery: Advances and opportunities. *Nature Reviews Drug Discovery* 20(3), 200–216. DOI: 10.1038/s41573-020-00114-z.

Azadnia, R., Al-Amidi, M.M., Mohammadi, H., Cifci, M.A., Daryab, A. *et al.* (2022) An AI based approach for medicinal plant identification using deep CNN based on global average pooling. *Agronomy* 12(11), 2723. DOI: 10.3390/agronomy12112723.

Bai, C., Yang, J., Cao, B., Xue, Y., Gao, P. *et al.* (2020) Growth years and post-harvest processing methods have critical roles on the contents of medicinal active ingredients of *Scutellaria baicalensis*. *Industrial Crops and Products* 158, 112985. DOI: 10.1016/j.indcrop.2020.112985.

Bai, R., Zhou, J., Wang, S., Zhang, Y., Nan, T. *et al.* (2024) Identification and classification of *Coix* seed storage years based on hyperspectral imaging technology combined with deep learning 13(3), 498. DOI: 10.3390/foods13030498.

Bi, S., Xu, L., Chen, S., Bu, S. and Xu, Y. (2021) Detection of herbal combinations and pharmacological mechanisms of clinical prescriptions for coronary heart disease using data mining and network pharmacology. *Evidence-Based Complementary and Alternative Medicine* 2021(1), 9234984. DOI: 10.1155/2021/9234984.

Bui, Q.M., Nguyen, H.N.M., Le, V.N., Nguyen, T.T., Truong, N.M. *et al.* (2023) AIQS-DB: Revolutionizing the simultaneous analysis of organic compounds. *Applied Sciences* 13(14), 8031. DOI: 10.3390/app13148031.

Cai, Z., Huang, Z., He, M., Li, C., Qi, H. *et al.* (2023) Identification of geographical origins of Radix Paeoniae Alba using hyperspectral imaging with deep learning-based fusion approaches. *Food Chemistry* 422, 136169. DOI: 10.1016/j.foodchem.2023.136169.

Chai, Y., Huang, Y., Han, J., Chen, F., Zhu, J. *et al.* (2023) To explore the herbal combinations and pharmacological mechanisms of yourong huang in the treatment of cervical Spondylosis based on data mining and network pharmacology. *Natural Product Communications* 18(6), 1934578X231180458. DOI: 10.1177/1934578X231180458.

Chang, S.-S., Huang, H.-J. and Chen, C.Y.-C. (2011) Two birds with one stone? Possible dual-targeting H1N1 inhibitors from traditional Chinese medicine. *PLOS Computational Biology* 7(12), e1002315. DOI: 10.1371/journal.pcbi.1002315.

Chen, H. and He, Y. (2022) Machine learning approaches in traditional Chinese medicine: A systematic review. *The American Journal of Chinese Medicine* 50(1), 91–131. DOI: 10.1142/S0192415X22500045.

Chen, W., Tong, J., He, R., Lin, Y., Chen, P. *et al.* (2021a) An easy method for identifying 315 categories of commonly-used Chinese herbal medicines based on automated image recognition using AutoML platforms. *Informatics in Medicine Unlocked* 25, 100607. DOI: 10.1016/j.imu.2021.100607.

Chen, W., Liu, C., Ji, X., Joseph, J., Tang, Z. *et al.* (2021b) Stanene-based nanosheets for β-elemene delivery and ultrasound-mediated combination cancer therapy. *Angewandte Chemie (International Ed. in English)* 60(13), 7155–7164. DOI: 10.1002/anie.202016330.

Chen, X., Ruan, C., Zhang, Y. and Chen, H. (2019) Heterogeneous information network based clustering for precision traditional Chinese medicine. *BMC Medical Informatics and Decision Making* 19(Suppl 6), 264. DOI: 10.1186/s12911-019-0963-0.

Chen, Z., Zhao, M., You, L., Zheng, R., Jiang, Y. *et al.* (2022) Developing an artificial intelligence method for screening hepatotoxic compounds in traditional Chinese medicine and Western medicine combination. *Chinese Medicine* 17(1), 58. DOI: 10.1186/s13020-022-00617-4.

Cheng, N., Chen, Y., Gao, W., Liu, J., Huang, Q. *et al.* (2021) An improved deep learning model: S-textBLCNN for traditional Chinese medicine formula classification. *Frontiers in Genetics* 12. DOI: 10.3389/fgene.2021.807825.

Cheng, Y., Xiao, M., Chen, J., Wang, D., Hu, Y. *et al.* (2022) Quality assessment and Q-markers discovery of Tongsaimai tablet by integrating serum pharmacochemistry and network pharmacology for anti-atherosclerosis benefit. *Chinese Medicine* 17(1), 103. DOI: 10.1186/s13020-022-00658-9.

Chu, Z., Li, F., Wang, D., Xu, S., Gao, C. *et al.* (2022) Research on identification method of tangerine peel year based on deep learning. *Food Science and Technology* 42, e64722. DOI: 10.1590/fst.64722.

Chung, M.-C., Su, L.-J., Chen, C.-L. and Wu, L.-C. (2024) AI-assisted literature exploration of innovative Chinese medicine formulas. *Frontiers in Pharmacology* 15, 1347882. DOI: 10.3389/fphar.2024.1347882.

Cruz-Tirado, J.P., Lima Brasil, Y., Freitas Lima, A., Alva Pretel, H., Teixeira Godoy, H. *et al.* (2023) Rapid and non-destructive cinnamon authentication by NIR-hyperspectral imaging and classification chemometrics tools. *Spectrochimica Acta. Part A: Molecular and Biomolecular Spectroscopy* 289, 122226. DOI: 10.1016/j.saa.2022.122226.

Daina, A., Michielin, O. and Zoete, V. (2017) SwissADME: A free web tool to evaluate pharmacokinetics, drug-likeness and medicinal chemistry friendliness of small molecules. *Scientific Reports* 7(1), 42717. DOI: 10.1038/srep42717.

Dou, Z., Xia, Y., Zhang, J., Li, Y., Zhang, Y. *et al.* (2021) Syndrome differentiation and treatment regularity in traditional Chinese medicine for type 2 diabetes: A text mining analysis. *Frontiers in Endocrinology* 12, 728032. DOI: 10.3389/fendo.2021.728032.

Duan, P., Yang, K., Su, X., Fan, S., Dong, X. *et al.* (2024) HTINet2: Herb–target prediction via knowledge graph embedding and residual-like graph neural network. *Briefings in Bioinformatics* 25(5). DOI: 10.1093/bib/bbae414.

Fan, L., Wu, L., Yu, X.-H., Chen, Y.-B., Lin, L. *et al.* (2021) The ethnopharmacology, phytochemistry, pharmacology and toxicology of the genus Erycibe (Convolvulaceae). *Journal of Ethnopharmacology* 278, 114312. DOI: 10.1016/j.jep.2021.114312.

Feng, G., Sun, Y., Liu, S., Song, F., Pi, Z. *et al.* (2019) Stepwise targeted matching strategy from *in vitro* to *in vivo* based on ultra-high performance liquid chromatography tandem mass spectrometry technology to quickly identify and screen pharmacodynamic constituents. *Talanta* 194, 619–626. DOI: 10.1016/j.talanta.2018.10.074.

Fang, S., Dong, L., Liu, L., Guo, J., Zhao, L. *et al.* (2021) HERB: A high-throughput experiment- and reference-guided database of traditional chinese medicine. *Nucleic Acids Research* 49(D1), D1197–D1206. DOI: 10.1093/nar/gkaa1063.

Fu, Y., Fang, Y., Gong, S., Xue, T., Wang, P. *et al.* (2023) Deep learning-based network pharmacology for exploring the mechanism of licorice for the treatment of COVID-19. *Scientific Reports* 13(1), 5844. DOI: 10.1038/s41598-023-31380-7.

Gan, X., Shu, Z., Wang, X., Yan, D., Li, J. *et al.* (2023) Network medicine framework reveals generic herb-symptom effectiveness of traditional Chinese medicine. *Science Advances* 9(43), eadh0215. DOI: 10.1126/sciadv.adh0215.

Gao, D., Niu, M., Wei, S.-Z., Zhang, C.-E., Zhou, Y.-F. *et al.* (2020) Identification of a pharmacological biomarker for the bioassay-based quality control of a thirteen-component TCM formula (Lianhua Qingwen) used in treating influenza A virus (H1N1) infection. *Frontiers in Pharmacology* 11, 746. DOI: 10.3389/fphar.2020.00746.

Gong, G.F., Chen, Y.Y., Zhang, Y.M., Fan, Y.Q., Zhou, Q. *et al.* (2019) A novel bis-component AIE smart gel with high selectivity and sensitivity to detect CN–, fe 3+ and h 2 PO 4–. *Soft Matter* 15(31), 6348–6352. DOI: 10.1039/C9SM01055A.

Gong, J.-N., Zhao, L., Chen, G., Chen, X., Chen, Z.-D. *et al.* (2021) A novel artificial intelligence protocol to investigate potential leads for diabetes mellitus. *Molecular Diversity* 25(3), 1375–1393. DOI: 10.1007/s11030-021-10204-8.

Guo, F., Tang, X., Zhang, W., Wei, J., Tang, S. *et al.* (2020a) Exploration of the mechanism of traditional Chinese medicine by AI approach using unsupervised machine learning for cellular functional similarity of compounds in heterogeneous networks, XiaoErFuPi granules as an example. *Pharmacological Research* 160, 105077. DOI: 10.1016/j.phrs.2020.105077.

Guo, R., Luo, X., Liu, J., Liu, L., Wang, X. *et al.* (2020b) Omics strategies decipher therapeutic discoveries of traditional Chinese medicine against different diseases at multiple layers molecular-level. *Pharmacological Research* 152, 104627. DOI: 10.1016/j.phrs.2020.104627.

Guo, S., Wang, J., Wang, Y., Zhang, Y., Bi, K. *et al.* (2019) Study on the multitarget synergistic effects of Kai-Xin-San against Alzheimer's disease based on systems biology. *Oxidative Medicine and Cellular Longevity* 2019(1), 1707218. DOI: 10.1155/2019/1707218.

Gupta, R., Srivastava, D., Sahu, M., Tiwari, S., Ambasta, R.K. *et al.* (2021) Artificial intelligence to deep learning: Machine intelligence approach for drug discovery. *Molecular Diversity* 25(3), 1315–1360. DOI: 10.1007/s11030-021-10217-3.

Han, M., Zhang, J., Zeng, Y., Hao, F. and Ren, Y. (2022) A novel method of chinese herbal medicine classification based on mutual learning. *Mathematics* 10(9), 1557. DOI: 10.3390/math10091557.

Hao, W., Han, M., Li, S. and Li, F. (2022) MTAL: A novel Chinese herbal medicine classification approach with mutual triplet attention learning. *Wireless Communications and Mobile Computing* 2022(1), 1–9. DOI: 10.1155/2022/8034435.

Hassanzadeh, P., Atyabi, F. and Dinarvand, R. (2019) The significance of artificial intelligence in drug delivery system design. *Advanced Drug Delivery Reviews* 151–152, 169–190. DOI: 10.1016/j.addr.2019.05.001.

He, Q., Tan, X., Geng, S., Du, Q., Pei, Z. *et al.* (2022b) Network analysis combined with pharmacological evaluation strategy to reveal the mechanism of Tibetan medicine Wuwei Shexiang pills in treating rheumatoid arthritis. *Frontiers in Pharmacology* 13, 941013. DOI: 10.3389/fphar.2022.941013.

He, S., Yi, Y., Hou, D., Fu, X., Zhang, J. *et al.* (2022a) Identification of hepatoprotective traditional Chinese medicines based on the structure–activity relationship, molecular network, and machine learning techniques. *Frontiers in Pharmacology* 13, 969979. DOI: 10.3389/fphar.2022.969979.

He, X., Zhao, L., Zhong, W., Chen, H.-Y., Shan, X. *et al.* (2020) Insight into potent leads for Alzheimer's disease by using several artificial intelligence algorithms. *Biomedicine & Pharmacotherapy* 129, 110360. DOI: 10.1016/j.biopha.2020.110360.

He, X., Huang, S., Wu, M., Wu, M., Zhang, Y. *et al.* (2021) Simultaneous quantitative analysis of ten bioactive flavonoids in Citri Reticulatae Pericarpium Viride (Qing Pi) by ultrahigh-performance liquid chromatography and high-resolution mass spectrometry combined with chemometric methods. *Phytochemical Analysis* 32(6), 1152–1161. DOI: 10.1002/pca.3056.

Hu, H., Wang, T., Wei, Y., Xu, Z., Cao, S. *et al.* (2023) Non-destructive prediction of isoflavone and starch by hyperspectral imaging and deep learning in Puerariae Thomsonii Radix. *Frontiers in Plant Science* 14, 1271320. DOI: 10.3389/fpls.2023.1271320.

Hu, J., Chen, Z., Yang, M., Zhang, R. and Cui, Y. (2018) A multiscale fusion convolutional neural network for plant leaf recognition. *IEEE Signal Processing Letters* 25(6), 853–857. DOI: 10.1109/LSP.2018.2809688.

Hu, L., Zhang, M., Hu, P., Zhang, J., Niu, C. *et al.* (2024) Dual-channel hypergraph convolutional network for predicting herb–disease associations. *Briefings in Bioinformatics* 25(2). DOI: 10.1093/bib/bbae067.

Husaini, A.M., Haq, S.A.U., Shabir, A., Wani, A.B. and Dedmari, M.A. (2022) The menace of saffron adulteration: Low-cost rapid identification of fake look-alike saffron using foldscope and machine learning technology. *Frontiers in Plant Science* 13, 945291. DOI: 10.3389/fpls.2022.945291.

Hussein, B.R., Malik, O.A., Ong, W.H. and Silk, J.W.F. (2020) Automated classification of tropical plant species data based on machine learning techniques and leaf trait measurements. In: *Computational Science and Technology: 6th ICCST 2019*, Springer, Singapore, Kota Kinabalu, Malaysia, August 29–30, 2019, pp. 85–94. DOI: 10.1007/978-981-15-0058-9_9.

Ianevski, A., Giri, A.K. and Aittokallio, T. (2022) SynergyFinder 3.0: An interactive analysis and consensus interpretation of multi-drug synergies across multiple samples. *Nucleic Acids Research* 50(W1), W739–W743. DOI: 10.1093/nar/gkac382.

Isert, C., Atz, K. and Schneider, G. (2023) Structure-based drug design with geometric deep learning. *Current Opinion in Structural Biology* 79, 102548. DOI: 10.1016/j.sbi.2023.102548.

Jia, Y., Sun, M., Shi, Y., Zhu, Z., Van Wijk, E. *et al.* (2020) A comparative study of aged and contemporary Chinese herbal materials by using delayed luminescence technique. *Chinese Medicine* 15(1), 1–12. DOI: 10.1186/s13020-020-0287-0.

Jiang, Z., Lv, A., Zhong, L., Yang, J., Xu, X. *et al.* (2023) Rapid prediction of adulteration content in *Atractylodis rhizoma* based on data and image features fusions from near-infrared spectroscopy and hyperspectral imaging techniques. *Foods* 12(15), 2904. DOI: 10.3390/foods12152904.

Jiménez-Luna, J., Grisoni, F., Weskamp, N. and Schneider, G. (2021) Artificial intelligence in drug discovery: Recent advances and future perspectives. *Expert Opinion on Drug Discovery* 16(9), 949–959. DOI: 10.1080/17460441.2021.1909567.

Jin, Y., Ji, W., Shi, Y., Wang, X. and Yang, X. (2023) Meta-path guided graph attention network for explainable herb recommendation. *Health Information Science and Systems* 11(1), 5. DOI: 10.1007/s13755-022-00207-6.

Jing, J., Zhu, C., Gong, R., Qi, X., Zhang, Y. *et al.* (2023) Research progress on the active ingredients of traditional Chinese medicine in the intervention of atherosclerosis: A promising natural immunotherapeutic adjuvant. *Biomedicine & Pharmacotherapy* 159, 114201. DOI: 10.1016/j.biopha.2022.114201.

Jumper, J., Evans, R., Pritzel, A., Green, T., Figurnov, M. *et al.* (2021) Highly accurate protein structure prediction with alphafold. *Nature* 596(7873), 583–589. DOI: 10.1038/s41586-021-03819-2.

Kabir, M.H., Guindo, M.L., Chen, R., Liu, F., Luo, X. *et al.* (2022) Deep learning combined with hyperspectral imaging technology for variety discrimination of *Fritillaria thunbergii*. *Molecules* 27(18), 6042. DOI: 10.3390/molecules27186042.

Kattenborn, T., Eichel, J. and Fassnacht, F.E. (2019) Convolutional neural networks enable efficient, accurate and fine-grained segmentation of plant species and communities from high-resolution UAV imagery. *Scientific Reports* 9(1), 17656. DOI: 10.1038/s41598-019-53797-9.

Ke, Z., Liu, M., Liu, J., Su, Z., Li, L. *et al.* (2024) The application of artificial intelligence in the research and development of traditional Chinese medicine. *International Journal of Drug Discovery and Pharmacology* 3(1), 100001. DOI: 10.53941/ijddp.2024.100001.

Kha, Q.-H., Le, V.H., Hung, T.N.K., Nguyen, N.T.K. and Le, N.Q.K. (2023) Development and validation of an explainable machine learning-based prediction model for drug–food interactions from chemical structures. *Sensors* 23(8), 3962. DOI: 10.3390/s23083962.

Kim, J.-H., Kim, B.-G., Roy, P.P. and Jeong, D.-M. (2019) Efficient facial expression recognition algorithm based on hierarchical deep neural network structure. *IEEE Access* 7, 41273–41285. DOI: 10.1109/ACCESS.2019.2907327.

Kim, T.J., Pyun, D.H., Park, S.Y., Lee, H.J., Abd El-Aty, A.M. *et al.* (2021) Patchouli alcohol improves wound healing in high fat diet-fed mice through AMPK-mediated suppression of inflammation and TGFb1 signaling. *Biochemical and Biophysical Research Communications* 561, 136–142. DOI: 10.1016/j.bbrc.2021.05.036.

Kuenzi, B.M., Park, J., Fong, S.H., Sanchez, K.S., Lee, J. *et al.* (2020) Predicting drug response and synergy using a deep learning model of human cancer cells. *Cancer Cell* 38(5), 672–684. DOI: 10.1016/j.ccell.2020.09.014.

Lai, X., Wang, X., Hu, Y., Su, S., Li, W. *et al.* (2020) Editorial: Network pharmacology and traditional medicine. *Frontiers in Pharmacology* 11. DOI: 10.3389/fphar.2020.01194.

Lee, M., Shin, H., Park, M., Kim, A., Cha, S. *et al.* (2022) Systems pharmacology approaches in herbal medicine research: A brief review. *BMB Reports* 55(9), 417–428. DOI: 10.5483/BMBRep.2022.55.9.102.

Lee, P., Bubeck, S. and Petro, J. (2023) Benefits, limits, and risks of GPT-4 as an AI chatbot for medicine. *New England Journal of Medicine* 388(13), 1233–1239. DOI: 10.1056/NEJMsr2214184.

Lee, S.H., Chan, C.S., Wilkin, P. and Remagnino, P. (2015) Deep-plant: Plant identification with convolutional neural networks. In: *2015 IEEE International Conference on Image Processing (ICIP)*, Quebec City, QC, Canada, pp. 452–456. DOI: 10.1109/ICIP.2015.7350839.

Li, D., Lv, B., Wang, D., Xu, D., Qin, S. *et al.* (2020a) Network pharmacology and bioactive equivalence assessment integrated strategy driven Q-markers discovery for Da-Cheng-Qi decoction to attenuate intestinal obstruction. *Phytomedicine* 72, 153236. DOI: 10.1016/j.phymed.2020.153236.

Li, D., Hu, J., Zhang, L., Li, L., Yin, Q. *et al.* (2022b) Deep learning and machine intelligence: New computational modeling techniques for discovery of the combination rules and pharmacodynamic characteristics of traditional Chinese medicine. *European Journal of Pharmacology* 933, 175260. DOI: 10.1016/j.ejphar.2022.175260.

Li, G., Nikolic, D. and Van Breemen, R.B. (2016) Identification and chemical standardization of licorice raw materials and dietary supplements using UHPLC-MS/MS. *Journal of Agricultural and Food Chemistry* 64(42), 8062–8070. DOI: 10.1021/acs.jafc.6b02954.

Li, J.-Y., Chen, H.-Y., Dai, W., Lv, Q.-J. and Chen, C.Y.-C. (2019) Artificial intelligence approach to investigate the longevity drug. *The Journal of Physical Chemistry Letters* 10(17), 4947–4961. DOI: 10.1021/acs.jpclett.9b02220.

Li, L., Wang, L., Fan, W., Jiang, Y., Zhang, C. *et al.* (2020b) The application of fermentation technology in traditional Chinese medicine: A review. *The American Journal of Chinese Medicine* 48(4), 899–921. DOI: 10.1142/S0192415X20500433.

Li, L., Zuo, Z. and Wang, Y. (2022c) Practical qualitative evaluation and screening of potential biomarkers for different parts of Wolfiporia cocos using machine learning and network pharmacology. *Frontiers in Microbiology* 13, 931967. DOI: 10.3389/fmicb.2022.931967.

Li, M., Shi, Y., Zhang, J., Wan, X., Fang, J. *et al.* (2023a) Rapid evaluation of Ziziphi Spinosae semen and its adulterants based on the combination of FT-NIR and multivariate algorithms. *Food Chemistry* 20, 101022. DOI: 10.1016/j.fochx.2023.101022.

Li, N., Yu, J., Mao, X., Zhao, Y. and Huang, L. (2022a) The research and development thinking on the status of artificial intelligence in traditional Chinese medicine. *Evidence-Based Complementary and Alternative Medicine* eCAM, 7644524. DOI: 10.1155/2022/7644524.

Li, S., Wang, B.-Y., Cao, L., Xiao, L.-H., Chen, P. *et al.* (2023b) Research and development practice of traditional Chinese medicine based on network target theory and technology. *Zhongguo Zhong Yao Za Zhi = Zhongguo Zhongyao Zazhi = China Journal of Chinese Materia Medica* 48(22), 5965–5976. DOI: 10.19540/j.cnki.cjcmm.20230923.701.

Li, X., Kong, W., Liu, X., Zhang, X., Wang, W. *et al.* (2021) Application of laser-induced breakdown spectroscopy coupled with spectral matrix and convolutional neural network for identifying geographical origins of Gentiana rigescens Franch. *Frontiers in Artificial Intelligence* 4, 735533. DOI: 10.3389/frai.2021.735533.

Liang, D., Yin, Y.-H., Miao, L.-Y., Zheng, X., Gao, W. *et al.* (2019) Integrating chemical similarity and bioequivalence: A pilot study on quality consistency evaluation of dispensing granule and traditional decoction of Scutellariae radix by a totality-of-the-evidence approach. *Journal of Pharmaceutical and Biomedical Analysis* 169, 1–10. DOI: 10.1016/j.jpba.2019.02.030.

Lin, L.-C., Chang, H.-Y., Lin, T.E., Lin, J.-R., Hsia, S.-M. *et al.* (2022) Structure-based virtual screening discovers novel kidney-type glutaminase inhibitors. *Biomedicine & Pharmacotherapy* 154, 113585. DOI: 10.1016/j.biopha.2022.113585.

Lin, Y.-C., Huang, W.-T., Ou, S.-C., Hung, H.-H., Cheng, W.-Z. *et al.* (2019) Neural network analysis of Chinese herbal medicine prescriptions for patients with colorectal cancer. *Complementary Therapies in Medicine* 42, 279–285. DOI: 10.1016/j.ctim.2018.12.001.

Lipinski, C.F., Maltarollo, V.G., Oliveira, P.R., da Silva, A.B.F. and Honorio, K.M. (2019) Advances and perspectives in applying deep learning for drug design and discovery. *Frontiers in Robotics and AI* 6, 108. DOI: 10.3389/frobt.2019.00108.

Liu, C. and Guo, D. (2020) Inheriting essence, keeping integrity and innovation. *Chinese Herbal Medicines* 12(1), 1–2. DOI: 10.1016/j.chmed.2020.01.001.

Liu, C., Xu, F., Zuo, Z. and Wang, Y. (2022a) An identification method of herbal medicines superior to traditional spectroscopy: Two-dimensional correlation spectral images combined with deep learning. *Vibrational Spectroscopy* 120, 103380. DOI: 10.1016/j.vibspec.2022.103380.

Liu, C., Jiang, Q., Zhou, Z., Lei, P., Zhang, P. *et al.* (2024) Studies on chemical profiling and pharmacokinetics of traditional Chinese medicine Formula Kang Shuai Lao Pian. *Arabian Journal of Chemistry* 17(1), 105398. DOI: 10.1016/j.arabjc.2023.105398.

Liu, J., Huang, Q., Yang, X. and Ding, C. (2022c) HPE-GCN: Predicting efficacy of tonic formulae via graph convolutional networks integrating traditionally defined herbal properties. *Methods* 204, 101–109. DOI: 10.1016/j.ymeth.2022.05.003.

Liu, K., Chen, X., Ren, Y., Liu, C., Lv, T. *et al.* (2022b) Multi-target-based polypharmacology prediction (mtTPP): An approach using virtual screening and machine learning for multi-target drug discovery. *Chemico-Biological Interactions* 368, 110239. DOI: 10.1016/j.cbi.2022.110239.

Liu, S., Dong, H., Zhang, M., Geng, W. and Wang, X. (2023a) Identification of different degrees of processed ginger using GC-IMS combined with machine learning. *Journal of Pharmaceutical Analysis* 14(1), 149–151. DOI: 10.1016/j.jpha.2023.10.005.

Liu, Y., Cui, X., Zhang, X., Xie, Z., Wang, W. *et al.* (2023b) Exploring the potential mechanisms of Tongmai Jiangtang capsules in treating diabetic nephropathy through multi-dimensional data. *Frontiers in Endocrinology* 14, 1172226. DOI: 10.3389/fendo.2023.1172226.

Liu, Z., Shen, T., Zhang, J., Li, Z., Zhao, Y. *et al.* (2021a) A novel multi-preprocessing integration method for the qualitative and quantitative assessment of wild medicinal plants: Gentiana rigescens as an example. *Frontiers in Plant Science* 12, 759248. DOI: 10.3389/fpls.2021.759248.

Liu, Z., Huang, D., Zheng, S., Song, Y., Liu, B. *et al.* (2021b) Deep learning enables discovery of highly potent anti-osteoporosis natural products. *European Journal of Medicinal Chemistry* 210, 112982. DOI: 10.1016/j.ejmech.2020.112982.

Lu, S.H., Zhai, H.L., Zhao, B.Q., Yin, B. and Zhu, L. (2020) Novel approach to the analysis of chemical third-order data. *Journal of Chemical Information and Modeling* 60(10), 4750–4756. DOI: 10.1021/acs.jcim.0c00554.

Ma, J., Zhou, X., Xie, B., Wang, C., Chen, J. *et al.* (2023) Application for identifying the origin and predicting the physiologically active ingredient contents of Gastrodia elata blume using visible–near-infrared spectroscopy combined with machine learning. *Foods* 12(22), 4061. DOI: 10.3390/foods12224061.

Mastropietro, A., Pasculli, G. and Bajorath, J. (2023) Learning characteristics of graph neural networks predicting protein–ligand affinities. *Nature Machine Intelligence* 5(12), 1427–1436. DOI: 10.1038/s42256-023-00756-9.

Mater, A.C. and Coote, M.L. (2019) Deep learning in chemistry. *Journal of Chemical Information and Modeling* 59(6), 2545–2559. DOI: 10.1021/acs.jcim.9b00266.

Matias, M., Pinho, J.O., Penetra, M.J., Campos, G., Reis, C.P. *et al.* (2021) The challenging melanoma landscape: From early drug discovery to clinical approval. *Cells* 10(11), 3088. DOI: 10.3390/cells10113088.

McGibbon, M., Shave, S., Dong, J., Gao, Y., Houston, D.R. *et al.* (2024) From intuition to AI: Evolution of small molecule representations in drug discovery. *Briefings in Bioinformatics* 25(1), bbad422. DOI: 10.1093/bib/bbad422.

Miao, J., Huang, Y., Wang, Z., Wu, Z. and Lv, J. (2023) Image recognition of traditional Chinese medicine based on deep learning. *Frontiers in Bioengineering and Biotechnology* 11, 1199803. DOI: 10.3389/fbioe.2023.1199803.

Mu, K., Liu, Y., Liu, G., Ran, F., Zhou, L. *et al.* (2023) A review of hemostatic chemical components and their mechanisms in traditional Chinese medicine and ethnic medicine. *Journal of Ethnopharmacology* 307, 116200. DOI: 10.1016/j.jep.2023.116200.

Mullowney, M.W., Duncan, K.R., Elsayed, S.S., Garg, N., Van der Hooft, J.J.J. *et al.* (2023) Artificial intelligence for natural product drug discovery. *Nature Reviews Drug Discovery* 22(11), 895–916. DOI: 10.1038/s41573-023-00774-7.

Nguyen, T.M., Quinn, T.P., Nguyen, T. and Tran, T. (2023) Explaining black box drug target prediction through model agnostic counterfactual samples. *IEEE/ACM Transactions on Computational Biology and Bioinformatics* 20(2), 1020–1029. DOI: 10.1109/TCBB.2022.3190266.

Ni, B., Liu, Y., Gao, X., Cai, M., Fu, J. *et al.* (2022) Isoliquiritigenin attenuates emodin-induced hepatotoxicity *in vivo* and *in vitro* through Nrf2 pathway. *Comparative Biochemistry and Physiology Part C* 261, 109430. DOI: 10.1016/j.cbpc.2022.109430.

Normile, D. (2003) The new face of traditional Chinese medicine. *Science* 299(5604), 188–190. DOI: 10.1126/science.299.5604.188.

Pan, C., Schoppe, O., Parra-Damas, A., Cai, R., Todorov, M.I. *et al.* (2019) Deep learning reveals cancer metastasis and therapeutic antibody targeting in the entire body. *Cell* 179(7), 1661–1676. DOI: 10.1016/j.cell.2019.11.013.

Pan, Y., Zhang, H., Chen, Y., Gong, Y., Yan, J. *et al.* (2024) Applications of hyperspectral imaging technology combined with machine learning in quality control of traditional Chinese medicine from the perspective of artificial intelligence: A review. *Critical Reviews in Analytical Chemistry* 8, 2850–2864. DOI: 10.1080/10408347.2023.2207652.

Park, S.-J., Kim, B.-G. and Chilamkurti, N. (2021) A robust facial expression recognition algorithm based on multi-rate feature fusion scheme. *Sensors (Basel, Switzerland)* 21(21), 6954. DOI: 10.3390/s21216954.

Peng, W., Li, N., Jiang, E., Zhang, C., Huang, Y. *et al.* (2022) A review of traditional and current processing methods used to decrease the toxicity of the rhizome of *Pinellia ternata* in traditional Chinese medicine. *Journal of Ethnopharmacology* 299, 115696. DOI: 10.1016/j.jep.2022.115696.

Picon, A., Seitz, M., Alvarez-Gila, A., Mohnke, P., Ortiz-Barredo, A. *et al.* (2019) Crop conditional convolutional neural networks for massive multi-crop plant disease classification over cell phone acquired images taken on real field conditions. *Computers and Electronics in Agriculture* 167, 105093. DOI: 10.1016/j.compag.2019.105093.

Preuer, K., Lewis, R.P., Hochrieter, S., Bender, A., Bulusu, K.C. *et al.* (2018) DeepSynergy: Predicting anti-cancer drug synergy with deep learning. *Bioinformatics* 34(9), 1538–1546. DOI: 10.1093/bioinformatics/btx806.

Pun, F.W., Liu, B.H.M., Long, X., Leung, H.W., Leung, G.H.D. *et al.* (2022) Identification of therapeutic targets for amyotrophic lateral sclerosis using Pandaomics - an AI-enabled biological target discovery platform. *Frontiers in Aging Neuroscience* 14, 914017. DOI: 10.3389/fnagi.2022.914017.

Rajpurkar, P., Chen, E., Banerjee, O. and Topol, E.J. (2022) AI in health and medicine. *Nature Medicine* 28(1), 31–38. DOI: 10.1038/s41591-021-01614-0.

Raza, A., Chohan, T.A., Buabeid, M., Arafa, E.-S.A., Chohan, T.A. *et al.* (2023) Deep learning in drug discovery: A futuristic modality to materialize the large datasets for cheminformatics. *Journal of Biomolecular Structure & Dynamics* 41(18), 9177–9192. DOI: 10.1080/07391102.2022.2136244.

Ren, A., Wu, T., Wang, Y., Fan, Q., Yang, Z. *et al.* (2023) Integrating animal experiments, mass spectrometry and network-based approach to reveal the sleep-improving effects of *Ziziphi spinosae* semen and γ-aminobutyric acid mixture. *Chinese Medicine* 18(1), 99. DOI: 10.1186/s13020-023-00814-9.

Ren, L., Zheng, X., Liu, J., Li, W., Fu, W. *et al.* (2019) Network pharmacology study of traditional Chinese medicines for stroke treatment and effective constituents screening. *Journal of Ethnopharmacology* 242, 112044. DOI: 10.1016/j.jep.2019.112044.

Romm, E.L. and Tsigelny, I.F. (2020) Artificial intelligence in drug treatment. *Annual Review of Pharmacology and Toxicology* 60(1), 353–369. DOI: 10.1146/annurev-pharmtox-010919-023746.

Ryu, J.Y., Kim, H.U. and Lee, S.Y. (2018) Deep learning improves prediction of drug–drug and drug–food interactions. *Proceedings of the National Academy of Sciences* 115(18), E4304–E4311. DOI: 10.1073/pnas.1803294115.

Shi, Y., He, T., Zhong, J., Mei, X., Li, Y. *et al.* (2024) Classification and rapid non-destructive quality evaluation of different processed products of *Cyperus rotundus* based on near-infrared spectroscopy combined with deep learning. *Talanta* 268(Pt 1), 125266. DOI: 10.1016/j.talanta.2023.125266.

Shi, Y.-H., Huang, Q.-W., Zhu, S.-M., Zhou, Y.-M., Zhang, L.-J. *et al.* (2022) Chemical profiling of *Fritillariae thunbergii* Miq prepared by different processing methods reveals two new quality markers: Zhebeininoside and imperialine-3-β-d-glucoside. *Journal of Ethnopharmacology* 283, 114670. DOI: 10.1016/j.jep.2021.114670.

Song, Z., Chen, G. and Chen, C.Y.-C. (2024) AI empowering traditional Chinese medicine? *Chemical Science* 15(41), 16844–16886. DOI: 10.1039/D4SC04107K.

Staub, B.N. and Sadrameli, S.S. (2019) The use of robotics in minimally invasive spine surgery. *Journal of Spine Surgery (Hong Kong)* 5(Suppl 1), S31–S40. DOI: 10.21037/jss.2019.04.16.

Sun, X. and Qian, H. (2016) Chinese herbal medicine image recognition and retrieval by convolutional neural network. *PLOS ONE* 11(6), e0156327. DOI: 10.1371/journal.pone.0156327.

Tan, C., Wu, C., Huang, Y., Wu, C. and Chen, H. (2020) Identification of different species of *Zanthoxyli pericarpium* based on convolution neural network. *PloS One* 15(4), e0230287. DOI: 10.1371/journal.pone.0230287.

Tang, M., Wen, C., Lin, J., Chen, H. and Ran, T. (2023) Discovery of novel A2AR antagonists through deep learning-based virtual screening. *Artificial Intelligence in the Life Sciences* 3, 100058. DOI: 10.1016/j.ailsci.2023.100058.

Tao, L., Zhu, F., Qin, C., Zhang, C., Xu, F. *et al.* (2014) Nature's contribution to today's pharmacopeia. *Nature Biotechnology* 32(10), 979–980. DOI: 10.1038/nbt.3034.

Tao, Y., Bao, J., Liu, Q., Liu, L. and Zhu, J. (2022) Application of deep-learning algorithm driven intelligent raman spectroscopy methodology to quality control in the manufacturing process of Guanxinning tablets. *Molecules (Basel, Switzerland)* 27(20), 6969. DOI: 10.3390/molecules27206969.

Thomford, N.E., Senthebane, D.A., Rowe, A., Munro, D., Seele, P. *et al.* (2018) Natural products for drug discovery in the 21st century: Innovations for novel drug discovery. *International Journal of Molecular Sciences* 19(6), 1578. DOI: 10.3390/ijms19061578.

Tian, D., Chen, W., Xu, D., Xu, L., Xu, G. *et al.* (2024) A review of traditional Chinese medicine diagnosis using machine learning: Inspection, auscultation-olfaction, inquiry, and palpation. *Computers in Biology and Medicine* 170, 108074. DOI: 10.1016/j.compbiomed.2024.108074.

Tran, T.T.V., Surya Wibowo, A., Tayara, H. and Chong, K.T. (2023) Artificial intelligence in drug toxicity prediction: Recent advances, challenges, and future perspectives. *Journal of Chemical Information and Modeling* 63(9), 2628–2643. DOI: 10.1021/acs.jcim.3c00200.

Tripathi, M.K., Nath, A., Singh, T.P., Ethayathulla, A.S. and Kaur, P. (2021) Evolving scenario of big data and artificial intelligence (AI) in drug discovery. *Molecular Diversity* 25(3), 1439–1460. DOI: 10.1007/s11030-021-10256-w.

Tsai, P.-W., Rogio, K.G.G., Hsieh, C.-Y., Cruz, K.A.D.C., Lee, C.-J. *et al.* (2023) Optimal stimulation of citrus reticulate for bioenergy extraction in MFCs and antioxidant activity via traditional Chinese medicine processing methods. *Journal of the Taiwan Institute of Chemical Engineers* 143, 104690. DOI: 10.1016/j.jtice.2023.104690.

Tu, Y., Li, L., Wang, Z. and Yang, L. (2021) Advances in analytical techniques and quality control of traditional Chinese medicine injections. *Journal of Pharmaceutical and Biomedical Analysis* 206, 114353. DOI: 10.1016/j.jpba.2021.114353.

Vo, T.H., Nguyen, N.T.K., Kha, Q.H. and Le, N.Q.K. (2022) On the road to explainable AI in drug-drug interactions prediction: A systematic review. *Computational and Structural Biotechnology Journal* 20, 2112–2123. DOI: 10.1016/j.csbj.2022.04.021.

Wakchaure, M., Patle, B.K. and Mahindrakar, A.K. (2023) Application of AI techniques and robotics in agriculture: A review. *Artificial Intelligence in the Life Sciences* 3, 100057. DOI: 10.1016/j.ailsci.2023.100057.

Walters, W.P. and Barzilay, R. (2021) Critical assessment of AI in drug discovery. *Expert Opinion on Drug Discovery* 16(9), 937–947. DOI: 10.1080/17460441.2021.1915982.

Wan, J., Zhang, Z., Wu, C., Tian, S., Zang, Y. *et al.* (2023) Astragaloside IV derivative HHQ16 ameliorates infarction-induced hypertrophy and heart failure through degradation of lncRNA4012/9456. *Signal Transduction and Targeted Therapy* 8(1), 414. DOI: 10.1038/s41392-023-01660-9.

Wang, J., Mo, W., Wu, Y., Xu, X., Li, Y. *et al.* (2022) Combined channel attention and spatial attention module network for Chinese herbal slices automated recognition. *Frontiers in Neuroscience* 16, 920820. DOI: 10.3389/fnins.2022.920820.

Wang, S., Hou, Y., Li, X., Meng, X., Zhang, Y. *et al.* (2021a) Practical implementation of artificial intelligence-based deep learning and cloud computing on the application of traditional medicine and western medicine in the diagnosis and treatment of rheumatoid arthritis. *Frontiers in Pharmacology* 12, 765435. DOI: 10.3389/fphar.2021.765435.

Wang, S., Yu, Y., Cui, M., Liu, K. and Liu, K. (2021d) Seed oil quality and cultivation of *Sambucus williamsii* hance as a new oil crop. *Frontiers in Nutrition* 8, 796175. DOI: 10.3389/fnut.2021.796175.

Wang, T.-H., Lee, C.-Y., Lee, T.-Y., Huang, H.-D., Hsu, J.B.-K. *et al.* (2021g) Biomarker identification through multiomics data analysis of prostate cancer prognostication using a deep learning model and similarity network fusion. *Cancers* 13(11), 2528. DOI: 10.3390/cancers13112528.

Wang, W.-Y., Xie, Y., Zhou, H. and Liu, L. (2021c) Contribution of traditional Chinese medicine to the treatment of COVID-19. *Phytomedicine: International Journal of Phytotherapy and Phytopharmacology* 85, 153279. DOI: 10.1016/j.phymed.2020.153279.

Wang, X., Wang, Z.-Y., Zheng, J.-H. and Li, S. (2021b) TCM network pharmacology: A new trend towards combining computational, experimental and clinical approaches. *Chinese Journal of Natural Medicines* 19(1), 1–11. DOI: 10.1016/S1875-5364(21)60001-8.

Wang, Y., Jafari, M., Tang, Y. and Tang, J. (2019) Predicting Meridian in Chinese traditional medicine using machine learning approaches. *PLOS Computational Biology* 15(11), e1007249. DOI: 10.1371/journal.pcbi.1007249.

Wang, Y., Lin, Y., Cheng, C., Chen, P., Zhang, P. *et al.* (2020) Corrigendum: NF-κb/TWIST1 mediates migration and phagocytosis of macrophages in the mice model of implant-associated *Staphylococcus aureus* osteomyelitis. *Frontiers in Microbiology* 11, 1301. DOI: 10.3389/fmicb.2020.01775.

Wang, Y., Shi, X., Li, L., Efferth, T. and Shang, D. (2021e) The impact of artificial intelligence on traditional Chinese medicine. *The American Journal of Chinese Medicine* 49(6), 1297–1314. DOI: 10.1142/S0192415X21500622.

Wang, Y., Yang, H., Chen, L., Jafari, M. and Tang, J. (2021f) Network-based modeling of herb combinations in traditional Chinese medicine. *Briefings in Bioinformatics* 22(5). DOI: 10.1093/bib/bbab106.

Wang, Y., Ren, J., Yao, Z., Wang, W., Wang, S. *et al.* (2023a) Clinical impact and risk factors of intensive care unit-acquired nosocomial infection: A propensity score-matching study from 2018 to 2020 in a teaching hospital in China. *Infection and Drug Resistance* 16, 569–579. DOI: 10.2147/IDR.S394269.

Wang, Y., Xiong, F., Zhang, Y., Wang, S., Yuan, Y. *et al.* (2023b) Application of hyperspectral imaging assisted with integrated deep learning approaches in identifying geographical origins and predicting nutrient contents of coix seeds. *Food Chemistry* 404(Pt A), 134503. DOI: 10.1016/j.foodchem.2022.134503.

Wang, Y.-C., Chen, T.-C.T. and Chiu, M.-C. (2023d) An improved explainable artificial intelligence tool in healthcare for hospital recommendation. *Healthcare Analytics* 3, 100147. DOI: 10.1016/j.health.2023.100147.

Wang, Z., Li, L., Song, M., Yan, J., Shi, J. *et al.* (2021h) Evaluating the traditional Chinese medicine (TCM) officially recommended in China for COVID-19 using ontology-based side-effect prediction framework (OSPF) and deep learning. *Journal of Ethnopharmacology* 272, 113957. DOI: 10.1016/j.jep.2021.113957.

Wang, Z., Li, W., Lu, J., Yuan, Z., Pi, W. *et al.* (2023c) Revealing the active ingredients of the traditional Chinese medicine decoction by the supramolecular strategies and multitechnologies. *Journal of Ethnopharmacology* 300, 115704. DOI: 10.1016/j.jep.2022.115704.

Wei, Z., Chen, J., Zuo, F., Guo, J., Sun, X. *et al.* (2023) Traditional Chinese medicine has great potential as candidate drugs for lung cancer: A review. *Journal of Ethnopharmacology* 300, 115748. DOI: 10.1016/j.jep.2022.115748.

Wongchai, A., Jenjeti, D. rao, Priyadarsini, A.I., Deb, N., Bhardwaj, A. *et al.* (2022) Farm monitoring and disease prediction by classification based on deep learning architectures in sustainable agriculture. *Ecological Modelling* 474, 110167. DOI: 10.1016/j.ecolmodel.2022.110167.

Wu, L., Xing, Y., Zheng, B. and Lin, A. (2016) A preliminary study of *Zanthoxylurn bungeanum* maxim varieties discriminating by computer vision. *Chinese Journal of Sensors and Actuators* 29(1), 136–140.

Wu, Q., Cai, C., Guo, P., Chen, M., Wu, X. *et al.* (2019) *In silico* identification and mechanism exploration of hepatotoxic ingredients in traditional Chinese medicine. *Frontiers in Pharmacology* 10, 458. DOI: 10.3389/fphar.2019.00458.

Xie, Y., Mai, C.-T., Zheng, D.-C., He, Y.-F., Feng, S.-L. *et al.* (2021) Wutou decoction ameliorates experimental rheumatoid arthritis via regulating NF-kb and Nrf2: Integrating efficacy-oriented compatibility of traditional Chinese medicine. *Phytomedicine: International Journal of Phytotherapy and Phytopharmacology* 85, 153522. DOI: 10.1016/j.phymed.2021.153522.

Xing, B., Li, S., Yang, J., Lin, D., Feng, Y. *et al.* (2021) Phytochemistry, pharmacology, and potential clinical applications of saffron: A review. *Journal of Ethnopharmacology* 281, 114555. DOI: 10.1016/j.jep.2021.114555.

Xiong, Y., Yan, L., Li, X., Duan, N., Lin, L. *et al.* (2021) Pattern classification in traditional chinese medicine and precision medicine: A narrative review. *Longhua Chinese Medicine* 4, 38–38. DOI: 10.21037/lcm-21-14.

Xu, H., Zhao, H., Ding, C., Jiang, D., Zhao, Z. *et al.* (2023) Celastrol suppresses colorectal cancer via covalent targeting peroxiredoxin 1. *Signal Transduction and Targeted Therapy* 8(1), 51. DOI: 10.1038/s41392-022-01231-4.

Xu, L., Zhang, Y., Zhang, P., Dai, X., Gao, Y. *et al.* (2019) Integrated metabolomics and network pharmacology strategy-driven active traditional Chinese medicine ingredients discovery for the alleviation of cisplatin nephrotoxicity. *Chemical Research in Toxicology* 32(12), 2411–2421. DOI: 10.1021/acs.chemrestox.9b00180.

Xu, Q., Guo, Q., Wang, C.-X., Zhang, S., Wen, C.-B. *et al.* (2021c) Network differentiation: A computational method of pathogenesis diagnosis in traditional Chinese medicine based on systems science. *Artificial Intelligence in Medicine* 118, 102134. DOI: 10.1016/j.artmed.2021.102134.

Xu, X., Chen, Y., Zhang, X., Zhang, R., Chen, X. *et al.* (2021a) Modular characteristics and the mechanism of Chinese medicine's treatment of gastric cancer: A data mining and pharmacology-based identification. *Annals of Translational Medicine* 9(24), 1777–1777. DOI: 10.21037/atm-21-6301.

Xu, Y., Wen, G., Hu, Y., Luo, M., Dai, D. *et al.* (2021b) Multiple attentional pyramid networks for Chinese herbal recognition. *Pattern Recognition* 110, 107558. DOI: 10.1016/j.patcog.2020.107558.

Xu, Y., Wu, W., Chen, Y., Zhang, T., Tu, K. *et al.* (2022) Hyperspectral imaging with machine learning for non-destructive classification of *Astragalus membranaceus* var. *mongholicus*, *Astragalus membranaceus*, and similar seeds. *Frontiers in Plant Science* 13, 1031849. DOI: 10.3389/fpls.2022.1031849.

Yang, J., Tian, S., Zhao, J. and Zhang, W. (2020b) Exploring the mechanism of TCM formulae in the treatment of different types of coronary heart disease by network pharmacology and machining learning. *Pharmacological Research* 159, 105034. DOI: 10.1016/j.phrs.2020.105034.

Xue, Q.-L., Miao, K.-H., Yu, Y. and Li, Z. (2023) Methodology for adaptive decision--making research on manufacturing process of traditional Chinese medicine based on deep reinforcement learning.

Zhongguo Zhong Yao Za Zhi = Zhongguo Zhongyao Zazhi = China Journal of Chinese Materia Medica 48(2), 562–568. DOI: 10.19540/j.cnki.cjcmm.20220705.304.

Yan, X., Zhang, S., Fu, H. and Qu, H. (2020) Combining convolutional neural networks and on-line raman spectroscopy for monitoring the Cornu Caprae Hircus hydrolysis process. *Spectrochimica Acta Part A: Molecular and Biomolecular Spectroscopy* 226, 117589. DOI: 10.1016/j.saa.2019.117589.

Yang, K., Long, X.-M., Cao, J.-J., Li, Y.-J., Wu, Y. *et al.* (2019b) An analytical strategy to explore the multicomponent pharmacokinetics of herbal medicine independently of standards: Application in *Gelsemium elegans* extracts. *Journal of Pharmaceutical and Biomedical Analysis* 176, 112833. DOI: 10.1016/j.jpba.2019.112833.

Yang, Q., Ji, H., Lu, H. and Zhang, Z. (2021) Prediction of liquid chromatographic retention time with graph neural networks to assist in small molecule identification. *Analytical Chemistry* 93(4), 2200–2206. DOI: 10.1021/acs.analchem.0c04071.

Yang, R., Liu, H., Bai, C., Wang, Y., Zhang, X. *et al.* (2020a) Chemical composition and pharmacological mechanism of Qingfei Paidu decoction and Ma Xing Shi Gan decoction against coronavirus disease 2019 (COVID-19): *In silico* and experimental study. *Pharmacological Research* 157, 104820. DOI: 10.1016/j.phrs.2020.104820.

Yang, S., Shen, Y., Lu, W., Yang, Y., Wang, H. *et al.* (2019c) Evaluation and identification of the neuroprotective compounds of Xiaoxuming decoction by machine learning: A novel mode to explore the combination rules in traditional Chinese medicine prescription. *BioMed Research International* 2019(1), 1–14. DOI: 10.1155/2019/6847685.

Yang, X., Wang, Y., Byrne, R., Schneider, G. and Yang, S. (2019a) Concepts of artificial intelligence for computer-assisted drug discovery. *Chemical Reviews* 119(18), 10520–10594. DOI: 10.1021/acs.chemrev.8b00728.

Yang, Y., Li, Q., Liu, Z., Ye, F. and Deng, K. (2019b) Understanding traditional Chinese medicine via statistical learning of expert-specific electronic medical records. *Quantitative Biology* 7(3), 210–232. DOI: 10.1007/s40484-019-0173-x.

Yu, X., Zhu, X., Zhang, L., Qin, J.-J., Feng, C. *et al.* (2022) In silico screening and validation of PDGFRA inhibitors enhancing radioiodine sensitivity in thyroid cancer. *Frontiers in Pharmacology* 13, 883581. DOI: 10.3389/fphar.2022.883581.

Yue, J., Li, W. and Wang, Y. (2021) Superiority verification of deep learning in the identification of medicinal plants: Taking *Paris polyphylla* var. Yunnanensis as an example. *Frontiers in Plant Science* 12, 752863. DOI: 10.3389/fpls.2021.752863.

Yue, J.Q., Huang, H.Y. and Wang, Y.Z. (2022) Extended application of deep learning combined with 2DCOS: Study on origin identification in the medicinal plant of *Paris polyphylla* var *yunnanensis*. *Phytochemical Analysis* 33(1), 136–150. DOI: 10.1002/pca.3076.

Zeng, J. and Jia, X. (2024) Systems theory-driven framework for AI integration into the holistic material basis research of traditional Chinese medicine. *Engineering* 40, 28–50. DOI: 10.1016/j.eng.2024.04.009.

Zhai, X., Wang, Q. and Li, M. (2016) Tu youyou's nobel prize and the academic evaluation system in China. *The Lancet* 387(10029), 1722. DOI: 10.1016/S0140-6736(16)30261-6.

Zhang, Y. and Wang, Y. (2023) Recent trends of machine learning applied to multi-source data of medicinal plants. *Journal of Pharmaceutical Analysis* 13(12), 1388–1407. DOI: 10.1016/j.jpha.2023.07.012.

Zhang, D., Wu, K., Zhang, X., Deng, S. and Peng, B. (2020b) In silico screening of shinese herbal medicines with the potential to directly inhibit 2019 novel coronavirus. *Journal of Integrative Medicine* 18(2), 152–158. DOI: 10.1016/j.joim.2020.02.005.

Zhang, H., Pan, J., Wu, X., Zuo, A.-R., Wei, Y. *et al.* (2019) Large-scale target identification of herbal medicine using a reverse docking approach. *ACS Omega* 4(6), 9710–9719. DOI: 10.1021/acsomega.9b00020.

Zhang, H., Ni, W., Li, J. and Zhang, J. (2020a) Artificial intelligence–based traditional Chinese medicine assistive diagnostic system: Validation study. *JMIR Medical Informatics* 8(6), e17608. DOI: 10.2196/17608.

Zhang, J., Fang, Y., Shao, X., Chen, H., Zhang, N. *et al.* (2024a) The future of molecular studies through the lens of large language models. *Journal of Chemical Information and Modeling* 64(3), 563–566. DOI: 10.1021/acs.jcim.3c01977.

Zhang, L., Zhang, C., Li, W., Li, L., Zhang, P. *et al.* (2023b) Rapid indentification of auramine O dyeing adulteration in *Dendrobium officinale*, saffron and *Curcuma* by SERS raman spectroscopy combined with SSA-BP neural networks model. *Foods* 12(22), 4124. DOI: 10.3390/foods12224124.

Zhang, S., Yang, K., Liu, Z., Lai, X., Yang, Z. *et al.* (2023c) DrugAI: A multi-view deep learning model for predicting drug–target activating/inhibiting mechanisms. *Briefings in Bioinformatics* 24(1), bbac526. DOI: 10.1093/bib/bbac526.

Zhang, W., Zheng, Q., Song, M., Xiao, J., Cao, Y. *et al.* (2021) A review on the bioavailability, bio-efficacies and novel delivery systems for piperine. *Food & Function* 12(19), 8867–8881. DOI: 10.1039/D1FO01971F.

Zhang, W., Zhang, C., Cao, L., Liang, F., Xie, W. *et al.* (2023a) Application of digital-intelligence technology in the processing of Chinese materia medica. *Frontiers in Pharmacology* 14. DOI: 10.3389/fphar.2023.1208055.

Zhang, X., Wang, L., Li, R., Wang, L., Fu, Z. *et al.* (2024b) Identification strategy of *Fructus gardeniae* and its adulterant based on UHPLC/Q-orbitrap-MS and UHPLC-QTRAP-MS/MS combined with PLS regression model. *Talanta* 267, 125136. DOI: 10.1016/j.talanta.2023.125136.

Zhang, Y., Wang, C., Wang, Y. and Cheng, P. (2022) Determining the stir-frying degree of *Gardeniae Fructus Praeparatus* based on deep learning and transfer learning. *Sensors* 8091. DOI: 10.3390/s22218091.

Zhao, J., Chen, X., Dou, X., Cao, Y., Wang, Y. *et al.* (2022) Automatic classification of medicinal materials based on three-dimensional point cloud and surface spectral information. In: *International Conference on Computer Graphics, Artificial Intelligence, and Data Processing (ICCAID 2021)*, Harbin, China, pp. 384–394. DOI: 10.1117/12.2631311.

Zhao, Z., Ren, X., Song, K., Qiang, Y., Zhao, J. *et al.* (2023) Pregenerator: TCM prescription recommendation model based on retrieval and generation method. *IEEE Access* 11, 103679–103692. DOI: 10.1109/ACCESS.2023.3316219.

Zhou, W. and Wang, Y. (2014) A network-based analysis of the types of coronary artery disease from traditional Chinese medicine perspective: Potential for therapeutics and drug discovery. *Journal of Ethnopharmacology* 151(1), 66–77. DOI: 10.1016/j.jep.2013.11.007.

Zhou, D., Zhao, Y., Chen, Z., Yan, X., Zhao, Y. *et al.* (2022a) Traditional processing increases biological activities of Dendrobium offificinale Kimura et Migo in Southeast Yunnan, China. *Scientific Reports* 12(1), 14814. DOI: 10.1038/s41598-022-17628-8.

Zhou, E., Shen, Q. and Hou, Y. (2024) Integrating artificial intelligence into the modernization of traditional Chinese medicine industry: A review. *Frontiers in Pharmacology* 15, 1181183. DOI: 10.3389/fphar.2024.1181183.

Zhou, L., Liu, S., Li, C., Sun, Y., Zhang, Y. *et al.* (2021b) Natural language processing algorithms for normalizing expressions of synonymous symptoms in traditional Chinese medicine. *Evidence-Based Complementary and Alternative Medicine* 2021(1), 6676607. DOI: 10.1155/2021/6676607.

Zhou, W., Yang, K., Zeng, J., Lai, X., Wang, X. *et al.* (2021a) FordNet: Recommending traditional Chinese medicine formula via deep neural network integrating phenotype and molecule. *Pharmacological Research* 173, 105752. DOI: 10.1016/j.phrs.2021.105752.

Zhou, Y., Zhang, Y., Lian, X., Li, F., Wang, C. *et al.* (2022b) Therapeutic target database update 2022: Facilitating drug discovery with enriched comparative data of targeted agents. *Nucleic Acids Research* 50(D1), D1398–D1407. DOI: 10.1093/nar/gkab953.

Zhou, Z., Chen, B., Chen, S., Lin, M. and Chen, Y. (2020) Applications of network pharmacology in traditional Chinese medicine research. *Evidence-Based Complementary and Alternative Medicine* 1, 1646905. DOI: 10.1155/2020/1646905.

Zhu, D., Han, J., Liu, C., Zhang, J. and Qi, Y. (2024) Modeling of flaxseed protein, oil content, linoleic acid, and lignan content prediction based on hyperspectral imaging. *Frontiers in Plant Science* 15, 1344143. DOI: 10.3389/fpls.2024.1344143.

Zhu, S., Liu, Q., Qiu, S., Dai, J. and Gao, X. (2022a) DNA barcoding: An efficient technology to authenticate plant species of traditional Chinese medicine and recent advances. *Chinese Medicine* 17(1), 112. DOI: 10.1186/s13020-022-00655-y.

Zhu, X., Zhu, M. and Ren, H. (2018) Method of plant leaf recognition based on improved deep convolutional neural network. *Cognitive Systems Research* 52, 223–233. DOI: 10.1016/j.cogsys.2018.06.008.

Zhu, Y., Ouyang, Z., Du, H., Wang, M., Wang, J. *et al.* (2022b) New opportunities and challenges of natural products research: When target identification meets single-cell multiomics. *Acta Pharmaceutica Sinica. B* 12(11), 4011–4039. DOI: 10.1016/j.apsb.2022.08.022.

Zuo, M.-T., Liu, Y.-C., Sun, Z.-L., Lin, L., Tang, Q. *et al.* (2021) An integrated strategy toward comprehensive characterization and quantification of multiple components from herbal medicine: An application study in *Gelsemium elegans*. *Chinese Herbal Medicines* 13(1), 17–32. DOI: 10.1016/j.chmed.2020.06.002.

Index

Printed and bound by CPI Group (UK) Ltd, Croydon, CR0 4YY

13/09/2026

14961088-0003